Spatial Deterministic Epidemics

Mathematical
Surveys
and
Monographs

Volume 102

Spatial Deterministic Epidemics

Linda Rass
John Radcliffe

American Mathematical Society

2000 *Mathematics Subject Classification.* Primary 92D30; Secondary 92D25.

ABSTRACT. This book uses rigorous analytic methods to determine the behaviour of spatial, deterministic models of certain multi-type epidemic processes where infection is spread by means of contact distributions. Results obtained include the existence of travelling wave solutions, the asymptotic speed of propagation and the spatial final size. The relationship with contact branching processes is also explored.

Library of Congress Cataloging-in-Publication Data

Rass, Linda, 1944–
Spatial deterministic epidemics / Linda Rass, John Radcliffe.
p. cm. — (Mathematical surveys and monographs, ISSN 0076-5376 ; no. 102)
Includes bibliographical references and index.
ISBN 0-8218-0499-5 (alk. paper)
1. Epidemiology—Mathematical models. 2. Medical geography—Mathematical models. I. Radcliffe, John, 1940– II. Title. III. Series.
RA652.2.M3 R375 2003
614.4′015′118–dc21 2002038456

---

10 9 8 7 6 5 4 3 2 1 08 07 06 05 04 03

Table of Contents

Preface

Mathematical biology has witnessed tremendous development in recent years. There is also an increasing realisation of the importance of spatial processes in all branches of population dynamics. One specialised area of research which has progressed to a point where a reasonably complete theory now exists concerns the rigorous mathematical analysis of spatial models of deterministic epidemics, in which individuals once infected cannot again become susceptible. Although several books have appeared in which spatial models are formulated and their behaviour described, the rigorous mathematical theory has until now only existed in a large number of papers scattered over many journals. The aim of this book is to collect together this material and unify it in order to make the results more accessible to researchers in this area and to make the mathematical techniques more readily available to mathematicians working in other branches of population dynamics.

This book concentrates on deterministic models of epidemics, with the sole exception of Chapter 9 where the connection with contact branching processes is explored. Epidemic models where infection is transferred between hosts and vectors, such as people and mosquitoes, or where infection can be transferred between several species of animals involve more than one population (or type). The general theory of these models of multi-type epidemics leads to the analysis of systems of non-linear integral equations, and requires the use of results on positivity and monotone techniques. These include the convexity and analyticity of the Perron-Frobenius root of a matrix whose entries are Laplace transforms of non-negative functions; and the existence of solutions of certain systems of equations involving a convex function. The appropriate mathematics is developed in two appendices.

The models include the $S \to L \to I \to R$ epidemic, in which an infected individual has a latent period before becoming infectious and eventually enters a removed state; and also more general models with infectivity varying with the time since infection. Chapter 1 contains an historical sketch. Chapter 2 sets up non-spatial models and treats the final size of an epidemic. Chapters 3, 4 and 5 are the core of the monograph, presenting a rigorous mathematical analysis of certain problems concerning the spatial spread of an epidemic, namely the pandemic theorem, the existence and uniqueness of wave solutions and the asymptotic speed of propagation of an epidemic; the central result of the monograph being that the asymptotic speed of propagation is in fact the minimum speed for which waves exist. Chapter 6 looks at epidemics on sites.

There is another approach, based on saddle point methodology, which can be used to obtain the speed of first spread in biological models. This is developed in Chapter 7 and applied to epidemic models. In all situations where an exact result has been proved, the result obtained by the saddle point method agrees with it. It is also used for $S \to I \to S$ epidemics and contact branching processes in Chapters 8 and 9 respectively. It is in fact a very powerful method which enables results to be obtained for much more complex situations, less tractable to exact analysis. There are applications to other areas of population dynamics such as genetics and evolutionary games. Although references are given, these applications are not included in this monograph.

The ideas developed have applications in other areas of population dynamics. Two of these which fit neatly into the general theme are included. The first application is the n-type $S \to I \to S$ deterministic epidemic, in which an infected individual can return to the susceptible state. The equilibrium solutions for the non-spatial model are determined and global convergence to the appropriate equilibrium is established. These results are then extended to a finite site model. The speed of first spread is obtained for the corresponding spatial model either in $\Re^N$ or on the N-dimensional integer lattice Z^N. The second application is contact branching processes. The connections between epidemics and contact branching processes are explained. The exact method is used for the contact birth process to obtain probabilistic convergence results. For more general models the saddle point method is applied to obtain the speed of first spread of the forward tail of the distribution of the furthest extent of the process.

In Chapter 1, links are given to certain other areas of population dynamics in which spatial models are used and where similar results to some of the results of this monograph have been obtained. Although in some ways we would have liked to have included some of these areas, we decided against it. The two most obvious omissions are spatial deterministic models in genetics and stochastic models for spatial epidemics.

The theory associated with the analogous problem in genetics of the spatial spread of a mutant gene uses techniques which overlap considerably with those used in this monograph. However, a comprehensive account of this subject would require a volume in its own right. In addition we decided to present the general n-type theory. The monotonicity of the functions involved in the epidemic models enabled the theory to be rigorously developed. No such general multi-type theory exists for the genetic models.

Deterministic and stochastic models are used to describe the behaviour of epidemics. Both only model the real world and each has its advantages and limitations in attempting to describe the possible modes of behaviour of epidemic systems. The reader might expect to find an account of both theories in a book on spatial epidemics. This is particularly true since both authors, although working at present in deterministic modelling, have a background in stochastic modelling. However, the theory associated with spatial stochastic epidemic models mainly concentrates on models in discrete space and uses quite different methodology. The continuous space stochastic models, analogous to the main models of this monograph, are less well developed. Whilst we are able to present a fairly complete account of the deterministic theory, this would not be possible at the present time for the corresponding models in the stochastic case.

We therefore decided in writing this monograph to restrict its scope to the theory of multi-type spatial deterministic epidemics, the requisite mathematics and a couple of immediate applications of the methodology. This enables us to present in a single volume a comprehensive and fairly complete account of an elegant general theory. Our choice of references, in similar fashion, has been deliberately selective and is mainly limited to the literature directly relevant to the text.

The approach adopted consistently throughout the monograph is to explain the ideas and methodology in the simple one-type case before proceeding to the rigorous analysis for the multi-type model. It is therefore possible on first reading to gain a good understanding of the material by confining attention to the more intuitive discussions of the simple models in each chapter.

Our interest in the development of a rigorous theory for deterministic epidemics was first triggered on reading a remarkable paper by Atkinson and Reuter two decades ago, and from subsequent discussions with the late Professor Reuter. Other strong influences came from the fundamental work of Diekmann and Thieme in epidemics and of Aronson, Weinberger and Lui in genetics. The methodology contained in their work provides both the one-type theory and the springboard for our development of the multi-type theory, which is presented herein.

We would like to dedicate the monograph to our respective families, Nicky and Sandy Rass and David, Paul and Rita Radcliffe for their continuous support and encouragement. We would also especially like to extend our warm appreciation to our erstwhile colleagues Professor Brian Connolly and Dr. James Gilson for their enthusiasm for the project and for their help with some of the technical problems. Thanks also are due to the late Professor Philip Holgate for some early useful discussions and insights and his support when getting the project off the ground.

John Radcliffe was supported in part during the writing of this monograph by a Leverhulme Research Fellowship and would like to express his gratitude to the Leverhulme Trust.

Linda Rass and John Radcliffe

CHAPTER 1

Introduction

1.1 An historical sketch and certain spatial problems

Early work providing rigorous methods in different areas of mathematical biology focused heavily on non-spatial models. For deterministic models of epidemics, the concern of this monograph, there are a number of recent books (e.g. Anderson and May [A1], Capasso [C1] and Diekmann and Heesterbeek [D8]) which provide a good account of the non-spatial theory.

Over a period of time there has been a growing awareness of the importance of including a spatial aspect when constructing realistic models of biological systems, with a consequent development of both approximate and mathematically rigorous methods of analysis. Spatial models are now extensively used throughout mathematical biology, and feature in the books mentioned above.

The spatial models may be deterministic or stochastic and can have the spatial aspect modelled by a diffusion term or by a continuous or discrete distribution. The space considered may be continuous ($\Re^N$) or discrete (a finite or infinite lattice). The earliest models developed were diffusion models in continuous space (see Fife [F2,F3]). An introduction to diffusion models in epidemics may be found in Murray [M8], further useful references being given in Fitzgibbon, Parrott and Webb [F5]. The use of a distribution, to model migration in genetics and contact in epidemics, developed at a later stage.

One area where a fairly complete, coherent body of spatial theory now exists is for deterministic models of certain epidemic processes which utilise contact distributions to model the spread of infection. Most work has concentrated on models in $\Re^N$ although some results have been obtained for lattice models. The rigorous methodology developed constitutes the core of this monograph. Early models involved only one type of individual, which may represent a human population or single species of animal or plant. However, many diseases involve more than one type of individual. These include tropical host-vector diseases, rabies with several different species and models in which the population has been stratified. These lead to multi-type models.

An historical outline of the development of the mathematical theory of epidemics up to 1975 is given in Chapter 2 of the encyclopaedic work of Bailey, [B1]. The germs of the ideas for the research on which this monograph is based appear in Chapter 9 of that book; which in particular contains an account of the heuristic derivations of the speed of propagation of a deterministic epidemic by Bartlett [B5] and the pandemic theorem of Kendall [K2], (see also Chapter 8 of Bartlett [B6]). In fact Bartlett [B5] also formulated stochastic models in continuous space. These were point process specifications which are essentially the stochastic analogues of the deterministic models in continuous space. Both deterministic and stochastic models used a contact distribution to model the spread of infection.

In the period covered by Bailey, work on these spatial models was quite limited in scope with heuristic methods paramount. The exception was pioneering work by Mollison [M4,M5] concerning wave solutions for deterministic models of a simple epidemic. However results showing the existence of wave solutions at certain speeds were obtained for a specific contact distribution only, with the methodology not easily adaptable to cover a more general setting. It is interesting to note that at this stage it was not envisaged that it would be possible to extend the exact treatment from the one-type model even to the relatively simple host-vector model (see Mollison [M6]).

In the latter half of the 1970's more general papers on spatial spread for deterministic models started to appear in the literature, commencing with a remarkable paper by Atkinson and Reuter [A7]. These papers used quite different techniques. Early work was confined to one-type models, with Diekmann and Thieme playing major roles in the development. Gradually over the next two decades a general theory of multi-type models was developed, primarily by the present authors. The results and relevant papers are described below.

This theory constitutes the main body of work presented in this monograph. In order to present the methodology in as simple a manner as possible, attention is restricted to the case where an infectious individual of any type can cause a susceptible of any specified type to become infected, possibly through a sequence of infections. This is termed a non-reducible epidemic. When this condition is not met the epidemic is said to be reducible. The extension to reducible epidemics uses mainly the same techniques but the degree of complexity is considerably increased. We therefore confine ourselves to referencing the relevant papers.

Although stochastic spatial models have also been developed, the main thrust of more recent work has focused on discrete space models rather than the continuous space point processes of Bartlett [B5]. Models with a single individual at each point of a lattice have been analysed using the methodology developed for percolation theory. An excellent survey is given by Metz, Mollison and van den Bosch [M2]. There are some similarities in results which have been obtained for the deterministic and stochastic models, but the stochastic methodology and models are too disjoint to be sensibly included in this monograph.

Consider the spatial spread of a deterministic epidemic in a homogeneous population, in which individuals once infected cannot return to the susceptible state. The spread of infection can be thought of as a wave travelling through the population. So a start would be to look at the possible wave solutions of the equations describing the spatial epidemic. Kendall [K3] considered wave solutions of a diffusion approximation to a one-type spatial epidemic. This indicated that there is a minimum speed below which wave solutions are not possible. Mollison [M4,M5] looked at this problem for an epidemic on the line and showed that in order for wave solutions to exist the contact distribution must be exponentially bounded in the forward tail. He obtained specific results when the contact distribution was double exponential, which were established for more general contact distributions by other authors. In fact waves are only possible if a certain threshold condition is met, in which case waves can only exist at speeds c greater than or equal to a critical speed c_0. The wave solution at a specific speed is unique modulo translation. Note however that this was only proved for the one-type model for speeds $c > c_0$. Atkinson and Reuter [A7] investigated the simple $S \to I$ and general $S \to I \to R$ epidemic and showed that no wave solutions exist for any speed c below a certain critical speed c_0. For each speed $c > c_0$ they constructed a wave solution. The existence of a wave solution for the case $c = c_0$ was proved by Brown and Carr [B11] using a limiting argument. A direct construction, when the contact distribution has compact support, has been used by Weinberger [W1] for a related problem in genetics. This method could clearly also be used in epidemics. The uniqueness modulo translation of the wave solution was proved by Barbour [B3] for all but the critical case using probabilistic methods. The uniqueness of the wave solution at speed $c = c_0$ can be proved by a method similar to the one used in genetics by Lui [L2].

Diekmann [D4-D7], Diekmann and Kaper [D9] and (independently) Thieme [T1-T4] set up a somewhat different one-type model in which the infection rate was allowed to vary with the time since infection. The $S \to I \to R$ model may be shown to be equivalent to a special case of such a model. Diekmann [D5] and Diekmann and Kaper [D9] considered wave solutions for the varying infectivity model and proved equivalent results to those obtained by Atkinson and Reuter [A7] and Barbour [B3]. Note however that their model did not cover the simple $S \to I$ epidemic, or situations when infectivity declined slowly, since an integrability constraint was imposed on the infection rate. Their papers used rather different methods based on complex analysis, which brought out the underlying structure clearly and provided the foundations upon which the multi-type theory was built.

Analogous results were shown to hold for a host-vector model with varying infectivity by Radcliffe, Rass and Stirling [R17] and subsequently for a non-reducible n-type model by Radcliffe and Rass [R1,R3]. The results obtained were more general since integrability constraints were not imposed and uniqueness was established for the critical wave speed. The reducible model was treated in Radcliffe and Rass [R8], some interesting results being obtained. In particular cases a multiplicity of waves was shown to be possible at certain speeds, in contrast to the non-reducible case where waves (if they exist at a specific speed) are unique modulo translation.

Note that results obtained concerning the existence of wave solutions at different speeds did not require the contact distributions to be radially symmetric. When there is radial symmetry, the direction independent critical wave speed was shown to be positive.

The next question concerns the speed of propagation of the epidemic. How fast must you run in order to avoid the epidemic? In order to perform a mathematical analysis it is first necessary to define precisely what is meant by the speed of propagation of the epidemic. A realistic assumption is to confine the initial infection to some compact region. Although the speed of spread of infection will at first depend on the quantity and configuration of the initial infection, intuition suggests that it will eventually settle down to a constant speed.

There are in fact several ways of defining the speed of propagation of an epidemic. One method for an epidemic on the line is to consider a point, such that the amount of infection ahead of this point is constant, and to define the speed as the rate at which this point moves out. For an epidemic in N-dimensional space, the analogue is to consider an $(N-1)$-dimensional hyperplane with the amount of infection beyond the hyperplane kept constant. The speed of propagation, in the direction orthogonal to the hyperplane, is then the rate at which the hyperplane moves out. The result obtained using this definition is termed the speed of first spread. This definition is used in the saddle point method to obtain an expression for the speed in certain models. Daniels [D1] used it to find the speed of first spread of the simple one-type epidemic on the line. The saddle point method is applied to the approximate linear equations which are valid in the forward front of the epidemic. For the $S \to I$ and $S \to I \to R$ models an explicit expression for the speed of first spread was obtained (Daniels [D1,D2], Mollison [M6]). Results were obtained for the n-type $S \to L \to I \to R$ (and subset models where the latent and removed stages may be omitted) for the non-reducible and reducible cases by Radcliffe and Rass [R5,R9]. The saddle point method does not require radial symmetry of the contact distributions and gives an explicit expression for the speed of first spread. Note that this method does in fact give precisely the same value for the speed of first spread as the value derived (using rigorous methodology) for the asymptotic speed of propagation discussed next. However this latter method requires the contact distributions to be radially symmetric. There is a conjecture (see van den Bosch, Metz and Diekmann [V1]) that conditions may be specified under which non-linear equations for spatial spread will have the same speed of spread as their linear approximation.

The saddle point method has also been used to obtain the speed of first spread for epidemics allowing a return to the susceptible state and for the speed of translation for contact branching processes (see Section 1.2). A discrete time saddle point method was derived in Radcliffe and Rass [R13]. This has been applied to a range of multi-type models in genetics, evolutionary game theory and branching processes (see Radcliffe and Rass [R13,R14,R16]).

Aronson and Weinberger [A4,A5] used the following idea to set up an alternative definition of the speed of spread for radially symmetric contact distributions, which is called the asymptotic speed of propagation. Suppose this speed is c^*. Then if you run faster than c^* you will leave the epidemic behind; whereas if you run slower than c^* then you will eventually be surrounded by the epidemic. Using a more precise formulation of this concept, a rigorous analysis can be performed.

Diekmann [D7] considered the one-type model with varying infectivity. Using the techniques of Aronson and Weinberger [A3,A4,A5,W1] he showed, in the case of a radially symmetric contact distribution which was exponentially dominated in the tail, that the asymptotic speed of propagation c^* was in fact the minimum speed, c_0, for which wave solutions exist. Results concerning the asymptotic speed of propagation were also obtained independently by Thieme [T3,T4]. The second of these papers provided results for models with non-constant population density; numerical studies appear in Shigesada and Kawasaki [S2]. Equivalent results to those of Diekmann were obtained for a host-vector and a non-reducible n-type model by Radcliffe and Rass [R6,R7]. For these models the asymptotic speed of propagation was shown to be the corresponding critical wave speed. For the reducible case the result is more complex (see Radcliffe and Rass [R12]). Consider only those types which can become infected as a result of the initial infection. Split these types into subsets of types, so that within each subset any type can cause an infection to occur to any other type but two-way transmission of infection is not possible between two types in different subsets. Let c_i be the critical wave speed if the ith subset of types is isolated from all other types. Then the asymptotic speed of propagation of infection for any type in subset j is the maximum of the c_i amongst all subsets i which can cause an infection to occur (through a series of infections) to the types in subset j.

Proofs deriving the asymptotic speed of propagation require radial symmetry of the contact distributions. Although an equivalent result is still presumably true in the non-rotationally symmetric case, as suggested by the saddle point method, a proof is still lacking (see van den Bosch, Metz and Diekmann [V1]).

When the contact distributions were radially symmetric, but not all exponentially dominated in the tail, the speed of propagation was shown to be zero if the threshold condition is not met and infinite otherwise. Note that exponential domination is the condition required for constant speed travelling wave solutions to be possible. Distributions when this condition does not hold have been considered in models for invasions by Kot, Lewis and van den Driessche [K6] and Clark [C2], (see also Mollison [M5]). When the redistribution kernels (the equivalent of the contact distributions for the epidemic models) are fat-tailed, they show that accelerating invasions can be generated. Such results may also to be obtainable for epidemic models.

The results concerning the asymptotic speed of propagation can be strengthened to give a shape theorem. When the threshold condition is met then the epidemic spreads out asymptotically like a sphere of radius $c_0 t$, where t is the time since the introduction of the initial infection. The speed of spread obtained by the saddle point method equals the asymptotic speed of propagation when there is radial symmetry. However the saddle point method and the wave solution results do not require this radial symmetry. The speed of spread in direction $\boldsymbol{\xi}$ is then the direction dependent minimum wave speed, so can be written as $c_0(\boldsymbol{\xi})$. This suggests that this will also be the speed of propagation in the non-radially symmetric case and that the shape of infection for large time t will have boundaries given by $c_0(\boldsymbol{\xi})t$, where $c_0(\boldsymbol{\xi})$ is a continuous function of its entries.

Shape theorems have also been obtained for stochastic models where individuals are located at the points of a two dimensional lattice (see Durrett [D13] and Cox and Durrett [C4]). The processes considered can be regarded as models for spatial epidemics or forest fires. An infected individual (burning tree) can only infect (set light to) any of its four immediate neighbours. An intuitive description of the shape theorems is as follows: conditional on the epidemic (forest fire) not dying out (i) the set of removed individuals (dead trees) grows linearly with time and has an asymptotic shape and (ii) the set of infected individuals (burning trees) will be near this growing boundary. For an account of this theory and a more precise statement of the shape theorems we refer the reader to chapter 9 of the book on particle systems and percolation theory by Durrett [D12].

One of the problems that arise in the study of non-spatial models concerns the final size of the epidemic i.e. the proportion of the individuals who eventually suffer the epidemic. This was treated by Kendall [K1] and has an analogue in the spatial case (see Diekmann [D5,D7], Thieme [T1,T2] and Radcliffe and Rass

[R2,R4,R6,R7,R12]). The pandemic theorem shows that, provided the threshold value of the epidemic parameters is exceeded and the contact distribution is radially symmetric, the epidemic spreads everywhere. It provides a positive uniform lower bound (for each type) on the proportion of individuals at each point who eventually suffer the epidemic. This lower bound was shown to be just the limit of the non-spatial final size (for that type) as the amount of initial infection tends to zero. The result was proved directly in 1 or 2 dimensions (Diekmann [D5], Radcliffe and Rass [R2,R4]) and later obtained for an N-dimensional space by sharpening the results obtained concerning the speed of propagation (Diekmann [D7], Radcliffe and Rass [R6,R7,R12]). The last of these papers concerns a reducible model. An upper bound on the spatial final size was also derived and hence the limit of the spatial final size was obtained as the amount of the initial infection tended to zero. One can also consider the behaviour at infinity, i.e. the proportion of individuals who eventually suffer the epidemic a long way from the initial source of infection. Kendall [K2] gave a heuristic treatment of the behaviour at infinity and obtained the result that the epidemic will eventually extend over the whole plane. A rigorous derivation for the more general varying infectivity model was given by Thieme [T2] for the one-type model and by Radcliffe and Rass [R7] for the non-reducible n-type model.

Finally consider running at speed c_0 in the epidemic on the line, and looking along its length at the proportions of infectives. It seems intuitively reasonable that the proportions would converge to the unique wave solution associated with the minimum speed c_0. Similar results have been proved for simple genetic models (Lui [L2,L3]) and branching diffusions (Bramson [B10]). Results on convergence to a wave form for diffusion models also appear in the book by Volpert, Volpert and Volpert [V2]. However this result has not yet been proved for spatial epidemic models, and remains an outstanding problem.

Results were also obtained by Radcliffe and Rass [R19] for a discrete space, n-type model with sites at the points of the N-dimensional infinite lattice. The sites represent conurbations such as towns or disconnected regions. Pandemic results were obtained which were analogous to those obtained for the continuous space n-type model. The proofs required symmetry conditions to be placed on both contact distributions and infection rates. The speed of first spread for the n-type $S \to L \to I \to R$ and subset models were obtained using a saddle point method (see Radcliffe and Rass [R11]). Symmetry constraints were not required.

At the same time as the theory for one-type models was being developed for epidemics, a parallel theory was being developed for two allele models in genetics using similar techniques.

Spatial models in genetics were initially formulated in continuous time. This originated with the works of Kolmogorov, Petrovsky and Piscounoff [K5] and Fisher [F4]. The problems investigated concerned the analysis of non-linear diffusion equations modelling the spatial spread of a mutant gene in a homogeneous population (see Fife [F2,F3]), Hadeler [H1] and Aronson and Weinberger [A4]). Results were obtained concerning the existence of wave solutions and the speed of propagation.

A discrete-time genetic model in which the migration process was represented by a random variable rather than diffusion was first introduced by Weinberger [W1,W2], and further analysed by Diekmann and Kaper [D9] and then subsequently in a series of papers by Lui [L2-L6] and Creegan and Lui [C5]. It is these discrete time genetic models that use similar techniques to the techniques described in this monograph. The migration distribution plays a role equivalent to the contact distribution in the epidemic models. There are three cases for the two allele model corresponding to when the fitness of the heterozygote is intermediate, superior or inferior respectively to the homozygote fitnesses. For the heterozygote intermediate and superior cases, analogous results are obtained to the results obtained for epidemic models in this monograph. These results concern the existence of wave solutions, the asymptotic speed of propagation of the mutant gene and the spatial final size. In addition the convergence to a wave form is established for the heterozygote intermediate case. Note that proofs for the heterozygote intermediate case do not cover all possible values of the fitness parameters. Results obtained for the heterozygote inferior case are somewhat different. Results for a system with two genes are discussed and extended in Radcliffe and Rass [R15] to cover a more general (non-symmetric) model for the spread of a strategy in evolutionary game theory.

In contrast to developments within epidemics, the rigorous results obtained for the case of two alleles have not been extended to multiple allele models although results have been obtained for simple models for sex-linked genes (see Lui [L7,L8]).

There are two related areas where there is a multi-type theory and where the methodology developed for spatial models of deterministic epidemics has direct application. These are included in this monograph. A brief discussion is given in the next section.

1.2 Multi-type problems in two related areas

Consider a multi-type, continuous space contact birth process starting with one type i individual at the origin at time $t = 0$. A birth occurs after an exponential time and the offspring can be of a different type. Let $U(t)$ denote the position of furthest spread in a specific direction, either of any type or of a specific type, by time t. Define $y_i(s,t)$ to be $P(U(t) > s)$. Then the $y_i(s,t)$ have been shown to satisfy the equations for the multi-type $S \to I$ epidemic on the real line. This connection was pointed out for the one-type model by Mollison [M6,M7] and for the multi-type model by Radcliffe and Rass [R10]. The initial conditions differ, so that proofs for the epidemic model must be adapted to obtain results concerning the speed of translation of the forward tail of the distribution of $U(t)$ for the contact birth process. This can then be extended to show that $U(t)/t$ converges in probability to the minimum wave speed for the associated $S \to I$ epidemic. The saddle point method has been used to obtain the front velocity for multi-type contact branching processes (see Daniels [D2] for the one-type model and Radcliffe and Rass [R10] for the multi-type model).

Note that probabilistic methods have also been used to obtain convergence results for contact branching process models, but only for the one-type case. The almost sure convergence of $U(t)/t$ was proved by Mollison [M6]; the value of the limit being obtained later by Biggins [B8]. Results for a specialised branching process where individuals die when they give birth have been obtained by Uchiyama [U1].

The second related area concerns deterministic models of epidemics where a return to the susceptible state is possible. One type $S \to I \to S$ models have been studied by a number of authors (e.g. Diekmann and Montijn [D10] and Gripenberg [G3]). Two type models have been considered by Cooke and Yorke [C3] and Roberts [R21].

The non-spatial multi-type $S \to I \to S$ model for a closed system has been considered by Hethcote and Thieme [H2] and Lajmanovich and Yorke [L1]. Local and global stability results were obtained. When a certain threshold condition is met the proportions of infectious individuals converge to the endemic equilibrium. When the condition is not met the epidemic dies out. Note that the determination of the equilibrium solutions gives an application of the main theorem of Appendix B, but with a different convex function to the one required for the models with permanent immunity. In Rass and Radcliffe [R20] the global asymptotic results were extended to a reducible system. It was shown that every solution of a multi-type model, which encompassed both the $S \to I$ and $S \to I \to S$ epidemics with stable population size, converges to an equilibrium even if the infection matrix is reducible.

Spatial multi-type models have also been considered. Corresponding convergence results can be derived for n-type epidemics on a finite number of sites by regarding the site and type combinations as new types. An infinite space model of an $S \to I$ $(\to S)$ epidemic was considered, with the space either $\Re^N$ or Z^N. The saddle point method was used (see Radcliffe and Rass [R11]) to obtain the speed of first spread of infection. However, no exact methodology has been developed at present.

Later work on multi-type models of $S \to I \to S$ type has concentrated on the non-spatial non-reducible case and has used monotonicity and sublinearity methods based on those of Krasnosel'skii [K7]. Simon and Jacquez [S3] use Liapunov functions to obtain results for multi-type models with many infection stages. Models with chronological age structure were considered in Busenberg, Iannelli and Thieme [B12] and convergence results derived. Time periodic forcing was considered by Aronsson and Melander [A6]; this work was then extended to arbitrary time-dependent forcing by Thieme ([T6] Section 4). Thieme [T5] also considered a population with distributed infection risk.

CHAPTER 2

The non-spatial epidemic

2.1 One-type models in a closed population

An infection such as measles is caused by a virus. A person who suffers a viral infection either dies or develops antibodies and recovers, and is then usually immune to the disease. This section considers a disease in one population where the infection is passed from individual to individual and individuals who recover from the infection cannot again become susceptible. The models are set up for a closed population, where no infection can enter from outside the population. Changes in the population size such as births into the system and deaths from causes other than the disease are neglected. This is reasonable for a model where the time scale for the epidemic is short relative to the life cycle of individuals.

Earlier models were specified by allowing different stages of infection, were conceptually simple and were formulated in terms of differential equations. A model with latent, infectious and removed stages is termed an $S \to L \to I \to R$ epidemic, with the obvious notation for sub-models with no latent and/or removed stage. A more general model allows the rate of infection to vary with the time since infection and is formulated in terms of integro-differential equations. Equations for the $S \to L \to I \to R$ epidemic can be shown to be equivalent to a special case of the equations for the varying infectivity model. The same is true for the sub-models.

Although models with different stages of infection have to a large extent been subsumed by the model with varying infectivity they are included in this chapter (albeit briefly) partly because they are of historical interest and provide simple examples, but mainly because of their importance in later chapters. The saddle point method of Chapter 7 can be applied only to these epidemic models and not to the general varying infectivity model. In addition it is the $S \to I$ and $S \to I \to R$ models which link to the $S \to I \to S$ model of Chapter 8 and the contact birth process of Chapter 9.

Model 1 (The $S \to L \to I \to R$ epidemic).

Consider an epidemic occurring over all real time. The population is divided into four classes: susceptible, latent, infectious and removed. An individual is said to be infected if he is carrying the virus or other infecting body; and infectious if he is capable of transmitting the disease. A latent individual is an individual who is infected but not yet infectious. A removed individual denotes an individual who has suffered the disease and has died or recovered or has been isolated from the population.

Let X(t), L(t), Y(t) and Z(t) denote the numbers (at time t) of susceptible, latent, infectious and removed individuals respectively. The rate of infection of a single susceptible individual by a single infectious individual is λ, which combines an infection rate per contact with a contact rate between two individuals. When first infected, an individual is latent. The rates at which latent individuals become infectious and infectious individuals become removed are represented by α and μ respectively. In fact μ is a combined recovery, removal by isolation and death rate (which it is usual to simply refer to as the removal rate).

Let $g'(t)$ denote the derivative of $g(t)$. The model is described by the equations,

$$\begin{aligned} X'(t) &= -\lambda X(t)Y(t), \\ L'(t) &= \lambda X(t)Y(t) - \alpha L(t), \\ Y'(t) &= \alpha L(t) - \mu Y(t), \\ Z'(t) &= \mu Y(t). \end{aligned} \tag{2.1}$$

Equations (2.1) can be rewritten in terms of the proportions $x(t) = X(t)/\sigma$, $l(t) = L(t)/\sigma$, $y(t) = Y(t)/\sigma$ and $z(t) = Z(t)/\sigma$, where σ is the population size. Equations (2.1) then become,

$$\begin{aligned} x'(t) &= -\sigma\lambda x(t)y(t), \\ l'(t) &= \sigma\lambda x(t)y(t) - \alpha l(t), \\ y'(t) &= \alpha l(t) - \mu y(t), \\ z'(t) &= \mu y(t), \end{aligned} \tag{2.2}$$

where $x(t)+l(t)+y(t)+z(t) = 1$. Now $X(t)$, $L(t)$, $Y(t)$ and $Z(t)$ are in fact discrete. However it is assumed that σ is large, so that it is not unreasonable to take $x(t)$, $l(t)$, $y(t)$ and $z(t)$ to be continuous and differentiable functions. The epidemic is considered over all time (so the equations hold for all t), with $\lim_{t\to-\infty} x(t) = 1$ and hence $\lim_{t\to-\infty} l(t) = \lim_{t\to-\infty} y(t) = \lim_{t\to-\infty} z(t) = 0$. For the solution to be non-trivial, necessarily $x(0) < 1$.

Essentially the same equations are obtained if the limit is taken as the population size tends to infinity. Take $\lambda = r\phi$, where r is the contact rate between two individuals and ϕ is the infection rate per contact. The number of contacts an individual makes per unit time will stay finite as the population size tends to infinity. Hence, with this assumption together with homogeneous mixing, σr tends to a positive constant as the population size σ tends to infinity, and so $\sigma\lambda$ tends to a positive constant also. We could therefore obtain equation (2.2) for this limiting case by replacing $\sigma\lambda$ by this constant.

When there is no removed stage, $\mu = 0$ and the last differential equation from equations (2.2) is omitted. This model is referred to as the $S \to L \to I$ model. For an $S \to I \to R$ epidemic, when there is no latent stage, $l(t) = 0$ and the middle two equations of equations (2.2) are replaced by $y'(t) = \sigma\lambda x(t)y(t) - \mu y(t)$. When $\mu = 0$ the equation for $z(t)$ is also omitted and the epidemic becomes an $S \to I$ epidemic. In the literature (see Bailey [B1]) the $S \to I \to R$ epidemic is referred to as the general epidemic, and the $S \to I$ epidemic (with no latent or removed stages) is called the simple epidemic.

The $S \to I$ epidemic has a simple explicit solution. If $x(0) = \alpha$, then $x(t) = 1/(1 + \theta e^{\sigma\lambda t})$, where $\theta = (1-\alpha)/\alpha$; and $y(t) = 1 - x(t)$. Shifting the time scale simply scales θ. Note that $x(t) \to 1$ as $t \to -\infty$ and (provided $\alpha < 1$) $x(t) \to 0$ as $t \to \infty$.

Model 2 (A model with varying infectivity).

Model 1 assumes that the rate of infection λ of an infectious individual is constant until the individual is removed by recovery, death or isolation. A model is now introduced which allows the rate of infection $\lambda(\tau)$ to be a function of the time τ since infection. This can be used to include a latent period of fixed duration where an infected individual is not infectious and may not show any symptoms. It also allows for a time beyond which an infected individual is no longer infectious.

Consider a population of size σ consisting of susceptible and infected individuals. Let x(t) be the proportion of susceptibles at time t and $I(t,\tau)d\tau$ be the proportion of individuals at time t who were infected in the time interval $(t-\tau-d\tau, t-\tau)$. When σ is large it is reasonable to consider $x(t)$ to be continuous and differentiable and for $I(t,\tau)$ to be a continuous function of τ. The epidemic is considered over all time with $\lim_{t\to-\infty} x(t) = 1$

In addition to the equation for the rate of decrease of the number of susceptibles (i.e. of $\sigma x(t)$), the total proportion of infected individuals at time t must equal $1-x(t)$. Also an individual who at time t was infected time τ ago was first infected at time $t-\tau$. Hence the model is described by the equations, for $\tau > 0$ and all t,

$$\begin{aligned} x'(t) &= -\sigma x(t) \int_0^\infty I(t,\tau)\lambda(\tau)d\tau, \\ x(t) &= 1 - \int_0^\infty I(t,\tau)d\tau, \\ I(t,\tau) &= I(t-\tau, 0). \end{aligned} \tag{2.3}$$

The relation for $I(t,\tau)$ implies continuity of $I(t,0)$. Substituting in the second equation gives $x(t) = 1 - \int_{-\infty}^{t} I(u,0)du$. Differentiating we obtain the relation $x'(t) = -I(t,0)$, so that $x'(t) \le 0$ (so $x(t)$ is monotone decreasing) and is continuous and bounded. From the definition of $x(t)$, $0 \le x(t) \le 1$.

Note that it is possible to write $\lambda(\tau) = r\phi(\tau)$ in model 2, where r is the contact rate per unit time between two individuals and $\phi(\tau)$ is the infection rate per contact by an infective who was infected time τ ago. As for model 1, a reasonable assumption is that σr tends to a limit as σ tends to infinity. Hence we may replace $\sigma\lambda(\tau)$ in equations (2.3) by this limit times $\phi(\tau)$, giving the same form of equation. If we write $\gamma(\tau) = \sigma\lambda(\tau)$ in equations (2.3), we can then interpret $\gamma(\tau)$ as a product between the number of contacts per unit time for an individual and the infection rate per contact by an individual infected time τ ago. This interpretation is also valid if the population size is allowed to become infinite.

In the $S \to I$ model, once a person is infected he has a constant infection rate λ. Thus the infection rate $\lambda(\tau) \equiv \lambda_1(\tau) = \lambda$. For the $S \to I \to R$ model, an infected individual again has a constant infection rate λ. However, the person is passing into a removed state at rate μ, so that infection is effectively disappearing exponentially at rate μ i.e. this model behaves like a special case of model 2, with $\lambda(\tau) \equiv \lambda_2(\tau) = \lambda e^{-\mu\tau}, 0 \le \tau < \infty$. Note that, if $\mu = 0$ then $\lambda_2(\tau) = \lambda_1(\tau)$. For the $S \to L \to I \to R$ epidemic, which allows for a latent period with rate α of moving from being latent to being infectious a similar result can be obtained; however this result is not intuitively obvious and is derived below.

The relation between models 1 and 2 can be established by direct comparison of the equations derived for $x(t)$ for each model. For model 2, consider a solution of equations (2.3) with the specified conditions. It follows from these equations that $I(t,\tau) = -x'(t-\tau)$ and hence that, for all t,

$$x'(t) = x(t) \int_0^\infty x'(t-\tau)\gamma(\tau)d\tau, \tag{2.4}$$

where $\gamma(\tau) = \sigma\lambda(\tau)$. Now suppose that $x(t)$ is a monotone decreasing solution (with a continuous, bounded derivative) to equation (2.4) with $0 \leq x(t) \leq 1$ and $\lim_{t\to-\infty} x(t) = 1$. Define $I(t,0) = -x'(t)$, and $I(t,\tau) = I(t-\tau,0)$. Then $x(t)$, $I(t,\tau)$ is a solution to equations (2.3) with the specified conditions.

Now consider model 1. Equations (2.2) imply that $x(t)$ is monotone decreasing with a continuous, bounded derivative. Consider a solution with $\lim_{t\to-\infty} x(t) = 1$ and hence $\lim_{t\to-\infty} l(t) = \lim_{t\to-\infty} y(t) = \lim_{t\to-\infty} z(t) = 0$. Then from these equations $x'(t) = -\sigma\lambda x(t)y(t)$, $y'(t) + \mu y(t) = \alpha l(t)$ and $l'(t) + \alpha l(t) = -x'(t)$. Hence $x'(t) = x(t)(-\sigma\lambda y(t))$, $y(t) = \int_{-\infty}^t \alpha l(u)e^{-\mu(t-u)}du$ and $l(u) = -\int_{-\infty}^u x'(\tau)e^{-\alpha(u-\tau)}d\tau$. Therefore

$$\begin{aligned}
x'(t) &= x(t) \int_{-\infty}^t \int_{-\infty}^u \sigma\lambda\alpha x'(\tau)e^{(\mu-\alpha)u+\alpha\tau-\mu t}d\tau du \\
&= x(t) \int_{-\infty}^t \int_{\tau}^t \sigma\lambda\alpha x'(\tau)e^{(\mu-\alpha)u+\alpha\tau-\mu t}du d\tau \\
&= x(t) \int_{-\infty}^t \frac{\sigma\lambda\alpha}{(\alpha-\mu)} x'(\tau)\left(e^{-\mu(t-\tau)} - e^{-\alpha(t-\tau)}\right) d\tau \\
&= x(t) \int_0^\infty x'(t-\tau)\gamma(\tau)d\tau,
\end{aligned}$$

where $\gamma(t) = \sigma\lambda_3(t)$ with $\lambda_3(t) = \lambda\alpha\left(e^{-\mu t} - e^{-\alpha t}\right)/(\alpha-\mu)$. This is just equation (2.4) with a specific $\gamma(\tau)$.

The steps may easily be reversed. Take $x(t)$ to be a non-negative, monotone decreasing solution (with a continuous derivative) to equation (2.4) with $\lim_{t\to-\infty} x(t) = 1$ and $\gamma(t) = \sigma\lambda_3(t)$. Define $l(t) = -\int_{-\infty}^t x'(u)e^{-\alpha(t-u)}du$, $y(t) = \int_{-\infty}^t \alpha l(u)e^{-\mu(t-u)}du$ and $z(t) = 1 - x(t) - l(t) - y(t)$. Then $x(t)$, $l(t)$, $y(t)$ and $z(t)$ satisfy equations (2.1). Hence the equations for model 1 are equivalent to those for model 2 for the special case when the infection rate $\lambda(t) = \lambda_3(t)$. Note that if we let $\alpha \to \infty$, then $\lambda_3(\tau) \to \lambda_2(\tau)$. An infinite rate of passing from the latent to the infectious stage is essentially equivalent to having no latent stage.

2.2 Epidemics initiated from outside

The models considered in Section 2.1 concern an epidemic in a closed population, where it is assumed that we can look back in time towards $t = -\infty$ to see how the epidemic developed. As $t \to -\infty$ everyone is assumed to be susceptible so that $\lim_{t\to-\infty} x(t) = 1$. This viewpoint is sometimes useful; for instance it will be used in Chapter 4 when looking at the possible wave solutions in the spatial case. The saddle point method is also applied to the spatial form of model 1 in Chapter 7. However it is often more convenient, and possibly more realistic, to look at the development of an epidemic which commences at time $t = 0$ by the introduction of infection from outside the population under consideration. These models only include these individuals in a term giving their infectious influence on the populations of susceptibles. An analogue of each of models 1 and 2 for this formulation is now considered.

Model 1* (The $S \to L \to I \to R$ model).

Consider a population as described in model 1, but where at time $t = 0$ everyone is susceptible. At this time infectious individuals from outside enter the population and can (at some time) infect susceptible individuals in the population. Let $\phi(t)$ and $\varepsilon(t)$ be the relative numbers (in relation to the population size σ) at time t of these latent and infectious individuals from outside. So for σ large these functions may be considered to be approximately continuous and differentiable. For such latent individuals the latent to infectious rate is α^* and the rate at which latent individuals leave is β^*. For the infectious individuals the infection rate is λ^* and the combined removal and leaving rate is μ^*. These infectious individuals from outside may be of the same type as those in the population, in which case $\lambda^* = \lambda$ and $\alpha^* = \alpha$, or may be individuals of some other type. Even if the individuals are of the same type as those in the population, μ^* may well be greater than μ since μ^* includes the rate at which individuals from outside leave the population and cease to have any infectious influence.

The model is described, for $t \geq 0$, by the equations

$$\begin{aligned}
x'(t) &= -x(t)\left(\sigma\lambda y(t) + \sigma\lambda^*\varepsilon(t)\right), \\
l'(t) &= x(t)\left(\sigma\lambda y(t) + \sigma\lambda^*\varepsilon(t)\right) - \alpha l(t), \\
y'(t) &= \alpha l(t) - \mu y(t), \\
z'(t) &= \mu y(t) \\
\phi'(t) &= -(\alpha^* + \beta^*)\phi(t), \\
\varepsilon'(t) &= \alpha^*\phi(t) - \mu^*\varepsilon(t).
\end{aligned}$$

The initial conditions are taken to be $x(0) = 1$, $l(0) = y(0) = z(0) = 0$, $\phi(0) = \phi$ and $\varepsilon(0) = \varepsilon$. From the last two equations $\phi(t) = \phi e^{-(\alpha^*+\beta^*)t}$ and hence

$$\varepsilon(t) = \varepsilon e^{-\mu^* t} + \frac{\alpha^*\phi}{\alpha^* + \beta^* - \mu^*}\left(e^{-\mu^* t} - e^{-(\alpha^*+\beta^*)t}\right). \tag{2.5}$$

Model 2* (A model with varying infectivity).

Consider a population as described in model 2, but where at time $t = 0$ all individuals are susceptible. Infected individuals are introduced from outside at this time. The number of these who were infected in the time interval $(-\tau - d\tau, -\tau)$ is $\sigma\varepsilon(\tau)d\tau$. The infection rate for such individuals is $\lambda^*(\tau)$. For σ large it is reasonable to take $\varepsilon(\tau)$ to be continuous. Then $\varepsilon = \int_0^\infty \varepsilon(\tau)d\tau$ is assumed to be finite (and is usually taken to be small), $\sigma\varepsilon$ represents the number of infected individuals to whom the population is exposed and $\sigma\lambda^*(\tau)$ is the number of contacts per unit time times the infection rate per contact.

As for model 1*, the infectious individuals from outside may be of the same type as those in the population, or may be individuals of some other type. Even if the individuals from outside are of the same type as those in the population, $\lambda^*(\tau)$ may well be different from $\lambda(\tau)$, since the former function includes an allowance for individuals from outside leaving the population.

The equations for this model, for $0 \leq \tau \leq t$ and $t \geq 0$, are

$$
\begin{aligned}
x'(t) &= -x(t)\left(\sigma \int_0^t I(t,\tau)\lambda(\tau)d\tau + \int_0^\infty \sigma\lambda^*(t+\tau)\varepsilon(\tau)d\tau\right), \\
x(t) &= 1 - \int_0^t I(t,\tau)d\tau, \\
I(t,\tau) &= I(t-\tau,0),
\end{aligned}
\tag{2.6}
$$

with the condition that $x(0) = 1$. As for model 2, $x(t)$ is differentiable and $I(t,\tau)$ is a continuous function of τ (and therefore $I(t,0)$ is continuous). From the equations above we then obtain the relation $x'(t) = -I(t,0) \leq 0$, so that $x(t)$ is monotone decreasing and $x'(t)$ is continuous.

This immediately gives an equivalent equation for $x(t)$, for $t \geq 0$, namely

$$
x'(t) = x(t)\left(\int_0^t x'(t-\tau)\gamma(\tau)d\tau - h(t)\right), \tag{2.7}
$$

where $\gamma(\tau) = \sigma\lambda(t)$ and $h(t) = \int_0^\infty \sigma\lambda^*(t+\tau)\varepsilon(\tau)d\tau$. The equivalence can be shown in the same manner as for equations (2.3) and (2.4) for model 2.

As in Section 2.1 it is easily shown that model 1* may be regarded as a special case of model 2* with $\gamma(t) = \sigma\lambda\alpha(e^{-\mu t} - e^{-\alpha t})/(\alpha - \mu)$ and a different function $h(t) = \sigma\lambda^*\varepsilon(t)$, where the expression for $\varepsilon(t)$ is given by equation (2.5). This function $h(t)$ is continuous and $a = \int_0^\infty h(t)dt$ will be finite if $\mu^* > 0$ and $\alpha^* + \beta^* > 0$. In this case a tends to zero as the relative number of individuals introduced from outside, $\varepsilon + \phi$, tends to zero. These are precisely the properties for $h(t)$ for model 2* which are needed to derive the results concerning the final size of the epidemic.

2.3 A multi-type model

So far we have considered epidemics in which the infection is transferred from individual to individual in a single population. Yellow fever is a viral disease which is usually transmitted between monkey and mosquito in the forest cycle or between man and mosquito in the urban cycle. Malaria is an asexual parasitic disease which is transmitted between man and mosquito and is an example of a disease in which reinfection occurs. Both of these diseases are examples of host-vector epidemics. In these 2-type epidemics the infection is transferred host $\rightarrow$ vector $\rightarrow$ host. Models for malaria are discussed in detail in Bailey [B2]. Another example of a 2-type epidemic is the carrier-borne epidemic. In this epidemic there are normal individuals and those who, once infected, become carriers. A carrier can carry the disease and infect other individuals but does not suffer from the disease so is less easy to detect. Removal rates by death or detection are therefore likely to be smaller for carriers than normal infectious individuals. A further application of the 2-type model is when some of the population have been vaccinated against a disease, the vaccination not conferring complete immunity. Infection rates will differ for vaccinated and non-vaccinated individuals. The severity of the disease, and hence the death rates, are also likely to differ.

An example of a multi-type epidemic is rabies involving several different animal populations with the infection being transmitted within and between populations. Sometimes it is convenient to stratify a population, with each stratum having a different contact distribution. The strata can then be considered as types, giving another example of a multi-type epidemic. We restrict our attention to multi-type epidemics where any type of individual may, possibly through a sequence of infections, infect any other type. This is termed a non-reducible epidemic. Since models of $S \to L \to I \to R$ type can be regarded as a special case of the model with varying infectivity we only consider the n-type version of the latter model.

Consider the n-type form of model 2^* of Section 2.2 where infection is initiated from outside. Let σ_i be the size of the ith population and take $x_i(t)$ and $I_i(t,\tau)d\tau$ to be the proportions in population i who at time t were respectively susceptible and infected in the time interval $(t-\tau-d\tau, t-\tau)$. Define $\varepsilon_j(\tau)d\tau$ to be the relative number (in relation to σ) of individuals of type j, who were introduced from outside at time $t=0$ and were infected in the time interval $(-\tau-d\tau,\tau)$. The scaling factor $\sigma = \sum_{j=1}^{n} \sigma_j$ is the total population size (excluding individuals from outside). Let $\lambda_{ij}(\tau)$ (and $\lambda^*_{ij}(\tau)$) represent the rate of infection of a susceptible in the ith population by an infected individual in the jth population (and j^{th} type from outside) who was infected time τ ago. Let $\varepsilon_j = \int_0^\infty \varepsilon_j(\tau)d\tau$; then $\sigma\varepsilon$ is the number of type j individuals introduced from outside. Clearly an alternative scaling factor related to the population sizes could be used. The types from outside may include different types to those in the n populations.

Equations (2.6) generalize to give, for $0 \le \tau \le t$ and $t \ge 0$,

$$(2.8)\qquad \begin{aligned} x_i'(t) &= -x_i(t)\left(\sum_{j=1}^{n} \sigma_j \int_0^t I_j(t,\tau)\lambda_{ij}(\tau)d\tau + h_i(t)\right), \\ x_i(t) &= 1 - \int_0^t I_i(t,\tau)d\tau, \\ I_i(t,\tau) &= I_i(t-\tau,0), \quad (i=1,...,n), \end{aligned}$$

where $h_i(t) = \sum_{j=1}^{k} \int_0^\infty \sigma\lambda^*_{ij}(t+\tau)\varepsilon_j(\tau)d\tau$. The initial conditions are that $x_i(0)=1$ for $i=1,...,n$. As for the one-type model, $0 \le x_i(t) \le 1$, the $x_i'(t)$ are differentiable and the $I_i(t,\tau)$ and $\varepsilon_i(\tau)$ are continuous functions of τ. The continuity of $I_i(t,0)$ and the relation $x_i'(t) = -I_i(t,0)$ are then obtained from equations (2.8). Hence each $x_i(t)$ is monotone decreasing with a continuous derivative. Substituting for $I_i(t,\tau)$ for the equation for $x_i'(t)$ in equations (2.8) and letting $\gamma_{ij}(\tau) = \sigma_j\lambda_{ij}(\tau)$ we obtain equivalent equations,

$$(2.9)\qquad x_i'(t) = x_i(t)\left(\sum_{j=1}^{n} \int_0^t x_j'(t-\tau)\gamma_{ij}(\tau)d\tau - h_i(t)\right), \quad (i=1,...,n).$$

Let $\{\mathbf{\Gamma}\}_{ij} = \gamma_{ij}$. Then the non-reducibility of the epidemic implies non-reducibility of the matrix $\mathbf{\Gamma}$ i.e. for every i, j there exists a sequence $i_i, ..., i_r$ with $i_1 = i$ and $i_r = j$ such that $\gamma_{i_s i_{s+1}} \neq 0$ for $s = 1, ..., (r-1)$. Equations (2.9) hold for $t \geq 0$ and are the starting point of a study of the final size of a non-spatial epidemic. Since the number of infected individuals from outside triggering the epidemic would usually be very small relative to σ, a natural problem to consider concerning the final size of each type is its limit as the ε_j tend to zero.

2.4 The solution of an equivalent system of equations

By imposing conditions on the scaled infection rates $\gamma_{ij}(\tau)$ and $\gamma_{ij}^*(\tau) = \sigma\lambda_{ij}^*(\tau)$ we can relate the solutions of equations (2.9) to those of the following equations and prove results concerning $\lim_{t\to\infty} x_i(t)$. For $t \geq 0$,

$$(2.10) \qquad -\log x_i(t) = \sum_{j=1}^{n} \int_0^t (1 - x_j(t-\tau))\gamma_{ij}(\tau)d\tau + H_i(t), \quad (i = 1, ..., n),$$

where $H_i(t) = \int_0^t h_i(s)ds$ is assumed to be positive for some i and t, so that the infection from outside has some effect. Each $\gamma_{ij}(\tau)$ and $\gamma_{ij}^*(\tau)$ is restricted to be bounded with continuous, bounded derivatives. We are interested in the development of an epidemic within the population, which is only triggered from outside. Hence we restrict $\gamma_{ij}^* = \int_0^\infty \gamma_{ij}^*(\tau)d\tau$ to be finite for all i, j. In addition each $\varepsilon_j(\tau)$ is taken to be continuous with ε_j finite.

It is simple to show that the solution we require (for the epidemic model described by equations (2.8)) has each $x_i(t) > 0$ for $t \geq 0$. Assume that this condition does not hold. From continuity and monotonicity of the $x_i(t)$ with $x_i(0) = 1$, this implies that there exists an i and $T > 0$ such that $x_i(t) > 0$ for $0 \leq t < T$ and $x_i(t) = 0$ for $t \geq T$. For $0 \leq t < T$, from equations (2.8) we obtain

$$0 \leq \frac{d(-\log x_i(t))}{dt} \leq \sum_{j=1}^{n} \sup_\tau \gamma_{ij}(\tau) \int_0^t I_i(t,\tau)d\tau + \sum_{j=1}^{k} \sup_\tau \gamma_{ij}^*(\tau)\varepsilon_j \leq C_i,$$

where $C_i = \sum_{j=1}^n \sup_\tau \gamma_{ij}(\tau) + \sum_{j=1}^k \sup_\tau \gamma_{ij}^*(\tau)\varepsilon_j$. Hence $x_i(t) \geq e^{-C_iT}$ for all $0 \leq t < T$ and continuity implies that $x_i(T) \geq e^{-C_iT} > 0$, which gives a contradiction.

LEMMA 2.1. *There exists a one to one correspondence between the positive monotone decreasing, differentiable solutions $x_i(t)$ $(i = 1, ..., n)$, with each $x_i(0) = 1$ and $x_i'(t)$ continuous, of equations (2.9) and (2.10). Any such solution has $x_i'(t)/x_i(t)$ bounded.*

PROOF. It is first shown that $h_i(t)$ is a continuous function of t. Now

$$|h_i(t+\delta t) - h_i(t)| \leq \sum_{j=1}^{k} \int_0^\infty |\gamma_{ij}^*(t+\tau+\delta t) - \gamma_{ij}^*(t+\tau)|\varepsilon_j(\tau)d\tau.$$

The result follows since $\gamma_{ij}^{*'}(t)$ is continuous and bounded and hence $\gamma_{ij}^*(t)$ is uniformly continuous for all i and j, and $\int_0^\infty \varepsilon_j(\tau)d\tau = \varepsilon_j$ exists for all $j = 1, ..., k$.

Consider a solution to equations (2.9) of the specified form. Integrating these equations from 0 to t we obtain

$$-\log x_i(t) = \sum_{j=1}^{n} \int_0^t \int_0^s (-x_j'(s-\tau))\gamma_{ij}(\tau)d\tau ds + H_i(t).$$

It follows that $x_i(t)$ satisfies equations (2.10) by noting that the jth term in the summation is equal to

$$\int_0^t \int_0^{t-\tau} (-x_j'(u))\gamma_{ij}(\tau)dud\tau = \int_0^t (1 - x_j(t-\tau))\gamma_{ij}(\tau)d\tau.$$

Now consider a solution to equations (2.10) of the specified form. Continuity of $h_i(t)$ implies that $H_i'(t) = h_i(t)$. Now each $\gamma_{ij}(t)$ is continuous, $x_i'(t)$ has bounded derivatives and $(1 - x_i(0)) = 0$. So we can differentiate equations (2.10) and use dominated convergence to obtain equations (2.9). From equations (2.9), $|x_i'(t)/x_i(t)| \leq \sum_{j=1}^{n} \sup_\tau \lambda_{ij}(\tau) + \sum_{j=1}^{k} \sup_\tau \lambda_{ij}^*(\tau)\varepsilon_j$.

□

Our analysis is therefore based on equations (2.10), which we term the epidemic equations. It is shown in Theorem 2.1 that there is a unique solution $x_i(t)$ $(i = 1, ..., n)$ to equations (2.10), which satisfies the conditions on $x_i(t)$ specified in Lemma 2.1. Therefore, from Lemma 2.1, the unique solution constructed in Theorem 2.1 provides the unique solution of the form required for the multi-type epidemic model described by equations (2.9), or equivalently equations (2.8).

THEOREM 2.1 (EXISTENCE AND UNIQUENESS OF SOLUTION). *There exists a solution $x_i(t)$ to equations (2.10), with $0 < x_i(t) \leq 1$ for $t \geq 0$ and $x_i(0) = 1$ $(i = 1, ..., n)$, which is unique. For this solution, each $x_i(t)$ is a monotone decreasing, differentiable function of t with continuous derivative.*

PROOF. Write $w_i(t) = -\log x_i(t)$, so that equations (2.10) become

$$w_i(t) = \sum_{j=1}^{n} \int_0^t \left(1 - e^{-w_j(t-\tau)}\right) \gamma_{ij}(\tau)d\tau + H_i(t), \quad (i = 1, ..., n). \tag{2.11}$$

The existence, uniqueness, monotonicity and continuity are first established.

Since each $\gamma_{ij}(t)$ is bounded, define $\theta = \max_{ij} \sup_{t\geq 0} \gamma_{ij}(t)$ and choose $k > 0$ such that $nk\theta < 1$. Let $\rho = nk\theta$ and define $T_s = sk$ for all non-negative integers s, so that $\lim_{s\to\infty} T_s = \infty$. A solution is constructed in the interval $[T_{s-1}, T_s]$ successively for each $s = 1, 2, ...$ We start with $s = 1$ and construct a solution in the interval $[0, T_1]$. For $t \in [0, T_1]$, define $w_i^{(0)}(t) = H_i(t)$ and define $w_i^{(m+1)}(t)$, for $m \geq 0$, recursively by

$$w_i^{(m+1)}(t) = \sum_{j=1}^{n} \int_0^t \left(1 - e^{-w_j^{(m)}(t-\tau)}\right) \gamma_{ij}(\tau)d\tau + H_i(t). \tag{2.12}$$

Since $H_i(t)$ is a monotone increasing function and (from the proof of Lemma 2.1) is continuous, $w_i^{(0)}(t)$ is monotone increasing and continuous. It is simple to show recursively that each $w_i^{(m)}(t)$ is also a monotone increasing, continuous function of t for $t \in [0, T_1]$.

From equations (2.12) $0 \leq w_i^{(m+1)}(t) \leq \rho + H_i(T_1)$ for all $m \geq 0$ and $|w_i^{(1)}(t) - w_i^{(0)}(t)| \leq \rho$. Also, since $|e^{-x} - e^{-y}| \leq |x - y|$ for all $x, y \geq 0$, for $m \geq 1$

$$|w_i^{(m+1)}(t) - w_i^{(m)}(t)| \leq \sum_{j=1}^{n} \int_0^t |w_j^{(m)}(t-\tau) - w_j^{(m-1)}(t-\tau)|\gamma_{ij}(\tau)d\tau.$$

Define $y_i^{(m)} = \sup_{t\in[0,T_1]} |w_i^{(m)}(t) - w_i^{(m-1)}(t)|$ for $m \geq 1$, let $\{\mathbf{y}^{(m)}\}_i = y_i^{(m)}$ and denote an n-vector of ones by $\mathbf{1}$. Then $y_i^{(1)} \leq \rho$ and, for $m \geq 1$, $\mathbf{y}^{(m+1)} \leq k\theta\mathbf{1}\mathbf{1}'\mathbf{y}^m$. Hence $y_i^{(m+1)} \leq k\theta\mathbf{1}'\mathbf{y}^{(m)}$ and $\mathbf{1}'\mathbf{y}^{(m+1)} \leq \rho\mathbf{1}'\mathbf{y}^{(m)}$ so that $\mathbf{1}'\mathbf{y}^{(m)} \leq n\rho^m$ and therefore $y_i^{(m+1)} \leq \rho^{m+1}$. Since $\rho < 1$, for any $l \geq 1$ and $m \geq 0$,

$$\sup_{t\in[0,T_1]} |w_i^{(m+l)}(t) - w_i^{(m)}(t)| \leq \sum_{j=m+1}^{m+l} y_i^{(j)} \leq \sum_{j=m+1}^{m+l} \rho^j \leq \rho^{m+1}/(1-\rho).$$

It immediately follows that the sequence of monotone increasing, continuous functions $w_i^{(m)}(t)$ is Cauchy convergent, uniformly for $t \in [0, T_1]$, to a monotone increasing, continuous limit $w_i(t)$ for $i = 1, ..., n$. Let $m \to \infty$ in equations (2.12). Using dominated convergence we obtain equations (2.11) for $t \in [0, T_1]$. Note that $w_i(0) = 0$ since $w_i^{(m)}(0) = 0$ for all m.

The uniqueness of the solution to equations (2.11) for $t \in [0, T_1]$ is now established by a contradiction argument. Let $w_i(t)$ and $w_i^*(t)$ be two distinct solutions. From equations (2.11), both solutions are bounded above by $\rho + H_i(T_1)$. Let

$$\{\mathbf{u}\}_i = ||w_i(t) - w_i^*(t)|| = \sup_{t\in[0,T_1]} |w_i(t) - w_i^*(t)|.$$

Then $\mathbf{u} \gneq \mathbf{0}$. From equations (2.11), $\mathbf{u} \leq k\theta\mathbf{1}\mathbf{1}'\mathbf{u}$. Then $\mathbf{1}'\mathbf{u} \leq \rho\mathbf{1}'\mathbf{u} < \mathbf{1}'\mathbf{u}$, and a contradiction is obtained. Hence the solution is unique.

This establishes the existence, uniqueness, monotonicity and continuity of the solution $w_i(t)$ for $i = 1, ..., n$ and $t \in [0, T_1]$. We now assume these results hold for $w_i(t)$ for $i = 1, ..., n$ and $t \in [0, T_s]$.

For $t \in [T_s, T_{s+1}]$ we define

$$w_i^{(0)}(t) = w_i(T_s) + \int_{T_s}^t h_i(u)du \equiv \sum_{j=1}^{n} \int_0^{T_s} \left(1 - e^{-w_j(\tau)}\right) \gamma_{ij}(t-\tau)d\tau + H_i(t),$$

and define $w_i^{(m)}(t)$ recursively by

$$w_i^{(m+1)}(t) = \sum_{j=1}^{n} \int_{T_s}^t \left(1 - e^{-w_j^{(m)}(\tau)}\right) \gamma_{ij}(t-\tau)d\tau + w_i(T_s) + \int_{T_s}^t h_i(u)du.$$

Then $w_i^{(m)}(T_s) = w_i(T_s)$ for all $m \geq 0$ and the proofs of existence, uniqueness, monotonicity and continuity for $t \in [T_s, T_{s+1}]$ are identical to those for $t \in [0, T_1]$. The existence, uniqueness, monotonicity and continuity of the solution $w_i(t)$ then follows for $i = 1, ..., n$ and $t \in [0, T_{s+1}]$.

Since $\lim_{s\to\infty} T_s = \infty$, by induction on s the global existence, uniqueness, monotonicity and continuity of the solution $w_i(t)$ is established for all $i = 1, ..., n$ and $t \geq 0$. Define $x_i(t) = e^{-w_i(t)}$, where $w_i(t)$ $(i = 1, ..., n)$ is the constructed solution. The existence, uniqueness, monotonicity and continuity for the $x_i(t)$ is then immediate.

Finally the differentiability results are established. Let $x_i(t)$, $(i = 1, ..., n)$, be a monotone decreasing, continuous solution of equations (2.10) with $x_i(0) = 1$. From the proof of Lemma 2.1, $H_i(t)$ is differentiable and its differential $h_i(t)$ is continuous.

Consider the jth term in the summation for the ith equation of (2.10). This may be written as $\int_0^t (1 - x_j(\tau))\gamma_{ij}(t - \tau)d\tau$. Now $x_j(t)$ is continuous with $x_j(0) = 1$ and $\gamma'_{ij}(t)$ is continuous and bounded. Using dominated convergence this term is differentiable and its differential is $\int_0^t (1 - x_j(\tau))\gamma'_{ij}(t - \tau)d\tau$, which is a continuous function of t. This then establishes the differentiability and continuity of $x'_i(t)$.

$\square$

2.5 The final size

Attention is now turned to the final size of the epidemic when the infection is introduced from outside. As a corollary the condition for a major epidemic to occur is obtained. This has been studied for the one type epidemic by Diekmann [D5] and Thieme [T1,T2]. The host-vector and n-type epidemics are considered in Radcliffe and Rass [R2, R4].

Consider the limiting behaviour of $x_i(t)$. Since $x_i(t)$ is monotone decreasing and $0 < x_i(t) \leq 1$, it tends to a limit as $t \to \infty$. Let $v_i = 1 - \lim_{t\to\infty} x_i(t)$. Then v_i is the proportion of population i individuals who eventually suffer from the epidemic, so is the final size of the epidemic for population i.

It is easily established that $a_i = \lim_{t\to\infty} H_i(t)$ exists and is finite. The assumption was made that $H_i(t) > 0$ for some i and $t > 0$. If this assumption did not hold, the infectives introduced from outside would not have any infectious influence on the populations under consideration. Since $H_i(t)$ is monotone increasing in t, this implies that $a_i > 0$ for at least one i.

The equations satisfied by v_i, $(i = 1, ..., n)$, are obtained. These are called the final size equations. These equations are shown to have a unique solution, so that the final size for each type is determined. A lower bound, not dependent on the amount of initial infection, is obtained. This bound is also the limit of the final size as the amount of initial infection tends to zero. This establishes the condition for a major epidemic. An epidemic is not major if each final size tends to zero as the amount of initial infection tends to zero. It is major otherwise.

Since ε_j is the relative number of infected individuals of type j initiating the epidemic, the amount of initial infection tends to zero if all the ε_j tend to zero. Since $a_i = \lim_{t\to\infty} H_i(t) \leq \sum_{j=1}^k \gamma_{ij}^* \varepsilon_j$, this is equivalent to considering the limit of each final size v_i as the a_j all tend to zero.

We are considering a non-reducible epidemic, therefore $\mathbf{\Gamma} = (\gamma_{ij})$ is a non-negative, non-reducible matrix. Unlike the γ^*_{ij}, the γ_{ij} were not restricted to be finite. The $S \to I$ epidemic provides an example when $\mathbf{\Gamma}$ has all its non-zero entries infinite. When $\mathbf{\Gamma}$ is finite, $\rho(\mathbf{\Gamma})$ denotes its Perron-Frobenius root (see Appendix A). If $\mathbf{\Gamma}$ has at least one infinite entry we define $\rho(\mathbf{\Gamma}) = \infty$.

It is instructive to first look at the one-type, host-vector and general 2-type epidemics in the case when $\mathbf{\Gamma}$ is finite. This is done in an intuitive manner. The rigorous proofs are then given for the n-type model.

The one-type epidemic.

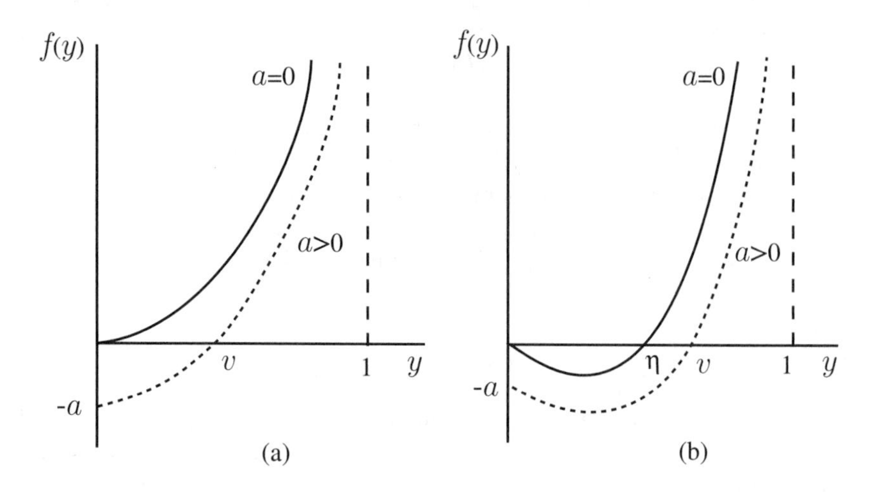

FIGURE 2.1. The function $f(y) = -\log(1-y) - \gamma y - a$ for the cases (a) $\gamma \leq 1$ and (b) $\gamma > 1$.

If $n = 1$ and the subscript notation is dropped, equations (2.10) become

$$-\log x(t) = \int_0^t (1 - x(t-\tau))\gamma(\tau)d\tau + H(t). \tag{2.13}$$

If γ is finite, equation (2.13) suggests that $y = v$ satisfies the equation

$$-\log(1-y) = \gamma y + a. \tag{2.14}$$

Consider the function $f(y) = -\log(1-y) - \gamma y - a$. Then $f(0) = -a$ and $f'(0) = 1-\gamma$, so is negative if $\gamma > 1$. Also $f''(y) = 1/(1-y)^2$, which is positive, so that f(y) is a convex function.

From figure 2.1 we see that for $a > 0$ equation (2.14) has a unique positive root v. Thus, for the one type epidemic with γ finite, v is the unique positive root of equation (2.14).

Note that if $\gamma \leq 1$, figure 2.1(a) shows that $v \to 0$ as $a \to 0$, so that the epidemic is not major . However if $\gamma > 1$, figure 2.1(b) shows that $v \to \eta$ as $a \to 0$, where η is the unique positive solution to equation (2.14) with $a = 0$. In this case the epidemic is major and η provides a lower bound for the proportion eventually suffering the epidemic, i.e. for the final size.

The condition for a major epidemic is therefore $\gamma > 1$. For the one-type model $\rho(\mathbf{\Gamma}) = \gamma$, hence this condition is equivalent to $\rho(\mathbf{\Gamma}) > 1$.

The host-vector epidemic.

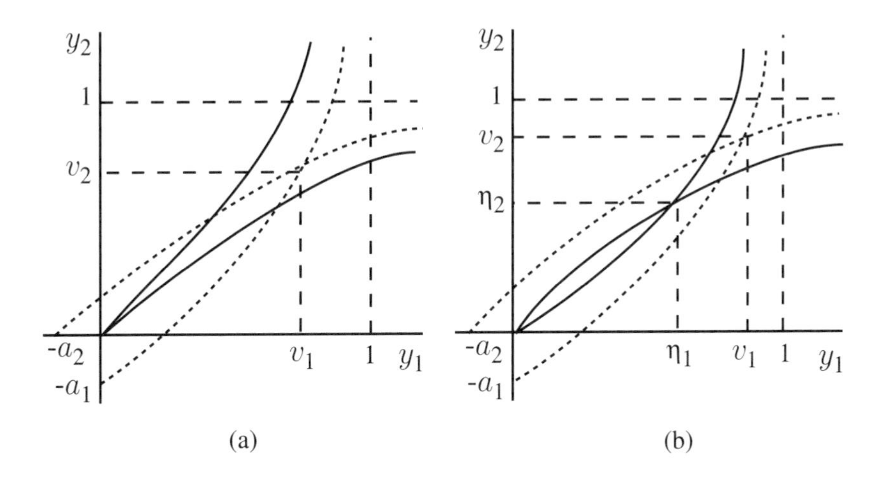

FIGURE 2.2. The pair of curves $y_2 = (-\log(1-y_1) - a_1)/\gamma_{12}$, and $y_1 = (-\log(1-y_2) - a_2)/\gamma_{21}$, for (a) $\gamma_{12}\gamma_{21} \le 1$; (b) $\gamma_{12}\gamma_{21} > 1$. The unbroken curves correspond to $a_1 = a_2 = 0$.

For the host-vector epidemic we need to take $n = 2$ in equations (2.10) and set $\gamma_{11} = \gamma_{22} = 0$. Then the corresponding pair of equations suggest that $y_i = v_i$, $i = 1, 2$ satisfy

$$-\log(1-y_1) = \gamma_{12}y_2 + a_1,$$
$$-\log(1-y_2) = \gamma_{21}y_1 + a_2.$$

These may be rewritten as

$$y_2 = -\frac{1}{\gamma_{12}}\log(1-y_1) - \frac{a_1}{\gamma_{12}}, \tag{2.15}$$

$$y_1 = -\frac{1}{\gamma_{21}}\log(1-y_2) - \frac{a_2}{\gamma_{21}}. \tag{2.16}$$

From equation (2.15) y_2 is a convex function of y_1 and, when $y_1 = 0$, $\dfrac{dy_2}{dy_1} = \dfrac{1}{\gamma_{12}}$. From equation (2.16) y_1 is a convex function of y_2 and, when $y_2 = 0$, $\dfrac{dy_2}{dy_1} = \gamma_{21}$. Thus by considering the tangents to the curves with $a_1 = 0$ and $a_2 = 0$ at (0,0), it is seen that when $\gamma_{12}\gamma_{21} \le 1$ we obtain figure 2.2(a), and when $\gamma_{12}\gamma_{21} > 1$ we obtain figure 2.2(b). Since at least one $a_i > 0$, we see from figure 2.2 that the curves intersect in the positive quadrant in a single point. This gives the unique solution to the final size equations.

Note that if $\gamma_{12}\gamma_{21} \le 1$, then v_1 and v_2 both tend to zero as a_1 and a_2 tend to zero, so that a major epidemic does not occur. However, if $\gamma_{12}\gamma_{21} > 1$, as a_1 and

a_2 tend to zero, $v_1 \to \eta_1$ and $v_2 \to \eta_2$, where $y_1 = \eta_1$ and $y_2 = \eta_2$ is the unique positive solution to equations (2.15) and (2.16) with $a_1 = a_2 = 0$. Here η_1 and η_2 provide lower bounds on the proportions eventually suffering the epidemic in the host and vector populations respectively.

The condition for a major epidemic is therefore $\gamma_{12}\gamma_{21} > 1$. Since $\rho(\mathbf{\Gamma}) = \sqrt{\gamma_{12}\gamma_{21}}$ for the host-vector epidemic, this condition is just $\rho(\mathbf{\Gamma}) > 1$.

The general non-reducible 2-type epidemic.

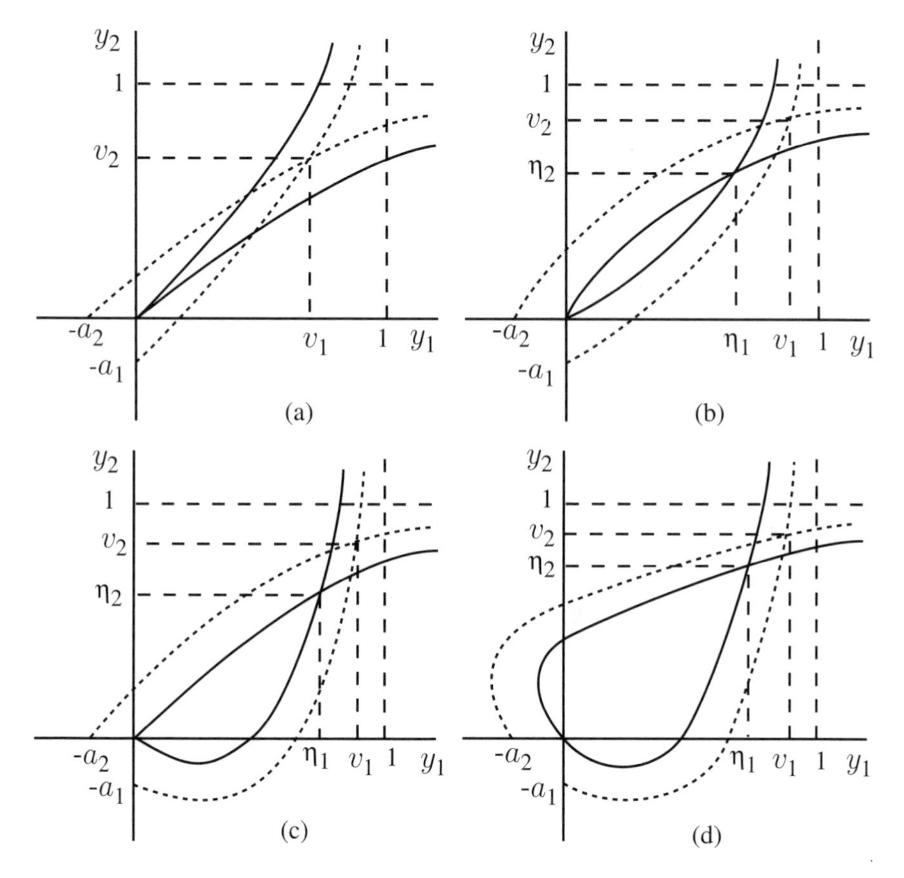

FIGURE 2.3. Plots of the curves $y_2 = \dfrac{-\log(1-y_1) - \gamma_{11}y_1 - a_1}{\gamma_{12}}$ and $y_1 = \dfrac{-\log(1-y_2) - \gamma_{22}y_2 - a_2}{\gamma_{21}}$, for each of four different cases. (a) $\gamma_{11} \le 1$, $\gamma_{22} \le 1$ and $\gamma_{12}\gamma_{21} \le (1-\gamma_{11})(1-\gamma_{22})$; (b) $\gamma_{11} \le 1$, $\gamma_{22} \le 1$ and $\gamma_{12}\gamma_{21} > (1-\gamma_{11})(1-\gamma_{22})$; (c) $\gamma_{11} > 1$ and $\gamma_{22} \le 1$; (d) $\gamma_{11} > 1$ and $\gamma_{22} > 1$. The unbroken curves correspond to $a_1 = a_2 = 0$.

Now consider the general case when $n = 2$, with infection not constrained to only be possible between individuals in different populations. Then equations (2.10) suggest that $y_i = v_i$, $i = 1, 2$ satisfy the equations

$$-\log(1-y_1) = \gamma_{11}y_1 + \gamma_{12}y_2 + a_1,$$
$$-\log(1-y_2) = \gamma_{21}y_1 + \gamma_{22}y_2 + a_2.$$

The pair of equations may be re-written in the form

$$y_2 = \frac{-\log(1-y_1) - \gamma_{11}y_1 - a_1}{\gamma_{12}}, \tag{2.17}$$

$$y_1 = \frac{-\log(1-y_2) - \gamma_{22}y_2 - a_2}{\gamma_{21}}. \tag{2.18}$$

From equation (2.17), y_2 is a convex function of y_1 and $\frac{dy_2}{dy_1} = \frac{(1-\gamma_{11})}{\gamma_{12}}$ when $y_1 = 0$. Similarly from equations (2.18), y_1 is a convex function of y_2 and $\frac{dy_1}{dy_2} = \frac{(1-\gamma_{22})}{\gamma_{21}}$ when $y_2 = 0$. Since at least one of a_1 and a_2 is positive, the curves meet in a unique point $y_1 = v_1$ and $y_2 = v_2$ in the positive quadrant.

Consider the case shown in figure 2.3 (a) and (b), when $\gamma_{11} \leq 1$ and $\gamma_{22} \leq 1$. When $a_1 = a_2 = 0$, the curves meet at $y_1 = y_2 = 0$ and will only also cross in the positive quadrant if $(1-\gamma_{11})(1-\gamma_{22}) < \gamma_{12}\gamma_{21}$. Hence if $\gamma_{11} \leq 1$, $\gamma_{22} \leq 1$ and $(1-\gamma_{11})(1-\gamma_{22}) \geq \gamma_{12}\gamma_{21}$ as in figure 2.3(a), v_1 and v_2 tend to zero as a_1 and a_2 tend to zero, and a major epidemic does not occur. If $\gamma_{11} \leq 1$, $\gamma_{22} \leq 1$ and $(1-\gamma_{11})(1-\gamma_{22}) < \gamma_{12}\gamma_{21}$ as in figure 2.3(b), then as a_1 and a_2 tend to zero, v_1 and v_2 tend to the unique positive solution $y_1 = \eta_1$ and $y_2 = \eta_2$ to equations (2.17) and (2.18) with $a_1 = a_2 = 0$.

Now consider the cases when $\gamma_{11} > 1$ and/or $\gamma_{22} > 1$. From figures 2.3(c) and 2.3(d) when $a_1 = a_2 = 0$ the curves cross at the origin and at a single point in the positive quadrant. As a_1 and a_2 tend to zero, $v_1 \to \eta_1$ and $v_2 \to \eta_2$, where $y_1 = \eta_1$ and $y_2 = \eta_2$ is the unique positive solution to equations (2.17) and (2.18) with $a_1 = a_2 = 0$. Figure 2.3(c) corresponds to the case when $\gamma_{11} > 1$ and $\gamma_{22} \leq 1$ while figure 2.3(d) corresponds to the case when $\gamma_{11} > 1$ and $\gamma_{22} > 1$.

For this two-type model the epidemic will not be major if $\gamma_{11} \leq 1$, $\gamma_{22} \leq 1$ and $\gamma_{12}\gamma_{21} \leq (1-\gamma_{11})(1-\gamma_{22})$. The condition $\rho(\mathbf{\Gamma}) \leq 1$ occurs precisely when $|\mathbf{\Gamma} - \mathbf{I}| \geq 0$ and $trace(\mathbf{\Gamma} - \mathbf{I}) \leq 0$, which occurs when $\gamma_{11} \leq 1$, $\gamma_{22} \leq 1$ and $(1-\gamma_{11})(1-\gamma_{22}) \leq \gamma_{12}\gamma_{21}$. So the condition for a major epidemic is $\rho(\mathbf{\Gamma}) > 1$. When this condition holds, η_1 and η_2 provide lower bounds on the final sizes.

The multi-type epidemic.

The possible non-negative solutions of certain systems of equations are derived in Appendix B. The final size equations for the n-type model are a special case. The results of Appendix B are thus used to prove the results for the final size equations for the n-type model.

THEOREM 2.2 (THE FINAL SIZE EQUATIONS).

1. *If $\mathbf{\Gamma}$ is finite, then $y_i = v_i, (i = 1, ..., n)$, is the unique positive solution to*

$$-\log(1-y_i) = \sum_{j=1}^{n} \gamma_{ij}y_j + a_i, \quad (i = 1, ..., n). \tag{2.19}$$

2. *If $\mathbf{\Gamma}$ has an infinite element in every row, then $v_i = 1$ for $i = 1, ..., n$.*
3. *The remaining case corresponds to a partitioning of $\mathbf{\Gamma}$ (by permutation of the indices) into*

$$\mathbf{\Gamma} = \begin{pmatrix} \mathbf{\Gamma}_{11} & \mathbf{\Gamma}_{12} \\ \mathbf{\Gamma}_{21} & \mathbf{\Gamma}_{22} \end{pmatrix},$$

where $(\mathbf{\Gamma}_{11}\ \mathbf{\Gamma}_{12})$ has no infinite element, and $(\mathbf{\Gamma}_{21}\ \mathbf{\Gamma}_{22})$ has at least one infinite element in each row. Let $\mathbf{v}$ and $\mathbf{a}$ be the n-vectors with $\mathbf{v}' = (\mathbf{v}_1'\ \mathbf{v}_2')$ and $\mathbf{a}' = (\mathbf{a}_1'\ \mathbf{a}_2')$. Then $\mathbf{v}_2 = \mathbf{1}$, and $\mathbf{y} = \mathbf{v}_1$ is the unique positive solution to

$$\mathbf{u} = \mathbf{\Gamma}_{11}\mathbf{y} + \mathbf{\Gamma}_{12}\mathbf{1} + \mathbf{a}_1, \tag{2.20}$$

where $\{\mathbf{u}\}_i = -\log(1 - \{\mathbf{y}\}_i)$.

PROOF. First observe that $H_i(t)$ is monotone increasing and bounded above by $\sum_{j=1}^{k} \gamma_{ij}^* \varepsilon_j$ so that $a_i = \lim_{t\to\infty} H_i(t)$ exists and is finite.

1. Consider $\int_0^t (1 - x_j(t-\tau))\gamma_{ij}(\tau)d\tau$ This can be expressed as

$$\int_0^\infty (1 - x_j(t-\tau))\gamma_{ij}(\tau)d\tau - \int_t^\infty (1 - x_j(t-\tau))\gamma_{ij}(\tau)d\tau.$$

The first term tends by monotone convergence to $\gamma_{ij}v_j$ as $t \to \infty$. Now $|\int_t^\infty (1 - x_j(t-\tau))\gamma_{ij}(\tau)d\tau| \le \int_t^\infty \gamma_{ij}(\tau)d\tau$, which tends to zero as t tends to infinity. Hence letting t tend to infinity in equations (2.10), we obtain

$$-\log(1 - v_i) = \sum_{j=1}^{n} \gamma_{ij}v_j + a_i, \ \ (i = 1, ..., n),$$

so that $y_i = v_i$, $i = 1, ..., n$, is a solution to equation (2.19). The uniqueness and positivity of the solution to equations (2.19) is an immediate consequence of Theorem B.2 of the Appendices, with $g(y) = -\log(1 - y)$.
2. If $v_i = 0$ for some i, it immediately follows that $x_i(t) \equiv 1$. From the ith equation of (2.10), for any j such that $\gamma_{ij} \ne 0$ it then follows that $x_j(t) \equiv 1$ and hence $v_j = 0$. Since $\mathbf{\Gamma}$ is non-reducible this then implies that $v_i = 0$ for all i. From equations (2.10) this is impossible since $H_i(t) > 0$ for some t and i. Hence $v_i > 0$ for all i.

   Suppose that γ_{ij} is infinite. Since $v_j > 0$, there exists a T_0 such that $(1 - x_j(t)) > \frac{1}{2}v_j$, for $t > T_0$. Hence

$$\int_0^t (1 - x_j(t-\tau))\gamma_{ij}(\tau)d\tau > \frac{1}{2}v_j \int_0^{t-T_0} \gamma_{ij}(\tau)d\tau,$$

for $t > T_0$. Therefore $\lim_{t\to\infty} \int_0^t (1 - x_j(t-\tau))\gamma_{ij}(\tau)d\tau = \infty$. Hence from equations (2.10) $\lim_{t\to\infty}(-\log x_i(t)) = \infty$, i.e. $v_i = 1$. Since $\mathbf{\Gamma}$ has an infinite element in each row, $v_i = 1$ for $i = 1, ..., n$.
3. It follows as in part 2 that $\mathbf{v}_2 = \mathbf{1}$. Suppose $\mathbf{\Gamma}_{11}$ is of order m. Using monotone convergence in equations (2.10) for $i = 1, ..., m$, we obtain equations

(2.20) with $\mathbf{y} = \mathbf{v}_1$. The uniqueness and positivity follow from Theorem B.2 of the Appendices with $g(y) = -\log(1-y)$. Note that $\mathbf{\Gamma}_{11}$ may be reducible, in which case $\mathbf{a} = \mathbf{\Gamma}_{12}\mathbf{1} + \mathbf{a}_1$ is of the form specified in part 4 of Theorem B.2.

□

The final theorem gives the condition for a major epidemic to occur. In addition when a major epidemic occurs it gives both lower bounds on the final sizes and the limits of the final sizes as the amount of initial infection tends to zero.

THEOREM 2.3 (CONDITION FOR A MAJOR EPIDEMIC).

1. *If $\rho(\mathbf{\Gamma}) \leq 1$, a sufficiently small amount of initial infection will not initiate a major epidemic, i.e. $v_i \to 0$, (i=1,...,n), as $\varepsilon_1, ..., \varepsilon_k$ all tend to zero.*
2. *If $\rho(\mathbf{\Gamma}) > 1$, a major epidemic will occur, no matter how little infection is introduced. In particular the following results hold:*
   (i) If $\rho(\mathbf{\Gamma})$ is finite, then $v_i \geq \{\boldsymbol{\eta}(\mathbf{0})\}_i$ and $v_i \to \{\boldsymbol{\eta}(\mathbf{0})\}_i$, as $\varepsilon_1, ..., \varepsilon_k$ all tend to zero; where $y_i = \{\boldsymbol{\eta}(\mathbf{0})\}_i$, is the unique positive solution to

$$-\log(1-y_i) = \sum_{j=1}^{n} \gamma_{ij} y_j, \quad (i = 1, ..., n).$$

   (ii) If $\mathbf{\Gamma}$ has at least one infinite element in each row, then $v_i = 1$ for all i.
   (iii) Suppose that $\mathbf{\Gamma}$ and $\mathbf{v}$ are partitioned as in Theorem 2.2, part 3. Let $\mathbf{y} = \boldsymbol{\eta}^*(\mathbf{b})$ be the unique solution of

$$-\log(1 - \{\mathbf{y}\}_i) = \{\mathbf{\Gamma}_{11}\mathbf{y}\}_i + \{\mathbf{b}\}_i, \quad (i = 1, ..., m),$$

   where $\mathbf{b}$ is non-negative and not identically zero. Then $\mathbf{v}_2 = \mathbf{1}$ and $\mathbf{v}_1 \geq \boldsymbol{\eta}^*(\mathbf{\Gamma}_{12}\mathbf{1})$ and $\mathbf{v}_1 \to \boldsymbol{\eta}^*(\mathbf{\Gamma}_{12}\mathbf{1})$ as $\varepsilon_1, ..., \varepsilon_k$ all tend to zero.

PROOF. Observe that each $a_i \to 0$ as $\varepsilon_j \to 0$ for all $j = 1, ..., k$. The results then follow immediately from Theorem B.2 of the Appendices and Theorem 2.2. □

A stochastic n-type model with constant infectivity rates λ_{ij} and removal rates μ_j, the stochastic equivalent of the $S \to I \to R$ model, was considered by Griffiths [G2]. He used a branching process approximation to describe the initial development of the epidemic. Eventual extinction of the population with probability one was shown to occur if and only if $\rho(\mathbf{\Gamma}) \leq 1$. This is precisely the condition obtained in Theorem 2.3 for the deterministic epidemic not to be major.

Note that the proofs of this chapter require the functions $h_i(t)$ to satisfy certain conditions, namely that each $h_i(t)$ is continuous and each $a_i = \int_0^\infty h_i(t)dt$ is finite and tends to zero as the amount of initial infection tends to zero. Hence the results are still be valid if infection is triggered in a somewhat different fashion, so that the specification of the $h_i(t)$ changes, provided these conditions are still met. In particular these conditions will hold for the $S \to L \to I \to R$ (and subset) models provided the individuals introduced from outside are being removed or leaving at some positive rate (so that their infectious influence decays with time).

CHAPTER 3

Bounds on the spatial final size

3.1 Specification of the model

The spatial analogue of model 2* of Chapter 2 is now described. This models the spread of infection through populations of susceptible individuals in $\Re^N$, when infection from outside the populations is introduced at time $t = 0$. This initial infection then triggers an epidemic within the populations.

Results analogous to those of the non-spatial models are derived in this chapter concerning the spatial final size. Equations for the final sizes for different types are obtained, and uniform upper and lower bounds derived. These give both the condition for a major epidemic to occur and enable a limiting result to be obtained concerning the spatial final size. Other results, specific to spatial models only, are derived in Chapters 4 and 5.

Consider n types of individuals, each type having uniform density in $\Re^N$. The density of type i individuals is σ_i. Then $\sigma = \sum_{i=1}^n \sigma_i$ is the total density for these types. Denote the proportion in population i at position s who are susceptible at time t by $x_i(\mathbf{s}, t)$. Let $I_i(\mathbf{s}, t, \tau)d\tau$ be the proportion of individuals in population i at position $\mathbf{s}$ who were infected by time t, that infection occurring in the time interval $(t-\tau-d\tau, t-\tau)$. Then $I_i(\mathbf{s}, t, \tau)d\tau\sigma_i d\mathbf{s}$ represents the number of infected individuals at time t who were infected in this time interval, and $x_i(\mathbf{s}, t)\sigma_i d\mathbf{s}$ represents the corresponding number of susceptibles at time t, who are in the region $\{\mathbf{r} \in \Re^N : \mathbf{s} \le \mathbf{r} \le \mathbf{s} + d\mathbf{s}\}$.

The infection rates will depend upon the vector distance between an infective and a susceptible individual. If infection is spread by contact, each contact rate varies with the distance apart and is likely to be higher for individuals who are spatially closer. Each $\lambda_{ij}(\tau)$ for the non-spatial model is therefore replaced by an infection rate for individuals at vector displacement $\mathbf{r}$. It is convenient to write this as $\lambda_{ij}(\tau) \times p_{ij}(\mathbf{r})$, where $\lambda_{ij}(\tau)$ represents the integrated contact rate (over $\mathbf{r}$) times the infection rate per contact of a type i susceptible by an individual in population j who was infected time τ ago and $p_{ij}(\mathbf{r})$ integrates to 1 and is termed the contact distribution. One would expect that $p_{ij}(\mathbf{r}) = p_{ji}(\mathbf{r})$; however this restriction is not imposed since it gives no simplification in the analysis. In general the infection rate by type i infectives of type j susceptibles differs from the corresponding rate of infection by type j infectives of type i susceptibles. For the spatial model we will refer to $\lambda_{ij}(\tau)$ as the infection rate for individuals infected time τ ago.

Infected individuals of k types are introduced from outside at time $t = 0$. As in Chapter 2 we consider the relative number of individuals of different types introduced from outside. For individuals introduced at position $\mathbf{s}$ this will be relative to the total population density at this position, which is constant over $\mathbf{s}$. Let $\sigma\varepsilon_j(\mathbf{s},\tau)d\mathbf{s}d\tau$ be the number of such individuals of type j in the region $(\mathbf{s},\mathbf{s}+d\mathbf{s})$ who were infected in the time interval $(-\tau-d\tau,-\tau)$. The rate of infection, by such individuals, of susceptibles from population i is $\lambda^*_{ij}(\tau)$ and the corresponding contact distribution when the vector displacement is $\mathbf{r}$ is $p^*_{ij}(\mathbf{r})$. Then $\sigma\varepsilon_j(\mathbf{s}) = \int_0^\infty \sigma\varepsilon_j(\mathbf{s},\tau)d\tau$ is the density of these type j infectives at position $\mathbf{s}$.

In similar manner to Chapter 2 the model is described, for $0 \le \tau \le t$ and $t \ge 0$, by the equations

(3.1)

$$\frac{\partial x_i(\mathbf{s},t)}{\partial t} = -x_i(\mathbf{s},t)\left(\sum_{j=1}^n \sigma_j \int_{\Re^N}\int_0^t I_j(\mathbf{s}-\mathbf{r},t,\tau)p_{ij}(\mathbf{r})\lambda_{ij}(\tau)d\tau d\mathbf{r} + h_i(\mathbf{s},t)\right),$$

$$x_i(\mathbf{s},t) = 1 - \int_0^t I_i(\mathbf{s},t,\tau)d\tau,$$

$$I_i(\mathbf{s},t,\tau) = I_i(\mathbf{s},t-\tau,0),\ (i=1,...,n),$$

where $h_i(\mathbf{s},t) = \sum_{j=1}^k \int_{\Re^N}\int_0^\infty \varepsilon_j(\mathbf{s}-\mathbf{r},\tau)p^*_{ij}(\mathbf{r})\sigma\lambda^*_{ij}(t+\tau)d\tau d\mathbf{r}$. The initial conditions are $x_i(\mathbf{s},0)\equiv 1$, for $i=1,...,n$. It is assumed that $h_i(\mathbf{s},t)>0$ for some i and for $\mathbf{s}$ and t in some open region, so that the individuals from outside have some infectious influence.

The interpretation of the $x_i(\mathbf{s},t)$ and $I_i(\mathbf{s},t,\tau)$, and equations (3.1), give the following conditions. Each $x_i(\mathbf{s},t)$ is non-negative and bounded above by 1, is jointly continuous in $\mathbf{s}$ and t and is partially differentiable with respect to t. The $\varepsilon_j(\mathbf{s},\tau)$ and $I_i(\mathbf{s},t,\tau)$, and hence the $I_i(\mathbf{s},\tau,0)$, are jointly continuous in $\mathbf{s}$ and τ. From equations (3.1) $x_i(\mathbf{s},t) = 1-\int_0^t I_i(\mathbf{s},u,0)du$ and hence $(\partial x_i(\mathbf{s},t)/\partial t) = -I_i(\mathbf{s},t,0) \le 0$. Therefore $x_i(\mathbf{s},t)$ is monotone decreasing in t and its partial derivative with respect to t is jointly continuous in $\mathbf{s}$ and t.

The conditions on the infection rates are the same as in the non-spatial case. As in Sections 2.3 and 2.4, define $\gamma_{ij}(\tau) = \sigma_j\lambda_{ij}(\tau)$ and $\gamma^*_{ij}(\tau) = \sigma\lambda^*_{ij}(\tau)$. The $\gamma_{ij}(t)$ and $\gamma^*_{ij}(t)$ are restricted to be bounded with continuous, bounded derivatives. Since the epidemic is only triggered by infection from outside, each integrated infection rate, $\gamma^*_{ij} = \int_0^\infty \gamma^*_{ij}(t)dt$, is taken to be finite. Also each $\varepsilon_i(\mathbf{s})$ is taken to be continuous and is assumed to be bounded (and would usually be small) with $\varepsilon_i = \int_{\Re^N}\varepsilon_i(\mathbf{s})d\mathbf{s}$ finite.

The contact distributions are assumed to have densities which are bounded continuous functions and are radially symmetric. The condition of radial symmetry is needed to obtain a lower bound on the proportion of individuals who eventually suffer the epidemic which is valid at all spatial positions for a given type. Note that general contact distributions could allow infection to spread in restricted directions only, so that there would be a region of $\Re^N$ in which no individuals were affected by the epidemic. The pandemic theorem, giving a uniform lower bound on the spatial final size, would then be vacuous.

It is then easy to show that each $x_i(\mathbf{s},t)$ is positive for all $\mathbf{s}$ and $t\ge 0$, and that its partial derivative with respect to t is uniformly bounded. From equations (3.1)

$$0 \leq -\frac{\partial x_i(\mathbf{s},t)}{\partial t} \leq x_i(\mathbf{s},t)\sum_{j=1}^{n}\sup_{\tau}\gamma_{ij}(\tau)\int_{\Re^N}(1-x_j(\mathbf{s}-\mathbf{r},t))p_{ij}(\mathbf{r})d\mathbf{r}$$

$$+\, x_i(\mathbf{s},t)\sum_{j=1}^{k}\sup_{\tau}\gamma_{ij}^*(\tau)\sup_{\mathbf{r}}\varepsilon_j(\mathbf{r}) \leq x_i(\mathbf{s},t)C_i,$$

where $C_i = \sum_{j=1}^{n}\sup_{\tau}\gamma_{ij}(\tau) + \sum_{j=1}^{k}\sup_{\tau}\gamma_{ij}^*(\tau)\sup_{\mathbf{r}}\varepsilon_j(\mathbf{r})$. It immediately follows that the partial derivative is uniformly bounded. Now suppose that $x_i(\mathbf{s},t) = 0$ for some i, $\mathbf{s}$ and t. Since $x_i(\mathbf{s},0) = 1$, for this i and $\mathbf{s}$ there exists a T such that $x_i(\mathbf{s},t) > 0$ for $0 \leq t < T$ and $x_i(\mathbf{s},T) = 0$. For $0 \leq t < T$, $-(\partial \log x_i(\mathbf{s},t)/\partial t) \leq C_i$ and hence $x_i(\mathbf{s},t) \geq e^{-C_iT} > 0$. Therefore, using continuity, $x_i(\mathbf{s},T) > 0$ and a contradiction is obtained. Hence $x_i(\mathbf{s},t) > 0$ for all i, $\mathbf{s}$ and $t \geq 0$.

Note that model 1* can also be extended to include a spatial aspect. Let $*$ denote convolution. In a similar manner to Section 2.1, with the obvious definitions, it can be shown that the spatial form of model 1* can be treated as a special case of the spatial form of model 2* with $\lambda_{ij}(t) = \sigma_j\lambda_{ij}\alpha_j\left(e^{-\mu_j t} - e^{-\alpha_j t}\right)/(\alpha_j - \mu_j)$ and

$$h_i(\mathbf{s},t) = \sum_{j=1}^{k}\sigma\lambda_{ij}^*\left(e^{-\mu_j^* t}\varepsilon_j * p_{ij}^*(\mathbf{s}) + \frac{\alpha_j^*\left(e^{-\mu_j^* t} - e^{-(\alpha_j^*+\beta_j^*)t}\right)}{\alpha_j^*+\beta_j^*-\mu_j^*}\phi_j * p_{ij}^*(\mathbf{s})\right),$$

provided $\alpha_i \neq \mu_i$ and $\alpha_i^* + \beta_i^* \neq \mu_i^*$ for all i, with a simple modification when equality occurs. The equivalent constraints on the $\varepsilon_i(\mathbf{s})$ and $\phi_i(\mathbf{s})$ are that they should be continuous, bounded and integrable (the integrals being denoted by ε_i and ϕ_i). For individuals from outside to have some infectious influence requires that $\varepsilon_i > 0$ for some i and/or both $\phi_i > 0$ and $\alpha_i^* > 0$ for some i. Since the infection from outside only triggers the epidemic, $\mu^* > 0$ for all i. It is then simple to show that $h_i(\mathbf{s},t)$ is jointly continuous and uniformly bounded for $\mathbf{s} \in \Re^N$ and $t \geq 0$. Also each $a_i(\mathbf{s}) = \int_0^\infty h_i(\mathbf{s},t)dt$ exists and is continuous and bounded and tends to zero as ε_j and ϕ_j tend to zero for all $j = 1, ..., k$. These are precisely the properties of $h_i(\mathbf{s},t)$ which are required to prove the results of this chapter.

3.2 The existence and uniqueness of solutions

By substituting for the $I_i(\mathbf{s},t,\tau)$ in the first set of equations (3.1) and changing the order of integration, the following equations for the $x_i(\mathbf{s},t)$ are obtained, which are easily shown to be equivalent to equations (3.1).

$$\frac{\partial x_i(\mathbf{s},t)}{\partial t} = x_i(\mathbf{s},t)\times$$

$$\left(\sum_{j=1}^{n}\int_0^t\int_{\Re^N}\frac{\partial x_j(\mathbf{s}-\mathbf{r},t-\tau)}{\partial t}p_{ij}(\mathbf{r})\gamma_{ij}(\tau)d\mathbf{r}d\tau - h_i(\mathbf{s},t)\right), \tag{3.2}$$

for $i = 1, ..., n$, where $h_i(\mathbf{s},t)$ is specified with equations (3.1). As in Chapter 2 define $\mathbf{\Gamma} = (\gamma_{ij})$, where $\mathbf{\Gamma}$ is non-reducible and may have some infinite entries.

The solutions of equations (3.2) can be linked to those of the following equations,

$$-\log x_i(\mathbf{s},t) = \sum_{j=1}^{n} \int_0^t \int_{\Re^N} (1 - x_j(\mathbf{s}-\mathbf{r}, t-\tau)) p_{ij}(\mathbf{r}) \gamma_{ij}(\tau) d\mathbf{r} d\tau \tag{3.3}$$
$$+ H_i(\mathbf{s},t), \ (i = 1, ..., n),$$

where $H_i(\mathbf{s},t) = \int_0^t h_i(\mathbf{s},w)dw$. A preliminary lemma is proved concerning the $H_i(\mathbf{s},t)$ before proving the equivalence of the solutions of equations (3.2) and (3.3).

LEMMA 3.1. *The functions $h_i(\mathbf{s},t)$ and $H_i(\mathbf{s},t)$ are jointly continuous and uniformly bounded for $\mathbf{s} \in \Re^N$ and $t \geq 0$. Also $H_i(\mathbf{s},t)$ tends to a limit $a_i(\mathbf{s})$ as t tends to infinity, which is continuous and bounded and is positive for some i and for $\mathbf{s}$ in some open ball.*

PROOF. Now $h_i(\mathbf{s},t) \leq \sum_{j=1}^{k} \sup_\tau \gamma_{ij}^*(\tau) \sup_{\mathbf{r}} p_{ij}^*(\mathbf{r}) \varepsilon_j$, so $h_i(\mathbf{s},t)$ is uniformly bounded. Also

$$|h_i(\mathbf{s}+\delta\mathbf{s}, t+\delta t) - h_i(\mathbf{s},t)| \leq |h_i(\mathbf{s}+\delta\mathbf{s}, t+\delta t) - h_i(\mathbf{s}+\delta\mathbf{s}, t)| + |h_i(\mathbf{s}+\delta\mathbf{s}, t) - h_i(\mathbf{s},t)|.$$

The first term on the right is bounded above by $|\delta t| \sum_{j=1}^{k} \sup_\tau {\gamma_{ij}^*}'(\tau) \sup_{\mathbf{r}} p_{ij}^*(\mathbf{r}) \varepsilon_j$. The second term is bounded above by

$$\sum_{j=1}^{k} \sup_\tau \gamma_{ij}^*(\tau) \int_{\Re^N} |p_{ij}^*(\mathbf{s}+\delta\mathbf{s}-\mathbf{r}) - p_{ij}^*(\mathbf{s}-\mathbf{r})| \varepsilon_j(\mathbf{r}) d\mathbf{r}.$$

The continuity then follows immediately since the $p_{ij}^*(\mathbf{r})$ are bounded and are uniformly continuous in a bounded interval and the $\varepsilon_j(\mathbf{r})$ are integrable.

Consider $H_i(\mathbf{s},t)$. This is uniformly bounded by $\sum_{j=1}^{k} \gamma_{ij}^* \sup_{\mathbf{r}} p_{ij}^*(\mathbf{r}) \varepsilon_j$. Also

$$\begin{aligned}
&|H_i(\mathbf{s}+\delta\mathbf{s}, t+\delta t) - H_i(\mathbf{s},t)| \\
&= |H_i(\mathbf{s}+\delta\mathbf{s}, t+\delta t) - H_i(\mathbf{s}+\delta\mathbf{s}, t)| + |H_i(\mathbf{s}+\delta\mathbf{s}, t) - H_i(\mathbf{s},t)| \\
&\leq |\delta t| \sup_{\mathbf{r}\in\Re^N, \ \tau\geq 0} h_i(\mathbf{r},\tau) + \sum_{j=1}^{k} \gamma_{ij}^* \int_{\Re^N} |p_{ij}^*(\mathbf{s}+\delta\mathbf{s}-\mathbf{r}) - p_{ij}^*(\mathbf{s}-\mathbf{r})| \varepsilon_j(\mathbf{r}) d\mathbf{r}.
\end{aligned}$$

The continuity again follows from the same properties of the $p_{ij}^*(\mathbf{r})$ and the $\varepsilon_j(\mathbf{r})$.

Since $H_i(\mathbf{s},t)$ is monotone increasing in t and is uniformly bounded above it tends to a limit $a_i(\mathbf{s})$ as t tends to infinity, which is bounded above by this uniform bound. Since $h_i(\mathbf{s},t) > 0$ for some i and for $\mathbf{s}$ and t in an open region the positivity result for $a_i(\mathbf{s}$ is immediate. It only remains to show continuity of $a_i(\mathbf{s})$. Now

$$|a_i(\mathbf{s}+\delta\mathbf{s}) - a_i(\mathbf{s})| \leq \sum_{j=1}^{k} \gamma_{ij}^* \int_{\Re^N} |p_{ij}^*(\mathbf{s}+\delta\mathbf{s}-\mathbf{r}) - p_{ij}^*(\mathbf{s}-\mathbf{r})| \varepsilon_j(\mathbf{r}) d\mathbf{r},$$

so that the continuity follows as before.

□

Lemma 3.2 proves, in similar fashion to Lemma 2.1, that there is a one-to-one correspondence between the solutions of equations (3.2) and (3.3) so that we can base the spatial analysis on equations (3.3). Theorem 3.1 then shows that equations (3.3) admit a unique solution which satisfies the required conditions.

LEMMA 3.2. *There exists a one to one correspondence between the positive, monotone decreasing (in t), jointly continuous solutions $x_i(\mathbf{s},t)$ $(i = 1,...,n)$ of equations (3.2) and (3.3) for which each $x_i(\mathbf{s},0) \equiv 1$ and the partial derivatives with respect to t are jointly continuous and are uniformly bounded for all $\mathbf{s}$ and for t in a finite interval. Any such solution has $\partial(-\log x_i(\mathbf{s},t))/\partial t$ uniformly bounded.*

PROOF. Let $x_i(\mathbf{s},t)$, $(i = 1,...,n)$, be a solution of equations (3.2) with the stated conditions. Integrating these equations from 0 to t and interchanging the integrals we obtain

$$-\log x_i(\mathbf{s},t) = \sum_{j=1}^{n} \int_{\Re^N} \int_0^t \int_0^u -\frac{\partial x_j(\mathbf{s}-\mathbf{r}, u-\tau)}{\partial u} p_{ij}(\mathbf{r})\gamma_{ij}(\tau) d\tau du d\mathbf{r} + H_i(\mathbf{s},t).$$

Taking $\theta = u - \tau$, interchanging the integrals over τ and θ and noting that $x_j(\mathbf{r},0) \equiv 1$ immediately gives equations (3.3).

Now consider a solution $x_i(\mathbf{s},t)$, for $i = 1,...,n$, of equations (3.3) with the stated conditions. We can re-write these equations as

$$-\log x_i(\mathbf{s},t) = \sum_{j=1}^{n} \int_0^t u_{ij}(\mathbf{s}, t-\tau)\gamma_{ij}(\tau)d\tau + H_i(\mathbf{s},t),$$

where $u_{ij}(\mathbf{s},t) = \int_{\Re^N}(1 - x_j(\mathbf{s}-\mathbf{r},t))p_{ij}(\mathbf{r})d\mathbf{r}$. Each $u_{ij}(\mathbf{s},t)$ is jointly continuous and uniformly bounded since the $p_{ij}(\mathbf{r})$ are continuous and integrable and the $x_j(\mathbf{s},t)$ are jointly continuous and uniformly bounded. Using dominated convergence and the continuity of, and uniform bound on, $\partial x_i(\mathbf{s},t)/\partial t$, $\partial u_{ij}(\mathbf{s},t)/\partial t$ exists and equals $\int_{\Re^N}(-\partial x_j(\mathbf{s}-\mathbf{r},t)/\partial t)p_{ij}(\mathbf{r})d\mathbf{r}$. The joint continuity and uniform bound on this partial derivative follows from the same properties of $\partial x_j(\mathbf{s},t)/\partial t$ and because $p_{ij}(\mathbf{r})$ is integrable.

Next consider $\int_0^t u_{ij}(\mathbf{s},t-\tau)\gamma_{ij}(\tau)d\tau$. Since $u_{ij}(\mathbf{s},0) = 1$, $\gamma_{ij}(t)$ is continuous and integrable over a finite range and from the properties just established for $u_{ij}(\mathbf{s},t)$, using dominated convergence the partial derivative of this integral exists and is equal to $\int_0^t(\partial u_{ij}(\mathbf{s},t-\tau)/\partial t)\gamma_{ij}(\tau)d\tau$. Finally observe that, from Lemma 3.1 and its definition, the function $H_i(\mathbf{s},t)$ is differentiable with respect to t and its derivative is $h_i(\mathbf{s},t)$. Differentiating the re-written form of equations (3.3) with respect to t, using the results obtained, then gives equations (3.2).

From equations (3.2) the following uniform bound is obtained,

$$\left|\frac{\partial(-\log x_i(\mathbf{s},t))}{\partial t}\right| \leq \sum_{j=1}^{n} \sup_{\tau} \gamma_{ij}(\tau) + \sum_{j=1}^{k} \sup_{\tau} \gamma_{ij}^*(\tau) \sup_{\mathbf{r}} p_{ij}^*(\mathbf{r})\varepsilon_j.$$

□

THEOREM 3.1 (EXISTENCE AND UNIQUENESS FOR THE SPATIAL CASE). *There exists a positive, monotone decreasing (in t) solution $x_i(\mathbf{s},t)$ to equations (3.3) with $x_i(\mathbf{s},0) \equiv 1$, $(i = 1,...,n)$, which is unique. The solution has $x_i(\mathbf{s},t)$ jointly continuous in $\mathbf{s}$ and t and partially differentiable with respect to t, with the partial derivative also jointly continuous and uniformly bounded for all $\mathbf{s}$ and t in a finite interval.*

PROOF. Write $w_i(\mathbf{s},t) = -\log x_i(\mathbf{s},t)$, so that equations (3.3) become

$$w_i(\mathbf{s},t) = \sum_{j=1}^{n} \int_0^t \int_{\Re^N} \left(1 - e^{-w_j(\mathbf{s}-\mathbf{r},t-\tau)}\right) \gamma_{ij}(\tau) p_{ij}(\mathbf{r}) d\mathbf{r} d\tau + H_i(\mathbf{s},t), \quad (i = 1,...,n).$$

We use the same sequence as in Theorem 2.1. This sequence has $0 = T_0 < T_1 < T_2 < ...$, such that $\lim_{s\to\infty} T_s = \infty$. The existence and uniqueness then follow (with the obvious definitions) if, for each positive integer s and $t \in [T_{s-1}, T_s]$, we let

$$||w_i^{(m+1)}(\mathbf{s},t) - w_i^{(m)}(\mathbf{s},t)|| = \sup_{\mathbf{s}\in\Re^N,\ t\in[T_{s-1},T_s]} |w_i^{(m+1)}(\mathbf{s},t) - w_i^{(m)}(\mathbf{s},t)|$$

and

$$||w_i(\mathbf{s},t) - w_i^*(\mathbf{s},t)|| = \sup_{\mathbf{s}\in\Re^N,\ t\in[T_{s-1},T_s]} |w_i(\mathbf{s},t) - w_i^*(\mathbf{s},t)|.$$

Observe that the convergence of $w_i^{(m)}(\mathbf{s},t)$ is uniform for all $\mathbf{s}$ and for all $t \in [T_{s-1}, T_s]$.

The monotonicity in t and the joint continuity of $w_i(\mathbf{s},t)$, and hence for $w_i(\mathbf{s},t)$, is proved by showing that each of the sequence of functions used in the construction satisfies these conditions. Cauchy convergence then ensures the conditions hold for the limit functions. Dominated convergence shows that the limit functions satisfy equations (3.3). The monotonicity of the sequence of functions is simple to establish. The continuity of the sequence is shown for $t \in [0, T_1]$, and can be established in similar fashion for $t \in [T_{s-1}, T_s]$.

From Lemma 3.1, $w_i^{(0)}(\mathbf{s},t) = H_i(\mathbf{s},t)$ is jointly continuous in $\mathbf{s}$ and t. To establish continuity of the sequence of functions it must be shown that if the $w_i^{(m)}(\mathbf{s},t)$ are jointly continuous, then so are the $w_i^{(m+1)}(\mathbf{s},t)$. For fixed i,j,m let $-\log x(\mathbf{s},t) = w_i^{(m)}(\mathbf{s},t)$, so that $x_i(\mathbf{s},t)$ is jointly continuous and uniformly bounded, and define $y(\mathbf{s},t) = \int_{\Re^N} x(\mathbf{s}-\mathbf{r},t) p_{ij}(\mathbf{r}) d\mathbf{r}$ and $z(\mathbf{s},t) = \int_0^t y(\mathbf{s},\tau)\gamma_{ij}(t-\tau)d\tau$. Then it suffices to establish the joint continuity of $z(\mathbf{s},t)$. As in Lemma 3.2, $y(\mathbf{s},t)$ is jointly continuous and uniformly bounded from the same properties for $x(\mathbf{s},t)$ and since $p_{ij}(\mathbf{r})$ is bounded and integrable. The joint continuity of $z(\mathbf{s},t)$ then follows from the results for $y(\mathbf{s},t)$ and the continuity of $\gamma_{ij}(t)$. The joint continuity of the sequence $w_i^{(m)}(\mathbf{s},t)$, $m = 0,1,2,...$, is therefore obtained.

Hence the existence, uniqueness, monotonicity and joint continuity of the limit function $w_i(\mathbf{s},t)$, and hence of $x_i(\mathbf{s},t) = e^{-w_i(\mathbf{s},t)}$, is established. It remains to show that $x_i(\mathbf{s},t)$ is partially differentiable with respect to t and that the derivative is a jointly continuous function which is uniformly bounded over all $\mathbf{s}$ and for t in a finite interval. We re-write the equation for $x_i(\mathbf{s},t)$ as

$$-\log x_i(\mathbf{s},t) = \sum_{j=1}^{n} \int_0^t u_{ij}(\mathbf{s},\tau)\gamma_{ij}(t-\tau)d\tau + H_i(\mathbf{s},t),$$

where $u_{ij}(\mathbf{s},t) = \int_{\Re^N}(1-x_j(\mathbf{s}-\mathbf{r},t))p_{ij}(\mathbf{r})d\mathbf{r}$. As in Lemma 3.2, $u_{ij}(\mathbf{s},t)$ is jointly continuous and uniformly bounded (by 1) and $H_i(\mathbf{s},t)$ is differentiable with respect to t. The differentiability of $-\log x_i(\mathbf{s},t)$, and hence of $x_i(\mathbf{s},t)$, then follows (using dominated convergence) from these results and the conditions on the $\gamma_{ij}(t)$, namely that they are continuous, differentiable and bounded with bounded derivatives. Then

$$-\frac{\partial \log x_i(\mathbf{s},t)}{\partial t} = \sum \left(\gamma_{ij}(0)u_{ij}(\mathbf{s},t) + \int_0^t u_{ij}(\mathbf{s},\tau)\gamma'_{ij}(t-\tau)d\tau\right) + h_i(\mathbf{s},t).$$

The joint continuity then follows as before, noting that $\gamma'_{ij}(t)$ is continuous and that, from Lemma 3.1, $h_i(\mathbf{s},t)$ is jointly continuous. Clearly for all $\mathbf{s}$,

$$\left|-\frac{\partial x_i(\mathbf{s},t)}{\partial t}\right| \leq \sum_{j=1}^{n}\left(\sup_\tau \gamma_{ij}(\tau) + t\sup_\tau \gamma'_{ij}(\tau)\right) + \sum_{j=1}^{k} \sup_\tau \gamma^*_{ij}(\tau)\sup_{\mathbf{r}} p^*_{ij}(\mathbf{r})\varepsilon_j.$$

Hence the partial derivative of $x_i(\mathbf{s},t)$ with respect to t is uniformly bounded for all $\mathbf{s}$ and t in a finite interval and is jointly continuous. This completes the proof of the theorem.

□

3.3 Results for a single population

It is instructive to look first at the case $n = 1$. For simplicity, the technical proofs will be omitted at this stage; proofs being given in later sections for general n. The same notation is used, but the subscripts are dropped here since they are unnecessary. A uniform lower bound for the final size was obtained by Diekmann [D5] and Thieme [T2].

Consider the limit as $t \to$ in equation (3.3). Since both $1-x(\mathbf{s},t)$ and $H(\mathbf{s},t)$ are monotone and bounded, they tend to limits as $t \to \infty$; these limits being denoted by $v(\mathbf{s})$ and $a(\mathbf{s})$ respectively. It was shown in Lemma 3.1 that $a(\mathbf{s})$ is a continuous function. There are two cases, depending upon whether γ is finite or infinite.

When γ is finite a simple argument shows that $v(\mathbf{s})$ is continuous and satisfies the equation

$$-\log(1-v(\mathbf{s})) = \gamma v * p(\mathbf{s}) + a(\mathbf{s}), \tag{3.4}$$

where $*$ denotes convolution. Equation (3.4) is referred to as the spatial final size equation. When γ is infinite, the argument is more complex. From Lemma 3.1, there exists an open ball $B \in \Re^N$ with $a(\mathbf{s}) > 0$ for $\mathbf{s} \in B$. An iterative argument, using the radial symmetry of $p(\mathbf{r})$, is used to show that $v * p(s) > 0$ for all $\mathbf{s}$; this result then being used to prove that $v(\mathbf{s}) \equiv 1$.

The case when γ is infinite immediately gives the result that, regardless of the amount of initial infection, everyone at all spatial positions will eventually suffer the epidemic.

When γ is finite, let $v = \inf_{\mathbf{s}\in\Re^N} v(\mathbf{s})$ and let $a = \inf_{\mathbf{s}\in\Re^N} a(\mathbf{s})$. From equation (3.4) we obtain

$$-\log(1 - v(\mathbf{s})) \geq \gamma v + a$$

for all $\mathbf{s}$, and so

$$-\log(1 - v) \geq \gamma v + a. \tag{3.5}$$

Note that if the initial infection for each type has bounded support, then $a = 0$. We now make use of the convexity of the function $f(y) = -\log(1 - y) - \gamma y - a$. The behaviour of this function is illustrated in Chapter 2, fig 2.1; the two possible cases when $\gamma \leq 1$ and when $\gamma > 1$ being shown. When $\gamma > 1$ and/or $a > 0$, let $y = \eta(a)$ be the unique positive solution to $f(y) = 0$. Define $\eta(0) = 0$ for $\gamma \leq 1$.

From fig 2.1 case (a), when $\gamma \leq 1$ the lower bound $v \geq \eta(a)$ is obtained. Hence $v(\mathbf{s}) \geq \eta(a)$ for all $\mathbf{s}$. Note that this lower bound is zero if $a = 0$ and also tends to zero as the amount of initial infection tends to zero. The lower bound is of little use in this case.

However when $\gamma > 1$ a useful bound is obtained. From fig 2.1 case (b), either both a and v are zero, or $v \geq \eta(a) > \eta(0) > 0$ and hence $v(\mathbf{s}) \geq \eta(0)$ for all $\mathbf{s}$. Provided the possibility that $v = 0$ can be excluded when $a = 0$, the result is obtained that, no matter how little infection is initially introduced, the proportion eventually suffering the epidemic at all spatial positions will be bounded below by $\eta(0)$, the same bound as in the non-spatial case.

The exclusion of $v = 0$ when $\gamma > 1$ and $a = 0$ is relatively simple provided we restrict the space to one or two dimensions and impose the condition that $\int_{\Re^N} |\mathbf{r}|^2 p(\mathbf{r}) d\mathbf{r} < \infty$. Consider equation (3.4), and let $l(\mathbf{s}) = \min(\eta(0), v(\mathbf{s}))$. Then $-\log(1 - l(\mathbf{s})) \geq \gamma l * p(\mathbf{s})$. Also, as is easily seen from fig 2.1, $-\log(1 - l(\mathbf{s})) \leq \gamma l(\mathbf{s})$. Hence

$$l(\mathbf{s}) \geq l * p(\mathbf{s}).$$

It follows from results on a convolution inequality, (Essén [E1] Theorem 3.1) that the inequality must be an equality. However, this result holds only for dimensions $N = 1$ and 2. Feller ([F1] Sections V1.10 and XV111.7) gives a probabilistic proof. This has an interesting interpretation in stochastic processes. The fact that this result is only true in dimensions $N = 1$ and 2 is related to a property of the symmetric random walk, namely that the random walk is persistent in $\Re^1$ and $\Re^2$, whereas it is transient in $\Re^N$, for $N \geq 3$. Note that the random walk is said to be persistent if it is certain to return to the starting point, otherwise it is said to be transient.

Thus $l(\mathbf{s})$ satisfies the equation

$$l(\mathbf{s}) = l * p(\mathbf{s}).$$

It is proved both in Essén (E[1] Theorem 2.1) and Feller ([F1] Section X1.2) that the only bounded continuous solutions of this equation are $l(\mathbf{s}) \equiv c$ for some constant $c \geq 0$. But $v = 0$ implies $c = 0$ and hence that $v(\mathbf{s}) \equiv 0$. From equation (3.4), this cannot possibly hold since $a(\mathbf{s}) > 0$ for $\mathbf{s} \in B$ and hence also $v(\mathbf{s}) > 0$ in this same region. Hence $v \neq 0$ and therefore $v(\mathbf{s}) \geq \eta(a) \geq \eta(0)$ for all $\mathbf{s}$.

A statement of Essén's results is given in Lemma 3.5. The proof of the pandemic theorem for general $\Re^N$ is much more complex, and is derived in Chapter 5 as a consequence following from the theorem on the speed of propagation (see Section 5.2).

An upper bound for $v(\mathbf{s})$, the final size of the epidemic at position $\mathbf{s}$, may also be obtained. This will then make it possible to identify more precisely what happens if the amount of initial infection is small. Note that when γ is infinite everyone at all points always suffers the epidemic, regardless of the amount of initial infection. We therefore only need to consider the case when γ is finite.

Consider equation (3.4), and let $w = \sup_{\mathbf{s} \in \Re^N} v(\mathbf{s})$ and $b = \sup_{\mathbf{s} \in \Re^N} a(\mathbf{s})$. Then we obtain the inequality $-\log(1 - v(\mathbf{s})) \leq \gamma w + b$ for all $\mathbf{s}$, and so $-\log(1 - w) \leq \gamma w + b$. It can be seen from fig 2.1 that $w \leq \eta(b)$. Hence bounds for the spatial final size for γ finite, which hold for all $\mathbf{s}$, are

$$\eta(0) \leq v(\mathbf{s}) \leq \eta(b).$$

From the bounds on $a(\mathbf{s})$ obtained in Lemma 3.1, b tends to zero as ε tends to zero. It is clear, as shown in fig 2.1, that $\eta(b) \to \eta(0)$ as $b \to 0$, i.e. as the amount of initial infection tends to zero.

Hence if $\gamma \leq 1$, since $\eta(0) = 0$, the final size tends everywhere to zero as the amount of initial infection tends to zero. When $\gamma > 1$, then the final size tends everywhere to $\eta(0) > 0$ as the amount of initial infection tends to zero. This is precisely the corresponding limit of the final size in the non-spatial case.

3.4 The spatial final size for the multi-type model

The same basic approach is used for general n, although of course more complexity is involved and some γ_{ij} may be finite and others may be infinite. An iterative method is utilised in order to make use of the result of Essén [E1]. The results were obtained in Radcliffe and Rass [R4].

First consider the limiting behaviour of $x_i(\mathbf{s}, t)$. Since $x_i(\mathbf{s}, t)$ is a monotone decreasing, bounded function of t for each $\mathbf{s}$, it will tend to a limit as $t \to \infty$. Define $v_i(\mathbf{s}) = 1 - \lim_{t \to \infty} x_i(\mathbf{s}, t)$. Then $v_i(\mathbf{s})$ represents the proportion of type i individuals at position $\mathbf{s}$ who eventually suffer the epidemic. This is termed the spatial final size.

In Lemma 3.1 it was shown that $H_i(\mathbf{s}, t)$ tends to a non-negative limit, denoted by $a_i(\mathbf{s})$, as $t \to \infty$. The continuity of $a_i(\mathbf{s})$ was also established.

THEOREM 3.2 (THE SPATIAL FINAL SIZE EQUATIONS).

1. *If $\rho(\mathbf{\Gamma})$ is finite, then $y_i(\mathbf{s}) = v_i(\mathbf{s})$ is a (continuous) solution to*

$$(3.6) \qquad -\log(1 - y_i(\mathbf{s})) = \sum_{j=1}^{n} \gamma_{ij} p_{ij} * y_j(\mathbf{s}) + a_i(\mathbf{s}), \quad (i = 1, ..., n, \ \mathbf{s} \in \Re^N).$$

2. *If* $\mathbf{\Gamma}$ *has an infinite element in every row, then* $v_i(\mathbf{s}) = 1$ *for all* $\mathbf{s}$ *and all* $i = 1, ..., n$.
3. *The remaining case corresponds to a partitioning of* $\mathbf{\Gamma}$ *(by permutation of the indices), as in Theorem 2.2 part 3, into*

$$\mathbf{\Gamma} = \begin{pmatrix} \mathbf{\Gamma}_{11} & \mathbf{\Gamma}_{12} \\ \mathbf{\Gamma}_{21} & \mathbf{\Gamma}_{22} \end{pmatrix},$$

where $(\mathbf{\Gamma}_{11}\ \mathbf{\Gamma}_{12})$ *contains* m *rows and has no infinite element, and* $(\mathbf{\Gamma}_{21}\ \mathbf{\Gamma}_{22})$ *has at least one infinite element in each row.*

Let $\mathbf{v}(\mathbf{s})$ *and* $\mathbf{a}(\mathbf{s})$ *be the* n*-vectors with* $\{\mathbf{v}(\mathbf{s})\}_i = v_i(\mathbf{s})$ *and* $\{\mathbf{a}(\mathbf{s})\}_i = a_i(\mathbf{s})$. *The corresponding partitioning of these vectors is given by* $(\mathbf{v}(\mathbf{s}))' = ((\mathbf{v}_1(\mathbf{s}))', (\mathbf{v}_2(\mathbf{s}))')$ *and* $(\mathbf{a}(\mathbf{s}))' = ((\mathbf{a}_1(\mathbf{s}))', (\mathbf{a}_2(\mathbf{s}))')$.

Then $\mathbf{v}_2(\mathbf{s}) = \mathbf{1}$, *and* $\mathbf{y}(\mathbf{s}) = (y_i(\mathbf{s})) = \mathbf{v}_1(\mathbf{s})$ *is a (continuous) solution to*

$$-\log(1 - y_i(\mathbf{s})) = \sum_{j=1}^{m} (\{\mathbf{\Gamma}_{11}\}_{ij} p_{ij} * y_j(\mathbf{s})) + \{\mathbf{\Gamma}_{12}\mathbf{1} + \mathbf{a}_1(\mathbf{s})\}_i, \tag{3.7}$$

for $i = 1, ..., m$, *and* $\mathbf{s} \in \Re^N$.

PROOF. Define $f_{ij}(\mathbf{s}, t) = \int_{\Re^N} (1 - x_j(\mathbf{r}, t)) p_{ij}(\mathbf{s} - \mathbf{r}) d\mathbf{r}$. By monotone convergence, $\lim_{t\to\infty} f_{ij}(\mathbf{s}, t) = \int_{\Re^N} (1 - v_j(\mathbf{r})) p_{ij}(\mathbf{s} - \mathbf{r}) d\mathbf{r}$. Denote this limit by $f_{ij}(\mathbf{s})$. Then $f_{ij}(\mathbf{s})$ is continuous since $v_j(\mathbf{r})$ is bounded and $p_{ij}(\mathbf{s})$ is continuous and integrable. The cases are now considered separately

1. Consider the ij^{th} term from the summation in the right hand side of equations (3.3), when $\mathbf{\Gamma}$ has all finite elements. Since the double integrals below exist, this may be written as

$$\int_0^t \int_{\Re^N} (1 - x_j(\mathbf{r}, t - \tau)) \gamma_{ij}(\tau) p_{ij}(\mathbf{s} - \mathbf{r}) d\mathbf{r} d\tau = \int_0^t f_{ij}(\mathbf{s}, t - \tau) \gamma_{ij}(\tau) d\tau$$
$$= \int_0^\infty f_{ij}(\mathbf{s}, t - \tau) \gamma_{ij}(\tau) d\tau - \int_t^\infty f_{ij}(\mathbf{s}, t - \tau) \gamma_{ij}(\tau) d\tau.$$

Consider the limit of the last two integrals as $t \to \infty$. The first tends by monotone convergence to $\int_0^\infty f_j(\mathbf{s}) \gamma_{ij}(\tau) d\tau = \gamma_{ij} f_j(\mathbf{s})$. The second integral is non-negative and bounded above by $\int_t^\infty \gamma_{ij}(\tau) d\tau$. It therefore tends to zero since γ_{ij} is finite. Hence equation (3.6), with $y_i(\mathbf{s}) = v_i(\mathbf{s})$ is obtained by letting $t \to \infty$ in equations (3.3). The continuity of $v_i(\mathbf{s})$ follows since the $p_{ij}(\mathbf{r})$ are continuous and differentiable.

2. Consider any i, j and $\mathbf{s}$ such that γ_{ij} is infinite and $f_{ij}(\mathbf{s}) > 0$. Then there exists a T such that $f_{ij}(\mathbf{s}, t) > f_{ij}(\mathbf{s})/2$ for $t > T$. Hence for $t > T$

$$\int_{\Re^N} \int_0^t (1 - x_j(\mathbf{s} - \mathbf{r}, t - \tau)) \gamma_{ij}(\tau) p_{ij}(\mathbf{r}) d\tau d\mathbf{r}$$
$$= \int_0^t f_{ij}(\mathbf{s}, \tau) \gamma_{ij}(t - \tau) d\tau > \frac{1}{2} f_{ij}(\mathbf{s}) \int_T^t \gamma_{ij}(t - \tau) d\tau.$$

The right hand side of this inequality tends to infinity as t tends to infinity. Hence, taking the limit in the ith equation of (3.3), we obtain the result that $v_i(\mathbf{s}) = 1$ for all i and $\mathbf{s}$ such that there exists a j with $f_{ij}(\mathbf{s}) > 0$ and γ_{ij} infinite. It then follows (using results from part 1 when γ_{ij} is finite) that $v_i(\mathbf{s}) > 0$ if $a_i(\mathbf{s}) > 0$ and/or there exists a j with $\gamma_{ij} > 0$ and $f_{ij}(\mathbf{s}) > 0$.

Since there is some initial infection, from Lemma 3.1 there exists an i and an open ball $B \in \Re^N$ such that $a_i(\mathbf{s}) > 0$, and hence also $v_i(\mathbf{s}) > 0$, for $\mathbf{s} \in B$. The non-reducibility of $\mathbf{\Gamma}$ implies that there exists a sequence $i = j_1, j_2, ..., j_r = i$ with $j_1, j_2, ..., j_{r-1}$ distinct and $\gamma_{j_k, j_k+1} > 0$ for $k = 1, ..., r-1$. Let $p(\mathbf{s}) = p_{j_1,j_2} * p_{j_2,j_3} * ... * p_{j_{r-1},j_r}(\mathbf{s})$. Since the contact distributions are continuous and radially symmetric, $p(\mathbf{s})$ is also continuous and radially symmetric. Hence there exists an open ball C centred on the origin, for which $p * p(\mathbf{s}) > 0$. Hence for any $\mathbf{s}$ there exists a m such that $a_i * p^{2m}(\mathbf{s}) > 0$, where $p^k(\mathbf{s})$ denotes the k-fold convolution of $p(\mathbf{s})$ with itself. Hence $v_i(\mathbf{s}) > 0$ for all $\mathbf{s}$ and for the specific i. The non-reducibility of $\mathbf{\Gamma}$ then implies that $v_i(\mathbf{s}) > 0$ for all i and all $\mathbf{s}$.

Hence $f_{ij}(\mathbf{s}) > 0$ for all i and j with $\gamma_{ij} \neq 0$ and for all $\mathbf{s}$. Since for each i there exists a j with γ_{ij} infinite and $-\log(x_i(\mathbf{s}, t)) > \frac{1}{2} f_{ij}(\mathbf{s}) \int_T^t \gamma_{ij}(t-\tau)d\tau$, we obtain the result that $v_i(\mathbf{s}) = 1$ for all i and all $\mathbf{s}$.

3. The result that $\mathbf{v}_2(\mathbf{s}) \equiv \mathbf{1}$ follows as for case 2. The equations (3.7) for $\mathbf{v}_1(\mathbf{s})$ are then derived as for case 1. The continuity of $v_i(\mathbf{s})$ follows as in part 1 for i such that γ_{ij} is finite for all j.

□

3.5 The pandemic theorem

For this section attention is confined to the case when $\rho(\mathbf{\Gamma}) > 1$ and the spatial dimension is restricted to $N \leq 2$. The second moment condition that $\int_{\Re^N} |\mathbf{r}|^2 p_{ij}(\mathbf{r}) d\mathbf{r} < \infty$ for all i and j is required in order to use the results of Essén [E1]. The results for general N can be derived using Lemma 3.3 together with non-zero bounds obtained in Chapter 5, which require more sophisticated mathematics. This is discussed in Section 5.7.

A preliminary lemma is proved, which establishes a lower bound when $\mathbf{\Gamma}$ has at least one infinite entry and provides a preliminary result when $\mathbf{\Gamma}$ has all finite entries. This lemma does not require the second moment conditions and restriction on the dimension.

LEMMA 3.3 (THE PANDEMIC LEMMA). *Suppose* $\rho(\mathbf{\Gamma}) > 1$.

1. *If* $\mathbf{\Gamma}$ *has all finite entries, then either* $v_i(\mathbf{s}) \geq \eta_i(\mathbf{0})$ *for all* $\mathbf{s}$ *and all* i, *or* $\inf_{\mathbf{s}} v_i(\mathbf{s}) = 0$ *for all* $i = 1, ..., n$. *Here* $y_i = \eta_i(\mathbf{0})$ *is the unique positive solution to*

$$-\log(1 - y_i) = \sum_{j=1}^{n} \gamma_{ij} y_j, \quad (i = 1, ..., n).$$

2. *If* $\mathbf{\Gamma}$ *has an infinite element in every row, then* $v_i(\mathbf{s}) = 1$ *for all* $\mathbf{s}$ *and all* $i = 1, ..., n$.

3. *The remaining case corresponds to a partitioning of* $\mathbf{\Gamma}$ *and* $\mathbf{v}(\mathbf{s})$ *as in Theorem 3.2, part 3. Then* $\mathbf{v}_2(\mathbf{s}) \equiv \mathbf{1}$ *and* $\mathbf{v}_1(\mathbf{s}) \geq \boldsymbol{\eta}^*(\mathbf{a}^*)$*; where* $\{\mathbf{a}^*\}_i = \sum_{j=m+1}^n \gamma_{ij}$ *for* $i = 1, ..., m$ *and* $\mathbf{y} = \boldsymbol{\eta}^*(\mathbf{a}^*)$ *is the unique positive solution to*

$$-\log(1 - \{\mathbf{y}\}_i) = \sum_{j=1}^{m} \gamma_{ij}\{\mathbf{y}\}_j + \{\mathbf{a}^*\}_i, \; (i = 1, ..., m).$$

PROOF. Define $v_i = \inf_{\mathbf{s}} v_i(\mathbf{s})$ and $a_i = \inf_{\mathbf{s}} a_i(\mathbf{s})$ for $i = 1, ..., n$.

1. Consider the final size equations,

$$-\log(1 - v_i(\mathbf{s})) = \sum_{j=1}^{n} \gamma_{ij} p_{ij} * v_j(\mathbf{s}) + a_i(\mathbf{s}), \; (i = 1, ..., n).$$

   It follows immediately that $-\log(1 - v_i) \geq \sum_{j=1}^{n} \gamma_{ij} v_j + a_i$. Hence $-\log(1 - v_i) = \sum_{j=1}^{n} \gamma_{ij} v_j + c_i$ for some non-negative constants $c_1, ..., c_n$. Then from Theorem B.2 either $v_i \geq \eta_i(\mathbf{0})$ for all i, and hence $v_i(\mathbf{s}) \geq \eta_i(\mathbf{0})$ for all i and $\mathbf{s}$, or $v_i = 0$ for all i. Note that this latter case is only possible when $c_i = a_i = 0$ for all i i.e. when the a_i are all zero and the original inequality for the v_i is an equality.
2. The result is immediate from the final size equations obtained in Theorem 3.2 part 2.
3. From Theorem 3.2 part (iii), we immediately obtain the result that $v_i(\mathbf{s}) = 1$ for all $\mathbf{s}$ and $i = m + 1, ..., n$. Also

$$-\log(1 - v_i) \geq \sum_{j=1}^{m} \gamma_{ij} v_j + \{\mathbf{a}^*\}_i, \; (i = 1, ..., m).$$

   The result then follows from Theorem B.2. Note that, since $\mathbf{\Gamma}$ is non-reducible, if $\mathbf{\Gamma}_{11}$ is reducible and is written in normal form, then $\mathbf{a}^*$ is of the form specified in Theorem B.2 part 1.

□

The pandemic theorem gives a lower bound for the spatial final size. The bound is obtained as an immediate consequence of the pandemic lemma when $\mathbf{\Gamma}$ has at least one infinite entry. When the infection matrix is finite, to obtain the result that $v_i(\mathbf{s}) \geq \eta_i(\mathbf{0})$ the case $v_i = \inf_{\mathbf{s}} v_i(\mathbf{s}) = 0$ for all i needs to be excluded. An iterative argument is needed which requires the proof of a preliminary lemma, Lemma 3.4.

LEMMA 3.4. *Let* $\mathbf{A} = (a_{ij})$ *be an* $n \times n$ *non-reducible, non-negative matrix with* $n > 1$ *and* $\rho(\mathbf{A}) > 1$. *Let*

$$a^*_{ij}(M) = a_{ij} + \frac{a_{in}a_{nj}(1 - (a_{nn}/\rho(\mathbf{A}))^M)}{(\rho(\mathbf{A}) - a_{nn})}, \; (i, j = 1, ..., n - 1). \tag{3.8}$$

Then $\mathbf{A}^*(M) = (a^*_{ij}(M))$ *is non-reducible for all integers* $M \geq 1$. *Also for sufficiently large* M, $\rho(\mathbf{A}^*(M)) > 1$.

PROOF. If $n = 2$, clearly $\mathbf{A}^*(M)$ is non-reducible. Consider the case where $n \geq 3$. Let i and j be any two distinct integers such that $1 \leq i, j \leq n-1$. Since $\mathbf{A}$ is non-negative and non-reducible, there exists a finite sequence of distinct integers $i_1, ..., i_r$ with $i_1 = i$ and $i_r = j$ such that $a_{i_s i_{s+1}} > 0$ for $s = 1, ..., r-1$. Note that since $n > 1$, by Theorem A.2 part 2, $\rho(\mathbf{A}) > a_{nn}$ and hence, for any $l < n$ and $m < n$, $a^*_{lm}(M) > 0$ if $a_{lm} > 0$ and/or $a_{ln}a_{nm} > 0$.

If $i_1, ..., i_r$ are all distinct from n, then $a^*_{i_s i_{s+1}} > 0$ for $s = 1, ..., r-1$. If, for some k, $i_k = n$, then $a^*_{i_{k-1} i_{k+1}} > 0$ and $a^*_{i_s i_{s+1}} > 0$ for any integer s such that $1 \leq k-2$ or $k+1 \leq s \leq r-1$. Hence $\mathbf{A}^*(M)$ is non-reducible.

Let $\mathbf{A}^* = (a^*_{ij})$, where

$$a^*_{ij} = a_{ij} + \frac{a_{in}a_{nj}}{(\rho(\mathbf{A}) - a_{nn})}.$$

We now show that $\rho(\mathbf{A}^*) = \rho(\mathbf{A})$. Let $\mathbf{A}$ and its left eigenvector $\mathbf{u} > \mathbf{0}$ corresponding to $\rho(\mathbf{A})$, be partitioned so that

$$\mathbf{A} = \begin{pmatrix} \mathbf{A}_{11} & \mathbf{a}_{12} \\ \mathbf{a}'_{21} & a_{nn} \end{pmatrix} \quad \text{and} \quad \mathbf{u} = \begin{pmatrix} \mathbf{u}_1 \\ u_2 \end{pmatrix}.$$

Then $\mathbf{u}'_1\mathbf{A}_{11} + u_2\mathbf{a}'_{21} = \rho(\mathbf{A})\mathbf{u}'_1$ and $\mathbf{u}'_1\mathbf{a}_{12} + u_2 a_{nn} = \rho(\mathbf{A})u_2$. Hence $u_2 = \mathbf{u}'_1\mathbf{a}_{12}/(\rho(\mathbf{A}) - a_{nn})$ and $\mathbf{u}'_1 > \mathbf{0}'$. Now

$$\mathbf{A}^* = \mathbf{A}_{11} + \frac{1}{(\rho(\mathbf{A}) - a_{nn})}\mathbf{a}_{12}\mathbf{a}'_{21}.$$

Therefore

$$\begin{aligned}\mathbf{u}'_1\mathbf{A}^* &= \mathbf{u}'_1\mathbf{A}_{11} + \frac{1}{(\rho(\mathbf{A}) - a_{nn})}\mathbf{u}'_1\mathbf{a}_{12}\mathbf{a}'_{21} \\ &= \mathbf{u}'_1\mathbf{A}_{11} + u_2\mathbf{a}'_{21} \\ &= \rho(\mathbf{A})\mathbf{u}'_1.\end{aligned}$$

Since the only eigenvector of a non-reducible square matrix $\mathbf{B}$ with all positive entries is the one corresponding to the maximum eigenvalue $\rho(\mathbf{B})$, (Theorem A2 part 7), it follows that $\rho(\mathbf{A}^*) = \rho(\mathbf{A})$.

Since $\rho(\mathbf{A}) > a_{nn}$, $a^*_{ij}(M) \uparrow a^*_{ij}$ as $M \to \infty$. Thus, from Lemma A.1, $\rho(\mathbf{A}^*(M)) \uparrow \rho(\mathbf{A}^*) = \rho(\mathbf{A})$ as $M \to \infty$. Hence $\rho(\mathbf{A}^*(M)) > 1$ if M is sufficiently large.

□

In order to prove the pandemic theorem we need to consider the bounded, continuous solutions $l(\mathbf{x})$ of the convolution inequality $l(\mathbf{x}) \geq l * p(\mathbf{x})$. A solution is said to be trivial if $l(\mathbf{x}) = l * p(\mathbf{x})$. Parts 1 and 2 of the following lemma follow from Theorem 3.1 of Essén [E1]. Part 3 follows from Theorem 2.1 of Essén [E1].

LEMMA 3.5. *Consider the bounded, continuous solutions* $l(\mathbf{x})$ *of the convolution inequality*

$$l(\mathbf{x}) \geq l * p(\mathbf{x}),$$

where $p(\mathbf{x})$ is a radially symmetric probability density function for $\mathbf{x} \in \Re^N$ with $\int_{\mathbf{x}\in\Re^N} |\mathbf{x}|^2 p(\mathbf{x})d\mathbf{x} < \infty$. Then the following results hold.

1. *If $N = 1$ or 2 no non-trivial solution exists.*
2. *If $N \geq 3$ non-trivial solutions always exist.*
3. *Any trivial solution is of the form $l(\mathbf{x}) = c$, where c is a constant.*

□

Hence, for the cases $N = 1$ and $N = 2$, if there is a bounded continuous solution $l(\mathbf{x})$ to the inequality $l(\mathbf{x}) \geq l * p(\mathbf{x})$, then $l(\mathbf{x}) \equiv c$ for some constant c. We can now prove the pandemic theorem, which gives a uniform lower bound for the final size for each type, provided the spatial dimension is restricted to be either one or two and $\int_{\mathbf{r}\in\Re^N} |\mathbf{r}|^2 p_{ij}(\mathbf{r})d\mathbf{r}$ is constrained to be finite for all i, j. This enables us to use Lemma 3.5.

THEOREM 3.3 (THE PANDEMIC THEOREM). *Suppose $\rho(\mathbf{\Gamma}) > 1$.*

1. *If $\mathbf{\Gamma}$ has all finite entries then $v_i(\mathbf{s}) \geq \eta_i(\mathbf{0})$ for all $\mathbf{s}$ and all i, where $y_i = \eta_i(\mathbf{0})$ is the unique positive solution to*

$$-\log(1 - y_i) = \sum_{j=1}^{n} \gamma_{ij} y_j, \quad (i = 1, ..., n). \tag{3.9}$$

2. *If $\mathbf{\Gamma}$ has an infinite element in every row, then $v_i(\mathbf{s}) = 1$ for all $\mathbf{s}$ and all $i = 1, ..., n$.*
3. *The remaining case corresponds to a partitioning of $\mathbf{\Gamma}$ and $\mathbf{v}(\mathbf{s})$ as in Theorem 2.2, part 3. Then $\mathbf{v}_2(\mathbf{s}) \equiv \mathbf{1}$ and $\mathbf{v}_1(\mathbf{s}) \geq \boldsymbol{\eta}^*(\mathbf{a}^*)$; where $\{\mathbf{a}^*\}_i = \sum_{j=m+1}^{n} \gamma_{ij}$ for $i = 1, ..., m$ and $\mathbf{y} = \boldsymbol{\eta}^*(\mathbf{a}^*)$ is the unique positive solution to*

$$-\log(1 - \{\mathbf{y}\}_i) = \sum_{j=1}^{m} \gamma_{ij} \{\mathbf{y}\}_j + \{\mathbf{a}^*\}_i, \quad (i = 1, ..., m).$$

PROOF. The results are mainly proved in the Lemma 3.3, (the pandemic lemma). It only remains to exclude, for case 1, the possibility that $v_i = 0$ for all $i = 1, ..., n$, where $v_i = \inf_{\mathbf{s}\in\Re^N} v_i(\mathbf{s})$. The proof therefore consists of showing that $v_i > 0$ for some i. The sufficiency of this condition was established in Lemma 3.3.

Consider the final size equations (3.6), namely

$$-\log(1 - v_i(\mathbf{s})) = \sum_{j=1}^{n} \gamma_{ij} p_{ij} * v_j(\mathbf{s}) + a_i(\mathbf{s}), \quad (i = 1, ..., n).$$

Define $k_i(\mathbf{s}) = \min(v_i(\mathbf{s}), \eta_i(0))$. From these equations we obtain

$$-\log(1 - k_i(\mathbf{s})) \geq \sum_{j=1}^{n} \gamma_{ij} p_{ij} * k_j(\mathbf{s}), \quad (i = 1, ..., n). \tag{3.10}$$

Let $y = \eta$ be the unique positive solution to $-\log(1-y) - \rho(\mathbf{\Gamma})y = 0$. The existence and uniqueness of the solution follows from Theorem B2. Take $\mathbf{u}' > \mathbf{0}'$ to be the left eigenvector of $\mathbf{\Gamma}$ corresponding to $\rho(\mathbf{\Gamma})$. Then, multiplying the ith equation of (3.9) (with $y_i = \eta_i(\mathbf{0})$) by $\{\mathbf{u}\}_i$ and summing over i, we obtain

$$\sum_{i=1}^{n} \{\mathbf{u}\}_i(-\log(1-\{\boldsymbol{\eta}(\mathbf{0})\}_i) - \rho(\mathbf{\Gamma})\{\boldsymbol{\eta}(\mathbf{0})\}_i) = 0.$$

Since $(-\log(1-\{\boldsymbol{\eta}(\mathbf{0})\}_i) - \rho(\mathbf{\Gamma})\{\boldsymbol{\eta}(\mathbf{0})\}_i) > 0$ if $\{\boldsymbol{\eta}(\mathbf{0})\}_i > \eta$, this implies that $\{\boldsymbol{\eta}(\mathbf{0})\}_i \leq \eta$ for some i. We may permute the suffices so that $\{\boldsymbol{\eta}(\mathbf{0})\}_n \leq \eta$. Since $k_n(\mathbf{s}) \leq \{\boldsymbol{\eta}(\mathbf{0})\}_n \leq \eta$ for all $\mathbf{s}$, we obtain

$$-\log(1-k_n(\mathbf{s})) \leq \rho(\mathbf{\Gamma})k_n(\mathbf{s}),$$

and hence

$$k_n(\mathbf{s}) \geq \sum_{j=1}^{n} \frac{\gamma_{nj}}{\rho(\mathbf{\Gamma})} p_{nj} * k_j(\mathbf{s}). \tag{3.11}$$

If $n = 1$, define $l_1^{(0)}(\mathbf{s}) = k_1(\mathbf{s})$ and $q_{11}^{(0)}(\mathbf{s}) = p_{11}(\mathbf{s})$. Then equation (3.11) becomes

$$l_1^{(0)}(\mathbf{s}) \geq q_{11}^{(0)} * l_1^{(0)}(\mathbf{s}). \tag{3.12}$$

When $n > 1$, for any non-negative function $u(\mathbf{s})$,

$$u * k_n(\mathbf{s}) \geq \sum_{j=1}^{n} \left[\frac{\gamma_{nj}}{\rho(\mathbf{\Gamma})}\right] u * p_{nj} * k_j(\mathbf{s}). \tag{3.13}$$

For each i, put $u(\mathbf{s}) = p_{in}(\mathbf{s})$ in inequality (3.13), and substitute in the ith inequality of (3.10), for each $i = 1, ..., n-1$. Then the following inequalities are obtained:

$$\begin{aligned} -\log(1-k_i(\mathbf{s})) &\geq \sum_{j=1}^{n-1} \left(\gamma_{ij}p_{ij} + \left[\frac{\gamma_{in}\gamma_{nj}}{\rho(\mathbf{\Gamma})}\right] p_{in} * p_{nj}\right) * k_j(\mathbf{s}) \\ &\quad + \left[\frac{\gamma_{in}\gamma_{nn}}{\rho(\mathbf{\Gamma})}\right] p_{in} * p_{nn} * k_n(\mathbf{s}). \end{aligned} \tag{3.14}$$

Now substitute inequality (3.13), with $u(\mathbf{s}) = p_{in}(\mathbf{s})$, into the ith inequality of (3.14), for each $i = 1, ..., n-1$. Defining $p_{nn}^t(\mathbf{s})$ to be the t-fold convolution of $p_{nn}(\mathbf{s})$ with itself, this gives the inequality

$$\begin{aligned} -\log(1-k_i(\mathbf{s})) \geq \sum_{j=1}^{n-1} \Bigg(&\gamma_{ij}p_{ij} + \gamma_{in}\left[\frac{\gamma_{nj}}{\rho(\mathbf{\Gamma})}\right] p_{in} * p_{nj} \\ &+ \gamma_{in}\left[\frac{\gamma_{nj}}{\rho(\mathbf{\Gamma})}\frac{\gamma_{nn}}{\rho(\mathbf{\Gamma})}\right] p_{in} * p_{nj} * p_{nn}\Bigg) * k_j(\mathbf{s}) + \gamma_{in}\left[\frac{\gamma_{nn}}{\rho(\mathbf{\Gamma})}\right]^2 p_{in} * p_{nn}^2 * k_n(\mathbf{s}). \end{aligned}$$

Repeating this process, we obtain for $M \geq 1$ the inequality

$$(3.15)\quad -\log(1-k_i(\mathbf{s})) \geq \sum_{j=1}^{n-1} \gamma_{ij}^*(M)\tilde{q}_{ij}^{(M)} * k_j(\mathbf{s}) + \gamma_{in}\left[\frac{\gamma_{nn}}{\rho(\mathbf{\Gamma})}\right]^M p_{in} * p_{nn}^M * k_n(\mathbf{s}),$$

for $i = 1, ..., n-1$, where

$$\gamma_{ij}^*(M) = \gamma_{ij} + \left[\frac{\gamma_{in}\gamma_{nj}\left(1 - \left[\frac{\gamma_{nn}}{\rho(\mathbf{\Gamma})}\right]^M\right)}{\rho(\mathbf{\Gamma}) - \gamma_{nn}}\right],$$

and

$$\tilde{q}_{ij}^{(M)}(\mathbf{s}) = \frac{1}{\gamma_{ij}^*(M)}\left(\gamma_{ij}p_{ij}(\mathbf{s}) + \left[\frac{\gamma_{in}\gamma_{nj}}{\rho(\mathbf{\Gamma})}\right]\right.$$
$$\left.\times\left(p_{in} * p_{nj} * \sum_{t=1}^{M-1}\left[\frac{\gamma_{nn}}{\rho(\mathbf{\Gamma})}\right]^t p_{nn}^t(\mathbf{s}) + p_{in} * p_{nj}(\mathbf{s})\right)\right).$$

Observe that $\int_{\mathbf{s}\in\Re^N} \tilde{q}_{ij}^{(M)}(\mathbf{s})d\mathbf{s} = 1$.

From Lemma 3.4, we choose M so that $\rho(\mathbf{\Gamma}^*(M)) > 1$, where $\mathbf{\Gamma}^*(M)$ is an $(n-1)\times(n-1)$ matrix with ij^{th} entry $\gamma_{ij}^*(M)$. For this M, define $\mathbf{\Gamma}^{(1)} = \mathbf{\Gamma}^*(M)$ and $q_{ij}^{(1)} = \tilde{q}_{ij}^{(M)}$. From inequality (3.15) we obtain

$$-\log(1-k_i(\mathbf{s})) \geq \sum_{j=1}^{n-1}\{\mathbf{\Gamma}^{(1)}\}_{ij}q_{ij}^{(1)} * k_j(\mathbf{s}),\ (i = 1, ..., n-1),$$

where, from Lemma 3.4, $\mathbf{\Gamma}^{(1)}$ is non-reducible with $\rho(\mathbf{\Gamma}^{(1)}) > 1$.

Now let $l_i^{(1)}(\mathbf{s}) = \min(k_i(\mathbf{s}), \eta_i^{(1)})$, where $y_i = \eta_i^{(1)}$ for $i = 1, ..., n-1$, is the unique positive solution to

$$-\log(1-y_i) = \sum_{j=1}^{n-1}\{\mathbf{\Gamma}^{(1)}\}_{ij}y_j,\ (i = 1, ..., n-1).$$

Then

$$-\log(1-l_i^{(1)}(\mathbf{s})) \geq \sum_{j=1}^{n-1}\{\mathbf{\Gamma}^{(1)}\}_{ij}q_{ij}^{(1)} * l_j^{(1)}(\mathbf{s}),\ (i = 1, ..., n-1).$$

This is the same as inequality (3.10) except for the following; the size of $\mathbf{\Gamma}^{(1)}$ is $(n-1)\times(n-1)$, the number of equations has been reduced to $n-1$ and $l_i^{(1)}(\mathbf{s})$, $q_{ij}^{(1)}(\mathbf{s})$ and $\mathbf{\Gamma}^{(1)}$ replace $k_i(\mathbf{s})$, $p_{ij}(\mathbf{s})$ and $\mathbf{\Gamma}$ respectively.

The argument can be repeated, so that after $n-1$ steps in all, we obtain the inequality

$$-\log(1-l_1^{(n-1)}(\mathbf{s})) \geq \{\mathbf{\Gamma}^{(n-1)}\}_{11}q_{11}^{(n-1)} * l_1^{(n-1)}(\mathbf{s}),$$

where $\mathbf{\Gamma}^{(n-1)}$ is a 1×1 matrix with $\mathbf{\Gamma}^{(n-1)} = \{\mathbf{\Gamma}^{(n-1)}\}_{11} > 1$. Also $l_1^{(n-1)}(\mathbf{s}) \leq \eta^{(n-1)}$ where $y = \eta^{(n-1)}$ is the unique positive solution to $-\log(1-y) = \rho(\mathbf{\Gamma}^{(n-1)})y$. Hence

$$-\log(1 - l_1^{(n-1)}(\mathbf{s})) \leq \{\mathbf{\Gamma}^{(n-1)}\}_{11} l_1^{(n-1)}(\mathbf{s}).$$

Therefore we obtain the inequality

$$l_1^{(n-1)}(\mathbf{s}) \geq q_{11}^{(n-1)} * l_1^{(n-1)}(\mathbf{s}). \tag{3.16}$$

Note that when $n = 1$ inequality (3.16) holds since it is just inequality (3.12) with $l_1^{(0)}(\mathbf{s}) = k_1(\mathbf{s})$ and $q_{11}^{(0)}(\mathbf{s}) = p_{11}(\mathbf{s})$. Hence inequality (3.16) holds for any value of $n \geq 1$.

Now apply the results on convolution inequalities given in Lemma 3.5. This implies that inequality (3.16) must be an equality, and that the only solution has $l_1^{(n-1)}(\mathbf{s}) \equiv c$, for some constant c.

But $l_1^{(n-1)}(\mathbf{s})$ is the minimum of $v_1(\mathbf{s})$ (for the re-ordered types) and n positive constants. Hence $c = 0$ would imply that $v_1(\mathbf{s}) \equiv 0$, which is impossible since it was shown in the proof of Theorem 3.2 that $v_i(\mathbf{s}) > 0$ for all i and $\mathbf{s}$. Therefore $c > 0$. Since $v_1(\mathbf{s}) \geq l_1^{(n-1)}(\mathbf{s}) \equiv c > 0$, this implies that $v_1 \geq c > 0$. We have therefore shown that $v_i = \inf_{\mathbf{s} \in \Re^N} v_i(\mathbf{s}) > 0$ for some i, completing the proof of the theorem.

□

3.6 Bounds and consequent limiting results

An upper bound can be obtained for the spatial final size. This result was obtained in Rass and Radcliffe [R18] and is proved here in Theorem 3.4. It can then be established that, if $\rho(\mathbf{\Gamma}) \leq 1$, the final size tends to zero everywhere as the amount of initial infection tends to zero. Thus no major epidemic occurs when $\rho(\mathbf{\Gamma})$ fails to exceed the threshold value of 1. When $\rho(\mathbf{\Gamma}) > 1$, the spatial final size is constrained to lie between two bounds; the lower of which comes from the pandemic theorem of Section 3.5. As the amount of infection triggering the epidemic tends to zero, the final size everywhere tends to the corresponding limit in the non-spatial case (see Theorem 2.3). These results are given in two corollaries.

THEOREM 3.4.

1. *If* $\mathbf{\Gamma}$ *has all finite entries then* $v_i(\mathbf{s}) \leq \eta_i(\mathbf{b})$ *for all* $\mathbf{s}$ *and* $i = 1, ..., n$, *where* $\{\mathbf{b}\}_i = \sup_{\mathbf{s} \in \Re^N} a_i(\mathbf{s})$ *and* $y_i = \eta_i(\mathbf{b})$ *is the unique positive solution (see Theorem B.2 part 1) to*

$$-\log(1 - y_i) = \sum_{j=1}^{n} \gamma_{ij} y_j + \{\mathbf{b}\}_i, \quad (i = 1, ..., n). \tag{3.17}$$

2. *If* $\mathbf{\Gamma}$ *has an infinite element in every row, then* $v_i(\mathbf{s}) = 1$ *for all* $\mathbf{s}$ *and* $i = 1, ..., n$.
3. *The remaining case corresponds to a partitioning of* $\mathbf{\Gamma}$ *and* $\mathbf{v}(\mathbf{s})$ *as in Theorem 3.2, part 3. Again take* $\{\mathbf{b}\}_i = \sup_{\mathbf{s} \in \Re^N} a_i(\mathbf{s})$. *Then* $\mathbf{v}_2(\mathbf{s}) \equiv \mathbf{1}$ *and*

$\mathbf{v}_1(\mathbf{s}) \leq \boldsymbol{\eta}^*(\mathbf{b}^*)$; *where* $\{\mathbf{b}^*\}_i = \sum_{j=m+1}^n \gamma_{ij} + \{\mathbf{b}\}_i$ *for* $i = 1, ..., m$ *and* $\mathbf{y} = \boldsymbol{\eta}^*(\mathbf{b}^*)$ *is the unique positive solution (see Theorem B.2 part 1) to*

$$-\log(1 - \{\mathbf{y}\}_i) = \sum_{j=1}^{m} \gamma_{ij}\{\mathbf{y}\}_j + \{\mathbf{b}^*\}_i, \quad (i = 1, ..., m). \tag{3.18}$$

PROOF. Define $b_i = \sup_{\mathbf{s}\in\Re^N} a_i(\mathbf{s})$, $u_i = \sup_{\mathbf{s}\in\Re^N} v_i(\mathbf{s})$ and $\mathbf{b} = (b_i)$. Note that $u_i > 0$ for all i.

1. From equation (3.6) we obtain the result that

$$-\log(1 - u_i) \leq \sum_{j=1}^{n} \gamma_{ij} u_j + b_i, \quad (i = 1, ..., n).$$

Let $\alpha = \max_i(u_i/\eta_i(\mathbf{b}))$. Choose i so that this maximum is achieved. Then for this i,

$$\begin{aligned} -\log(1 - \alpha\eta_i(\mathbf{b})) = -\log(1 - u_i) &\leq \sum_{j=1}^{n} \gamma_{ij}\alpha\eta_j(\mathbf{b}) + b_i \\ &= \alpha(-\log(1 - \eta_i(\mathbf{b})) + b_i(1 - \alpha). \end{aligned}$$

Therefore, since $\alpha > 0$ and $\eta_i(\mathbf{b}) > 0$,

$$\left(\frac{-\log(1 - \alpha\eta_i(\mathbf{b}))}{\alpha\eta_i(\mathbf{b})}\right) - \left(\frac{-\log(1 - \eta_i(\mathbf{b}))}{\eta_i(\mathbf{b})}\right) \leq \frac{b_i(1 - \alpha)}{\alpha\eta_i(\mathbf{b})}. \tag{3.19}$$

Suppose that $\alpha > 1$. Since $-\log(1-x)/x$ is a strictly monotone increasing function of x for $0 < x < 1$, the left hand side of the inequality (3.19) is positive. But the right hand side of the inequality cannot be positive. Hence a contradiction is obtained. Hence $\alpha \leq 1$. Hence $v_i(\mathbf{s}) \leq u_i \leq \eta_i(\mathbf{b})$ for all $\mathbf{s}$ and all $i = 1, ..., n$; thus giving the upper bound required for the spatial final size.

2. When $\boldsymbol{\Gamma}$ has an infinite element in every row, then the result follows immediately from Theorem 3.3.
3. From Theorem 3.3 we immediately obtain the result that $\mathbf{v}_2(\mathbf{s}) \equiv \mathbf{1}$. Also

$$-\log(1 - u_i) \leq \sum_{j=1}^{m} \gamma_{ij} u_j + \sum_{j=m+1}^{n} \gamma_{ij} + b_i = \sum_{j=1}^{m} \gamma_{ij} u_j + \{\mathbf{b}^*\}_i, \tag{3.20}$$

where $\{\mathbf{b}^*\}_i = \sum_{j=m+1}^n \gamma_{ij} + \{\mathbf{b}\}_i$ for $i = 1, ..., m$.

Define $\{\boldsymbol{\Gamma}_{11}\}_{ij} = \gamma_{ij}$ for $i, j = 1, ..., m$. The populations may be ordered so that $\boldsymbol{\Gamma}_{11}$ is in normal form. Then

$$\mathbf{\Gamma}_{11} = \begin{pmatrix} \mathbf{A}_{11} & \mathbf{0} & \dots & \mathbf{0} & \mathbf{0} & \dots & \mathbf{0} \\ \vdots & \vdots & \ddots & \vdots & \vdots & \ddots & \vdots \\ \mathbf{0} & \mathbf{0} & \dots & \mathbf{A}_{gg} & \mathbf{0} & \dots & \mathbf{0} \\ \mathbf{A}_{g+1,1} & \mathbf{A}_{g+1,2} & \dots & \mathbf{A}_{g+1,g} & \mathbf{A}_{g+1,g+1} & \dots & \mathbf{0} \\ \vdots & \vdots & \ddots & \vdots & \vdots & \ddots & \vdots \\ \mathbf{A}_{k,1} & \mathbf{A}_{k,2} & \dots & \mathbf{A}_{k,g} & \mathbf{A}_{k,g+1} & \dots & \mathbf{A}_{k,k} \end{pmatrix},$$

where the matrix $\mathbf{A}_{ij}$ is of size $r_i \times r_j$. Each matrix $\mathbf{A}_{ii}$ is non-reducible, $\mathbf{A}_{ij} = \mathbf{0}$ for $j > i$ and for $j < i \leq g$. For each $i > g$ there exists a $j < i$ with $\mathbf{A}_{ij} \neq \mathbf{0}$.

Let $\mathbf{u}$ and $\mathbf{z}$ be m vectors with i^{th} entries u_i and $-\log(1-u_i)$ respectively. Then $\mathbf{u}$, $\mathbf{z}$ and $\mathbf{b}^*$ may be equivalently partitioned into

$$\mathbf{u} = \begin{pmatrix} \mathbf{u}_1 \\ \vdots \\ \mathbf{u}_k \end{pmatrix}, \quad \mathbf{z} = \begin{pmatrix} \mathbf{z}_1 \\ \vdots \\ \mathbf{z}_k \end{pmatrix} \text{ and } \mathbf{b}^* = \begin{pmatrix} \mathbf{b}_1^* \\ \vdots \\ \mathbf{b}_k^* \end{pmatrix},$$

where $\mathbf{u}_i$, $\mathbf{z}_i$ and $\mathbf{b}_i^*$ are all r_i-vectors. Note that $\mathbf{b}^* \neq \mathbf{0}$ for $i = 1, ..., g$, since $\mathbf{\Gamma}$ is non-reducible.

Hence inequality (3.20) can be rewritten as

$$\begin{aligned} \mathbf{z}_i &\leq \mathbf{A}_{ii}\mathbf{u}_i + \mathbf{b}_i^*, & (i = 1, ..., g), \\ \mathbf{z}_i &\leq \sum_{j=1}^{i} \mathbf{A}_{ij}\mathbf{u}_j + \mathbf{b}_i^*, & (i = g+1, ..., k). \end{aligned} \tag{3.21}$$

Note that the solution $\mathbf{y} = \boldsymbol{\eta}^*(\mathbf{b}^*)$ to equations (3.18) may be equivalently partitioned into

$$\boldsymbol{\eta}^*(\mathbf{b}^*) = \begin{pmatrix} \boldsymbol{\eta}_1^*(\mathbf{b}^*) \\ \vdots \\ \boldsymbol{\eta}_k^*(\mathbf{b}^*) \end{pmatrix}.$$

Let $\{\boldsymbol{\phi}_i^*(\mathbf{b}^*)\}_j = -\log(1 - \{\boldsymbol{\eta}_i^*(\mathbf{b}^*)\}_j)$. Then, from Theorem B.2 part 1, $\boldsymbol{\eta}_i^*(\mathbf{b}^*)$ is the unique positive vector satisfying

$$\boldsymbol{\phi}_i^*(\mathbf{b}^*) = \mathbf{A}_{ii}\boldsymbol{\eta}_i^*(\mathbf{b}^*) + \mathbf{b}_i^*,$$

for $i = 1, ..., g$; while for $i = g+1, ..., k$,

$$\boldsymbol{\phi}_i^*(\mathbf{b}^*) = \mathbf{A}_{ii}\boldsymbol{\eta}_i^*(\mathbf{b}^*) + \sum_{j=1}^{i-1} \mathbf{A}_{ij}\boldsymbol{\eta}_j^*(\mathbf{b}^*) + \mathbf{b}_i^*.$$

For each $i = 1, ...g$, if we apply the proof used in part 1, we immediately obtain the result that $\mathbf{u}_i \leq \boldsymbol{\eta}_i^*(\mathbf{b}^*)$ for $i = 1, ..., g$. For $i > g$ a sequential argument may be used. Consider equations (3.21) for $i = g+1$.

$$\mathbf{z}_i \le \mathbf{A}_{ii}\mathbf{u}_i + \sum_{j=1}^{i-1} \mathbf{A}_{ij}\mathbf{u}_j + \mathbf{b}_i^* \le \mathbf{A}_{ii}\mathbf{u}_i + \sum_{j=1}^{i-1} \mathbf{A}_{ij}\boldsymbol{\eta}_j^*(\mathbf{b}^*) + \mathbf{b}_i^*.$$

Using the proof of part 1, we obtain $\mathbf{u}_{g+1} \le \boldsymbol{\eta}_{g+1}^*(\mathbf{b}^*)$. Proceeding sequentially for $i = g+2, ..., k$ we obtain the result that $\{\mathbf{v}_1(\mathbf{s})\}_i \le \mathbf{u}_i \le \boldsymbol{\eta}_i^*(\mathbf{b}^*)$ for all $\mathbf{s}$ and $i = 1, ..., m$.

□

Two corollaries can now be given. The first concerns the case when $\rho(\mathbf{\Gamma}) \le 1$. It shows that when the amount of initial infection tends to zero the final size of the epidemic also tends to zero everywhere. The second corollary covers the case when $\rho(\mathbf{\Gamma}) > 1$. When the amount of initial infection tends to zero, the final size tends everywhere to the corresponding spatial lower bound. This positive bound is also the limiting value of the non-spatial final size when the amount of initial infection tends to zero.

COROLLARY 3.1. *Let $\rho(\mathbf{\Gamma}) \le 1$. If $\sup_{\mathbf{s}} \varepsilon_i(\mathbf{s}) \to 0$ for all i, then $v_i(\mathbf{s}) \to 0$ for all $\mathbf{s}$ and all i.*

PROOF. The condition $\sup_{\mathbf{s}} \varepsilon_i(\mathbf{s}) \to 0$ for all i implies that $b_i \to 0$ for all i. From Theorem 3.4 the bounds on the final size are

$$\mathbf{0} = \eta_i(\mathbf{0}) \le v_i(\mathbf{s}) \le \eta_i(\mathbf{b}).$$

From Theorem B.2, $\eta_i(\mathbf{b})$ is a continuous function of $\mathbf{b}$. Hence the result immediately follows.

□

The next corollary uses the pandemic theorem, Theorem 3.3, to obtain the limit of the spatial final size. This required the conditions that $N \le 2$ and each $p_{ij}(\mathbf{r})$ is radially symmetric with $\int_{\mathbf{r}\in\Re^N} |\mathbf{r}|^2 p_{ij}(\mathbf{r}) d\mathbf{r}$ finite. However in Chapter 5 an alternative proof of the pandemic theorem is given, which is valid for all N and only requires radial symmetry of the contact distributions. Therefore the restriction on the dimension N and the second moment conditions on the contact distributions are not required for the proof.

COROLLARY 3.2. *Let $\rho(\mathbf{\Gamma}) > 1$. There are three cases to consider.*

1. *$\mathbf{\Gamma}$ has all finite entries. If $\sup_{\mathbf{s}} \varepsilon_i(\mathbf{s}) \to 0$ for all i, then $v_i(\mathbf{s}) \to \eta_i(\mathbf{0})$ for all $\mathbf{s}$ and all i, where $y_i = \eta_i(\mathbf{0})$ is the unique positive solution to*

$$-\log(1 - y_i) = \sum_{j=1}^{n} \gamma_{ij} y_j, \quad (i = 1, ..., n).$$

2. *$\mathbf{\Gamma}$ has an infinite element in every row. If $\sup_{\mathbf{s}} \varepsilon_i(\mathbf{s}) \to 0$ for all i, then $v_i(\mathbf{s}) \to 1$ for all $\mathbf{s}$ and all $i = 1, ..., n$.*
3. *The remaining case corresponds to a partitioning of $\mathbf{\Gamma}$ and $\mathbf{v}(\mathbf{s})$ as in Theorem 3.2, part 3. If $\sup_{\mathbf{s}} \varepsilon_i(\mathbf{s}) \to 0$ for all i, then $\mathbf{v}_2(\mathbf{s}) \to \mathbf{1}$ and $\mathbf{v}_1(\mathbf{s}) \to \boldsymbol{\eta}^*(\mathbf{a}^*)$; where $\{\mathbf{a}^*\}_i = \sum_{j=m+1}^{n} \gamma_{ij}$ for $i = 1, ..., m$ and $\mathbf{y} = \boldsymbol{\eta}^*(\mathbf{a}^*)$ is the unique positive solution to*

$$-\log(1-\{\mathbf{y}\}_i)=\sum_{j=1}^{m}\gamma_{ij}\{\mathbf{y}\}_j+\{\mathbf{a}^*\}_i,\ (i=1,...,m).$$

PROOF.

1. The upper and lower bounds on the final size obtained in Theorems 3.3 and 3.4 are
$$\eta_i(\mathbf{0})\le v_i(\mathbf{s})\le\eta_i(\mathbf{b}),$$
for all $\mathbf{s}$ and all i. Observe that the condition $\sup_{\mathbf{s}}\varepsilon_i(\mathbf{s})\to 0$ for all i implies that $b_i\to 0$ for all i. Using the continuity in $\mathbf{b}$ of $\eta_i(\mathbf{b})$ the result immediately follows.
2. If $\mathbf{\Gamma}$ has an infinite element in every row, then from Theorem 3.3, $v_i(\mathbf{s})=1$ for all $\mathbf{s}$ and all $i=1,...,n$ regardless of the amount of initial infection. The result is immediate.
3. From Theorem 3.3 $\mathbf{v}_2(\mathbf{s})\equiv\mathbf{1}$ regardless of the amount of initial infection. Also from Theorems 3.3 and 3.4,
$$\boldsymbol{\eta}^*(\mathbf{a}^*)\le\mathbf{v}_1(\mathbf{s})\le\boldsymbol{\eta}^*(\mathbf{b}^*),$$
where $\{\mathbf{a}^*\}_i=\sum_{j=m+1}^{n}\gamma_{ij}$ and $\{\mathbf{b}^*\}_i=\sum_{j=m+1}^{n}\gamma_{ij}+\{\mathbf{b}\}_i$ for $i=1,...,m$.
Observe that the condition $\sup_{\mathbf{s}}\varepsilon_i(\mathbf{s})\to 0$ for all i implies that $\mathbf{b}\to\mathbf{0}$ and hence $\mathbf{b}^*\to\mathbf{a}^*$. Using the continuity in $\mathbf{c}$ of $\boldsymbol{\eta}^*(\mathbf{c})$ the result immediately follows.

□

3.7 The behaviour at infinity

The pandemic result stated by Kendall [K2] is somewhat different from the form so far considered. The model he considered was an $S\to I\to R$ epidemic in a homogeneous population in the plane $\Re^2$. The result which was obtained concerned the limit of the final size of the epidemic as the distance from the initial focus of infection tended to infinity. Conditions were imposed on the initial configuration of the epidemic and the contact distributions. Kendall assumed that the final size at position $\mathbf{s}$ was eventually monotone decreasing in $|\mathbf{s}|$ and obtained the limit of the final size at infinity. The behaviour at infinity was also considered by Thieme [T1] for the one-type model.

The analogous result may be derived for the more general n-type models we have considered provided suitable conditions are imposed on the initial infection from outside and on the contact distributions. Specifically $\varepsilon_i(\mathbf{s},\tau)$, $p_{ij}(\mathbf{s})$ and $p^*_{ij}(\mathbf{s})$ are taken to be radially symmetric, monotone decreasing, continuous functions of $|\mathbf{s}|$ for all i,j. In addition, the contact distributions are assumed to be differentiable (except possibly at a finite number of points) with bounded derivatives. They are also taken to be convex functions of the radial distance $|\mathbf{r}|$ for $|\mathbf{r}|$ sufficiently large. This latter condition implies that each contact distribution is a convex function in the tails of the first co-ordinate $\{\mathbf{r}\}_1$ when the remaining co-ordinates are held fixed.

LEMMA 3.6. *The functions $H_i(\mathbf{s},t)$ and $w_i(\mathbf{s},t)$ are radially symmetric and monotone decreasing in $|\mathbf{s}|$ for $i = 1, ..., n$.*

PROOF. The proof makes use of the construction of $w_i(\mathbf{s},t)$ in Theorem 3.1 as the limit of a sequence of functions $w_i^{(m)}(\mathbf{s},t)$, where $w_i^{(0)}(\mathbf{s},t) = H_i(\mathbf{s},t)$.

First consider the function $H_i(\mathbf{s},t)$. We need only show the radial symmetry and monotonicity of $\int_{\mathbf{r}\in\Re^N} \varepsilon_j(\mathbf{r},u)p^*_{ij}(\mathbf{s}-\mathbf{r})d\mathbf{r}$, since

$$H_i(\mathbf{s},t) = \sum_{j=1}^{n} \int_0^t \int_0^\infty \int_{\mathbf{r}\in\Re^N} \varepsilon_j(\mathbf{r},u)p^*_{ij}(\mathbf{s}-\mathbf{r})\gamma^*_{ij}(\tau+u)d\mathbf{r}dud\tau.$$

The radial symmetry is easily seen. Rotate the axes so that the first co-ordinate of $\mathbf{r}$ is in the direction of $\mathbf{s}$. The functions in the integrand remain unaltered because of radial symmetry. The same integral is obtained for any $\mathbf{s}^*$ which has $|\mathbf{s}^*| = |\mathbf{s}|$.

Now consider the monotonicity of the integral. We can rotate the axes so that we need only consider a change in the first entry of $\mathbf{s}$, when that entry is positive. Let $'$ here represent the derivative with respect to the first entry. It suffices to show that

$$\frac{\partial}{\partial s_1} \int_{-\infty}^{\infty} \varepsilon_j((r_1, ...r_N),u)p^*_{ij}(s_1-r_1,...,s_N-r_N)dr_1 < 0.$$

Note that $p^{*\prime}_{ij}(\theta, r_2, ..., r_N) = -p^{*\prime}_{ij}(-\theta, r_2, ..., r_N)$. Thus, splitting the integral range and using dominated convergence and monotone convergence in the tails to justify the first step,

$$\begin{aligned}
&\frac{\partial}{\partial s_1} \int_{-\infty}^{\infty} \varepsilon_j((r_1,...,r_N),u)p^*_{ij}(s_1-r_1,...,s_N-r_N)dr_1 \\
&= \int_{-\infty}^{\infty} \varepsilon_j((r_1,...,r_N),u)p^{*\prime}_{ij}(s_1-r_1,...,s_N-r_N)dr_1 \\
&= \int_{s_1}^{\infty} \varepsilon_j((r_1,...,r_N),u)p^{*\prime}_{ij}(s_1-r_1,...,s_N-r_N)dr_1 \\
&+ \int_{-\infty}^{s_1} \varepsilon_j((r_1,...,r_N),u)p^{*\prime}_{ij}(s_1-r_1,...,s_N-r_N)dr_1 \\
&= -\int_0^\infty \varepsilon_j((s_1+\theta,r_2,...,r_N),u)p^{*\prime}_{ij}(\theta,s_2-r_2,...,s_N-r_N)d\theta \\
&+ \int_0^\infty \varepsilon_j((s_1-\theta,r_2,...,r_N),u)p^{*\prime}_{ij}(\theta,s_2-r_2,...,s_N-r_N)d\theta \\
&= \int_0^\infty [\varepsilon_j((s_1-\theta,r_2,...,r_N),u) - \varepsilon_j((s_1+\theta,r_2,...,r_N),u)] \\
&\times p^{*\prime}_{ij}(\theta,s_2-r_2,...,s_N-r_N)d\theta.
\end{aligned}$$

Since $s_1 > 0$ the first term in the last integrand is positive and the second term is negative for all $\theta > 0$. Hence the integral is negative. This establishes that $H_i(\mathbf{s},t)$ is a monotone decreasing function of $|\mathbf{s}|$.

The result for $w_i(\mathbf{s},t)$ is obtained by proving the radial symmetry and monotonicity of each of the iterative sequence of functions $w_i^{(m)}(\mathbf{s},t)$ defined in Theorem

3.1. Now $w_i^{(0)}(\mathbf{s},t) = H_i(\mathbf{s},t)$, which has already been shown to be radially symmetric and monotone decreasing in $|\mathbf{s}|$. Induction can then be used to prove the same result for $w_i^{(m)}(\mathbf{s},t)$ for all $m \geq 0$. Since the inductive step uses the same method previously used for $H_i(\mathbf{s},t)$ we omit the details. It therefore immediately follows that the limit function $w_i(\mathbf{s},t)$ is also radially symmetric and monotone decreasing in $|\mathbf{s}|$.

□

It follows from Lemma 3.6 that each of the final sizes $v_i(\mathbf{s})$ and every $a_i(\mathbf{s})$ is radially symmetric and monotone decreasing in $|\mathbf{s}|$ and is bounded below by zero. Hence $v_i(\mathbf{s})$ and $a_i(\mathbf{s})$ tend to limits v_i and a_i as $|\mathbf{s}|$ tends to infinity, and v_i and a_i must be the infimum of $v_i(\mathbf{s})$ and $a_i(\mathbf{s})$ taken over all $\mathbf{s} \in \Re^N$. Note that v_i was the notation used in the proof of Theorem 3.3 for $\inf_{\mathbf{s}\in\Re^N} v_i(\mathbf{s})$.

Define $\mathbf{a}$ and $\mathbf{v}$ to be the vectors with i^{th} entries a_i and v_i. Theorem 3.5 uses the pandemic theorem to obtain the limit, v_i, of each final size as the distance from the initial focus of infection tends to infinity. The proof given in this chapter (see Theorem 3.3) requires the conditions that $N \leq 2$ and that each contact distribution is radially symmetric with $\int_{\mathbf{r}\in\Re^N} |\mathbf{r}|^2 p_{ij}(\mathbf{r})d\mathbf{r}$ finite. However the alternative proof of the pandemic theorem, given in Chapter 5, is valid for all N. Therefore Theorem 3.5 does not require the restriction that the dimension should be one or two; it is valid for any dimension. In addition it is not required that the second moment about the origin for each contact distributions should be finite.

THEOREM 3.5.
1. *If* $\mathbf{\Gamma}$ *has all finite entries then* $\mathbf{v} = \boldsymbol{\eta}(\mathbf{a})$. *When* $\rho(\mathbf{\Gamma}) \leq 1$ *and* $\mathbf{a} = \mathbf{0}$, *then* $\mathbf{v} = \mathbf{0}$.
2. *If* $\mathbf{\Gamma}$ *has an infinite element in every row, then* $\mathbf{v} = \mathbf{1}$.
3. *The remaining case corresponds to a partitioning of* $\mathbf{\Gamma}$ *as in Theorem 3.2, part 3. Equivalently partition* $\mathbf{v}$ *and* $\mathbf{a}$ *so that*

$$\mathbf{v} = \begin{pmatrix} \mathbf{v}_1 \\ \mathbf{v}_2 \end{pmatrix} \text{ and } \mathbf{a} = \begin{pmatrix} \mathbf{a}_1 \\ \mathbf{a}_2 \end{pmatrix}.$$

Then $\mathbf{v}_2 = \mathbf{1}$ *and* $\mathbf{v}_1 = \boldsymbol{\eta}^*(\mathbf{a}_1 + \mathbf{\Gamma}_{12}\mathbf{1})$, *where* $\mathbf{y} = \boldsymbol{\eta}^*(\mathbf{b})$ *is the unique positive solution to*

$$-\log(1 - \{\mathbf{y}\}_i) = \{\mathbf{\Gamma}_{11}\mathbf{y} + \mathbf{b}\}_i, \quad (i = 1, ..., m).$$

PROOF.
1. From Lemma 3.6 it follows that the $v_i(\mathbf{s})$, are radially symmetric. Therefore $v_i(\mathbf{s}) = v_i((|\mathbf{s}|\ \mathbf{0}')')$. Define $\tilde{p}_{ij}(\{\mathbf{r}\}_1)$ to be the marginal density function obtained by integrating $p_{ij}(\mathbf{r})$ over all but the first entry of $\mathbf{r}$. From the final size equations (3.6), derived in Theorem 3.2 part 1, it follows that for $i = 1, ..., n$

$$-\log(1 - v_i((|\mathbf{s}|\ \mathbf{0}')')) = \sum_{j=1}^{n} \gamma_{ij} \int_{-\infty}^{\infty} v_j(([|\mathbf{s}| - u]\ \mathbf{0}')')\tilde{p}_{ij}(u)du + a_i((|\mathbf{s}|\ \mathbf{0}')'). \tag{3.22}$$

Using dominated convergence we obtain $-\log(1-v_i) = \sum_{j=1}^{n} \gamma_{ij} v_j + a_i$ for all i. The result immediately follows from Theorem B.2 when $\mathbf{a} \neq \mathbf{0}$ and/or $\rho(\mathbf{\Gamma}) \leq 1$. When $\mathbf{a} = \mathbf{0}$ and $\rho(\mathbf{\Gamma}) > 1$ then from Theorem B.2 either $\mathbf{v} = \mathbf{0}$ or $\mathbf{v} = \boldsymbol{\eta}(\mathbf{a})$. The case $\mathbf{v} = \mathbf{0}$ is excluded by Theorem 3.3.

2. This follows immediately from Theorem 3.2 part 2.
3. From Theorem 3.2 part 3, the result that $\mathbf{v}_2 = \mathbf{1}$ is immediate. From equations (3.7), proceeding in an identical manner to the proof of part 1, we obtain $-\log(1-v_i) = \{\mathbf{\Gamma}_{11}\mathbf{v}_1\}_i + \{\mathbf{\Gamma}_{12}\mathbf{1}\}_i + a_i$ for $i = 1, ..., m$. The result then follows from Theorem B.2.

□

When the initial infection is confined to a bounded region, as assumed in Chapter 5 when considering the asymptotic speed of propagation, then it can be shown that $\mathbf{a} = \mathbf{0}$ (see Theorem 5.5). Hence when $\rho(\mathbf{\Gamma}) \leq 1$, $\mathbf{v} = \mathbf{0}$. So, for any type, the proportion eventually suffering the epidemic at a point in $\Re^N$ tends to zero as the distance of that point from the region containing the initial infection tends to infinity. Also when $\rho(\mathbf{\Gamma}) > 1$ then v_i is equal to the lower bound obtained in the pandemic theorem, Theorem 3.3. In this case, for type i, the proportion eventually suffering the epidemic at a point in $\Re^N$ tends to $\eta_i(\mathbf{0})$ as the distance of the point from the initial focus of infection tends to infinity.

CHAPTER 4

Wave solutions

4.1 Specification and discussion

Consider a homogeneous population (of constant density on a line) in which all individuals are susceptible. This is the one-dimensional form of the one-type spatial models described in Chapter 3. Now imagine how a small amount of infection in this population will spread in a positive direction along the line. In the early stages of the epidemic, the proportion infected at different points will depend upon the initial proportion of infectives at different positions on the line. However, after a considerable time t has passed, the initial configuration is likely to have little effect. It seems intuitively reasonable that the proportion of individuals at position s who are still susceptible at time t, $x(s,t)$, will progressively behave more and more like a wave. This wave will travel through the population, with everyone ahead of the wave susceptible. As the wave advances individuals become infected and the proportion of susceptible individuals declines. Once the wave has passed, there will be a proportion of susceptibles left behind who did not get infected.

A wave travelling at speed c through the population will have $x(s+ct_1, t+t_1) = x(s,t)$ for all s, t and t_1. If you run at speed c in a positive direction, the proportion you observe where you are at any time will stay the same. Looking forwards along the line at a fixed point in time, the values of $x(s,t)$ will have the shape of a monotone increasing wave, tending to 1 in the front tail, as $s \to \infty$. In the back tail, as $s \to -\infty$, $x(s,t)$ will be shown to be $1 - \eta(0)$. Here $\eta(0)$ is both the limit of the non-spatial final size (obtained in Chapter 2) and the limit of the spatial final size (see Chapter 3) when the amount of initial infection tends to zero.

For an epidemic occurring in $\Re^N$, wave solutions may also be considered. Let $x(\mathbf{s},t)$ be the proportion of individuals at position $\mathbf{s}$ who are still susceptible at time t. If a wave travels at speed c in a direction specified by direction cosines $\boldsymbol{\xi}$ then $x(\mathbf{s} + ct_1\boldsymbol{\xi}, t + t_1) = x(\mathbf{s}, t)$ for all $\mathbf{s}$, t and t_1. A wave front will pass through the population in the direction specified by $\boldsymbol{\xi}$ if the function $x(\mathbf{s},t)$ is constant on the hyperplane orthogonal to the direction $\boldsymbol{\xi}$. Note that in this case $x(\mathbf{s},t)$ will be a function of $\boldsymbol{\xi}'\mathbf{s} - ct$ only. At time t, as you look along any line with direction cosines $\boldsymbol{\xi}$, $x(\mathbf{s} + k\boldsymbol{\xi}, t)$ will be monotone increasing in k and will tend to 1 as $k \to \infty$ and to $1 - \eta(0)$ as $k \to -\infty$.

The population need not be homogeneous. Spatial models where there are n types are discussed in Chapter 3. Take $x_i(\mathbf{s},t)$ to be the proportion of type i at position $\mathbf{s}$ and time t who are still susceptible. For a wave front to pass through the population in direction $\boldsymbol{\xi}$, we require $x_i(\mathbf{s},t)$ to be constant for all $\mathbf{s}$ for which $\boldsymbol{\xi}'\mathbf{s}$ is constant. Also, for each i, $x_i(\mathbf{s} + k\boldsymbol{\xi}, t)$ must be monotone increasing in k and

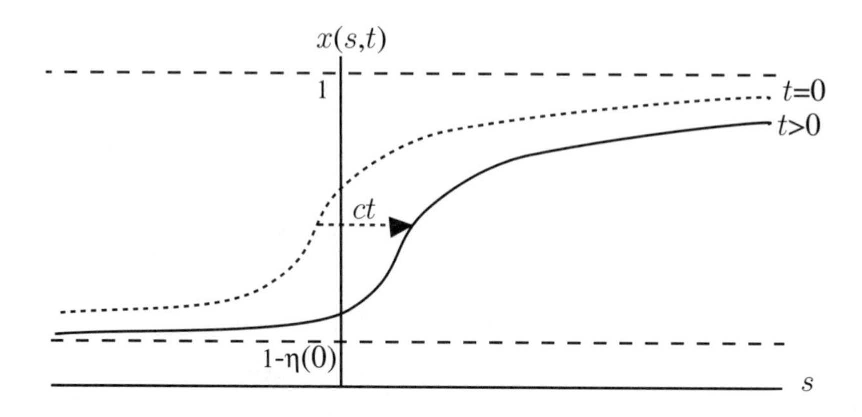

FIGURE 4.1. The travelling wave for a homogeneous population when $N = 1$.

must tend to 1 as $k \to \infty$ and to $1 - \eta_i(\mathbf{0})$ as $k \to -\infty$. Here the $\eta_i(\mathbf{0})$ can be shown to be the limiting final sizes obtained in Chapters 2 and 3.

4.2 The wave equations

For the spatial model considered in Chapter 3, infection was triggered at time $t = 0$ by the introduction of infection from outside the populations. This model is not appropriate for this chapter. In order to look at the possible speeds for which wave solutions exist, we think of the epidemic as existing for all time. Hence we need to consider the spatial analogue of the n-type form of model 2 of Chapter 2. This model is described and equations are obtained which must be satisfied by a wave travelling in a specified direction. We are led to a study of a system of integral equations. The speeds for which wave solutions exist can then be determined.

Consider n types of individuals, each type having uniform density in $\Re^N$. The density of type i individuals is σ_i. Let the rate of infection of an individual in population i by an individual in population j, who was infected time τ ago, be $\lambda_{ij}(\tau)$. The contact distribution, representing the distance $\mathbf{r}$ over which infection occurs between type j infective and type i susceptible individuals, has density $p_{ij}(\mathbf{r})$. The proportion of individuals in population i at position $\mathbf{s}$, who are still susceptible at time t is $x_i(\mathbf{s}, t)$. Let $I_i(\mathbf{s}, t, \tau)d\tau$ be the proportion of individuals in population i at position $\mathbf{s}$ who at time t were infected in the time interval $(t - \tau - d\tau, t - \tau)$. The model is described by the equations

$$\begin{aligned} \frac{\partial x_i(\mathbf{s}, t)}{\partial t} &= -x_i(\mathbf{s}, t) \sum_{j=1}^{n} \sigma_j \int_{\Re^N} \int_0^\infty I_j(\mathbf{s} - \mathbf{r}, t, \tau) p_{ij}(\mathbf{r}) \lambda_{ij}(\tau) d\tau d\mathbf{r}, \\ x_i(\mathbf{s}, t) &= 1 - \int_0^\infty I_i(\mathbf{s}, t, \tau) d\tau, \\ I_i(\mathbf{s}, t, \tau) &= I_i(\mathbf{s}, t - \tau, 0), \ (i = 1, ..., n). \end{aligned} \tag{4.1}$$

As in Chapter 3, we consider a solution with each $x_i(\mathbf{s}, t)$ jointly continuous and partially differentiable with respect to t. Also $0 \leq x_i(\mathbf{s}, t) \leq 1$ and $\lim_{t \to -\infty} x_i(\mathbf{s}, t) = 1$. In addition $I_i(\mathbf{s}, t, \tau)$, and hence $I_i(\mathbf{s}, \tau, 0)$, is jointly continuous in $\mathbf{s}$ and τ. From equations (4.1), $x_i(\mathbf{s}, t) = 1 - \int_{-\infty}^{t} I_i(\mathbf{s}, u, 0)du$, so that $I_i(\mathbf{s}, t, 0)$ is uniformly bounded over t and $(\partial/\partial t)x_i(\mathbf{s}, t) = -I_i(\mathbf{s}, t, 0)$. Hence the partial derivative is also jointly continuous and uniformly bounded over all t and $x_i(\mathbf{s}, t)$ is monotone decreasing in t.

Equations are now obtained which must be satisfied by a wave travelling in a specified direction. We look for wave solutions travelling with velocity $c > 0$ in a direction with direction cosines $\boldsymbol{\xi}$ and let $p_{ij}(r)$ be the projection of the contact distribution in that direction. Let $x_i(\mathbf{s}, t) = x_i(u)$ and $I_i(\mathbf{s}, t, \tau) = I_i(u, \tau)$, where $u = \boldsymbol{\xi}'\mathbf{s} - ct$. Since $I_i(\mathbf{s}, t, \tau) = I_i(\mathbf{s}, t - \tau, 0)$, it immediately follows that $I_i(u, \tau) = I_i(u + c\tau, 0)$. Hence equations (4.1) become

$$
\begin{aligned}
x_i'(u) &= c^{-1}x_i(u)\sum_{j=1}^{n}\sigma_j \int_{-\infty}^{\infty}\int_{0}^{\infty} I_j(u - r, \tau)p_{ij}(r)\lambda_{ij}(\tau)d\tau dr, \\
x_i(u) &= 1 - \int_0^{\infty} I_i(u, \tau)d\tau, \\
I_i(u, \tau) &= I_i(u + c\tau, 0).
\end{aligned} \tag{4.2}
$$

Any solution must have each $0 \leq x_i(u) \leq 1$ and $x_i(u)$ monotone increasing and differentiable and $I_i(u, 0)$ continuous. We specify that all individuals are susceptible before the wave, i.e. $\lim_{u \to \infty} x_i(u) = 1$, for $i = 1, \ldots, n$. From these equations $x_i(u) = 1 - \int_0^{\infty} I_i(u + c\tau, 0)d\tau = 1 - c^{-1}\int_u^{\infty} I_i(r, 0)dr$. Since $x_i(u) \leq 1$ for all u, $I_i(u, 0)$ is bounded. We obtain the relation $cx_i'(u) = I_i(u, 0)$, so that $x_i'(u)$ is also continuous and bounded.

From equations (4.2), $I_i(u - r, \tau) = I_i(u - r + c\tau) = cx_i'(u - r + c\tau)$. Making this substitution and changing variable so that $s = c\tau$ and $t = c\tau - r$, we obtain a system of differential equations in the $x_i(u)$,

$$
x_i'(u) = x_i(u)\sum_{j=1}^{n}\int_{-\infty}^{\infty} x_j'(u + t)v_{ij}(t)dt, \; (i = 1, \ldots, n), \tag{4.3}
$$

where, if we define $\gamma_{ij}(\tau) = \sigma_j\lambda_{ij}(\tau)$ as in previous chapters,

$$
v_{ij}(t) = c^{-1}\int_0^{\infty} p_{ij}(s - t)\gamma_{ij}(s/c)ds. \tag{4.4}
$$

Clearly $v_{ij}(t)$ depends upon c, however the dependence is suppressed to avoid complicating the notation at this stage. Any non-negative, monotone increasing solution $x_i(u)$ $(i = 1, \ldots, n)$ to equations (4.3) with continuous, bounded derivatives and with $\lim_{u \to \infty} x_i(u) = 1$, for $i = 1, \ldots, n$, is a solution to equations (4.2) if we define $I_i(u, 0) = cx_i'(u) \geq 0$ and $I_i(u, \tau) = I_i(u + c\tau, 0)$.

We now relate the solutions of equations (4.3) to the solutions of each of two systems of equations; firstly the equations which will be used when considering waves at the critical speed

$$(4.5) \qquad x_i'(u) = x_i(u) \sum_{j=1}^{n} \int_{-\infty}^{\infty} (1 - x_j(u+t)) v_{ij}'(t) dt, (i = 1, ..., n);$$

secondly equations which are used for the major part of the analysis

$$(4.6) \qquad w_i(u) = \sum_{j=1}^{n} \int_{-\infty}^{\infty} (1 - e^{-w_j(u+t)}) v_{ij}(t) dt, (i = 1, ..., n),$$

with $w_i(u) = -\log x_i(u)$ for all i, so that each $w_i(u)$ is monotone decreasing and $\lim_{u\to\infty} w_i(u) = 0$.

In order to do this we need to impose some conditions on the $p_{ij}(\mathbf{r})$ and $\gamma_{ij}(\tau)$. The $p_{ij}(\mathbf{r})$ are assumed to be continuous for all $\mathbf{r}$. However they are not assumed to be radially symmetric; an assumption which is necessary in Chapters 3 and 5 to prove the results on the pandemic theorem and the asymptotic speed of propagation. Hence the projections of the contact distribution in the direction specified by $\boldsymbol{\xi}$, the $p_{ij}(r)$, are continuous functions of r but may not be symmetric about zero.

As in Chapters 2 and 3, the $\gamma_{ij}(\tau)$ are assumed to be bounded with continuous bounded derivatives. In addition $\gamma_{ij}'(\tau)$ is assumed to change sign at most a finite number of times, the number of changes being denoted by k_{ij}. This last condition on the $\gamma_{ij}'(\tau)$ was not required in Chapters 3 and 5. So this present chapter requires an additional constraint on the infection rates but less constraints on the contact distributions.

These constraints imply conditions on the $v_{ij}(t)$. For each positive speed c, $v_{ij}(u)$ is bounded (by $\sup_t \gamma_{ij}(t)$) and differentiable, since $\gamma_{ij}(t)$ is bounded and differentiable with continuous, bounded derivatives and $p_{ij}(r)$ is continuous and integrable. The derivative is

$$(4.7) \qquad v_{ij}'(s) = c^{-1} p_{ij}(-s)\gamma_{ij}(0) + c^{-2} \int_{-s}^{\infty} p_{ij}(t)\gamma_{ij}'((s+t)/c) dt.$$

This is easily shown to be continuous and bounded with $|v_{ij}'(t)| \le c^{-1} \sup_t \gamma_{ij}(t) + c^{-2} \sup_t \gamma_{ij}'(t)$. In addition $v_{ij}'(t)$ is absolutely integrable since

$$(4.8) \quad \int_{-\infty}^{\infty} |v_{ij}'(s)| ds \le c^{-1}\gamma_{ij}(0) + c^{-1} \int_{0}^{\infty} |\gamma_{ij}'(u)| du \le c^{-1}(k_{ij} + 2) \sup_t \gamma_{ij}(t).$$

Before proving the correspondence between the solutions of the three systems of equations we show that any solution to equations (4.2) for a specific $c > 0$ must have each $x_i(u)$ positive and $x_i'(u)/x_i(u)$ bounded. The solution required to equations (4.3) must therefore satisfy these conditions. From equations (4.2),

$$0 \le x_i'(u) \le c^{-1} x_i(u) \sum_{j=1}^{n} \sup_{\tau} \gamma_{ij}(\tau) \int_{-\infty}^{\infty} (1 - x_j(u-r)) p_{ij}(r) dr \le A_i x_i(u),$$

where $A_i = c^{-1}\sum_{j=1}^{n}\sup_\tau \gamma_{ij}(\tau)$. Suppose that $x_i(u) = 0$ for some i and u. Then, since $x_i(u)$ is monotonic and $\lim_{u\to\infty} x_i(u) = 1$, for this i there exists a U such that $x_i(u) > 0$ for $u > U$ but $x_i(u) = 0$ for $u \leq U$. Thus for $U < u < U + 1$, $x_i(u) \geq x_i(U+1)e^{-A_i} > 0$ and hence from continuity $x_i(U) > 0$ and a contradiction is obtained. Also $x_i'(u)/x_i(u) \leq c^{-1}\sum_{j=1}^{n}\sup_\tau \gamma_{ij}(\tau)$, so is bounded.

LEMMA 4.1. *There is a one to one correspondence between the positive, monotone increasing, differentiable solutions $x_i(u)$ $(i = 1, ..., n)$ of equations (4.3) and (4.5) for which $\lim_{u\to\infty} x_i(u) = 1$ and $w_i(u) = -\log x_i(u)$ has continuous bounded derivatives. Any such solution provides a solution to equations (4.6) with this definition of $w_i(u)$.*

PROOF. It is simple to show the equivalence of solutions $x_i(u)$ to equations (4.3) and (4.5). Integrating by parts the ij^{th} term in the summation in either system of equations and noting that $\lim_{t\to-\infty} v_{ij}(t) = 0$ and $\lim_{t\to\infty} v_{ij}(t)$ is finite we obtain

$$\int_{-\infty}^{\infty} x_j'(u+t)v_{ij}(t)dt = \int_{-\infty}^{\infty} (1 - x_j(u+t))v_{ij}'(t)dt.$$

Hence the solutions $x_i(u)$, with the stated conditions, to (4.3) and (4.5) are identical.

It is now sufficient to prove that, if $x_i(u)$, $(i = 1, ...n)$, is a solution to equations (4.3) with the stated conditions, then $w_i(u) = -\log(x_i(u))$ $(i = 1, ..., n)$ is a solution to equations (4.6).

Consider a positive, monotone increasing, differentiable solution $x_i(u)$ $(i = 1, ..., n)$ to equations (4.3), where each $x_i'(u)/x_i(u)$ is continuous and bounded and $\lim_{u\to\infty} x_i(u) = 1$. Then each $w_i(u) = -\log(x_i(u))$ is monotone decreasing and differentiable, with continuous bounded derivatives, and tends to zero as $u \to \infty$. Equations (4.3) may be rewritten as,

$$-w_i'(u) = \sum_{j=1}^{n}\int_{-\infty}^{\infty} x_j'(u+s)v_{ij}(s)ds, \; (i = 1, ..., n).$$

Integrating from u to R, and interchanging the integrals we obtain

$$w_i(u) - w_i(R) = \sum_{j=1}^{n}\int_{-\infty}^{\infty} (x_j(s+R) - x_j(s+u))v_{ij}(s)ds.$$

Since $w_i(R) \downarrow 0$ and $x_i(R) \uparrow 1$ as $R \to \infty$, using monotone convergence and replacing $x_j(s+u)$ by $e^{-w_j(s+u)}$, we obtain equations (4.6).

□

Note that it is possible to also show that any solution of equations (4.6) must be a solution of equations (4.5) and hence of equations (4.3), however an additional constraint needs to be imposed, namely that each infection rate is convex in the tail.

We can avoid imposing the constraint for the following reason. The analysis of wave solutions for the spatial models is based on the fundamental integral equations (4.6) together with the conditions that each $w_i(u)$ is non-negative, monotone decreasing and differentiable with continuous bounded derivatives and

$\lim_{u\to\infty} w_i(u) = 0$. From Lemma 4.1, if no solution to equations (4.6) is shown to be possible at a specific speed, then equations (4.3) and (4.5) cannot have a solution at that speed. For speeds c for which a solution to equations (4.6) exists, the solution can be shown to be unique and its behaviour in the upper tail established. This is then used to show that the solution is also a solution (and hence from Lemma 4.1 is the unique solution) to equations (4.3) and (4.5) with the specified conditions. The proof of existence of a wave solution at the critical speed uses equations (4.5) directly.

A brief indication of the alternative direct proof, assuming convexity of the infection rates in the tail is now given out of interest. It is not necessary to assume that the solution is differentiable.

Consider any monotone decreasing, continuous solution $w_i(u)$ to (4.6) with $\lim_{u\to\infty} w_i(u) = 0$, $(i = 1, ..., n)$. Let $x_i(u) = e^{-w_i(u)}$. Then $x_i(u)$ is positive, monotone increasing and continuous with $\lim_{u\to\infty} x_i(u) = 1$, $(i = 1, ..., n)$. We can re-write equations (4.6) in the form

$$(4.9) \qquad -\log x_i(u) = c^{-1}\sum_{j=1}^{n}\int_u^\infty D_{ij}(t)\gamma_{ij}[(t-u)/c]dt, (i = 1, ..., n),$$

where $D_{ij}(t) = \int_{-\infty}^{\infty}(1-x_j(s))p_{ij}(t-s)ds$ is easily shown to be continuous. Consider the integral in the jth term of equations (4.9). For any given u

$$\frac{d}{du}\int_u^\infty D_{ij}(t)\gamma_{ij}[(t-u)/c]dt = -D_{ij}(u)\gamma_{ij}(0)$$
$$+ \lim_{\delta u\to 0}\int_u^\infty D_{ij}(t)\left\{\frac{\gamma_{ij}[(t-u-\delta u)/c] - \gamma_{ij}[(t-u)/c]}{\delta u}\right\}dt.$$

Split the range of integration into (u, a) and (a, ∞), where a is sufficiently large such that $\gamma_{ij}(\tau)$ is convex for $\tau > (a-u-1)/c$, then apply dominated convergence and monotone convergence respectively to the integrals over these ranges. Hence $x_i(u)$ is differentiable and

$$\frac{x_i'(u)}{x_i(u)} = \frac{d}{du}\left\{-c^{-1}\sum_{j=1}^{n}\int_u^\infty D_{ij}(t)\gamma_{ij}[(t-u)/c]dt\right\}$$
$$= c^{-1}\sum_{j=1}^{n}\left\{D_{ij}(u)\gamma_{ij}(0) + c^{-1}\int_u^\infty D_{ij}(t)\gamma_{ij}'[(t-u)/c]dt\right\}$$
$$= \sum_{j=1}^{n}\int_{-\infty}^{\infty}(1-x_i(t))v_{ij}'(t-u)dt.$$

So the $x_i(u)$ are differentiable and satisfy equations (4.5). The result that $x_i'(u)/x_i(u)$ is continuous and uniformly bounded follows since $v_{ij}'(u)$ is bounded and absolutely integrable and $x_i(s)$ is continuous and bounded.

4.3 A discussion of the single population case

The case when $n = 1$ will now be treated in a heuristic way to explain how a certain characteristic equation enables us to determine for which speeds wave solutions are possible. An indication is also given of the method of construction of waves and proofs of uniqueness and non-existence at certain speeds.

Consider equation (4.6) when $n = 1$, where again the subscripts are dropped as unnecessary. Then we have

$$w(u) = \int_{-\infty}^{\infty} (1 - e^{-w(u+t)})v(t)dt, \tag{4.10}$$

where $w(u)$ is continuous and monotone decreasing with $\lim_{u\to\infty} w(u) = 0$. Note that $\int_{-\infty}^{\infty} v(t)dt = \int_0^{\infty} \gamma(\tau)d\tau = \gamma$, γ being so defined in Section 2.5 of Chapter 2.

A constraint on γ and a limit.

We first find the limit of the solution to the wave equation at $-\infty$ and obtain the threshold value $\gamma = 1$, waves only being possible when $\gamma > 1$.

Let $m = \lim_{u\to-\infty} w(u)$. If γ is finite, then by monotone convergence we obtain the result that $m = \gamma(1 - e^{-m})$. If we write $l = 1 - e^{-m}$, so that $l = \lim_{u\to-\infty}(1 - x(u))$, then $-\log(1 - l) = \gamma l$. When $\gamma \leq 1$, from Figure 2.1(a) with $a = 0$ the only solution l to this equation is $l = 0$, which corresponds to $x(u) \equiv 1$, or equivalently to $m = 0$ and $w(u) \equiv 0$. This is the trivial wave solution when there is no infection. We immediately obtain the condition that no non-trivial wave solution is possible unless $\gamma > 1$.

When $\gamma > 1$ and is finite, from Figure 2.1(b) with $a = 0$, in addition to the zero solution there is a unique positive value of l satisfying $-\log(1 - l) = \gamma l$. The solution is $l = \eta(0)$, where $\eta(0)$ is the lower bound on the spatial final size obtained in Section 3.4 for the one-type model. Hence for a non-trivial wave solution, $w(u)$ tends to $-\log(1 - \eta(0))$ and $x(u)$ tends to $1 - \eta(0)$ as $u \to \infty$. Therefore it is possible to have a non-trivial wave solution when $\gamma > 1$. Once the wave has passed through the population, $1 - \eta(0)$ gives the proportion who remained unaffected by its passage.

When γ is infinite, if there is a non-trivial wave solution then $m \neq 0$. Then the right hand side of equations (4.10) tends to infinity as u tends to minus infinity. Hence m must be infinite and therefore $\lim_{u\to-\infty} x(u) = 0$. Thus it is possible to have a wave solution. The wave affects everyone when it passes through the population.

A condition on the contact distribution.

If $w(u)$ is a non-trivial solution to equation (4.10), it is possible to show that its Laplace transform, $W(\lambda) = \int_{-\infty}^{\infty} w(u)e^{\lambda u}du$, exists for $0 < Re(\lambda) < \delta$ for some $\delta > 0$. A simple proof was given for the $S \to I$ epidemic by Atkinson and Reuter [A7] and is briefly indicated below..

For the $S \to I$ epidemic $v(t) = c^{-1}\sigma \int_0^{\infty} p(s - t)\lambda ds$, so that equation (4.10) becomes

$$w(u) = \frac{\sigma\lambda}{c} \int_{-\infty}^{\infty} (1 - e^{-w(u+t)})Q(-t)dt,$$

where $Q(s) = \int_s^\infty p(r)dr$. Since $\lim_{t\to\infty} Q(-t) = 1$, we can find k so that $k\sigma\lambda/(4c) > 1$ and $Q(-t) \geq 1/2$ for $t > k$. Since $w(u)$ is monotone decreasing, $\lim_{x\downarrow 0}(1 - e^{-x})/x = 1$ and $\lim_{u\to\infty} w(u) = 0$, there exists a U such that $(1 - e^{-w(u)}) > \frac{1}{2}w(u)$ for $u > U$. Hence for $u > U$,

$$w(u) = \frac{\sigma\lambda}{c}\int_{-\infty}^{\infty}(1 - e^{-w(u+t)})Q(-t)dt \geq \frac{\sigma\lambda}{2c}\int_k^{2k}(1 - e^{-w(u+t)})dt$$
$$\geq \frac{k\sigma\lambda}{2c}(1 - e^{-w(u+2k)}) \geq \frac{k\sigma\lambda}{4c}w(u+2k).$$

Therefore $w(u + 2rk) \leq e^{-2rkA}w(u)$ for $u > U$ and any positive integer r, where $A = (1/(2k))\log(k\sigma\lambda/(4c)) > 0$. Hence $w(u) = O(e^{-Au})$ for u large. A suitable δ to use is then A, since the Laplace transform $W(\lambda)$ exists for $0 < Re(\lambda) < A$.

This method of proof does not generalise even to the $S \to I \to R$ epidemic and a much more complex proof is needed in general, this being given in Theorem 4.2 for the general n-type model.

We now consider the implication of taking the Laplace transform of equation (4.10). This enables us to obtain the characteristic equation, and also provides a condition on the contact distribution which is necessary for wave solutions to be possible.

Since $1 - e^{-w(u)}$ behaves like $w(u)$ for $w(u)$ small, its Laplace transform also exists for $0 < Re(\lambda) < \delta$. Now consider $V(\lambda) = \int_{-\infty}^{\infty} v(u)e^{-\lambda u}du$. Taking the Laplace transform of equation (4.10) implies that $V(\lambda)$ also exists for $0 < Re(\lambda) < \delta$. Hence if we denote the abscissae of convergence of $W(\lambda)$ and $V(\lambda)$ in the positive half of the complex plane by Δ_w and Δ_v respectively, then necessarily $\Delta_w \leq \Delta_v$. It is easily seen that $V(\lambda) = \sigma P(\lambda)\Lambda(c\lambda)$, where $P(\lambda) = \int_{-\infty}^{\infty} p(r)e^{\lambda r}dr$ and $\Lambda(\lambda) = \int_0^\infty \lambda(r)e^{-\lambda r}dr$. Note that $\Lambda(c\lambda)$ exists for all $c > 0$ and all $Re(\lambda) > 0$. A condition on the contact distribution is thus obtained for the existence of a non-trivial wave solution. Since $P(\lambda)$ must exist for $0 < \lambda < \delta$, no non-trivial wave solution is possible at any positive speed c unless the projection of the contact distribution in direction $\boldsymbol{\xi}$, $p(r)$, is exponentially dominated in the forward tail (i.e. $p(r)e^{\lambda r} \to 0$ as $r \to \infty$, where λ may be taken to be any value such that $0 < \lambda < \delta$).

The characteristic equation.

We therefore only need consider the case when $\gamma > 1$ and $p(r)$ is exponentially dominated in the forward tail. Then

$$W(\lambda) = V(\lambda)\int_{-\infty}^{\infty} e^{\lambda r}(1 - e^{-w(r)})dr. \tag{4.11}$$

This equation is valid for $0 < Re(\lambda) < \Delta_w$. The speed c is not used in the notation, but clearly $V(\lambda) = \sigma\Lambda(c\lambda)P(\lambda)$ is a function of c. However Δ_v is just the abscissa of convergence of $P(\lambda)$, so is independent of c. We define $K_c(\lambda) = V(\lambda)$ so that the dependence on c is made explicit. Equation (4.11) may be rewritten in the form

$$(K_c(\lambda) - 1)W(\lambda) = K_c(\lambda)T(\lambda), \tag{4.12}$$

where $T(\lambda) = \int_{-\infty}^{\infty} e^{\lambda r}(e^{-w(r)} - 1 + w(r))dr > 0$ for all real λ. For $w(r)$ small, $(e^{-w(r)} - 1 + w(r))$ behaves like $(w(r))^2/2$, so that when Δ_w is finite, $T(\lambda)$ is finite and positive for λ real with $0 < \lambda < 2\Delta_w$.

Now $K_c(0) = \gamma > 1$. Then for any $0 < \lambda^* < \Delta_v$, if $K_c(\lambda) > 1$ for $0 \leq \lambda \leq \lambda^*$, it then follows that $W(\lambda)$ exists in some open region of λ^*. Hence either $\Delta_w = \Delta_v$, or $K_c(\lambda) = 1$ has a solution in the range $(0, \Delta_v)$. In the latter case, if $\lambda = \alpha_c$ is the smallest positive root of $K_c(\lambda) = 1$, then $K_c(\alpha_c)T(\alpha_c)$ exists and is positive whilst $(K_c(\alpha_c) - 1) = 0$. Hence from equation (4.12), $W(\alpha_c) = \infty$ and so $\Delta_w = \alpha_c$.

We refer to $K_c(\lambda) = 1$ as the characteristic equation. This equation is central to the identification of the speeds $c > 0$ for which wave solutions exist. The properties of $K_c(\lambda)$ when λ is real are easily established and are used to prove the existence, uniqueness and non-existence of waves at different speeds.

The properties of $K_c(\lambda)$ are now described for $c > 0$ and $0 \leq \lambda \leq \Delta_v$; these being illustrated in Figure 4.2.

1. First observe that the Laplace transforms $P(\theta)$ and $\Lambda(\theta)$ are continuous functions of θ within their region of convergence, which tend to one and γ/σ respectively as $\theta \downarrow 0$. Hence $K_c(\lambda) = V(\lambda) = \sigma P(\lambda)\Lambda(c\lambda)$ is a jointly continuous function of c and λ for $c > 0$ and $0 < \lambda < \Delta_v$. Also $K_c(0) = \lim_{\theta \downarrow 0} K_c(\theta) = \gamma > 1$ for all $c > 0$.
2. The abscissa of convergence of $K_c(\lambda)$ is just Δ_v, and is the same for all $c > 0$. When Δ_v is finite, since $\Lambda(c\lambda)$ exists and is positive for all $0 < \lambda < \infty$, the limit of $K_c(\lambda)$ as $\lambda \uparrow \Delta_v$ is determined by $P(\Delta_v)$. This limit $K_c(\Delta_v)$ is finite for all $c > 0$ when $P(\Delta_v)$ is finite and is infinite for all $c > 0$ when $P(\Delta_v)$ is infinite. Now consider the case when $\Delta_v = \infty$. Since $K_c(\lambda) = \int_{-\infty}^{\infty} e^{-\lambda u} v(u)du$, its limit as $\lambda \to \infty$ is either infinity or zero, the latter limit occurring precisely when $v(u) = 0$ for all $u < 0$.
3. Clearly $\Lambda(c\lambda) = \displaystyle\int_0^{\infty} \lambda(\tau)e^{-c\lambda\tau}d\tau$ is a strictly decreasing function of c for each $\lambda > 0$. Hence $K_c(\lambda)$ is also a strictly decreasing function of c for each $0 < \lambda < \Delta_v$. When both Δ_v and $P(\Delta_v)$ are finite, then $K_c(\Delta_v)$ is also a strictly decreasing function of c.
4. Now consider $K_c(\lambda)$ for each fixed $c > 0$. For any such c, the second derivative of $K_c(\lambda)$ with respect to λ is given by $K_c''(\lambda) = \int_{-\infty}^{\infty} u^2 e^{\lambda u} v(u)du > 0$, so that $K_c(\lambda)$ is a strictly convex function of λ for all $0 < \lambda < \Delta_v$. Hence, for a specific value of c, $K_c(\lambda) = 1$ can have at most two solutions in the range $0 < \lambda \leq \Delta_v$.

Connection with wave speeds.

For each $0 < \lambda^* < \Delta_v$, $\lim_{c \to \infty} K_c(\lambda^*) = 0$. Then there exists a c^* sufficiently large such that $K_{c^*}(\lambda^*) < 1$. Since $K_{c^*}(0) = \gamma > 1$, continuity in λ then implies that $K_{c^*}(\lambda) = 1$ has a solution for some $0 < \lambda < \lambda^*$. So the characteristic equation has a solution for some positive speeds.

If $K_{c^*}(\lambda^*) < 1$ for some $c^* > 0$ and $0 < \lambda^* \leq \Delta_v$, then $K_c(\lambda^*) < 1$ for all $c \geq c^*$. The same properties of $K_c(\lambda)$ then imply that, for each $c \geq c^*$, the characteristic equation $K_c(\lambda) = 1$ has a solution for some $0 < \lambda < \lambda^*$. In addition, if $K_{c^*}(\lambda) \geq 1$ for all $0 < \lambda \leq \Delta_v$, then for each $0 < c < c^*$ it follows that $K_c(\lambda) > 1$ for all $0 \leq \lambda \leq \Delta_v$.

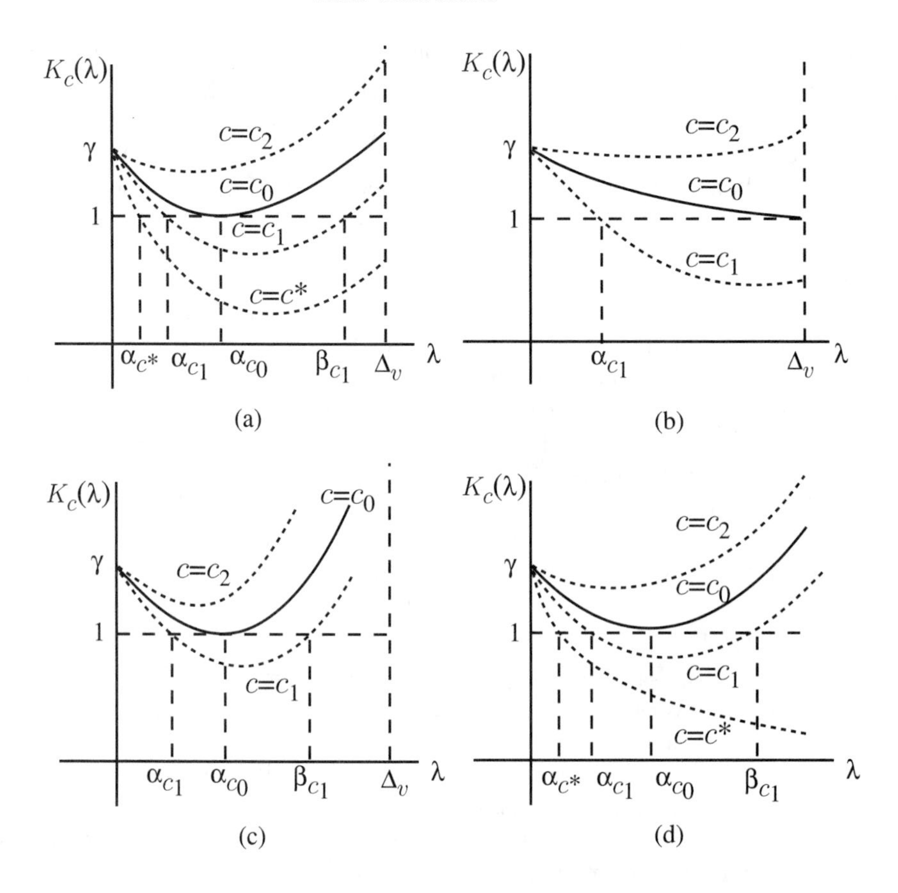

FIGURE 4.2. The function $K_c(\lambda)$ when γ is finite and $c_0 > 0$. Cases (a), (b) and (c) have Δ_v finite and case (d) has Δ_v infinite. Case (c) has $P(\Delta_v)$ infinite. Cases (a) and (b) both have $P(\Delta_v)$ finite, the special case when $K_{c_0}(\Delta_v) = 1$ being illustrated in (b). In the figures, $0 < c_2 < c_0 < c_1 < c^*$. Note that in case (d) the curve for $c = c^*$ will only occur for some choices of contact distribution and infection rate.

Define c_0 to be the infimum of positive values of c for which $K_c(\lambda) = 1$ has a positive real solution. Consider the case when $c_0 > 0$. Then if $K_{c_0}(\lambda) < 1$ for some λ, continuity and strict monotonicity in c implies that there exists a $0 < c < c_0$ for which $K_c(\lambda) < 1$. This gives a contradiction. Hence $K_{c_0}(\lambda) \geq 1$ for $0 \leq \lambda \leq \Delta_v$. Therefore, for each $0 < c < c_0$, $K_c(\lambda) > 1$ for $0 \leq \lambda \leq \Delta_v$, so that the characteristic equation has no solution. If it is assumed that a non-trivial wave solution exists at speed c, where $0 < c < c_0$, this result is used to obtain a contradiction, which establishes that no wave solution is possible at any speed below c_0.

For both the cases $c_0 > 0$ and $c_0 = 0$, for each $c > c_0$ there is either one or two solutions to the characteristic equation. This depends upon whether Δ_v and $P(\Delta_v)$ are finite or infinite. The different cases are shown in Figure 4.2 for the case when $c_0 > 0$ and γ is finite. For $c > c_0$ we define α_c to be the smallest positive root of $K_c(\lambda) = 1$. Also we define β_c to be the other positive root if there is one and to be Δ_v otherwise. Then $K_c(\alpha_c) = 1$ for each $c > c_0$ and, from the convexity in λ, $K_c(\lambda) < 1$ for all $\alpha_c < \lambda < \beta_c$, so that the root α_c is simple. These results are used to construct a solution and establish uniqueness for each speed $c > c_0$.

Finally consider the solutions of the characteristic equation at speed c_0 when $c_0 > 0$. It has already been shown that $K_{c_0}(\lambda) \geq 1$ for all $0 \leq \lambda \leq \Delta_v$. Equality for some λ is now shown. Monotonicity in c implies that α_c is a monotone decreasing function of c for $c > c_0$. When Δ_v and $P(\Delta_v)$ are finite, then $\alpha_c \leq \Delta_v$, so that α_c tends to a limit α as $c \downarrow c_0$ and hence, using joint continuity in c and λ, $K_{c_0}(\alpha) = \lim_{c \downarrow c_0} K_c(\alpha_c) = 1$. Note that α may equal Δ_v.

When Δ_v is infinite, since $K_{c_0}(\lambda) \geq 1$ for all λ, necessarily $K_{c_0}(\Delta_v)$ is infinite. When Δ_v is finite and $P(\Delta_v)$ is infinite, then $K_{c_0}(\Delta_v)$ is also infinite, so that both cases may be treated simultaneously. If γ is infinite we can choose $0 < A < B < \Delta_v$ such that $1 < K_{c_0}(A) < K_{c_0}(B)$ and when γ is finite we take $A = 0$ and choose $0 < B < \Delta_v$ so that $1 < \gamma = K_{c_0}(A) < K_{c_0}(B)$. Then there exists a $c^* > c_0$ such that $1 < K_{c^*}(A) < K_{c_0}(A) < K_{c^*}(B)$, and hence for each $c_0 < c < c^*$, $1 < K_c(A) < K_c(B)$. But $K_c(\lambda)$ is a convex function of λ which is less than one for some λ. Hence the characteristic equation has two roots and $0 < \alpha_c < \beta_c < B < \Delta_v$. Therefore α_c tends to a limit $\alpha \leq B < \Delta_v$ as $c \downarrow c_0$ and $K_{c_0}(\alpha) = \lim_{c \downarrow c_0} K_c(\alpha_c) = 1$.

Hence in all cases $K_{c_0}(\alpha) = 1$ for some $0 < \alpha \leq \Delta_v$, with strict inequality except in the case when Δ_v and $P(\Delta_v)$ are finite when α may be equal to Δ_v. The strict convexity in λ then implies that α is the unique root of the characteristic equation when $c = c_0$, which is of even multiplicity if $\alpha < \Delta_v$. Denote this α by α_{c_0}.

When $c = c_0$ a limiting argument is used to show that a wave solution exists at speed c_0. The proof does not cover the case when γ is infinite but $\lim_{u \to \infty} v(u) = 0$. Provided $\alpha_{c_0} < \Delta_v$ it has been shown that the root has even multiplicity. This is used to prove uniqueness of the wave solution at speed c_0. Note that for the special case when Δ_v and $P(\Delta_v)$ are both finite with $K_{c_0}(\Delta_v) = 1$ uniqueness is not established.

The next subsection gives examples concerning the different forms the plots of $K_c(\lambda)$ can take, both in the case when γ is finite and $c_0 > 0$ (as illustrated in Figure 4.2) and when γ is infinite and/or $c_0 = 0$. The results for $K_c(\lambda)$ contained in this present subsection are then used in the remaining subsections to show that when $c_0 = 0$ then a non-trivial wave solution exists at each speed $c > 0$, the wave at each speed being unique modulo translation (i.e. shift of origin). When $c_0 > 0$ then no non-trivial wave solution is possible at any speed $0 < c < c_0$. A wave solution exists at each speed $c \geq c_0$, with the wave solution at each speed unique modulo translation. However existence and uniqueness at speed c_0 is not established for all cases.

Some illustrative examples.

A brief discussion, with examples, is now given in this subsection to illustrate how the different plots of $K_c(\lambda)$ can arise. Although this is interesting, it is not essential to understanding the one-type case. It can therefore be omitted on first reading the section.

Figure 4.2 only shows the case when $c_0 > 0$ and γ is finite. Note that for case (d) (i.e. $\Delta_v = \infty$) with $c_0 > 0$, curves of the form $c = c^*$ will not always occur. When $c_0 = 0$ the illustrations in Figure 4.2 remain the same except that the curves for $c = c_0$ and $c = c_2$ should be deleted. For case (d) it is possible to have both forms of curve $c = c_1$ and $c = c^*$, or to have only curves of one type which can be either of the form $c = c_1$ or $c = c^*$. These two curves correspond to the cases when the limit as $\lambda \to \infty$ of $K_c(\lambda)$ is infinity or zero respectively. When γ is infinite the vertical axis shown in Figure 4.2 simply becomes the asymptote as $\lambda \downarrow 0$. An example when γ is infinite is the simple $S \to I$ epidemic.

Examples are first given for the case when $c_0 > 0$. When γ is finite with $\gamma P(\lambda) > 1$ for all $0 \leq \lambda \leq \Delta_v$ then it is easy to see that $c_0 > 0$. Assume that the characteristic equation has a solution for all $c > 0$ and consider the monotone function α_c. If $\Delta_v = \infty$ necessarily $P(\Delta_v) = \infty$. For $P(\Delta_v)$ infinite, choose B such that $P(\lambda) > 2\gamma$ for $\lambda > B$. Then there exists a $c^* > 0$ such that $K_c(B) > \gamma$ for $0 < c < c^*$. When Δ_v and $P(\Delta_v)$ are both finite, take $B = \Delta_v$. Then in either case α_c is monotone increasing and bounded above by B, with $P(B)$ finite. Hence α_c tends to a limit $\alpha \leq B$ as $c \downarrow 0$ and joint continuity of $K_c(\lambda)$ in c and λ then gives $\gamma P(\alpha) = 1$. This is a contradiction and hence $c_0 > 0$ when $\gamma > 1$ is finite and $\gamma P(\lambda) > 1$ for all $0 \leq \lambda \leq \Delta_v$. Note that this case must occur when the contact distribution is symmetric about zero, since $\gamma P(\lambda)$ is convex and $\gamma P'(0) = 0$ so that $\gamma P(\lambda) \geq \gamma > 1$ for all $0 \leq \lambda \geq \Delta_v$.

The double exponential distribution with $p(u) = (\alpha/2)e^{-\alpha|u|}$ and normal contact distribution with $p(u) = e^{-u^2/(2\sigma^2)}/\sqrt{2\pi\sigma^2}$ (so mean zero) provide examples of contact distributions corresponding to Δ_v finite (and equal to α) and infinite respectively. These are illustrated in Figure 4.2(c) and 4.2(d). An $S \to I \to R$ epidemic with normal contact distribution with mean zero provides an example of a case when only curves of type $c = c_1$ occur. The same epidemic, but with a symmetric contact distribution with mean zero behaving like $e^{-|u|}/u^2$ in the tails, provides an example of Δ_v finite (and equal to 1) with $K_c(\Delta_v)$ finite, so gives an example of the case illustrated in Figure 4.2 (a) (or possibly (b)) .

Now consider the special case illustrated in Figure 4.2 (b) when Δ_v and $P(\Delta_v)$ are both finite with $K_{c_0}(\Delta_v) = 1$, for which uniqueness of the wave solution at speed c_0 is not established. An example of when this special case can occur is easily constructed. First observe that if there exists a $c > 0$ such that $K_c(\Delta_v) = 1$ and $K_c'(\Delta_v) \leq 0$, the convexity in λ implies that $c = c_0$ and hence $K_{c_0}(\Delta_v) = 1$ as required. Consider a simple $S \to I$ epidemic. Then $K_c(\theta) = \dfrac{\sigma\lambda P(\theta)}{c\theta}$ and hence the derivative of $K_c(\theta)$ with respect to θ is given by $K_c'(\theta) = \dfrac{\sigma\lambda(\theta P'(\theta) - P(\theta))}{c\theta^2}$. The condition $K_c(\Delta_v) = 1$ then simply specifies that $c_0 = \sigma\lambda P(\Delta_v)/\Delta_v$, which is positive. The condition $K_c'(\Delta_v) \leq 0$ is equivalent to the condition $P(\Delta_v) - \Delta_v P'(\Delta_v) \geq 0$. Define

$$p(u) = \begin{cases} A\frac{1}{u^4}e^{-|u|}, & \text{for } |u| \geq 1, \\ \frac{A}{2e}(7 - 5u^2), & \text{for } |u| < 1, \end{cases}$$

where $A > 0$ is chosen so that $p(u)$ integrates to one. Then $\Delta_v = 1$ and

$$P(1) - P'(1) = \int_{-\infty}^{\infty} (1-u)e^u p(u)du$$

$$> \int_0^1 (1-u)e^u p(u)du - \int_1^{\infty} (u-1)e^u p(u)du = A(13e^{-1} - 25/6) > 0.$$

Hence this example has $\Delta_v = 1$, $c_0 = \sigma\lambda P(1) > 0$, $K_{c_0}(\Delta_v) = 1$ and $K'_{c_0}(\Delta_v) < 0$.

We now provide examples for the case when $c_0 = 0$. When $\gamma > 1$ is finite it is easy, as for the case $c_0 > 0$, to consider the $P(\lambda)$ for which c_0 will be zero. Now $K_c(\lambda) \uparrow \gamma P(\lambda)$ as $c \downarrow 0$ for all $0 \leq \lambda < \Delta_v$, with the right hand inequality not strict if $P(\Delta_v)$ is finite. Then if $\gamma P(\lambda^*) \leq 1$ for some λ^*, strict monotonicity in c implies that $K_c(\lambda^*) < 1$ for all $c > 0$ and hence $c_0 = 0$.

Consider the adjusted Figure 4.2 for $c_0 = 0$. An example for case (c) occurs when the contact distribution is a shifted double exponential with negative mean $-a$ and scale parameter $\alpha = 1$. Then $P(\lambda) = e^{-\lambda a}/(1-\lambda^2)$, so that $\inf_\lambda \gamma P(\lambda) < 1$ for a sufficiently large choice of a.

The cases (a) and (b) correspond to $\gamma P(\lambda) = 1$ for some $0 < \lambda \leq \Delta_v$ with $P(\Delta_v)$ finite, where $\gamma P(\Delta_v) > 1$ for case (a) and $\gamma P(\Delta_v) \leq 1$ for case (b). An example for case (b) is simple to construct when $\gamma > 1$ is finite using the contact distribution $p(u)$ given in the previous example for case (b) when $c_0 > 0$, but shifted by an amount $-a$. The contact distribution is then symmetric about $-a$ and not zero and $P(\theta) = e^{-a\theta} P^*(\theta)$, where

$$P^*(\theta) = \int_{|u|\geq 1} e^{\theta u} A \frac{1}{u^4} e^{-|u|} du + \int_{|u|<1} e^{\theta u} \frac{A}{2e}(7 - 5u^2)du.$$

We need only choose a so that $\gamma P(\Delta_v) \leq 1$, since then necessarily $c_0 = 0$. Now $\Delta_v = 1$ and the condition $\gamma P(\Delta_v) \leq 1$ is just $\gamma e^{-a} P^*(1) \leq 1$. So an example of case (b) when $c_0 = 0$ is obtained if we take $a \geq \log(\gamma P^*(1))$. Note that $P^{*\prime}(0) = 0$, as it is the Laplace transform of a contact distribution which is symmetric about zero. Then the convexity of $P^*(\theta)$ implies that $P^*(1) > P^*(0) = 1$, and hence that the value of a required is positive.

Figure 4.2(d) shows the case when Δ_v is infinite. An example with $c_0 = 0$ clearly occurs when the contact distribution is one-sided, so that $p(u) = 0$ for $u > 0$. Then, as $\lambda \to \infty$, $P(\lambda) \to 0$ and hence for all $c > 0$ $K_c(\lambda) \to 0$. Hence this is an example when only curves of type $c = c^*$ occur.

If the support of $\lambda(t)$ is $[D, \infty)$ and the support of $p(u)$ is $[A, B]$, where $A < B$, $B > 0$ and $D > 0$, then the support of $v(u)$ for speed $c > 0$ is $[cD - B, \infty)$. Hence

$$\lim_{\lambda\to\infty} K_c(\lambda) = \begin{cases} \infty, & \text{for } 0 < c < B/D, \\ 0, & \text{for } c \geq B/D. \end{cases}$$

Then Δ_v is infinite and both forms of curve corresponding to $c = c_1$ and $c = c^*$ occur.

An $S \to I \to R$ epidemic with normal contact distribution with mean a and variance b gives an example of when Δ_v is infinite and only curves of type $c = c_1$ occur. The normal contact distribution with zero mean, i.e. $a = 0$, has $c_0 > 0$. In fact when $a \geq 0$, $P'(0) = a \geq 0$, so that $\gamma P(\theta) \geq \gamma P(0) = \gamma > 1$ for all $\theta \geq 0$, which implies that $c_0 > 0$. An example with $c_0 = 0$ can be obtained by choosing a negative value of a such that $\inf_{\theta\geq 0}\gamma P(\theta)\leq 1$. Since $P(\theta)=\exp(a\theta+b\theta^2/2)$, the infimum is obtained at $\theta=-a/b$ for $a<0$. Hence we need to choose $a<0$ and b such that $\gamma\exp(-a^2/(2b))\leq 1$, i.e. $a^2\geq 2b\log(\gamma)$, to have $c_0=0$. Observe that when $a<0$ and this condition is not met, then $c_0>0$.

Construction of waves for positive speeds $c\geq c_0$.
In all cases, for $c>c_0$, $K_c(\alpha_c)=1$ and $K_c(\lambda)<1$ for $\alpha_c<\lambda<\beta_c$. This is used to construct a non-trivial wave solution for each $c>c_0$ and to establish uniqueness. It can be shown that $w(u)$ behaves like $Ae^{-\alpha_c u}$ for u large. An iterative method of construction is used.

A brief sketch of the construction proof is given for any specified $c>c_0$. Define $w^{(0)}(u)=Ae^{-\alpha_c u}$ for $A\geq 2$, a condition required later in the existence proof. Then, for $m\geq 0$, define

$$w^{(m+1)}(u)=\int_{-\infty}^{\infty}(1-e^{-w^{(m)}(u+t)})v(t)dt. \tag{4.13}$$

It is shown that $w(u)=\lim_{m\to\infty}w^{(m)}(u)$ satisfies the wave equation (4.11).

Since $1-e^{-x}\leq x$ for $x\geq 0$, then

$$w^{(1)}(u)\leq\int_{-\infty}^{\infty}w^{(0)}(u+t)v(t)dt=Ae^{-\alpha_c u}V(\alpha_c)=Ae^{-\alpha_c u}K_c(\alpha_c)=w^{(0)}(u).$$

We can then use induction. Assume that $w^{(m)}(u)\leq w^{(m-1)}(u)$ for $1\leq m\leq M$. Then consider equation (4.13) with $m=M$. Since $w^{(M)}(u)\leq w^{(M-1)}(u)$, it immediately follows that

$$w^{(M+1)}(u)\leq\int_{-\infty}^{\infty}(1-e^{-w^{(M-1)}(u+t)})v(t)dt=w^{(M)}(u).$$

Hence by induction the constructed sequence $\{w^{(m)}(u)\}$ is monotone decreasing in m for all u. It is easy to show that each member of the constructed sequence is non-negative and monotone decreasing in u with $\lim_{u\to\infty}w^{(m)}(u)=0$ for all m. Hence the sequence tends to a limit $w(u)$ as $m\to\infty$, where $w(u)$ is a non-negative, monotone decreasing solution to the wave equation (4.10) with $\lim_{u\to\infty}w(u)=0$.

The construction automatically gives the upper bound $w(u)\leq Ae^{-\alpha_c u}$. A lower bound on the constructed wave solution is now obtained in order to show that the solution is not the trivial wave solution $w(u)\equiv 0$. This requires the condition that $A\geq 2$. Since $c>c_0$, there is a $\delta>0$ with $\delta<\alpha_c$ such that $k_c(\alpha_c+\delta)<1$. Again induction is used to show that $w^{(m)}(u)\geq Ae^{-\alpha_c u}-Be^{-(\alpha_c+\delta)u}$ for this δ and a suitable choice of B, namely $B=A^2K_c(\alpha_c+\delta)/(2(1-K_c(\alpha_c+\delta)))$. Note that B is positive since $K_c(\alpha_c+\delta)<1$.

The result that $w^{(m)}(u)\geq Ae^{-\alpha_c u}-Be^{-(\alpha_c+\delta)u}$ obviously holds for $m=0$. Assume that it holds for $0\leq m\leq M$. From equation (4.10), since $1-e^{-w}\geq w-w^2/2$ for $w\geq 0$,

$$w^{(M+1)}(u)\geq\int_0^{\infty}\left(1-e^{-w^{(M)}(t)}\right)v(t-u)dt\geq\int_0^{\infty}\left(w^{(M)}(t)-\frac{(w^{(M)}(t))^2}{2}\right)v(t-u)dt.$$

Now using both upper and lower bounds on $w^{(M)}(t)$,

$$w^{(M+1)}(u) \geq \int_0^\infty (Ae^{-\alpha_c t} - Be^{-(\alpha_c+\delta)t} - \frac{1}{2}A^2 e^{-2\alpha_c t})v(t-u)dt.$$

Since $\delta < \alpha_c$, then

$$\begin{aligned} w^{(M+1)}(u) &\geq \int_0^\infty (Ae^{-\alpha_c t} - (B + \frac{1}{2}A^2)e^{-(\alpha_c+\delta)t})v(t-u)dt \\ &\geq \int_{-\infty}^\infty (Ae^{-\alpha_c t} - (B + \frac{1}{2}A^2)e^{-(\alpha_c+\delta)t})v(t-u)dt \\ &= Ae^{-\alpha_c u}K_c(\alpha_c) - (B + \frac{1}{2}A^2)K_c(\alpha_c+\delta)e^{-(\alpha_c+\delta)u} \\ &= Ae^{-\alpha_c u} - Be^{-(\alpha_c+\delta)u}. \end{aligned}$$

Note that the second step is valid since the integrand for all $t < 0$ is non-positive as $A \geq 2$ and $B > 0$. Therefore the inequality holds for $m = M + 1$, and hence by induction it holds for all non-negative integers m. It must also hold for the limit function $w(u)$. Hence

$$Ae^{-\alpha_c u} - Be^{-(\alpha_c+\delta)u} \leq w(u) \leq Ae^{-\alpha_c u}, \tag{4.14}$$

and so $w(u) > 0$ for u such that $e^{\delta u} > B/A$. Thus $w(u)$ is a non-trivial solution of the wave equation.

Continuity of $w(u)$ is established using uniform continuity of $v(u)$ in a bounded interval, the bound $1 - e^{-w(u)} \leq w(u) \leq Ae^{-w(u)}$ and the easily derived properties that $v(u)$ is bounded for all u and is integrable for $u \leq 0$. Differentiability is shown by splitting the range in the right hand side of equation (4.10) into $t \geq 0$ and $t < 0$ and writing the lower integral as

$$\int_{-\infty}^0 \int_0^\infty (1 - e^{-w(t)})c^{-1}\gamma(s/c)p(s+u-t)dsdt = c^{-1}\int_u^\infty \gamma((r-u)/c)R(r)dr,$$

where $R(r) = \int_{-\infty}^0 (1 - e^{-w(t)})p(r-t)dt$ is continuous. It is then simple to show that $x(u) = e^{-w(u)}$ satisfies equations (4.5) with $n = 1$ and that $x'(u)/x(u)$ is continuous and bounded.

When $c = c_0 > 0$ consider a monotone decreasing sequence $\{c_k\}$ which tends to c_0 as k tends to infinity. A wave solution $w^{(k)}(u)$ can be constructed for each c_k and any translate used. If γ is finite the proof is based on equation (4.10). The translate is chosen so that $w^{(k)}(0)$ is fixed for all k, which anchors the limit function so that it is not identically zero. The sequence $\{w^{(k)}(u)\}$ can be shown to tend to a limit $w(u)$ as $k \to \infty$, where $w(u)$ is non-trivial and differentiable and satisfies the wave equation with $c = c_0$. This equation can then be used to show that $\lim_{u\to\infty} w(u) = 0$. Since each $x^{(m)}(u) = e^{-w^{(m)}(u)}$ satisfies equation (4.5) with $c = c_m$ and $n = 1$, it is also shown that $x(u) = e^{-w(u)}$ satisfies this equation with $c = c_0$. It is then simple to show that $x'(u)/x(u)$ is continuous and bounded.

The proof requires the $w^{(k)}(u)$ to be uniformly bounded, which does not hold when γ is infinite. In this case the one-type form of equations (4.5), with $c = c_k$, for $x^{(k)}(u) = e^{-w^{(k)}(u)}$ is used. The translate chosen has $x^{(k)}(0) = 1/2$. The sequence $\{x_k(u)\}$ can be shown to tend to a limit $x(u)$ as $k \to \infty$, where $x(u)$ is differentiable and $x(0) = 1/2$, and to satisfy equation (4.5) with $n = 1$ and $c = c_0$. Provided $\lim_{u\to\infty} v(u) > 0$, using monotone convergence it is simple to show that $\lim_{u\to\infty} x(u) = 1$. Since $v'(t)$ is absolutely integrable and $x(u)$ is monotone increasing and continuous with $x(0) > 0$ it is also easily proved that $x(u) > 0$ for all u and $x'(u)/x(u)$ is bounded.

From Lemma 4.1, since $x(u)$ is a solution to equations (4.5) with $n = 1$ and $c = c_0$, it is also a solution to the corresponding equation (4.3) and $w(u) = -\log x(u)$ is a solution to equation (4.10) with $c = c_0$.

Existence at speed $c_0 > 0$ has not been shown when $\lim_{u\to\infty} v(u) = 0$ but γ is infinite. It is easy to show that this corresponds to the case when $\gamma(t)$ tends to zero as t tends to infinity, but too slowly for the scaled infection rate to be integrable. Let $Q(s)$ be the distribution function for the contact distribution. Then, integrating by parts,

$$v(u) = \int_o^\infty \gamma\left(\frac{s}{c_0}\right) p(s-u)ds = \lim_{t\to\infty} \gamma(t) - \gamma(0)Q(-u) - \frac{1}{c_0}\int_0^\infty \gamma'\left(\frac{s}{c_0}\right) Q(s-u)ds.$$

Since $\gamma'(s)$ is absolutely integrable and $\lim_{u\to-\infty} Q(u) = 0$, using dominated convergence we obtain the result that $\lim_{u\to\infty} v(u) = \lim_{u\to\infty} \gamma(u)$. A simple example when $\lim_{t\to\infty} \gamma(t) = 0$ and $\gamma = \infty$ occurs when $\gamma(t) = \sigma/(1+t)$.

Uniqueness results.

Now consider the uniqueness of the wave solution modulo translation at each speed for which a solution exists. Uniqueness for all speeds $c > c_0$ was established for the $S \to I \to R$ model by Barbour [B3], using a martingale, and for the model with varying infectivity by Diekmann and Kaper [D9]. The method of proof given here uses the methods of the latter authors. Uniqueness for the case $c = c_0$ was not established until results were obtained for multi-type models.

For $c > c_0$, the bounds obtained in (4.14) for the constructed wave solution $w(u)$ indicate that $\lim_{u\to\infty} w(u)e^{\alpha_c u} = A$, where the only restriction on A is that $A \geq 2$. It can be shown that any solution $w(u)$ has to satisfy this limiting result. By choosing a particular translate of a wave solution (which has the effect of scaling this limit) we can ensure that $\lim_{u\to\infty} w(u)e^{\alpha_c u} = A$, for any fixed $A > 0$. This may then be used to show that the wave solution is unique modulo translation.

For $c_0 > 0$ with $K_{c_0}(\Delta_v) > 1$, the function $K_{c_0}(\lambda) = 1$ has a root of multiplicity $2s$, for some $s > 0$, at $\lambda = \alpha_{c_0}$, and $K_{c_0}(\lambda) > 1$ for $\lambda \neq \alpha_{c_0}$. It may then be shown, for speed c_0, that any solution $w(u)$ to the wave equation has $\lim_{u\to\infty} w(u)e^{\alpha_{c_0} u}/u^{2s-1} = A$. Note that A may regarded as fixed since we may take a suitable translate of $w(u)$. Uniqueness is again established by contradiction, although a somewhat different argument is required.

A sketch of the simpler proof for the case $c > c_0$ is given. Assume that there are two distinct continuous solutions $w_1(u)$ and $w_2(u)$ with $\lim_{u\to\infty} w_i(u)e^{\alpha_c u} = A$ for $i = 1, 2$, so that one is not a translate of the other. Define $D(u) = (w_1(u) - w_2(u))e^{\alpha_c u}$. This is a continuous function of u which tends to zero as u tends to either plus or minus infinity. Let $D = \sup_u |D(u)| > 0$. This must be achieved for some finite u. For this value of u from equation (4.10),

$$D \leq \int_{-\infty}^{\infty} |e^{-w_1(t)} - e^{-w_2(t)}| e^{\alpha_c u} v(t-u)dt$$
$$< \int_{-\infty}^{\infty} |w_1(t) - w_2(t)| e^{\alpha_c t} e^{-\alpha_c (t-u)} v(t-u)dt \leq DK_c(\alpha_c) = D.$$

A contradiction is obtained, so that the solution is unique modulo translation.

Non-existence of waves at speeds $0 < c < c_0$.

The condition that $K_c(\lambda) > 1$ for all $0 < \lambda \leq \Delta_v$ when $c < c_0$ is used to establish non-existence of waves for these speeds. In this case $\Delta_w = \Delta_v$. A proof by contradiction is again used; a sketch of the proof being given below. This proof is much simpler than the proof for the general multi-type case, and is based on the work of Atkinson and Reuter [A7].

If $K_c(\Delta_v)$ is finite, then also Δ_v must be finite. It can then be shown that $W(\lambda)$ exists in an open region about $\lambda = \Delta_w$, which contradicts the definition of Δ_w. However there is no simple derivation of this result; the result for the general n-type case is contained in Theorem 4.2 part 2, which is proved in Section 4.5. The contradiction argument shows that no non-trivial wave solution is possible in this case

When $K_c(\Delta_v)$ is infinite a simple contradiction argument is possible for the one-type case. Clearly there exists a $0 < \theta < \Delta_v$ with $K_c(\theta) > \gamma$ and $K_c^{'}(\theta) > 0$. Consider equation (4.11). For any constant E, multiply the equation by $e^{-\lambda E}$, differentiate with respect to λ, multiply by $e^{\lambda E}$ and put $\lambda = \theta$. Then

$$\int_{-\infty}^{\infty} (u-E)w(u)e^{\theta u}du = K_c(\theta)\int_{-\infty}^{\infty} (u-E)(1-e^{-w(u)})e^{\theta u}du$$
$$+ K_c^{'}(\theta)\int_{-\infty}^{\infty} (1-e^{-w(u)})e^{\theta u}du.$$

Rearranging the equation we obtain

$$\int_{-\infty}^{\infty} (u-E)w(u)e^{\theta u}\left(1 - K_c(\theta)\frac{1-e^{-w(u)}}{w(u)}\right)du = K_c^{'}(\theta)\int_{-\infty}^{\infty} (1-e^{-w(u)})e^{\theta u}du \tag{4.15}$$

.

The right hand side of equation (4.15) is positive, regardless of the choice of E. Now consider the left hand side. Let $r(u) = (1-e^{-w(u)})/w(u)$. Then

$$\frac{dr(u)}{du} = \frac{dw(u)}{du}\frac{1+w(u)-e^{w(u)}}{e^{w(u)}w(u)^2} > 0$$

for all u, since $w^{'}(u) < 0$. Therefore $r(u)$ is an increasing function of u which tends to 1 as $u \to -\infty$ and to $\eta(0)/(-\log(1-\eta(0))) = 1/\gamma$ as $u \to \infty$. Then $1 - [K_c(\theta)(1-e^{-w(u)})/w(u)]$ is a strictly decreasing negative function of u. We can therefore always take E sufficiently large so that the left hand side of equation

(4.15) is negative. Hence a contradiction is obtained. No non-trivial wave solution at speed c is therefore possible for any $0 < c < c_0$.

4.4 Some preliminary results for the multi-type model

Now consider n populations or types. Conditions on the infection matrix are obtained and the limits of the wave solution at minus infinity are derived. Define $\gamma_{ij} = \int_{-\infty}^{\infty} v_{ij}(t)dt = \int_0^{\infty} \gamma_{ij}(\tau)d\tau$. This does not depend on c and may be infinite. Let $\mathbf{\Gamma} = (\gamma_{ij})$. We call $\mathbf{\Gamma}$ the infection matrix and restrict it in this chapter to be non-reducible.

A necessary condition on the infection matrix for the existence of a non-trivial solution to equations (4.6) is established very simply in Lemma 4.2. Theorem 4.1 then shows that if $\rho(\mathbf{\Gamma}) > 1$ and a non-trivial solution $w_i(u)$, $(i = 1, ..., n)$, exists to equations (4.6) for some speed $c > 0$, then $l_i = \lim_{u\to-\infty} e^{-w_i(u)}$ exists, for $(i = 1, ..., n)$, and the l_i are the unique solutions to certain equations.

LEMMA 4.2. *If $\rho(\mathbf{\Gamma}) \leq 1$, then no non-trivial, continuous, monotone decreasing wave solution to equations (4.6) is possible at any speed.*

PROOF. If $\rho(\mathbf{\Gamma}) \leq 1$, since $\mathbf{\Gamma}$ is non-reducible it must necessarily have finite elements. Assume that a non-trivial solution $w_i(u)$ $(i = 1, ..., n)$, to (4.6) exists and let $m_i = \lim_{u\to-\infty} w_i(u)$ so that $e^{-m_i} = l_i$. Let $u \to -\infty$ in equations (4.6). By monotone convergence we obtain $\mathbf{m} = \mathbf{\Gamma}(\mathbf{1} - \mathbf{l})$ where $\{\mathbf{l}\}_i = l_i$, $\{\mathbf{m}\}_i = m_i = -\log l_i$ and $\mathbf{0} \leq \mathbf{l} \leq \mathbf{1}$. Note that $\mathbf{m} \neq \mathbf{0}$ and hence $\mathbf{l} \neq \mathbf{1}$ since $w_i(u)$ is a non-trivial solution of (4.6).

It follows that

$$-\mathbf{m} + \mathbf{1} - \mathbf{l} = (\mathbf{I} - \mathbf{\Gamma})(\mathbf{1} - \mathbf{l}).$$

Since $\mathbf{\Gamma}$ is non-negative and non-reducible, by Theorem A.1 there exists a $\mathbf{u} > \mathbf{0}$ such that $\mathbf{u}'\mathbf{\Gamma} = \rho(\mathbf{\Gamma})\mathbf{u}'$. Hence if $u_j = \{\mathbf{u}\}_j$,

$$\sum_{j=1}^{n} u_j \left(-m_j + 1 - e^{-m_j}\right) = (1 - \rho(\mathbf{\Gamma})) \sum_{j=1}^{n} u_j \left(1 - e^{-m_j}\right).$$

Since $\mathbf{u} > \mathbf{0}$ and $\mathbf{m} \neq \mathbf{0}$, the right-hand side of this equation is non-negative while the left-hand side of the equation is negative, which is a contradiction. Hence a non-trivial wave solution to (4.6) cannot exist at any speed when $\rho(\mathbf{\Gamma}) \leq 1$.

□

THEOREM 4.1. *If $\rho(\mathbf{\Gamma}) > 1$ and there exists a non-trivial, continuous, monotone decreasing solution $w_i(u)$ $i = 1, ..., n$, to equations (4.6) with $\lim_{u\to\infty} w_i(u) = 0$, then $l_i = \lim_{u\to-\infty} e^{-w_i(u)}$ $(i = 1, ..., n)$ exists and $\mathbf{l}' = (l_1, ..., l_n)$ is uniquely defined.*

1. *When $\mathbf{\Gamma}$ has all finite elements, $\mathbf{l}$ is the unique solution $\mathbf{0} < \mathbf{l} < \mathbf{1}$ to*

$$\mathbf{m} = \mathbf{\Gamma}(\mathbf{1} - \mathbf{l}),$$

where $\{\mathbf{m}\}_i = -\log(l_i)$.

2. *If $\mathbf{\Gamma}$ has an infinite element in each row, then $\mathbf{l} = \mathbf{0}$.*

3. *When* $\mathbf{\Gamma}$ *has no infinite element in exactly* k *rows,* $1 \leq k < n$*, we may permute the index set so that*

$$\mathbf{\Gamma} = \begin{pmatrix} \mathbf{\Gamma}_{11} & \mathbf{\Gamma}_{12} \\ \mathbf{\Gamma}_{21} & \mathbf{\Gamma}_{22} \end{pmatrix},$$

where $\mathbf{\Gamma}_{11}$ *is a square matrix of order* k*,* $\mathbf{\Gamma}_{11}$ *and* $\mathbf{\Gamma}_{12}$ *have no infinite elements and* $(\mathbf{\Gamma}_{21}, \mathbf{\Gamma}_{22})$ *has an infinity in every row. Then* $l_i = 0$ *for* $i = k+1, ..., n$ *and* $\mathbf{l}^{*\prime} = (l_1, ..., l_k)$ *is the unique solution,* $\mathbf{0} < \mathbf{l}^* < \mathbf{1}$ *to*

$$\mathbf{m}^* = \mathbf{\Gamma}_{11}(\mathbf{1} - \mathbf{l}^*) + \mathbf{\Gamma}_{12}\mathbf{1},$$

where $\{\mathbf{m}^*\}_i = -\log(l_i)$.

PROOF. From equations (4.6) it is seen that if $l_i = 1$ then for any j such that $\gamma_{ij} \neq 0$, $l_j = 1$. Since $\mathbf{\Gamma}$ is non-reducible for every pair of distinct suffices i, j there is a distinct sequence $i_1, ...i_r$ with $i_1 = i$ and $i_r = j$ such that $\gamma_{i_s i_{s+1}} \neq 0$ for $s = 1, ..., r-1$. Hence if $l_i = 1$ for some i, $l_j = 1$ for all $j = 1, ..., n$. Thus either $\mathbf{l} = \mathbf{1}$ or $\mathbf{l} < \mathbf{1}$. Note that $\mathbf{l} = \mathbf{1}$ corresponds to the trivial solution $w_i(u) = 0$ all u and i. Hence necessarily $\mathbf{l} < \mathbf{1}$.

1. Here $\mathbf{\Gamma}$ is finite. Let $u \to -\infty$ in equations (4.6). Then by monotone convergence we obtain $\mathbf{m} = \mathbf{\Gamma}(\mathbf{1} - \mathbf{l})$, where $\{\mathbf{m}\}_i = -\log(\mathbf{l}_i)$ is finite and so $\mathbf{l} > \mathbf{0}$. From Theorem B.2 there is a solution $\mathbf{0} < \mathbf{l} < \mathbf{1}$ to this equation precisely when $\rho(\mathbf{\Gamma}) > 1$, and this solution is unique.
2. If $\gamma_{ij} = \infty$ for some j, since $\mathbf{l} < \mathbf{1}$ when the wave solution is non-trivial, from equations (4.6) it follows that $l_i = 0$. Hence, when every row of $\mathbf{\Gamma}$ contains at least one infinite element, $\mathbf{l} = \mathbf{0}$. Note that since $\mathbf{\Gamma}$ is non-reducible with an infinite element $\rho(\mathbf{\Gamma}) = \infty$.
3. When exactly k rows do not contain an infinite element, $1 \leq k < n$, we may partition $\mathbf{\Gamma}$ as in part 3 of the statement of the theorem. As in case 2, the last $(n-k)$ equations of (4.6) then give $l_i = 0$ for $i = k+1, ..., n$. Applying monotone convergence to the first k of equations (4.6) we obtain

$$\log(m_i) = \sum_{j=1}^{k} \{\mathbf{\Gamma}_{11}\}_{ij}(1 - l_j) + \{\mathbf{\Gamma}_{12}\mathbf{1}\}_i, \text{ for } i = 1, .., k.$$

Writing $\{\mathbf{l}^*\}_i = l_i$, for $i = 1, ..., k$, and putting $\mathbf{a} = \mathbf{\Gamma}_{12}\mathbf{1}$ we obtain

$$\mathbf{m}^* = \mathbf{\Gamma}_{11}(\mathbf{1} - \mathbf{l}^*) + \mathbf{a},$$

where $\{\mathbf{m}*\}_i = -\log(\{\mathbf{l}^*\}_i)$. Since $\mathbf{\Gamma}$ is non-reducible, writing $\mathbf{\Gamma}_{11}$ in normal form and using Theorem B.2 with $\mathbf{B} = \mathbf{\Gamma}_{11}$, $\mathbf{a}$ is of the form described in Theorem B.2 part 3. Hence there exists a unique solution $\mathbf{l}^*$ to these equations, and $\mathbf{0} < \mathbf{l}^* < \mathbf{1}$. As in part 2, $\rho(\mathbf{\Gamma}) = \infty$.

□

Observe that the equations satisfied by $\mathbf{l}$ are identical to those satisfied by $\mathbf{1} - \boldsymbol{\eta}(\mathbf{0})$, where $\boldsymbol{\eta}(\mathbf{0})$ is the lower bound for the vector of final sizes obtained for both the spatial and non-spatial epidemic in Chapters 2 and 3. The vector $\boldsymbol{\eta}(\mathbf{0})$

is also the limit of the vector of final sizes as the amount of initial infection tends to zero. Hence $\{\mathbf{l}\}_i = 1 - \boldsymbol{\eta}(\mathbf{0})$ represents the amount remaining susceptible in population i once the epidemic wave has passed through the system.

We assume for the remainder of this chapter that $\rho(\mathbf{\Gamma}) > 1$, since otherwise no wave solutions are possible.

4.5 The regions of convergence of certain transforms

Let $W_i(\lambda) = \int_{-\infty}^{\infty} e^{\lambda u} w_i(u)du$ and $V_{ij}(\lambda) = \int_{-\infty}^{\infty} e^{-\lambda u} v_{ij}(u)du$, and define $\mathbf{V}(\lambda)$ to be the matrix with (ij)th element $V_{ij}(\lambda)$.

We prove theorems concerning the regions of convergence of these transforms in the complex plane. These results are crucial to the proofs of existence, uniqueness and non-existence of wave solutions at different speeds in Sections 4.7 to 4.10. The convergence of $W_i(\lambda)$ also implies that, for non-trivial wave solutions to be possible $p_{ij}(r)$, $(i, j = 1, ..., n)$ must have exponentially bounded tails at $+\infty$.

The following lemma is a modified form of Lemma 3.4 of Diekmann [D5]. Let N_+ denote the set of positive integers, and $\Re_+$ denote the non-negative half of the real line.

LEMMA 4.3. *Let $f : \Re \to \Re$ be non-negative and $f \in L_1(\Re_+)$. Define $f^{(0)}(x) = f(x)$ and recursively define $f^{(k)}(x) = \int_x^{\infty} f^{(k-1)}(y)dy$ for those values $k \in N_+$ for which $f^{(k-1)} \in L_1(\Re_+)$. Then for any $r \in \Re_+$ if $\int_r^{\infty} f^{(k)}(x)dx$ converges for all non-negative integers k, then so does $\int_r^{\infty}(x-r)^k f(x)dx$. Also, for each non-negative integer k,*

$$\int_r^{\infty} (x-r)^k f(x)dx = k! \int_r^{\infty} f^{(k)}(x)dx.$$

PROOF. Assume that $\int_s^{\infty} f^{(k)}(x)dx$ converges for some $s = r \in \Re_+$, and hence for all $s \geq r$, and for all non-negative integers k. Induction on k is then used to show that, for any non-negative integer k, $\int_s^{\infty}(x-s)^k f(x)dx$ converges and is equal to $k! \int_s^{\infty} f^{(k)}(x)dx$ for all $s \geq r$.

First observe that the result trivially holds when $k = 0$, since

$$\int_s^{\infty} (x-s)^0 f(x)dx = \int_s^{\infty} f(x)dx = 0! \int_s^{\infty} f^{(0)}(x)dx.$$

Now assume that $\int_s^{\infty}(x-s)^k f(x)dx$ exists and equals $k! \int_s^{\infty} f^{(k)}(x)dx$ for all $s \geq r$ and all integers $0 \leq k \leq K$. Then, for any $u \geq r$,

$$\frac{1}{K!} \int_u^{\infty} (x-u)^K f(x)dx = \int_u^{\infty} f^{(K)}(x)dx = f^{(K+1)}(u).$$

For any $s \geq r$, we now integrate both sides of this equality over u taking values from s to ∞. The integral of the right hand side exists and is just $\int_s^{\infty} f^{(K+1)}(u)du$. Hence the integral of the left hand side also exists, and

$$\int_s^\infty f^{(K+1)}(x)dx = \frac{1}{K!}\int_s^\infty \int_u^\infty (x-u)^K f(x)dxdu$$
$$= \frac{1}{K!}\int_s^\infty \int_s^x (x-u)^K f(x)dudx = \frac{1}{(K+1)!}\int_s^\infty (x-s)^{K+1} f(x)dx.$$

Hence, for all $s \geq r$, $\int_s^\infty (x-s)^k f(x)dx$ exists and equals $k! \int_s^\infty f^{(k)}(x)dx$ for $k = K+1$. Therefore by induction this result holds for all positive integers k, which completes the proof of the lemma.

□

This lemma is required in the proof of the Theorem 4.2. The first half of this theorem proves that, if a non-trivial wave solution $w_i(u)$ $(i = 1, ..., n)$ exists, then the Laplace transforms must exist in some region in the right hand half of the complex plane. The second part of the theorem is required for the proof of Theorem 4.4, which concerns the relation between the abscissae of convergence of the $W_i(\lambda)$ and the roots of the characteristic equation. It is also required in Section 4.8 when proving the non-existence of waves at subcritical speeds.

THEOREM 4.2. *Let $w_i(u)$, for $i = 1, ..., n$, be a non-trivial solution of equations (4.6) with $w_i(u)$ differentiable, with its derivative continuous and bounded, and with $w_i(u)$ decreasing monotonically to zero as $u \to \infty$. Then the following results hold:*

1. *There exists a positive real δ such that each $W_i(\lambda)$, $i = 1, ..., n$, converges for $\lambda \in S_\delta$ where $S_\delta = \{\lambda \in \mathbb{C} : 0 < Re(\lambda) < \delta\}$.*
2. *If for a real positive β, the condition $\rho(\mathbf{V}(\beta)) > 1$ holds and $W_i(\beta)$ converges for each $i = 1, ..., n$, then the abscissa of convergence of each $W_i(\lambda)$ is to the right of β.*

PROOF. Since $w_i'(u)$ is bounded it immediately follows that $w_i(u)$ is at most $O(|u|)$ as $u \to -\infty$. When $\mathbf{\Gamma}$ is a finite matrix, in fact $w_i(u)$ tends to a finite limit as $u \to -\infty$. This is easily seen from Theorem 4.1. Hence, for any non-negative β and each $i = 1, ..., n$, the integral $\displaystyle\int_{-\infty}^{r} e^{(\lambda+\beta)u} w_i(u)du$ converges for $Re(\lambda) > 0$ and any finite r. In order to prove the theorem, it suffices to prove the convergence of $\displaystyle\int_r^\infty e^{(\lambda+\beta)u} w_i(u)du$, $i = 1, ..., n$, for some finite r and $0 < Re(\lambda) < \delta$. The cases $\beta = 0$ and $\beta > 0$ prove parts 1 and 2 respectively.

If $\beta = 0$, since $w_i(u)$ is a non-trivial solution then $\rho(\mathbf{V}(0)) = \rho(\mathbf{\Gamma}) > 1$, and if $\beta > 0$ then $\rho(\mathbf{V}(\beta)) > 1$. Hence, for all $\beta \geq 0$, $\mathbf{V}(\beta)$ is non-negative and non-reducible with $\rho(\mathbf{V}(\beta)) > 1$. Observe that $\rho(\theta\mathbf{B}) = \theta\rho(\mathbf{B})$ for any real positive θ. From Theorem A.2, the Perron-Frobenius root of a non-reducible matrix is a continuous, strictly increasing function of its entries. Hence there exist finite positive reals α and θ, with $0 < \theta < 1$, such that $\mathbf{\Gamma}^*$ is non-reducible and $\rho(\mathbf{\Gamma}^*) > 1$, where $\mathbf{\Gamma}^* = (\theta\gamma_{ij}^*)$ and $\gamma_{ij}^* = \int_{-\alpha}^{\alpha} e^{-\beta u} v_{ij}(u)du$.

Define $f_i(u) = e^{\beta u} w_i(u)$,

$$v_{ij}^*(u) = \begin{cases} e^{-\beta u} v_{ij}(u), & \text{for } |u| \leq \alpha, \\ 0, & \text{for } |u| > \alpha, \end{cases}$$

$$\Psi_{ij}(u) = \int_{-\infty}^{\infty} e^{\beta(u+s)}(1 - e^{-w_j(u+s)})(e^{-\beta s}v_{ij}(s) - v_{ij}^*(s))ds$$
$$+ \int_{-\alpha}^{\alpha} e^{\beta(u+s)}(1 - e^{-w_j(u+s)} - \theta w_j(u+s))v_{ij}^*(s)ds,$$

and

$$n_{ij}(x) = \begin{cases} \int_x^\infty v_{ij}^*(y)dy, & \text{for } x \geq 0, \\ -\int_{-\infty}^x v_{ij}^*(y)dy = \int_x^\infty v_{ij}^*(y)dy - \gamma_{ij}{}^*, & \text{for } x < 0. \end{cases}$$

Note that $n_{ij}(x) = 0$ if $|x| > \alpha$.

Equations (4.6) may be written in the form

$$f_i(u) = \sum_{j=1}^n \left[\theta \int_{-\infty}^{\infty} f_j(u+s)v_{ij}^*(s)ds + \Psi_{ij}(u)\right].$$

Now $\theta < 1$ and $w_i(u) \to 0$ as $u \to \infty$, so that $\left(1 - e^{-w_i(u)} - \theta w_i(u)\right) \geq 0$ for u sufficiently large. Hence there exists a positive L sufficiently large such that $\Psi_{ij}(u) \geq 0$, for all i, j and all $u > L + \alpha$.

Choose $r > L + \alpha$ and integrate these equations from r to x. We obtain

$$\begin{aligned} &\int_r^x f_i(u)du - \sum_{j=1}^n \int_r^x \Psi_{ij}(u)du \\ &= \theta \sum_{j=1}^n \int_r^x \int_{-\infty}^{\infty} f_j(u+s)v_{ij}^*(s)dsdu \\ &= \sum_{j=1}^n \left(\theta\gamma_{ij}^* \int_r^x f_j(u)du - \theta \int_{-\infty}^{\infty} [f_j(u+r) - f_j(u+x)]n_{ij}(u)du\right), \end{aligned} \tag{4.16}$$

and hence

$$(\mathbf{\Gamma}^* - \mathbf{I})\mathbf{f}^{(1)}(r,x) + \mathbf{\Psi}^{(1)}(r,x)\mathbf{1} = \theta\mathbf{b}^{(1)}(x), \tag{4.17}$$

where $\mathbf{f}^{(1)}(r,x)$, and $\mathbf{b}^{(1)}(r,x)$ are the vectors with i^{th} elements $\int_r^x f_i(u)du$ and $\sum_{j=1}^n \int_{-\infty}^{\infty} [f_j(u+r) - f_j(u+x)]n_{ij}(u)du$ respectively. Also $\mathbf{\Psi}^{(1)}(r,x)$ is the matrix with ij^{th} entry $\int_r^x \Psi_{ij}(u)du$.

We now consider the two cases $\beta = 0$ and $\beta > 0$ separately. In each case $\mathbf{f}^{(1)}(r,\infty)$ and $\mathbf{\Psi}^{(1)}(r,x)$ are shown to exist and an upper bound is obtained for $\mathbf{f}^{(1)}(r)$, where $\mathbf{f}^{(1)}(r) = \mathbf{f}^{(1)}(r,\infty)$. Note that we also define $\mathbf{\Psi}^{(1)}(r) = \mathbf{\Psi}^{(1)}(r,\infty)$. The cases are then recombined for the remainder of the proof. Let $f_i^{(1)}(r)$ and $\Psi_{ij}^{(1)}(r)$ denote the ith entry of $\mathbf{f}^{(1)}(r)$ and the ij^{th} entry of $\mathbf{\Psi}^{(1)}(r)$ respectively.

1. Consider the case when $\beta = 0$.

   Since $\mathbf{\Gamma}^*$ is finite, non-negative and non-reducible, from Theorem A.1 it has a positive left eigenvector $\mathbf{a}'$ corresponding to the maximum eigenvalue $\rho(\mathbf{\Gamma}^*)$. Multiplying (4.17) on the left by $\mathbf{a}'$ we obtain

$$(\rho(\mathbf{\Gamma}^*) - 1)\mathbf{a}'\mathbf{f}^{(1)}(r, x) + \mathbf{a}'\mathbf{\Psi}^{(1)}(r, x)\mathbf{1} = \theta\mathbf{a}'\mathbf{b}^{(1)}(r, x).$$

   Since all the terms on the left-hand side are positive for $r > L + \alpha$, and the right-hand side tends by monotone convergence to the finite limit

$$C = \theta \sum_{i=1}^{n} \{\mathbf{a}\}_i \sum_{j=1}^{n} \int_{-\infty}^{\infty} f_j(u + r) n_{ij}(u) du$$

   as $x \to \infty$, we can conclude that $f_i^{(1)}(r) = \lim_{x\to\infty} \int_r^x f_i(u)du$ and $\mathbf{\Psi}_{ij}^{(1)}(r) = \lim_{x\to\infty} \int_r^x \Psi_{ij}(u)du$ exist for all $i, j = 1, ..., n$. In addition, since $f_j(u)$ is a decreasing function of u when $\beta = 0$ and $n_{ij}(u) = 0$ for $|u| > \alpha$, it follows that

$$\{\mathbf{a}\}_i(\rho(\mathbf{\Gamma}^*) - 1) \int_r^{\infty} f_i(u)du \leq C \leq \theta\mathbf{a}'\mathbf{N}\mathbf{f}(r - \alpha),$$

   where $\mathbf{N} = (N_{ij})$, $N_{ij} = \int_{-\infty}^{\infty} |n_{ij}(u)|du$ and $\{\mathbf{f}(r - \alpha)\}_i = f_i(r - \alpha)$.

   Let $\mathbf{f}^{(0)}(u)$ be the vector with i^{th} element $f_i^{(0)}(u)$, where $f_i^{(0)}(u) = f_i(u)$. Hence we obtain the bounds

$$\mathbf{f}^{(1)}(r) \leq \mathbf{M}\mathbf{f}^{(0)}(r - \alpha),$$

   where $\mathbf{M} = (M_{ij})$, and $M_{ij} = \theta(\rho(\mathbf{\Gamma}^*) - 1)^{-1}(\{\mathbf{a}\}_i)^{-1}\{\mathbf{a}'\mathbf{N}\}_j$. This is the required bound on $\mathbf{f}^{(1)}(r)$.

2. Now consider the case when $\beta > 0$.

   Let $F_i = \displaystyle\int_{-\infty}^{\infty} f_i(u)du$. This integral exists and hence $\int_r^{\infty} f_i(u)du$ exists and is bounded above by F_i. Thus we immediately obtain the result that $f_i^{(1)}(r) \leq F_i$, for $i = 1, ..., n$.

   From equations (4.16), for all $x \geq r$ we obtain the inequality

$$\begin{aligned}\sum_{j=1}^{n} \int_r^x \Psi_{ij}(u)du &= \int_r^{\infty} f_i(u)du - \theta \sum_{j=1}^{n} \int_{-\infty}^{\infty} v_{ij}^*(s) \int_{x+s}^{u+s} f_j(t)dtds \\ &\leq F_i.\end{aligned}$$

   Consider any $r > L+\alpha$. Since all the terms on the left-hand side are positive it follows that $0 \leq \displaystyle\int_r^x \Psi_{ij}(u)du \leq F_i$ and hence $\Psi_{ij}^{(1)}(r) = \displaystyle\int_r^{\infty} \Psi_{ij}(u)du$ exists for all i, j.

The proof now treats cases 1 and 2 concurrently, so that $\beta \geq 0$. We have already defined $f_i^{(0)}(r)$, $f_i^{(1)}(r)$ and $\Psi_{ij}^{(1)}(r)$ for $r > L + \alpha$. For such an r, let

$\mathbf{b}^{(1)}(r) = \mathbf{b}^{(1)}(r, \infty)$. Sequences $\mathbf{f}^{(k)}(u, x) = (f_i^{(k)}(u, x))$, $\mathbf{\Psi}^{(k)}(u, x) = (\Psi_{ij}^{(k)}(u, x))$ and $\mathbf{b}^{(k)}(u, x) = (b_i^{(k)}(u, x))$ are now defined recursively by

$$f_i^{(k+1)}(u, x) = \int_u^x f_i^{(k)}(s)ds, \quad \Psi_{ij}^{(k+1)}(u, x) = \int_u^x \Psi_{ij}^{(k)}(s)ds \text{ and}$$

$$b_i^{(k+1)}(r, x) = \int_r^x b_i^{(k)}(u)du = \sum_{j=1}^n \int_{-\infty}^{\infty} [f_j^{(k-1)}(u + r) - f_j^{(k-1)}(u + x)]n_{ij}(u)du,$$

for $k = 1, 2, ...$ and $i, j = 1, ..., n$. The existence of $b_i^{(1)}(u)$, $f_i^{(1)}(u)$ and $\Psi_{ij}^{(1)}(u)$ have already been established for $u > L + \alpha$. These are all decreasing functions of u.

Let $x \to \infty$ in equations (4.16) so that the equation becomes

$$f_i^{(1)}(r) - \sum_{j=1}^n \Psi_{ij}^{(1)}(r) = \theta \sum_{j=1}^n \int_{-\infty}^{\infty} f_j^{(1)}(r + s)v_{ij}^*(s)ds.$$

Replace r by u and integrate over u from r to x, for any $r > L + \alpha$. Then identical equations to equations (4.16) are obtained with $f_i^{(1)}(u)$ and $\Psi_{ij}^{(1)}(u)$ replacing $f_i(u)$ and $\Psi_i(u)$. This then gives an equivalent equation to equation (4.17) with $\mathbf{f}^{(2)}(r, x) = (f_i^{(2)}(r, x))$, $\mathbf{\Psi}^{(2)}(r, x) = (\Psi_{ij}^{(2)}(r, x))$ and $\mathbf{b}^{(2)}(r, x) = (b_i^{(2)}(r, x))$ replacing $\mathbf{f}^{(1)}(r, x)$, $\mathbf{\Psi}^{(1)}(r, x)$ and $\mathbf{b}^{(1)}(r, x)$.

As in case 1, since $b_i^{(2)}(r) = b_i^{(2)}(r, \infty)$ exists, $f_i^{(2)}(r) = f_i^{(2)}(r, \infty)$ and $\Psi_{ij}^{(2)}(r) = \Psi_{ij}^{(2)}(r, \infty)$ can be shown to exist for $r > L + \alpha$ and $\mathbf{f}^{(2)}(r) \leq \mathbf{M}\mathbf{f}^{(1)}(r - \alpha)$, where $\mathbf{f}^{(k)}(r)$ is a vector with i^{th} element $f_i^{(k)}(r)$ for $k = 0, 1, ...$

Using induction, for all $k \in N_+$, we can prove the existence of $f_i^{(k)}(r)$ and $\Psi_{ij}^{(k)}(r)$ and establish the bounds $\mathbf{f}^{(k+1)}(r) \leq \mathbf{M}\mathbf{f}^{(k)}(r - \alpha)$ for $k = 1, 2, ...$ Since $\mathbf{f}^{(k)}(u)$ has monotone decreasing entries for $k \geq 1$, the following inequality holds,

$$f_i^{(k+1)}(r - \alpha) = f_i^{(k+1)}(r) + \int_{r-\alpha}^r f_i^{(k)}(u)du \leq f_i^{(k+1)}(r) + \alpha f_i^{(k)}(r - \alpha).$$

Hence, for $k = 1, 2, ...,$

$$\mathbf{f}^{(k+1)}(r - \alpha) \leq \mathbf{f}^{(k+1)}(r) + \alpha\mathbf{f}^{(k)}(r - \alpha) \leq (\mathbf{M} + \alpha\mathbf{I})\mathbf{f}^{(k)}(r - \alpha).$$

Now Lemma 4.2 is applied, and we obtain the inequality

$$\begin{aligned} \frac{1}{k!}\int_r^{\infty} (u - r)^k f_i(u)du &= \int_r^{\infty} f_i^{(k)}(u)du \\ &\leq f_i^{(k+1)}(r - \alpha) \leq \{(\mathbf{M} + \alpha\mathbf{I})^k\mathbf{f}^{(1)}(r - \alpha)\}_i, \end{aligned}$$

for $i = 1, ..., n$ and $k = 2, 3, ...$ Note that

$$\mathbf{f}^{(1)}(r - \alpha) \leq \begin{cases} (\mathbf{M} + \alpha\mathbf{I})\mathbf{f}^{(0)}(r - \alpha), & \beta = 0, \\ \mathbf{F}, & \beta > 0, \end{cases}$$

where $\mathbf{F}$ has the ith element F_i.

Define δ to be the minimum of the reciprocals of the absolute values of the eigenvalues of $(\mathbf{M}+\alpha\mathbf{I})$, i.e. $\delta = 1/(\rho(\mathbf{M})+\alpha)$. Then

$$\left|\int_r^\infty e^{\lambda(u-r)} f_i(u)du\right| \le \int_r^\infty e^{Re(\lambda)(u-r)} f_i(u)du = \sum_{k=0}^\infty \frac{(Re(\lambda))^k}{k!} \int_r^\infty (u-r)^k f_i(u)du,$$

which converges when $r > L+\alpha$ and $Re(\lambda) < \delta$. Thus, for each $i = 1, ..., n$, the integral $\int_r^\infty e^{\lambda(u-r)+\beta u} w_i(u)du$ converges for $r > L+\alpha$ and $Re(\lambda) < \delta$. This completes the proof of the theorem.

□

If the Laplace transforms of $w_i(u)$ and $v_{ij}(u)$ exist for some positive real entry, denote their abscissae of convergence in the right-hand half of the complex plane by Δ_{w_i} and $\Delta_{v_{ij}}$. Then $\Delta_{w_i} = \sup\{\lambda \in \Re_+ : W_i(\lambda) < \infty\}$ and $\Delta_{v_{ij}} = \sup\{\lambda \in \Re_+ : V_{ij}(\lambda) < \infty\}$. Hence $W_i(\lambda)$ is analytic in the strip $\{\lambda \in \mathbb{C} : 0 < Re(\lambda) < \Delta_{w_i}\}$ and $V_{ij}(\lambda)$ is analytic in the strip $\{\lambda \in \mathbb{C} : 0 \le Re(\lambda) < \Delta_{v_{ij}}\}$, since both are transforms of non-negative functions. Define $\Delta_w = \min_i \Delta_{w_i}$ and $\Delta_v = \min_{i,j} \Delta_{v_{ij}}$.

THEOREM 4.3. *If $w_i(u)$, for $i = 1, ..., n$, is a non-trivial, monotone decreasing, differentiable solution of equations (4.6) with continuous bounded derivatives and with $\lim_{u\to\infty} w_i(u) = 0$, then $\Delta_{w_i} = \Delta_w$ for all i and $\Delta_w \le \Delta_v$.*

PROOF. From Theorem 4.2 part 1, Δ_{w_i} is defined. Since the region of convergence of $\int_{-\infty}^\infty (1-e^{-w_i(s)})e^{\lambda s}ds$ is the same as that of $W_i(\lambda)$, and both transforms are finite and positive for λ real with $0 < \lambda < \Delta_{w_i}$, from equations (4.6) $V_{ij}(\lambda)$ exists for $0 < \lambda \le \Delta_{w_i}$, so that $\Delta_{v_{ij}}$ is defined. Taking transforms of equations (4.6) we obtain

$$W_i(\lambda) = \sum_{j=1}^n V_{ij}(\lambda) \int_{-\infty}^\infty (1-e^{-w_j(s)})e^{\lambda s}ds. \tag{4.18}$$

Now $V_{ij}(\lambda) > 0$ when $\gamma_{ij} \ne 0$ for $0 < \lambda < \Delta_v$. It immediately follows that $\Delta_{w_i} \le \Delta_{v_{ij}}$ and $\Delta_{w_i} \le \Delta_{w_j}$ if $\gamma_{ij} \ne 0$.

Since $\mathbf{\Gamma}$ is non-reducible, for each $i \ne j$ there exists an integer k and a sequence of non-zero elements $\gamma_{i_s i_{s+1}}$ for $s = 1, ..., k$, with $i_1 = i$ and $i_{k+1} = j$. Hence $\Delta_{w_i} \le \Delta_{w_j}$, for all $i \ne j$, and thus $\Delta_{w_i} = \Delta_w$ for all i. Then $\Delta_w = \Delta_{w_i} \le \Delta_{v_{ij}}$ holds for all i, j from which we obtain the result $\Delta_w \le \Delta_v$.

□

COROLLARY 4.1. *Necessary conditions for the existence of a non-trivial solution to equations (4.6), with the conditions specified in Theorem 4.3, are that $p_{ij}(r)e^{\delta^* r} \to 0$ as $r \to \infty$, $(i, j = 1, ..., n)$, for some positive real δ^*.*

PROOF. From Theorem 4.2, there exists a $\delta > 0$ such that $W_i(\lambda)$ exists for $0 < Re(\lambda) < \delta$ for all i. Hence $\Delta_w \geq \delta$. From Theorem 4.3, $\Delta_v \geq \Delta_w$. Therefore $\Delta_{v_{ij}} \geq \delta$ for all i, j. But $V_{ij}(\lambda) = \sigma_j P_{ij}(\lambda)\Lambda_{ij}(c\lambda)$, where $P_{ij}(\lambda) = \int_{-\infty}^{\infty} e^{\lambda u} p_{ij}(u)du$ and $\Lambda_{ij}(\lambda) = \int_0^\infty e^{-\lambda u}\lambda_{ij}(u)du$. Now $\Lambda_{ij}(c\lambda)$ is finite for all $c > 0$ and all $0 < Re(\lambda) < \infty$. Hence if we take any $0 < \delta^* < \delta$, $P_{ij}(\delta^*)$ is finite and so $p_{ij}(r)e^{\delta^* r} \to 0$ as $r \to \infty$ for all i, j as required.

□

4.6 The characteristic equation

For any speed $c > 0$, $\mathbf{V}(\lambda)$ is a finite, non-negative, non-reducible matrix for each $0 < \lambda < \Delta_v$. Let $K_c(\lambda) = \rho(\mathbf{V}(\lambda))$, the Perron-Frobenius root of $\mathbf{V}(\lambda)$. This is the generalisation of the function $K_c(\lambda)$ defined in Section 4.2 for the case $n = 1$. We define the characteristic equation to be $K_c(\lambda) = 1$. Note that $\mathbf{V}(\lambda) = (V_{ij}(\lambda))$, where $V_{ij}(\lambda) = \sigma_j P_{ij}(\lambda)\Lambda_{ij}(c\lambda)$, so that $\mathbf{V}(\lambda)$ depends upon c.

The behaviour of the function $K_c(\lambda)$ for general n can be shown to be essentially the same as the behaviour for $n = 1$. In order to establish this behaviour, some preliminary results need to be proved.

THEOREM 4.4. *Let $w_i(u)$, $i = 1, ..., n$, be a non-trivial, monotone decreasing, differentiable solution of equations (4.6) with continuous bounded derivatives and with $\lim_{u\to\infty} w_i(u) = 0$. Then the following results hold.*

1. *If $\Delta_w < \Delta_v$, then Δ_w is the smallest positive root of $K_c(\lambda) = 1$.*
2. *If $K_c(\lambda) = 1$ has a solution in the range $0 < \lambda < \Delta_v$, and if α is the smallest positive root, then $\Delta_w = \alpha < \Delta_v$.*

PROOF. We first prove part 1 so that $\Delta_w < \Delta_v$. From Theorem 4.2, $\Delta_w > 0$. Also equations (4.6) can be written in the form

$$(\mathbf{V}(\lambda) - \mathbf{I})\mathbf{W}(\lambda) = \mathbf{R}(\lambda), \tag{4.19}$$

where $\{\mathbf{W}(\lambda)\}_i = W_i(\lambda)$, and $\mathbf{R}(\lambda) = \mathbf{V}(\lambda)\mathbf{T}(\lambda)$, with

$$\{\mathbf{T}(\lambda)\}_i = \int_{-\infty}^{\infty} \left(w_i(s) - (1 - e^{-w_i(s)})\right) e^{\lambda s} ds.$$

First the abscissa of convergence of $R_i(\lambda)$ is shown to be to greater than Δ_w for all $i = 1, .., n$. Take any real ν and θ such that $0 \leq \nu < \theta < \Delta_w$, with $\nu \leq \Delta_v - \Delta_w$ if $\mathbf{V}(\Delta_v)$ has all finite entries, and $\nu < \Delta_v - \Delta_w$ otherwise. Observe that $\sup_{u\in\Re}(e^{\theta u}w_i(u))$ is finite for all i. Then we obtain the inequality

(4.20)

$$
\begin{aligned}
0 \leq R_i(\Delta_w+\nu) &= \sum_{j=1}^{n} V_{ij}(\Delta_w+\nu)\int_{-\infty}^{\infty}\left(w_j(u)-(1-e^{-w_j(u)})\right)e^{(\Delta_w+\nu)u}du \\
&\leq \frac{1}{2}\sum_{j=1}^{n} V_{ij}(\Delta_w+\nu)\int_{-\infty}^{\infty} w_j^2(u)e^{(\Delta_w+\nu)u}du \\
&= \frac{1}{2}\sum_{j=1}^{n} V_{ij}(\Delta_w+\nu)\int_{-\infty}^{\infty}\left(w_j(u)e^{\theta u}\right)\left(w_j(u)e^{(\Delta_w+\nu-\theta)u}\right)du \\
&\leq \frac{1}{2}\sum_{j=1}^{n} V_{ij}(\Delta_w+\nu)\sup_{u\in\Re}(e^{\theta u}w_j(u))W_j(\Delta_w+\nu-\theta)).
\end{aligned}
$$

The right hand side of this inequality is finite, and hence the abscissa of convergence of $R_i(\lambda)$ is to the right of Δ_w for $i = 1,..,n$.

Now $\mathbf{V}(\lambda)$ is a non-reducible matrix with finite, non-negative entries for each $0 < \lambda \leq \Delta_w$. From Theorem A.1, there exists a left eigenvector $(\mathbf{u}(\lambda))' > \mathbf{0}'$ corresponding to the Perron-Frobenius eigenvalue $K_c(\lambda)$ of $\mathbf{V}(\lambda)$. From Theorem A.3, both $K_c(\lambda)$ and $\mathbf{u}(\lambda)$ are continuous for $0 < \lambda \leq \Delta_w$. Multiplying equations (4.19) on the left by $(\mathbf{u}(\lambda))'$ we obtain

(4.21) $$(K_c(\lambda)-1)(\mathbf{u}(\lambda))'\mathbf{W}(\lambda) = (\mathbf{u}(\lambda))'\mathbf{R}(\lambda), \text{ for } 0 < \lambda < \Delta_w.$$

For all λ in this range, $(K_c(\lambda)-1)$ must stay positive since the vectors in equation (4.21) are all positive. Hence $K_c(\lambda) = 1$ has no root in the range $0 < \lambda < \Delta_w$. It only remains to show that $K_c(\Delta_w) = 1$.

Suppose that $W_i(\Delta_w)$ is finite for some i. From equation (4.19) we have

$$\lim_{\lambda\uparrow\Delta_w}\left(\sum_{j\neq i} V_{ij}(\lambda)W_j(\lambda)\right) = R_i(\Delta_w) + (1-V_{ii}(\Delta_w))W_i(\Delta_w),$$

which is finite. Since $\mathbf{V}(\lambda)$ is non-reducible, this implies that if $W_i(\Delta_w)$ is finite for some i, then all the entries of $\mathbf{W}(\Delta_w)$ are finite. Since $\mathbf{R}(\Delta_w) > \mathbf{0}$ and $\mathbf{u} > \mathbf{0}$, taking the limit of equations (4.21) as $\lambda \to \Delta_w$ we obtain

$$(K_c(\Delta_w)-1)(\mathbf{u}(\Delta_w))'\mathbf{W}(\Delta_w) = (\mathbf{u}(\Delta_w))'\mathbf{R}(\Delta_w) > 0.$$

Hence $K_c(\Delta_w) > 1$ and is finite. From Theorem 4.2 part 2, with $\beta = \Delta_w$, this then implies that the abscissa of convergence of the $W_i(\lambda)$ is to the right of Δ_w, which is a contradiction.

Hence $\lim_{\lambda\uparrow\Delta_w} W_i(\lambda)$ is infinite for all i. Therefore $\lim_{\lambda\uparrow\Delta_w}\mathbf{u}'(\lambda)\mathbf{W}(\lambda)$ is also infinite. Taking the limit as $\lambda \uparrow \Delta_w$ in equation (4.21), the right hand side stays finite. Hence $K_c(\Delta_w) = 1$.

Therefore $\lambda = \Delta_w$ is the smallest positive root of $K_c(\lambda) = 1$, which concludes the proof of part 1.

Now consider part 2, so that $K_c(\lambda) = 1$ has a solution in the range $(0, \Delta_v)$, and the smallest positive root is α. If $\Delta_w < \alpha$, then from part 1 $K_c(\Delta_w) = 1$, which is a contradiction. Hence $\Delta_w \geq \alpha$. From equation (4.21) with $\lambda = \alpha$, as in part 1 since the right hand side is positive and $K_c(\alpha) - 1 = 0$, necessarily $(\mathbf{u}(\alpha))'\mathbf{W}(\alpha)$ is infinite. Again as in part 1 this implies that $W_i(\alpha)$ is infinite for all i and hence that $\Delta_w = \alpha$. This completes the proof of the theorem.

□

Results concerning the Perron Frobenius root contained in Appendix A are now used to establish the properties of $K_c(\lambda)$. These properties were obtained in Radcliffe and Rass [R1,R3]. The definition of $K_c(\lambda)$, for λ real, also needs to be extended to complex λ in a region about the interval $(0, \Delta_v)$ of the real axis.

Consider any real $\lambda_0 \in (0, \Delta_v)$. The entries of $\mathbf{V}(\lambda)$ are analytic in the region $0 < Re(\lambda) < \Delta_v$. From Theorem A.3 part 1, for any $\lambda_0 \in (0, \Delta_v)$, the definition of $K_c(\lambda)$ may be extended to λ in an open ball in the complex plane centred on λ_0. In this ball there is a unique eigenvalue of $\mathbf{V}(\lambda)$ which has largest real part. We define $K_c(\lambda)$ to be this eigenvalue.

LEMMA 4.4 (PROPERTIES OF $K_c(\lambda)$).

1. *For a fixed speed $c > 0$, the abscissa of convergence, Δ_v, of $\mathbf{V}(\lambda)$ in the right hand half of the complex plane does not depend on c, and $K_c(\lambda)$ is finite for $0 < \lambda < \Delta_v$.*
2. $\lim_{\lambda \downarrow 0} K_c(\lambda) = K_c(0) = \rho(\mathbf{\Gamma})$. *This does not depend on c and must be greater than one for wave solutions to be possible.*
3. *For $0 < \lambda < \Delta_v$, $K_c(\lambda)$ is continuous in λ, and for each λ is a continuous strictly decreasing function of c ($c > 0$) with $\inf_{c>0} K_c(\lambda) = 0$.*
4. *Suppose Δ_v is finite. If $\mathbf{V}(\Delta_v)$ has all finite elements for some $c > 0$, then the same is true for all $c > 0$; and $K_c(\Delta_v) = \lim_{\lambda \uparrow \Delta_v} K_c(\lambda)$ is finite (and necessarily non-zero) for all $c > 0$, and is a continuous decreasing function of c. If $\mathbf{V}(\Delta_v)$ has an infinite element for some $c > 0$, then the same is true for all $c > 0$; and $K_c(\Delta_v) = \lim_{\lambda \uparrow \Delta_v} K_c(\lambda)$ is infinite for all $c > 0$.*
5. *If Δ_v is infinite, $K_c(\lambda)$ either tends to zero or infinity as $\lambda \to \infty$ for each $c > 0$. If $\lim_{\lambda\to\infty} K_{c^*}(\lambda) = \infty$ then $\lim_{\lambda\to\infty} K_c(\lambda) = \infty$ for all $c < c^*$. If $\lim_{\lambda\to\infty} K_{c^*}(\lambda) = 0$ then $\lim_{\lambda\to\infty} K_c(\lambda) = 0$ for all $c > c^*$.*
6. *For each $0 < \lambda_0 < \Delta_v$, there exists a neighbourhood of λ_0 in which $K_c(\lambda)$ is analytic.*
7. *Except in the degenerate case when $K_c(\lambda) \equiv \rho(\mathbf{\Gamma})$, $K_c(\lambda)$ is a strictly convex function of λ for $0 < \lambda < \Delta_v$. If $K_c'(\lambda) \neq 0$, then $K_c''(\lambda) > 0$. For the non-degenerate case, if $K_c'(\lambda) = 0$, then there exists a positive integer m such that the derivative $K_c^{2m}(\lambda) > 0$ and $K_c^s(\lambda) = 0$ for $s < 2m$.*

PROOF.

1. The abscissa of convergence of $\mathbf{V}(\lambda)$ is Δ_v, where $\Delta_v = \min_{ij} \Delta_{v_{ij}}$. Now $V_{ij}(\lambda) = \sigma_j P_{ij}(\lambda)\Lambda_{ij}(c\lambda)$, where

$$P_{ij}(\lambda) = \int_{-\infty}^{\infty} e^{\lambda r} p_{ij}(r)dr \text{ and } \Lambda_{ij}(\lambda) = \int_0^{\infty} e^{-\lambda s}\lambda_{ij}(s)ds.$$

When $\gamma_{ij} \neq 0$, since $\Lambda_{ij}(\lambda)$ is finite and positive for all $0 < \lambda < \infty$, $\Delta_{v_{ij}}$ is the abscissa of convergence of $P_{ij}(\lambda)$ and does not depend on c. Also $P_{ij}(\lambda)$

is finite and positive for all $0 \leq \lambda < \Delta_{v_{ij}}$. If $\gamma_{ij} = 0$, then $V_{ij}(\lambda) \equiv 0$ so that trivially $\Delta_{v_{ij}} = \infty$. Hence Δ_v does not depend on c, and $K_c(\lambda)$ is finite and positive for $0 < \lambda < \Delta_v$.

2. It follows from Theorem A.3 part 3 that $\lim_{\lambda \downarrow 0} K_c(\lambda) = K_c(0)$. Now $K_c(0) = \rho(\Gamma)$, which does not depend on c, and has been assumed to be greater than one. If $\rho(\Gamma) \leq 1$ then no non-trivial solution to (4.6) exists, by Lemma 4.2, for any $c > 0$.
3. The continuity of $K_c(\lambda)$ in λ follows from Theorem A.3 part 1. Now $V_{ij}(\lambda) = \sigma_j P_{ij}(\lambda)\Lambda_{ij}(c\lambda)$ is a continuous strictly decreasing function of c, if $\gamma_{ij} \neq 0$. Since $\mathbf{V}(\lambda)$ is non-reducible, and its entries are decreasing functions of c, the result that $K_c(\lambda)$ is a strictly decreasing function of c for each $\lambda \in (0, \Delta_v)$ follows from Theorem A.2 part 1.

   For any λ with $0 < \lambda < \Delta_v$, $\Lambda_{ij}(c\lambda) \to 0$ as $c \to \infty$, and $P_{ij}(\lambda)$ is finite for all j. By Theorem A.2 part 2,

$$0 < K_c(\lambda) \leq \max_i \sum_{j=1}^{n} \sigma_j \Lambda_{ij}(c\lambda) P_{ij}(\lambda),$$

   which tends to zero as $c \to \infty$. Therefore $\inf_{c>0} K_c(\lambda) = 0$, for $0 < \lambda < \Delta_v$.
4. $V_{ij}(\lambda) = \sigma_j P_{ij}(\lambda)\Lambda_{ij}(c\lambda)$, where $\Lambda_{ij}(\lambda)$ is finite for all λ such that $0 < \lambda < \infty$. The behaviour of $V_{ij}(\lambda)$ at $\lambda = \Delta_v < \infty$ only depends on that of $P_{ij}(\lambda)$, which is independent of c. Thus if $\mathbf{V}(\Delta_v)$ has all its elements finite for some c, the same is true for all $c > 0$, and hence $K_c(\Delta_v)$ is finite for all $c > 0$.

   Since Δ_v is finite, $V_{ij}(\Delta_v) \neq 0$ if $\gamma_{ij} \neq 0$. Thus $\mathbf{V}(\Delta_v)$ is nonreducible. Then, from Theorem A.3 part 3, $\lim_{\lambda \uparrow \Delta_v} K_c(\lambda) = K_c(\Delta_v)$.

   When $\mathbf{V}(\Delta_v)$ has all finite elements, observing that $\mathbf{V}(\Delta_v) \neq 0$ when $n = 1$ since $\rho(\mathbf{\Gamma}) > 1$, by Theorem A.1 we obtain the result that $K_c(\Delta_v) > 0$. Since $\mathbf{V}(\Delta_v)$ is non-reducible, the proof that $K_c(\Delta_v)$ is a continuous strictly decreasing function of c then follows as for property 3.

   As the convergence of of $V_{ij}(\lambda)$ depends only on that of $P_{ij}(\lambda)$, which is independent of c, it either converges for all c or diverges for all c. Therefore if $\mathbf{V}(\Delta_v)$ has an infinite element for some c, it has an infinite element for all c and so from Theorem A.3 part 3 $\lim_{\lambda \uparrow \Delta_v} K_c(\lambda)$ is infinite for all $c > 0$.
5. For $\lambda > 0$ and $c > 0$, define

$$T_c(\lambda; i_1, ..., i_r) = \prod_{j=1}^{r} \left[\sigma_{i_{j+1}} P_{i_j, i_{j+1}}(\lambda) \Lambda_{i_j, i_{j+1}}(c\lambda)\right],$$

   where $i_{r+1} = i_1$. Since $\prod_{j=1}^{r} \Lambda_{i_j, i_{j+1}}(c\lambda)$ is a decreasing function of c, it immediately follows that $T_c(\lambda; i_1, ..., i_r)$ is also a decreasing function of c.

   When Δ_v is infinite we make use of the fact that $T_c(\lambda; i_1, ..., i_r)$ is the two-sided Laplace transform of the r-fold convolution of $v_{i_s i_{s+1}}(x)$ for $s = 1, ..., r$. Hence $\lim_{\lambda \to \infty} T_c(\lambda; i_1, ..., i_r)$ is zero if the convolution is zero for all negative entries and is infinite otherwise. Then from Theorem A.3 part 4, $K_c(\lambda)$ either tends to zero or infinity as $\lambda \to \infty$ for each $c > 0$. This theorem may also be used to obtain the other results.

If $\lim_{\lambda\to\infty} K_{c^*}(\lambda) = \infty$, then there exists a sequence $i_1, ..., i_r$ such that $\lim_{\lambda\to\infty} T_{c^*}(\lambda; i_1, ..., i_r) = \infty$. Using the monotonicity in c, for all $c < c^*$ it then follows that $\lim_{\lambda\to\infty} T_c(\lambda; i_1, ..., i_r) = \infty$, and therefore $\lim_{\lambda\to\infty} K_c(\lambda) = \infty$.

Similarly if $\lim_{\lambda\to\infty} K_{c^*}(\lambda) = 0$, then for all distinct sequences $i_1, ..., i_r$ $\lim_{\lambda\to\infty} T_{c^*}(\lambda; i_1, ..., i_r) = 0$. Hence, for all $c > c^*$, $\lim_{\lambda\to\infty} T_c(\lambda; i_1, ..., i_r) = 0$ for all such sequences, and therefore $\lim_{\lambda\to\infty} K_c(\lambda) = 0$.

6. Since the $V_{ij}(\lambda)$ are non-negative analytic functions of λ for $0 < Re(\lambda) < \Delta_V$, the result immediately follows from Theorem A.3 part 1.
7. From Theorem A.3 part 6, $K_c(\lambda)$ is superconvex for $0 < \lambda < \Delta_v$, i.e. $\log(K_c(\lambda))$ is convex. Since $K_c(\lambda)$ is analytic and non-zero for $0 < \lambda < \Delta_v$, this then implies that $K_c''(\lambda) \geq [K_c'(\lambda)]^2/K_c(\lambda)$. Hence $K_c''(\lambda) > 0$ if $K_c'(\lambda) \neq 0$. Since $K_c(\lambda)$ is analytic, it then follows that $K_c(\lambda)$ is a strictly convex function of λ except in the degenerate case when $K_c(\lambda)$ is constant for all $\lambda \in (0, \Delta_v)$. In the non-degenerate case, from the strict convexity it follows immediately that, if $K_c'(\lambda) = 0$, there exists a positive integer m such that $K_c^{2m}(\lambda) > 0$ and $K_c^s(\lambda) = 0$ for $s < 2m$.

   Note that if the degenerate case occurs for some $c^* > 0$ then $K_{c^*}(\lambda) \equiv \rho(\mathbf{\Gamma})$ is finite. Then $K_c(\lambda)$ cannot be constant for any other positive value of c, since $K_c(0) = \rho(\mathbf{\Gamma})$ and $K_c(\lambda)$ is a strictly monotone decreasing function of c for $\lambda \in (0, \Delta_v)$.

□

Classification of possible speeds.

The implications of Lemma 4.4 are now discussed.

Wave solutions are only possible if $\rho(\mathbf{\Gamma}) > 1$ and the contact distributions are exponentially dominated in the tail so that $\Delta_v > 0$; in which case $K_c(0) > 1$ for all $c > 0$. Note that $K_c(\lambda) < 1$ for some $c > 0$ since $\inf_c K_c(\lambda) = 0$. We define

$$c_0 = \inf\{c > 0 : K_c(\lambda) < 1 \text{ for some } \lambda \in (0, \Delta_v)\}.$$

For each $c > c_0$, $K_c(\lambda) = 1$ has a positive root. Let $\lambda = \alpha$ be the smallest such root; then $K_c(\alpha) = 1$ is simple and $K_c(\lambda) < 1$ for $\lambda \in (\alpha, \alpha + \delta)$, for some positive δ. This is used to construct a wave solution for each speed $c > c_0$ and to establish uniqueness of the wave at each speed.

When $c = c_0 > 0$, $K_{c_0}(\lambda) = 1$ has a unique positive root at $\lambda = \alpha$. Provided $\alpha \neq \Delta_v$, the root is a zero of even multiplicity. The direct construction method cannot be used for $c = c_0$. Instead a wave solution is obtained as the limit of a sequence of wave solutions at speeds tending from above to c_0. The proof of uniqueness at speed c_0 utilises the even multiplicity of the zero of $K_{c_0}(\lambda) = 1$.

For each $c > c_0$, $K_c(\lambda) > 1$ for all $\lambda \in (0, \Delta_v]$. A contradiction argument uses this result to show non-existence of waves at all speeds $c > c_0$.

Lemma 4.4 implies that if $K_c(\lambda)$ has its smallest positive real root simple at speed c, then the same is true for any higher speed. Additionally if it has no positive real root at speed c, then the same is true at any lower positive speed.

These results are summarised in Theorem 4.5. The proofs of these results are contained in Sections 4.7 to 4.10.

THEOREM 4.5 (RESULTS CONCERNING WAVE SPEEDS). *If either $\rho(\mathbf{\Gamma}) \leq 1$ or some $p_{ij}(r)$ is not exponentially dominated in the tail, then no wave solution is possible at any speed $c > 0$.*

When $\rho(\mathbf{\Gamma}) > 1$ and the $p_{ij}(r)$ are all exponentially dominated in the tail, define $c_0 = \inf\{c : K_c(\lambda) < 1 \text{ for some } \lambda \in (0, \Delta_v)\}$. The following results then hold.

1. *If $c_0 = 0$, then a non-trivial wave solution exists at each speed $c > 0$, which is unique modulo translation.*
2. *If $c_0 > 0$, then no non-trivial wave solution is possible at any positive speed $c < c_0$. A non-trivial wave solution exists at each speed $c \geq c_0$, however the existence proof for the case $c = c_0$ does not cover all cases. When some of the scaled infection rates $\gamma_{ij}(t)$ tend to zero as t tends to infinity, but too slowly for $\gamma_{ij}(t)$ to be integrable, the conditions required for the proof of Theorem 4.8 (concerning existence at speed c_0) may not be satisfied. The wave at each speed $c \geq c_0$ is unique modulo translation, except in an exceptional case which occurs when $c = c_0$ and $K_c(\Delta_v) = 1$. Uniqueness has not been established in this case.*

□

An example when the exceptional case occurs has been given for the case $n = 1$ in Section 4.3.

4.7 Existence of waves at supercritical speeds

The existence of waves at each speed $c > c_0$ is established. The method of proof was developed for the one-type model by Diekmann [D5] and extended to multi-type models by Radcliffe and Rass [R1]. For each such speed c, there is a smallest positive root of $K_c(\lambda) = 1$, which depends upon c and which we denote by $\lambda = \alpha$. Also there exists a δ such that $0 < \delta < \min(\alpha, \Delta_v - \alpha)$ for which $K_c(\lambda) < 1$ for $\alpha < \lambda < \alpha + \delta < \Delta_v$. The existence theorem is framed in terms of these conditions.

THEOREM 4.6 (EXISTENCE OF WAVES AT SUPERCRITICAL SPEEDS). *If $K_c(\alpha) = 1$ and $K_c(\lambda) < 1$ for $\alpha < \lambda < \alpha + \delta < \min(2\alpha, \Delta_v)$ for some positive δ, then equations (4.6) have a non-trivial solution $w_i(u)$ $(i = 1, ..., n)$, which is non-negative, monotone decreasing and differentiable with continuous bounded derivatives and with $\lim_{u\to\infty} w_i(u) = 0$. Also $x_i(u) = e^{-w_i(u)}$ $(i = 1, ..., n)$ satisfy equations (4.5).*

PROOF. By Theorem A.1 and Theorem A.2 part 7, since $\mathbf{V}(\alpha)$ is a finite non-negative non-reducible matrix, its maximum eigenvalue $K_c(\alpha)$ is simple and has corresponding right eigenvector $\mathbf{A} > \mathbf{0}$, where $\mathbf{A}$ is unique up to a multiple. We may take $\mathbf{A}$ so that $\{\mathbf{A}\}_i > 2$ for $i = 1, ..., n$.

Define $w_i^{(0)}(u) = \{\mathbf{A}\}_i e^{-\alpha u}$ for $i = 1, ..., n$. Now recursively define $w_i^{(k)}(u)$, for all positive integers k, by

$$w_i^{(k)}(u) = \sum_{j=1}^{n} \int_{-\infty}^{\infty} \left(1 - e^{-w_j^{(k-1)}(u+t)}\right) v_{ij}(t)dt, \quad (i = 1, ..., n). \tag{4.22}$$

The sequence $\{w_i^{(k)}(u)\}$ so constructed is now shown to tend to a limit $w_i(u)$ as $k \to \infty$, for $i = 1, ..., n$, which is a monotone decreasing solution to equations (4.6) with $w_i(\infty) = 0$ for all i.

Consider equation (4.22), with $k = 1$. Then, since $(1 - e^{-w}) \leq w$ for all $w \geq 0$,

$$w_i^{(1)}(u) \leq \sum_{j=1}^{n} \int_{-\infty}^{\infty} w_j^{(0)}(u+t) v_{ij}(t) dt = \sum_{j=1}^{n} \{\mathbf{A}\}_j e^{-\alpha u} V_{ij}(\alpha).$$

Defining $\mathbf{w}^{(k)}(u)$ to be the vector with i^{th} entry $w_i^{(k)}(u)$, for all u we obtain

$$\mathbf{w}^{(1)}(u) \leq e^{-\alpha u} \mathbf{V}(\alpha) \mathbf{A} = e^{-\alpha u} \mathbf{A} = \mathbf{w}^{(0)}(u).$$

Now suppose that $\mathbf{w}^{(k)}(u) \leq \mathbf{w}^{(k-1)}(u)$ for all u and all integers $1 \leq k \leq K$. Since $(1 - e^{-w})$ is an increasing function of w, for each i we immediately obtain from equations (4.22) with $k = K + 1$

$$\begin{aligned} w_i^{(K+1)}(u) &= \sum_{j=1}^{n} \int_{-\infty}^{\infty} \left(1 - e^{-w_j^{(K)}(u+t)}\right) v_{ij}(t) dt \\ &\leq \sum_{j=1}^{n} \int_{-\infty}^{\infty} \left(1 - e^{-w_j^{(K-1)}(u+t)}\right) v_{ij}(t) dt = w_i^{(K)}(u). \end{aligned}$$

Hence $\mathbf{w}^{(K+1)}(u) \leq \mathbf{w}^{(K)}(u)$ for all u. Therefore, by induction on k, $\{w_i^{(k)}(u)\}$ is a monotone decreasing (in k) sequence for each u and all $i = 1, ..., n$.

A simple induction argument shows that $w_i^{(k)}(u)$ is non-negative and monotone decreasing in u for all non-negative integers k and all $i = 1, ..., n$.

The function $w_i^{(0)}(u) = \{\mathbf{A}\}_i e^{-\alpha u}$ is non-negative and monotone decreasing in u. We assume that $w_i^{(k)}(u)$ is non-negative and monotone decreasing in u for all integers $0 \leq k \leq K$. Note that $(1 - e^{-w})$ is a monotone increasing function of w, and is non-negative if $w \geq 0$. From equations (4.22), with $k = K+1$, for any $s > 0$,

$$\begin{aligned} w_i^{(K+1)}(u+s) &= \sum_{j=1}^{n} \int_{-\infty}^{\infty} \left(1 - e^{-w_j^{(K)}(u+s+t)}\right) v_{ij}(t) dt \\ &\leq \sum_{j=1}^{n} \int_{-\infty}^{\infty} \left(1 - e^{-w_j^{(K)}(u+t)}\right) v_{ij}(t) dt = w_i^{(K+1)}(u), \end{aligned}$$

so that $w_i^{(K+1)}(u)$ is monotone decreasing in u. The non-negativity trivially follows from equation (4.22) and the non-negativity of $w^{(K)}(u)$. Hence by induction $w_i^{(k)}(u)$ is non-negative and monotone decreasing in u for all non-negative integers k and all $i = 1, ..., n$.

Thus, for each i, $\{w_i^{(k)}(u)\}$ is a monotone decreasing sequence of monotone decreasing functions of u, which is bounded below by the zero function. The sequence therefore tends to a limit $w_i(u)$ as $k \to \infty$, which is monotone decreasing and non-negative. Using monotone convergence the $w_i(u)$ satisfy the wave equations (4.6). Since $0 \leq w_i^{(k)}(u) \leq w_i^{(0)}(u)$ for all k, with $\lim_{u\to\infty} w_i^{(0)}(u) = 0$, it immediately follows that $\lim_{u\to\infty} w_i(u) = 0$ for all $i = 1, ..., n$.

We need to exclude the possibility that the solution constructed is the trivial solution with $w_i(u) \equiv 0$ for all i. There exists a δ such that $0 < \delta < \alpha$ and $0 < K_c(\alpha+\delta) < 1$. For this δ we obtain a lower bound on the vector $\mathbf{w}^{(k)}(u)$ with i^{th} entry $w_i^{(k)}(u)$. The bound, which holds for all u and all non-negative integers k, is

$$\mathbf{w}^{(k)}(u) \geq \mathbf{A}e^{-\alpha u} - \mathbf{B}e^{-(\alpha+\delta)u}, \tag{4.23}$$

where $\mathbf{B} = \frac{1}{2}(\mathbf{I}-\mathbf{V}(\alpha+\delta))^{-1}\mathbf{V}(\alpha+\delta)\mathbf{C}$ and $\{\mathbf{C}\}_i = \{\mathbf{A}\}_i^2$. Note that $\mathbf{B} > \mathbf{0}$ since $\mathbf{C} > \mathbf{0}$ and $(\mathbf{I}-\mathbf{V}(\alpha+\delta))^{-1} > \mathbf{0}$. The latter result follows from Theorem A.2 part 4, since $\mathbf{V}(\alpha+\delta)$ is non-reducible.

Inequality (4.23) clearly holds for $k = 0$. Assume that the inequality is valid for all integers $0 \leq k \leq K$. From equation (4.22) with $k = K+1$, since $(1-e^{-w}) \geq w - \frac{1}{2}w^2$ for any non-negative w we have

$$\begin{aligned} w_i^{(K+1)}(u) &\geq \sum_{j=1}^{n} \int_0^\infty \left(1 - e^{-w_j^{(K)}(t)}\right) v_{ij}(t-u)dt \\ &\geq \sum_{j=1}^{n} \int_0^\infty \left(w_j^{(K)}(t) - \frac{1}{2}(w_j^{(K)}(t))^2\right) v_{ij}(t-u)dt. \end{aligned}$$

Note that $(w_j^{(K)}(t))^2 \leq (w_j^{(0)}(t))^2 = \{\mathbf{A}\}_j^2 e^{-2\alpha t} \leq \{\mathbf{A}\}_j^2 e^{-(\alpha+\delta)t}$ for $t \geq 0$. From the inductive hypothesis with $k = K$, $w_j^{(K)}(u) \geq \{\mathbf{A}\}_j e^{-\alpha u} - \{\mathbf{B}\}_j e^{-(\alpha+\delta)u}$. Hence for each i we obtain

$$\begin{aligned} w_i^{(K+1)}(u) &\geq \sum_{j=1}^{n} \int_0^\infty \left[\{\mathbf{A}\}_j e^{-\alpha t} - (\{\mathbf{B}\}_j + \frac{1}{2}\{\mathbf{A}\}_j^2)e^{-(\alpha+\delta)t}\right] v_{ij}(t-u)dt \\ &= \sum_{j=1}^{n} \int_{-\infty}^\infty \left[\{\mathbf{A}\}_j e^{-\alpha t} - (\{\mathbf{B}\}_j + \frac{1}{2}\{\mathbf{A}\}_j^2)e^{-(\alpha+\delta)t}\right] v_{ij}(t-u)dt \\ &- \sum_{j=1}^{n} \int_{-\infty}^0 \left[\{\mathbf{A}\}_j e^{-\alpha t} - (\{\mathbf{B}\}_j + \frac{1}{2}\{\mathbf{A}\}_j^2)e^{-(\alpha+\delta)t}\right] v_{ij}(t-u)dt. \end{aligned}$$

The second term on the right hand side will be positive if $(\{\mathbf{B}\}_j + \frac{1}{2}\{\mathbf{A}\}_j^2) \geq \{\mathbf{A}\}_j$, i.e. if $\{\mathbf{B}\}_j \geq \{\mathbf{A}\}_j(1 - \frac{1}{2}\{\mathbf{A}\}_j)$, for all $j = 1, ..., n$. This is trivially true since $\mathbf{B} > \mathbf{0}$ and $\mathbf{A}$ has been scaled so that $\{\mathbf{A}\}_j \geq 2$ for all $j = 1, ..., n$.

Therefore

$$\begin{aligned} w_i^{(K+1)}(u) &\geq \sum_{j=1}^{n} \int_{-\infty}^\infty \left[\{\mathbf{A}\}_j e^{-\alpha t} - (\{\mathbf{B}\}_j + \frac{1}{2}\{\mathbf{A}\}_j^2)e^{-(\alpha+\delta)t}\right] v_{ij}(t-u)dt \\ &= \sum_{j=1}^{n} \left[\{\mathbf{A}\}_j e^{-\alpha u} V_{ij}(\alpha) - (\{\mathbf{B}\}_j + \frac{1}{2}\{\mathbf{A}\}_j^2)e^{-(\alpha+\delta)u} V_{ij}(\alpha+\delta)\right]. \end{aligned}$$

Hence

$$\begin{aligned}\mathbf{w}^{(K+1)}(u) &\geq e^{-\alpha u}\mathbf{V}(\alpha)\mathbf{A} - e^{-(\alpha+\delta)u}\mathbf{V}(\alpha+\delta)(\mathbf{B}+\frac{1}{2}\mathbf{C}) \\ &= e^{-\alpha u}\mathbf{A} - e^{-(\alpha+\delta)u}\left[\mathbf{V}(\alpha+\delta)\mathbf{B} + (\mathbf{I}-\mathbf{V}(\alpha+\delta))\mathbf{B}\right] \\ &= e^{-\alpha u}\mathbf{A} - e^{-(\alpha+\delta)u}\mathbf{B}.\end{aligned}$$

Thus the inequality (4.23) holds for $k = K+1$, and so by induction it must hold for all non-negative integers k.

It immediately follows that $\mathbf{w}(u) \geq \mathbf{A}e^{-\alpha u} - \mathbf{B}e^{-(\alpha+\delta)u}$ for all u. For u sufficiently large this vector lower bound has all positive entries. This then proves that the constructed solution is not the trivial solution.

It is now shown that each $w_i(u)$ is a continuous function of u. First observe that

$$\int_{-\infty}^{0} v_{ij}(s)ds = c^{-1}\int_{-\infty}^{0}\int_{0}^{\infty}\gamma_{ij}(t/c)p_{ij}(t-s)dtds \leq c^{-1}\sup\gamma_{ij}(u)\int_{0}^{\infty} rp_{ij}(r)dr,$$

which is finite since $p_{ij}(r)$ is exponentially dominated in the tail. Now for any $R > 0$ and $|\delta u| < 1$, since the $v_{ij}(t)$ are bounded, $w_i(t) \leq \{\mathbf{A}\}_i e^{-\alpha t}$ and $1 - e^{-w} \leq w$, hence

$$\begin{aligned}|w_i(u+\delta u) - w_i(u)| &\leq \sum_{j=1}^{n}\int_{-\infty}^{\infty}(1-e^{-w_j(t)})|v_{ij}(t-u-\delta u) - v_{ij}(t-u)|dt \\ &\leq \sum_{j=1}^{n}\left(\int_{-R}^{R}|v_{ij}(t-u-\delta u) - v_{ij}(t-u)|dt\right. \\ &\left. +2\sup_{s} v_{ij}(s)\int_{R}^{\infty}\{\mathbf{A}\}_j e^{-\alpha t}dt + 2\int_{-\infty}^{-R+1+u} v_{ij}(s)ds\right).\end{aligned}$$

Continuity then follows since $\alpha > 0$, $v_{ij}(t)$ is uniformly continuous in a bounded interval and $\int_{-\infty}^{0} v_{ij}(s)ds$ is finite.

In order to establish differentiability we split the range of $\int_{-\infty}^{\infty}(1-e^{-w_j(t)})v_{ij}(t-u)dt$ into $t > 0$ and $t < 0$. Differentiability of the first integral is then immediate, using dominated convergence, since $v_{ij}(t-u)$ is differentiable with continuous bounded derivatives and $1 - e^{-w_j(t)} \leq w_j(t) \leq \{\mathbf{A}\}_j e^{-\alpha t}$, so that $(1 - e^{-w_j(t)})$ is integrable over $t > 0$. Also

$$\frac{d}{du}\int_{0}^{\infty}(1-e^{-w_j(t)})v_{ij}(t-u)dt = -\int_{0}^{\infty}(1-e^{-w_j(t)})v'_{ij}(t-u)dt.$$

Let $R_{ij}(r) = \int_{-\infty}^{0} p_{ij}(r-t)(1-e^{-w_j(t)})dt$, which is continuous since the contact distributions are continuous and integrable. Then

$$\int_{-\infty}^{0}(1-e^{-w_j(t)})v_{ij}(t-u)dt = c^{-1}\int_{u}^{\infty}\gamma_{ij}((r-u)/c)R_{ij}(r)dr.$$

Now $\gamma'_{ij}(t)$ is bounded and continuous and $\int_u^\infty R_{ij}(r)dr \leq \int_u^\infty (r-u)p_{ij}(r)dr$ is finite, hence the integral is differentiable and

$$\begin{aligned}\frac{d}{du}\int_{-\infty}^{0}(1-e^{-w_j(t)})v_{ij}(t-u)dt &= \frac{-1}{c}\left(\gamma_{ij}(0)R_{ij}(u)+\int_u^\infty \gamma'_{ij}\left(\frac{r-u}{c}\right)R_{ij}(r)dr\right)\\ &= -\int_{-\infty}^{0}(1-e^{-w_j(t)})v'_{ij}(t-u)dt.\end{aligned}$$

Hence each $w_i(u)$ is differentiable, with its derivative bounded since $v'_{ij}(t)$ is absolutely integrable, and so $x_i(u)$ is differentiable and the $x_i(u)$ satisfy equations (4.5).

□

A wave solution $w_i(u)$ $(i = 1, ..., n)$ to equations (4.6) has therefore been constructed at all speeds $c > c_0$. This wave also provides a solution (with the specified conditions) to equations (4.5), and hence to equations (4.3), if we take $x_i(u) = e^{-w_i(u)}$. Note that, with this definition, $x_i(u)$ is positive and monotone increasing with $\lim_{u\to\infty} x_i(u) = 1$ since $w_i(u) > 0$ and is monotone decreasing with $\lim_{u\to\infty} w_i(u) = 0$.

The solution constructed at speed c in Theorem 4.6 is wedged between bounds, so that

$$e^{-\alpha u}\mathbf{A} - e^{-(\alpha+\delta)u}\mathbf{B} \leq \mathbf{w}(u) \leq e^{-\alpha u}\mathbf{A}.$$

This enables us to determine the behaviour of the constructed solution when u is large. Specifically, $\mathbf{w}(u)$ behaves like $e^{-\alpha u}\mathbf{A}$ as $u \to \infty$, where $\alpha = \Delta_w$. Note that $\mathbf{A}$ is the eigenvector corresponding to $K_c(\alpha)$, so that $\mathbf{A}$ is unique up to a multiple. Shifting the origin for u will scale the vector $\mathbf{A}$. It is shown in Section 4.10 that any solution to equations (4.6) has the same behaviour for u large, which enables uniqueness (modulo translation) to be established. From Lemma 4.1 this proves that there is a unique solution to equations (4.3) and (4.5) at each speed $c > c_0$, which is the constructed solution.

4.8 Non-existence of waves at subcritical speeds

When $c_0 > 0$, for any $0 < c < c_0$ $K_c(\lambda) > 1$ for all $\lambda \in [0, \Delta_v]$. This condition on $K_c(\lambda)$ is used to show that no wave solution is possible at any speed $0 < c < c_0$. The proof makes use of the result contained in Theorem 4.2 part 2. Note that $K_c(0) = \rho(\mathbf{\Gamma})$, which has been assumed to be greater than one. The proofs are based on the work of Diekmann and Kaper [D9], Radcliffe, Rass and Stirling [R17] and Radcliffe and Rass [R1].

THEOREM 4.7 (NON-EXISTENCE OF WAVES AT SUBCRITICAL SPEEDS). *If, for some positive c,* $K_c(\lambda) > 1$ *for all* $0 \leq \lambda < \Delta_v$ *and* $\lim_{\lambda\uparrow\Delta_v} K_c(\lambda) > 1$, *then for this value of c no non-trivial, monotone decreasing, differentiable solution* $w_i(u)$ *(with continuous, bounded derivatives) exists to equations (4.6) with* $\lim_{u\to\infty} w_i(u) = 0$ *for* $i = 1, ..., n$.

PROOF. Consider a positive c such that there is no solution to $K_c(\lambda) = 1$ for $\lambda \in [0, \Delta_v]$. Now assume that, for this value of c, there exists a non-trivial solution $w_i(u)$, $i = 1, ..., n$, to equations (4.6). Theorem 4.3 shows that $\Delta_w \leq \Delta_v$. From Theorem 4.4 the possibility that $\Delta_w < \Delta_v$ is excluded since there is no solution to $K_c(\lambda) = 1$. Hence $\Delta_w = \Delta_v$.

From Lemma 4.4 part 5, If $\Delta_v = \infty$, then $\lim_{\lambda\to\infty} K_c(\lambda)$ is either zero or infinity. Since the limit is greater than one, only the infinite limit is possible. There are therefore two cases to consider:

(i) $\mathbf{V}(\Delta_v)$ has all finite entries, $K_c(\Delta_v) > 1$ and Δ_v is finite.

(ii) $\mathbf{V}(\Delta_v)$ has at least one infinite entry and $\lim_{\lambda\uparrow\Delta_v} K_c(\lambda)$ is infinite.

Both cases are shown to lead to a contradiction; the proof in case (i) being much simpler than in case (ii). This then establishes the result that no non-trivial solution to equations (4.6) is possible for any speed c such that $K_c(\lambda) > 1$ for all $0 \leq \lambda < \Delta_v$ and $\lim_{\lambda\uparrow\Delta_v} K_c(\lambda) > 1$.

Case (i)

Rewriting equation (4.21) from Theorem 4.4, we have

$$(4.24) \qquad (\mathbf{u}(\lambda))'\mathbf{W}(\lambda) = \frac{1}{(K_c(\lambda) - 1)}(\mathbf{u}(\lambda))'\mathbf{R}(\lambda), \text{ for } 0 < \lambda < \Delta_w.$$

Now consider inequality (4.20) in Theorem 4.4 and take Δ_v and $\mathbf{V}(\Delta_v)$ finite, $\Delta_w = \Delta_v$ and $\nu = 0$. This gives

$$0 \leq R_i(\Delta_v) \leq \frac{1}{2}\sum_{j=1}^{n} V_{ij}(\Delta_v) \sup\{w_j(s)e^{\theta s} : s \in \Re\} W_j(\Delta_w - \theta) \text{ for } 0 < \theta < \Delta_v.$$

Hence $R_i(\Delta_v)$ is finite for all $i = 1, ..., n$. Since $\mathbf{V}(\lambda)$ is non-reducible with entries continuous functions of λ for $\lambda \in (0, \Delta_v]$, by Theorem A.3 parts 2 and 3 $(\mathbf{u}(\lambda))'$ has entries which are positive and continuous functions of λ and $(\mathbf{u}(\Delta_v))' = \lim_{\lambda\uparrow\Delta_v}(\mathbf{u}(\lambda))'$ has finite positive entries.

This implies, from equation (4.24), that $\lim_{\lambda\uparrow\Delta_v}(\mathbf{u}(\lambda))'\mathbf{W}(\lambda)$ exists. Since $\mathbf{u}(\lambda) > \mathbf{0}$ for $\lambda \in (0, \Delta_v]$ it immediately follows that $W_i(\Delta_v)$ is finite for all $i = 1, ..., n$. As $K_c(\Delta_v) > 1$, Theorem 4.2 part 2 with $\beta = \Delta_v = \Delta_w$ can be applied to obtain the result that the abscissa of convergence of $W_i(\lambda)$ is to the right of Δ_w for $i = 1, ..., n$. This contradicts the definition of Δ_w.

Hence a contradiction has been obtained. Therefore for case (i) no non-trivial solution $w_i(u)$, $i = 1, ..., n$, to equations (4.6) is possible.

Case (ii)

Since $1 - e^{-x} < x$ for all $x > 0$ and $\lim_{u\to\infty}\left(1 - e^{-w_i(u)}\right)/w_i(u) = 1$ for each $i = 1, ..., n$, for each fixed α with $0 < \alpha < 1$ there exists an X_0 such that $\left(1 - e^{-w_i(u)}\right) \geq \alpha w_i(u)$ for all $i = 1, ..., n$ and all $u \geq X_0$. Choose such an α and obtain the corresponding X_0. From equation (4.6), for any $b > 0$ we then have

$$(4.25) \qquad w_i(u) \geq \alpha \sum_{j=1}^{n} \int_{-b}^{\infty} w_j(u+s)v_{ij}(s)ds \text{ for } u \geq X_0 + b \text{ and } i = 1, ..., n.$$

As $\lim_{\lambda\uparrow\Delta_v} K_c(\lambda) = \infty$, there exists a distinct subsequence $i_1, ..., i_r$, with $i_{r+1} = i_1$, such that $\lim_{\lambda\uparrow\Delta_v} \prod_{j=1}^r V_{i_j i_{j+1}}(\lambda) = \infty$. Define $v(s) = v_{i_1 i_2} * v_{i_2 i_3} * ... * v_{i_r i_1}(s)$, i.e. the convolution of $v_{i_j i_{j+1}}(s)$ for $j = 1, ..., r$. Then this implies that

$$\lim_{\lambda\uparrow\Delta_v} \int_{-\infty}^0 v(s)e^{-\lambda s} ds = \lim_{\lambda\uparrow\Delta_v} \int_{\sum s_j<0}\cdots\int \prod_{j=1}^r \left(v_{i_j i_{j+1}}(s_j)e^{-\lambda s_j}\right) ds_1...ds_r = \infty.$$

Note that $\int_{-\infty}^0 v(s)e^{-\lambda s}ds$ is a continuous function of λ for $0 < \lambda < \Delta_v$. Hence for any $K > 0$ there exists a $\lambda_0 \in (0, \Delta_v)$ such that $\int_{-\infty}^0 v(s)e^{-\lambda_0 s}ds > K$. Choose K such that $K\alpha^r > 1$ and find the corresponding λ_0.

Let $S_b = \{\mathbf{s} = (s_1, ..., s_r) \in \Re^n : \sum s_j < 0 \text{ and } s_j > -b \text{ for } j = 1, ..., r\}$. Define

$$V(b, \lambda_0) = \int_{\mathbf{s}\in S_b}\cdots\int \prod_{j=1}^r \left(v_{i_j i_{j+1}}(s_j)e^{-\lambda_0 s_j}\right) ds_1...ds_r.$$

This function $V(b, \lambda_0)$ is a monotone increasing, continuous function of b with $\lim_{b\to\infty} V(b, \lambda_0) = \int_{-\infty}^0 v(s)e^{-\lambda_0 s}ds > K$. Hence for any $0 < K^* < K$ there exists a b such that $V(b, \lambda_0) > K^*$. Take $K^* < \alpha^{-r}$ and find the corresponding b. Then $V(b, \lambda_0)\alpha^r > 1$. From equation (4.25) we have, for $u \geq X_0 + rb$,

$$w_{i_j}(u) \geq \alpha \int_{-b}^\infty w_{i_{j+1}}(u+s)v_{i_j i_{j+1}}(s)ds \text{ for } j = 1, ..., r.$$

Hence, for $u \geq X_0 + rb$,

(4.26)

$$\begin{aligned} &e^{\lambda_0 u}w_{i_1}(u) \\ &\geq \alpha^r \int_{-b}^\infty ... \int_{-b}^\infty e^{\lambda_0(u+\sum s_j)} w_{i_1}(u + \sum s_j) \prod_{j=1}^r \left(v_{i_j i_{j+1}}(s_j)e^{-\lambda_0 s_j}\right) ds_1...ds_r \\ &\geq \alpha^r \int_{\mathbf{s}\in S_b}\cdots\int e^{\lambda_0(u+\sum s_j)} w_{i_1}(u + \sum s_j) \prod_{j=1}^r \left(v_{i_j i_{j+1}}(s_j)e^{-\lambda_0 s_j}\right) ds_1...ds_r. \end{aligned}$$

We next prove that $w_{i_1}(u) > 0$ for all u. Now $w_{i_1}(u)$ is a continuous, decreasing function of u. Equations (4.6) and the non-reducibility of $\mathbf{\Gamma}$ imply that $w_{i_1}(u) > 0$ for some u. Hence we need only exclude the possibility that there exists a finite value C such that $w_{i_1}(u) > 0$ for $u < C$ but $w_{i_1}(u) = 0$ for $u \geq C$.

From equation (4.6) $w_i(u) > 0$ if $\int_{-\infty}^\infty w_j(u+t)v_{ij}(t)dt > 0$. Then using this result with $i = i_s$ and $j = i_{s+1}$, for each $s = 1, ..., r$, we obtain the condition that $w_{i_1}(u) > 0$ if $\int_{-\infty}^\infty w_{i_1}(u+t)v(t)dt > 0$. But the Laplace transform, $V(\lambda) = \int_{-\infty}^\infty e^{-\lambda u}v(u)du \to \infty$ as $\lambda \uparrow \Delta_v$. Hence there exist constants $0 \leq A_1 < A_2$ such that $v(t) > 0$ for $-A_2 \leq t \leq -A_1$. Hence $w_{i_1}(C + (A_1 + A_2)/2) > 0$ which is a contradiction. Therefore $w_{i_1}(u) > 0$ for all u.

Since $w_{i_1}(u) > 0$ all u and $W_{i_1}(\lambda_0)$ is finite, then $e^{\lambda_0 u}w_{i_1}(u) > 0$ for all u and $\lim_{u\to\infty} e^{\lambda_0 u}w_{i_1}(u) = 0$. Consider a strictly monotone decreasing sequence $\{y_j\}$, where $y_1 < w_{i_1}(0)$ and $\lim_{j\to\infty} y_j = 0$. Define

$$x_j = \sup\{x \in \Re : w_{i_1}(u)e^{\lambda_0 u} \geq y_j \text{ for all } u \in (0,x)\}.$$

Since $w_{i_1}(u)e^{\lambda_0 u}$ is positive and continuous in u with $\lim_{u\to\infty} w_{i_1}(u)e^{\lambda_0 u} = 0$, clearly $w_{i_1}(x_j)e^{\lambda_0 x_j} = y_j$. Also x_j tends monotonically to infinity as j tends to infinity. Take j sufficiently large so that $x_j > X_0 + rb$. Now consider inequality (4.26) with $u = x_j$ for this j. Since $y_j > 0$,

$$y_j = w_{i_1}(x_j)e^{\lambda_0 x_j} \geq \alpha^r y_j V(b, \lambda_0) > y_j.$$

This is a contradiction. Hence, for case (ii) also, no non-trivial solution $w_i(u)$, $i = 1, ..., n$, to equations (4.6) is possible.

□

4.9 Existence of waves at critical speed

A wave solution has been constructed for each speed $c > c_0$. A direct construction at speed c_0 may be possible, using the the same approach as Weinberger [W1] for a related problem in population genetics. However this approach would constrain the contact distributions to have bounded support.

In order to prove the existence of a wave solution at speed c_0 for more general contact distributions a limiting argument is used. This approach was first used by Brown and Carr [B11] for the one-type $S \to I$ epidemic.

Wave solutions can be constructed corresponding to a monotone decreasing sequence $\{c_m\}$ of wave speeds, where $\lim_{m\to\infty} c_m = c_0$. In each case an appropriate translate of the wave solution can be selected so that $\mathbf{w}^{(m)}(u)$ is the selected vector translate corresponding to speed c_m. When $\mathbf{\Gamma}$ is finite $\mathbf{w}(u) = \lim_{m\to\infty} \mathbf{w}^{(m)}(u)$ can be shown to be a non-trivial solution to the wave equations (4.6) corresponding to critical speed c_0. This approach is not possible if γ_{ij} is infinite for some i, j because in this case $\lim_{u\to-\infty} w_i^{(m)}(u)$ is infinite so that the $w_i^{(m)}(u)$ are not uniformly bounded.

When the infection matrix has all non-zero entries infinite, with the corresponding $v_{ij}(u)$ tending to a positive limit as u tends to infinity, then the proof is based on $x_i^{(m)}(u) = -\log w_i^{(m)}(u)$, and uses equations (4.5). Then $\mathbf{x}(u) = \lim_{m\to\infty} \mathbf{x}^{(m)}(u)$ can be shown to be a non-trivial solution to the wave equations (4.5) corresponding to critical speed c_0, and hence from Theorem 4.1 to equations (4.3) and (4.6). An example of this case is the multi-type $S \to I$ epidemic.

A proof can still be given in many cases when some non-zero γ_{ij} are infinite and some are finite. The difficulty arises when, for some i, j, γ_{ij} is infinite but $\lim_{u\to\infty} v_{ij}(u) = 0$. This latter condition was shown in Section 4.3 to correspond to an equivalent condition on the scaled infection rate, namely $\lim_{u\to\infty} \gamma_{ij}(u) = 0$. The precise conditions under which the existence of a wave solution at the critical speed can be established are specified in Theorem 4.8.

THEOREM 4.8. *When $c_0 > 0$ and $\mathbf{\Gamma}$ is finite, there exists a non-trivial, continuous, monotone decreasing solution $w_i(u)$ $(i = 1, ..., n)$, with $\lim_{u\to\infty} w_i(u) = 0$, to equations (4.6) with speed $c = c_0$. Also $x_i(u) = e^{-w_i(u)}$ is differentiable and satisfies equations (4.5) and $x_i'(u)/x_i(u)$ is continuous and bounded.*

When $c_0 > 0$ and $\mathbf{\Gamma}$ has some infinite entries, then (provided a certain condition holds) there exists a non-trivial, monotone decreasing, differentiable solution $x_i(u)$ $(i = 1, ..., n)$, with each $x_i'(u)/x_i(u)$ continuous and bounded and $\lim_{u\to\infty} x_i(u) = 1$, to equations (4.5) with speed $c = c_0$. The condition required is that a type l can be chosen for which:

1. *γ_{lj} is infinite with $\lim_{u\to\infty} v_{lj}(u) > 0$ for at least one j.*
2. *For this l and each $i = 1, ..., n$ there exists a positive integer r and a sequence $j_1, ..., j_r = i$ for which γ_{l,j_1} is infinite with $\lim_{u\to\infty} v_{l,j_1}(u) > 0$ and for each $1 \le s \le r-1$ one of two conditions hold; either $\gamma_{j_s,j_{s+1}}$ is finite and non-zero or $\gamma_{j_s,j_{s+1}}$ is infinite with $\lim_{u\to\infty} v_{j_s,j_{s+1}}(u) > 0$.*

PROOF. When $\mathbf{\Gamma}$ has some infinite entries and there are k rows with no infinite entries, re-order the types so that the first k rows of $\mathbf{\Gamma}$ have no infinite entries and the type l specified in the conditions is now labeled as type n. When $\mathbf{\Gamma}$ is finite, no re-ordering is necessary and $k = n$.

For any $c > c_0$ a solution $w_i(u)$, for $i = 1, ..., n$, has been constructed in Theorem 4.6 to equations (4.6). This was also shown to give a solution $x_i(u) = e^{-w_i(u)}$, for $i = 1, ..., n$, to equations (4.5). We work with $x_i(u)$ since, when some entries of $\mathbf{\Gamma}$ are infinite, the $w_i(u)$ will not all be bounded. We will use equations (4.5) for all $i = 1, ..., n$ and also equations (4.6), but only for $i = 1, ..., k$. From Theorem 4.2, for $i \le k$, all $c > c_0$ and all u, $x_i(u) \ge 1 - \eta_i(\mathbf{0}) > 0$ and hence $w_i(u) \le e^{-1-\eta_i(\mathbf{0})} < \infty$.

Equation (4.5) may be rewritten in the form

$$cx_i'(u) = x_i(u) \sum_{j=1}^{n} (\gamma_{ij}(\infty) - \gamma_{ij}(0) y_{ij}(u) - s_{ij}(u)),\ (i = 1, ..., n), \tag{4.27}$$

where $y_{ij}(u) = \int_{-\infty}^{\infty} p_{ij}(-s) x_j(u+s)ds$ and $s_{ij}(u) = \int_0^{\infty} y_{ij}(u+cr)\gamma_{ij}'(r)dr$.
The first k of equations (4.6) can similarly be written as

$$-\log(x_i(u)) = \sum_{j=1}^{n} (\gamma_{ij} - t_{ij}(u)),\ (i = 1, ..., k), \tag{4.28}$$

where $t_{ij}(u) = \int_0^{\infty} \gamma_{ij}(r) y_{ij}(u+cr)dr$.

Consider a monotone decreasing sequence $\{c_m\}$ such that $\lim_{m\to\infty} c_m = c_0$. Let $x_i^{(m)}(u)$, $d_i^{(m)}(u)$, $y_{ij}^{(m)}(u)$ and $s_{ij}^{(m)}(u)$ respectively be the selected wave translate $x_i(u)$, the derivative $x_i'(u)$ and functions $y_{ij}(u)$ and $s_{ij}(u)$ corresponding to speed c_m. For $i \le k$ also let $t_{ij}^{(m)}(u)$ be the function $t_{ij}(u)$ corresponding to speed c_m. The translate selected for each c_m has $x_n(0) = (1 + (1 - \eta_n(0)))/2$, where $1 - \eta_n(0) = \lim_{u\to-\infty} x_n^{(m)}(u)$ and is independent of c_m. This will anchor the limiting solution so that it is not the trivial solution.

We now show that the sequences $\{x_i^{(m)}(u)\}$, $\{d_i^{(m)}(u)\}$, $\{y_{ij}^{(m)}(u)\}$ and $\{s_{ij}^{(m)}(u)\}$, for all i and j, and $\{t_{ij}^{(m)}(u)\}$ for $i \le k$ and all j, are uniformly bounded and equicontinuous. To show that they are equicontinuous, the derivatives are shown to be uniformly bounded.

Clearly $|x_i^{(m)}(u)| \leq 1$ and $|y_{ij}^{(m)}(u)| \leq \int_{-\infty}^{\infty} p_{ij}(-s)ds = 1$. Using the same notation as in Theorem 4.2, let k_{ij} be the number of changes of sign of $\gamma'_{ij}(t)$ for $t \in [0, \infty)$, and let $L_{ij} = \sup_{t\in[0,\infty)} \gamma_{ij}(t)$. Then

$$|s_{ij}^{(m)}(u)| \leq \int_0^{\infty} |\gamma'_{ij}(\theta)| d\theta \leq (k_{ij}+1)L_{ij}.$$

and hence also, using equations (4.27),

$$|d_i^{(m)}(u)| \leq \frac{1}{c_0} \sum_{j=1}^{n} (L_{ij} + L_{ij} + (k_{ij}+1)L_{ij}).$$

Also for $i \leq k$, $|t_{ij}^{(m)}(u)| \leq \int_0^{\infty} \gamma_{ij}(s)ds = \gamma_{ij}$. Hence the sequences are uniformly bounded.

Now consider the derivatives. Note that we already have obtained a bound for $x'_{ij}(u)$. Differentiating $y_{ij}(u)$ and $s_{ij}(u)$ for all i, j and $t_{ij}(u)$ for $i \leq k$ and all j, and using dominated convergence, we obtain the following:

$$|y_{ij}^{(m)'}(u)| \leq \int_{-\infty}^{\infty} |x_j^{(m)'}(u+s)| p_{ij}(-s) ds \leq \frac{1}{c_0} \sum_{l=1}^{n} (k_{jl}+3)L_{jl},$$

$$|s_{ij}^{(m)'}(u)| \leq \int_0^{\infty} |y_{ij}^{(m)'}(u + c_m\theta)||\gamma'_{ij}(\theta)| d\theta \leq \frac{1}{c_0}(k_{ij}+1)L_{ij} \sum_{l=1}^{n} (k_{jl}+3)L_{jl},$$

$$|t_{ij}^{(m)'}(u)| \leq \int_0^{\infty} |y_{ij}^{(m)'}(c_m s + u)||\gamma_{ij}(s)| ds \leq \frac{\gamma_{ij}}{c_0} \sum_{l=1}^{n} (k_{jl}+3)L_{jl}.$$

Finally, differentiating equation (4.27),

$$\left| \frac{d^2}{du^2} x_i^{(m)}(u) \right| \leq \frac{1}{c_0} \sum_{j=1}^{n} (|x_i^{(m)'}(u)| [\gamma_{ij}(\infty) + \gamma_{ij}(0)|y_{ij}^{(m)}(u)| + |s_{ij}^{(m)}(u)|]$$
$$+ x_i^{(m)}(u)[\gamma_{ij}(0)|y_{ij}^{(m)'}(u)| + |s_{ij}^{(m)'}(u)|]),$$

where the right hand side of this inequality is bounded, since each component has already been shown to be bounded. Hence the sequences of functions are equicontinuous.

Now use Ascoli's theorem and a nested subsequence argument. There exists a subsequence of $\{c_m\}$, which we will again denote by $\{c_m\}$, such that $c_m \downarrow c_0$ as $m \to \infty$ and the corresponding sequences $x_i^{(m)}(u)$, $d_i^{(m)}(u)$, $y_{ij}^{(m)}(u)$ and $s_{ij}^{(m)}(u)$ for all i, j, and $t_{ij}^{(m)}(u)$ for $i \leq k$ and all j, converge uniformly on every bounded interval and hence pointwise on $\Re$ to $x_i(u)$, $d_i(u)$, $y_{ij}(u)$, $s_{ij}(u)$ and $t_{ij}(u)$. Note that $x_n(0) = (1 + (1 - \eta_n(0)))/2$. It therefore follows that $x_i(u)$ is differentiable and $d_i(u) = x_i{}'(u)$, (see Bartle [B4], Theorem 19.12).

Using the dominated convergence theorem,

$$y_{ij}(u) = \lim_{m\to\infty} \int_{-\infty}^{\infty} x_j^{(m)}(u+s)p_{ij}(-s)ds = \int_{-\infty}^{\infty} \lim_{m\to\infty} x_j^{(m)}(u+s)p_{ij}(-s)ds$$
$$= \int_{-\infty}^{\infty} x_j(u+s)p_{ij}(-s)ds,$$
$$s_{ij}(u) = \lim_{m\to\infty} \int_0^{\infty} y_{ij}^{(m)}(u+c_m\theta)\gamma'_{ij}(\theta)d\theta = \int_0^{\infty} \lim_{m\to\infty} y_{ij}^{(m)}(u+c_m\theta)\gamma'_{ij}(\theta)d\theta,$$

and for $i \le k$,

$$t_{ij}(u) = \lim_{m\to\infty} \int_0^{\infty} y_{ij}^{(m)}(u+c_m\theta)\gamma_{ij}(\theta)d\theta = \int_0^{\infty} \lim_{m\to\infty} y_{ij}^{(m)}(u+c_m\theta)\gamma_{ij}(\theta)d\theta.$$

Clearly $y_{ij}(u)$ is continuous. Using this and the dominated convergence theorem it is easily shown that $\lim_{m\to\infty} y_{ij}^{(m)}(u+c_m\theta) = y_{ij}(u+c_0\theta)$, so that

$$s_{ij}(u) = \int_0^{\infty} y_{ij}(u+c_0\theta)\gamma'_{ij}(\theta)d\theta = \int_0^{\infty}\int_{-\infty}^{\infty} x_j(u+c_0\theta+s)p_{ij}(-s)\gamma'_{ij}(\theta)dsd\theta,$$

and for $i \le k$,

$$t_{ij}(u) = \int_0^{\infty} y_{ij}(u+c_0\theta)\gamma_{ij}(\theta)d\theta = \int_0^{\infty}\int_{-\infty}^{\infty} x_j(u+c_0\theta+s)p_{ij}(-s)\gamma_{ij}(\theta)dsd\theta.$$

For $i \le k$, $x_i^{(m)}(u) \ge (1-\eta_i(\mathbf{0})$ and hence $x_i(u) \ge (1-\eta_i(\mathbf{0}) > 0$ for all u. Therefore $\lim_{m\to\infty} -\log(1-x^{(m)}(u)) = -\log(1-x_i(u))$. Hence $x_i(u)$, $(i=1,...,n)$, satisfy equations (4.27) and (4.28) with $c = c_0$. They therefore also satisfy equation (4.5) with $c = c_0$.

Each $x_i(u)$ has been shown to be differentiable. It is bounded and monotone increasing with bounded derivative, since it is the limit of monotone increasing functions which are uniformly bounded and the derivatives are uniformly bounded. It is not the trivial solution with $x_i(u) \equiv 1$ for all u and i since $x_n(0) = (1+(1-\eta_n(\mathbf{0})))/2 < 1$. It is now shown that $\lim_{u\to\infty} x_i(u) = 1$ for all $i = 1,...,n$.

First consider the case when $\mathbf{\Gamma}$ is finite, so that $k = n$. Then $x_i(u)$, $(i = 1,...,n)$, satisfy equations (4.28) with $c = c_0$. Now $x_i(u)$ is monotone increasing and also $(1-\eta_i(\mathbf{0})) \le x_i(u) \le 1$. Using monotone convergence as $u \to \infty$, $y_i = \lim_{u\to\infty} -\log x_i(u)$ for $i = 1,...,n$ is a solution to

$$y_i = \sum_{j=1}^{n} \gamma_{ij}\left(1-e^{-y_j}\right) \text{ for } j = 1,...,n.$$

This has only two possible solutions, $y_i = 0$ all i or $y_i = -\log(1-\eta_i(\mathbf{0}))$ all i. Since $\lim_{u\to\infty} x_n(u) \ge x_n(0) = (1+(1-\eta_n(\mathbf{0}))/2 > (1-\eta_n(\mathbf{0}))$, necessarily we have $y_i = 0$ and so $\lim_{u\to\infty} x_i(u) = 1$, for all $i = 1,...,n$.

Now consider the case when $k < n$ with the conditions specified in the statement of the theorem. Since $x_n(0) > 0$ and $x_n(u)$ is monotone increasing, $x_n(u) > 0$ for all $u \ge 0$. Also $x_n(u) \le 1$, so that $0 < \lim_{u\to\infty} x_n(u) \le 1$ and $\lim_{u\to\infty} x'_n(u) = 0$.

Consider the last equation with $i = n$ of equations (4.5) with $u \geq 0$ and $c = c_0$. Let $u \to \infty$. It immediately follows using dominated convergence that, for any j such that $\lim_{u\to\infty} v_{nj}(u) > 0$ (so that necessarily γ_{nj} is infinite), $\lim_{u\to\infty} x_j(u) = 1$.

Assume that it has been shown that $\lim_{u\to\infty} x_i(u) = 1$. From continuity, there exists a U such that $x_i(u) > 0$ for $u \geq U$. If $i > k$ the same argument can be used as for the case $i = n$ to show that $\lim_{u\to\infty} x_j(u)$ for any j such that γ_{ij} is infinite and $\lim_{u\to\infty} v_{ij}(u) > 0$. If $i \leq k$ we can use dominated convergence in the ith equation of (4.6) with $c = c_0$ to obtain

$$0 = \sum_{j=1}^{n} \gamma_{ij}(1 - \lim_{u\to\infty} x_j(u)).$$

it immediately follows that $\lim_{u\to\infty} x_j(u) = 1$ for all j such that $\gamma_{ij} \neq 0$.

Now for any i take the sequence $j_1, ..., j_r = i$ specified in the statement of the theorem. These results then successively show that $\lim_{u\to\infty} x_{j_s}(u) = 1$ for $s = 1, ..., r$, so that $\lim_{u\to\infty} x_i(u) = 1$.

We have therefore obtained a non-trivial solution $x_i(u)$, for $i = 1, .., n$, to equations (4.5) with $c = c_0$, which are each non-negative, differentiable and monotone increasing functions of u with $\lim_{u\to\infty} x_i(u) = 1$. From equations (4.5) as in Section 4.2 it is simple to show that $x_i(u) > 0$ for all u and $i = 1, ..., n$ and that each $x_i'(u)/x_i(u)$ is bounded.

□

Since there is a solution $x_i(u)$ $(i = 1, ..., n)$ to equations (4.5) at speed $c = c_0$, from Lemma 4.1 the $x_i(u)$ provide a non-trivial solution to equations (4.3) and to equations (4.6) if we take $w_i(u) = -\log x_i(u)$.

4.10 Uniqueness of waves modulo translation

Theorem 4.6 constructs a wave solution to equations (4.6) for each speed $c > c_0$, which leads to a solution of equations (4.5) and hence equations (4.3). In Theorem 4.8 when $c_0 > 0$ a wave solution has also been shown to exist to these systems of equations, subject to certain constraints, at the critical speed $c = c_0$. This section now proves that the wave solution to equations (4.6) is unique modulo translation at each positive speed $c \geq c_0$, except in an exceptional case. From Lemma 4.1, the solution constructed at each speed $c > c_0$ therefore provides the unique solution to equations (4.3) and (4.5) at that speed modulo translation. Also from that lemma, the limiting solution obtained in Theorem 4.8 provides the unique solution modulo translation to equations (4.3) and (4.5) at speed $c = c_0$.

The behaviour of $w_i(u)$ as u tends to infinity is established in Theorem 4.9. This is then used in Theorem 4.10 to prove the uniqueness results. Note that, from Theorem 4.4, if $K_c(\lambda) = 1$ has a solution in the range $(0, \Delta_v)$ then the smallest positive root $\alpha = \Delta_w < \Delta_v$. It is possible to have no solution to the characteristic equation in the range $(0, \Delta_v)$, but to have a solution when $\lambda = \Delta_v$, in which case necessarily $c = c_0$. An example is given in Section 4.3. The behaviour of $w_i(u)$ as u tends to infinity cannot then be obtained, and so uniqueness has not been proved.

Two lemmas are first proved concerning the derivatives of $|\mathbf{I} - \mathbf{V}(\lambda)|$ at $\lambda = \Delta_w$ and the number of zeros this determinant has in a strip of the complex plane centred on the line $Re(\lambda) = \Delta_w$. From these the form of the Laurent expansion for $\mathbf{W}(\lambda)$ can be determined.

LEMMA 4.5. *Let $\lambda = \Delta_w < \Delta_v$ be the smallest positive root of $K_c(\lambda) = 1$.*

1. *When $c > c_0$, then $\lambda = \Delta_w$ is a simple zero of $|\mathbf{I} - \mathbf{V}(\lambda)| = 0$ and $(d/d\lambda)|\mathbf{I} - \mathbf{V}(\lambda)| > 0$ at $\lambda = \Delta_w$.*
2. *When $c = c_0 > 0$, then $\lambda = \Delta_w$ is a zero of multiplicity $2r$ of $|\mathbf{I} - \mathbf{V}(\lambda)|$ for some positive integer r, and $(d^{2r}/d\lambda^{2r})|\mathbf{I} - \mathbf{V}(\lambda)| < 0$ at $\lambda = \Delta_w$.*

PROOF. Let the eigenvalues of $\mathbf{V}(\lambda)$ be $\mu_i(\lambda)$ for $i = 1, .., n$, where $\mu_1(\lambda) = K_c(\lambda)$. Then $|\mathbf{I} - \mathbf{V}(\lambda)| = (1 - K_c(\lambda))(1 - \mu_2(\lambda))...(1 - \mu_n(\lambda))$.

1. When $c > c_0$, then $K_c'(\Delta_w) < 0$. Also $|\mathbf{I} - \mathbf{V}(\lambda)|$ and $(1 - K_c(\lambda))$ are analytic in an open region about $\lambda = \Delta_w$ with $(1 - \mu_2(\Delta_w))...(1 - \mu_n(\Delta_w)) > 0$. Since $\lambda = \Delta_w$ is a simple root of $K_c(\lambda) = 1$, it is also a simple root of $|\mathbf{I} - \mathbf{V}(\lambda)| = 0$. The function $|\mathbf{I} - \mathbf{V}(\lambda)|/(1 - K_c(\lambda))$ is analytic in an open region about $\lambda = \Delta_w$. Hence at $\lambda = \Delta_w$,

$$\frac{d}{d\lambda}|\mathbf{I} - \mathbf{V}(\lambda)| = -K_c'(\Delta_w)(1 - \mu_2(\Delta_w))...(1 - \mu_n(\Delta_w)) > 0.$$

2. Let $K_c^j(\lambda)$ denote the jth derivative of $K_c(\lambda)$ with respect to λ. When $c = c_0 > 0$, from Lemma 4.4 part 7, there exists a positive integer r such that $K_c^j(\Delta_w) = 0$ for $j = 1, ..., 2r - 1$ and $K_c^{2r}(\Delta_w) > 0$. By a similar argument to that of the proof of part (i), it follows that $\lambda = \Delta_w$ is a zero of multiplicity $2r$ of $|\mathbf{I} - \mathbf{V}(\lambda)|$ and at $\lambda = \Delta_w$,

$$\frac{d^{2r}}{d\lambda^{2r}}|\mathbf{I} - \mathbf{V}(\lambda)| = -K_c^{2r}(\Delta_w)\prod_{j=2}^{n}(1 - \mu_j(\Delta_w)) < 0.$$

□

LEMMA 4.6.

1. *Define $\mathbf{C}(\lambda)$ to be the matrix with ij^{th} entry $|V_{ij}(\lambda)|$. For any real $x \in (0, \Delta_v)$ and any $\varepsilon_1 > 0$ there exists a $\delta_1 > 0$ and a positive A sufficiently large such that $\rho(C(\lambda)) \leq \varepsilon_1$ for all λ such that $|Re(\lambda) - x| \leq \delta_1$ and $|Im(\lambda)| \geq A$.*
2. *If $\rho(\mathbf{V}(x)) = 1$ for some $x \in (0, \Delta_v)$, then there exists a $\delta_2 > 0$ such that the only zero of $|\mathbf{I} - \mathbf{V}(\lambda)|$ in the strip $|Re(\lambda) - x| \leq \delta_2$ is at $\lambda = x$.*

PROOF.

1. It is first shown that for every $\varepsilon^* > 0$ and $x \in (0, \Delta_v)$ there exists a $\delta_{ij} > 0$ and a positive A_{ij} sufficiently large such that $|V_{ij}(x + dx + iy)| \leq \varepsilon^*$ for $|dx| \leq \delta_{ij}$ and $|y| \geq A_{ij}$.

   Consider any $x \in (0, \Delta_v)$ and any $\varepsilon^* > 0$. From the Riemann-Lebesgue lemma (Apostol [A2]), $|V_{ij}(x + iy)| \to 0$ as $|y| \to \infty$. Hence there exists an $A_{ij} > 0$ such that $|V_{ij}(x + iy)| \leq \varepsilon^*/2$ for $|y| \geq A_{ij}$.

   Now choose δ^* such that $0 < \delta^* < \min(x, \Delta_v - x)$. Then for any $B > 0$ and $|dx| \leq \delta^*$, since $1 - e^{-|w|} \leq |w|$,

$$\begin{aligned}
&|V_{ij}((x+dx+iy)-V_{ij}((x+iy)| \\
&\le \int_0^\infty e^{(x+|dx|)u}\left(1-e^{-|dx|u}\right)v_{ij}(u)du+\int_{-\infty}^0 e^{(x-|dx|)u}\left(1-e^{|dx|u}\right)v_{ij}(u)du \\
&\le \int_B^\infty e^{(x+\delta^*)u}v_{ij}(u)du+\int_{-\infty}^{-B} e^{(x-\delta^*)u}v_{ij}(u)du \\
&+\int_0^B e^{(x+\delta^*)u}\left(1-e^{-|dx|u}\right)v_{ij}(u)du+\int_{-B}^0 e^{(x-\delta^*)u}\left(1-e^{|dx|u}\right)v_{ij}(u)du \\
&\le \int_B^\infty e^{(x+\delta^*)u}v_{ij}(u)du+\int_{-\infty}^{-B} e^{(x-\delta^*)u}v_{ij}(u)du \\
&+|dx|B\left(\int_0^B e^{(x+\delta^*)u}v_{ij}(u)du+\int_{-B}^0 e^{(x-\delta^*)u}v_{ij}(u)du\right).
\end{aligned}$$

Since $V_{ij}(x+\delta^*)$ and $V_{ij}(x-\delta^*)$ are finite, we can choose B sufficiently large so that $\int_B^\infty e^{(x+\delta^*)u}v_{ij}(u)du \le \varepsilon^*/8$ and $\int_{-\infty}^{-B} e^{(x-\delta^*)u}v_{ij}(u)du \le \varepsilon^*/8$. For this B take $\delta_{ij} = \min(\delta^*, \varepsilon^*/\{4B[V_{ij}(x+\delta^*)+V_{ij}(x-\delta^*)]\})$. Then $|V_{ij}(x+dx+iy)-V_{ij}(x+iy)| \le \varepsilon^*/2$ for $y \in \Re$ and $|dx| \le \delta_{ij}$. Hence, for $|dx| \le \delta_{ij}$ and $|y| \ge A_{ij}$,

$$V_{ij}(x+dx+iy) \le V_{ij}(x+iy)+\varepsilon^*/2 \le \varepsilon^*.$$

Now consider any $\varepsilon_1 > 0$. For each i,j, take $\varepsilon^* = \varepsilon_1/n$ and find the corresponding δ_{ij} and A_{ij}. Define $A = \max_{i,j} A_{ij}$ and $\delta_1 = \min_{i,j}\delta_{ij}$. For all λ such that $|Re(\lambda)-x| \le \delta_1$ and $|Im(\lambda)| \ge A$, $\mathbf{C}(\lambda) \le \frac{\varepsilon_1}{n}\mathbf{11}'$ and hence (from Theorem A.2 part 1) $\rho(\mathbf{C}(\lambda)) \le \rho\left(\frac{\varepsilon_1}{n}\mathbf{11}'\right) = \varepsilon_1$. This completes the proof of part 1.

2. Consider a real value $x \in (0,\Delta_v)$ such that $\rho(\mathbf{V}(x)) = 1$. Choose $0 < \varepsilon_1 < 1$ and find δ_1 and A from part 1 so that $\rho(C(\lambda)) \le \varepsilon_1 < 1$ for $|Re(\lambda)-x| \le \delta_1$ and $|Im(\lambda)| \ge A$. Now from Theorem A.2 part 3, if μ is an eigenvalue of $\mathbf{V}(\lambda)$ then $|\mu| \le \rho(\mathbf{C}(\lambda))$. Hence, for $|Re(\lambda)-x| \le \delta_1$ and $|Im(\lambda)| \ge A$, all of the eigenvalues of $\mathbf{V}(\lambda)$ have modulus less than one so that $|\mathbf{I}-\mathbf{V}(\lambda)| \ne 0$. Hence the zeros of $|\mathbf{I}-\mathbf{V}(\lambda)|$ in the strip $x-\delta_1 \le Re(\lambda) \le x+\delta_1$ in fact lie in the rectangle with $x-\delta_1 \le Re(\lambda) \le x+\delta_1$ and $-A \le Im(\lambda) \le A$.

   But $|\mathbf{I}-\mathbf{V}(\lambda)|$ is analytic in this closed rectangle and hence has only a finite number of zeros in this region. Hence by reducing the width of the rectangle the only zeros of the determinant which are in the rectangle, and hence in the corresponding strip, will be on the line $Re(\lambda) = x$.

   Now $\rho(\mathbf{V}(x)) = 1$ and $|V_{ij}(x+iy)| \le V_{ij}(x)$ for $y \ne 0$, with the inequality strict when $\gamma_{ij} \ne 0$. Therefore, from Theorem A.2 part 3, the moduli of the eigenvalues of $\mathbf{V}(x+iy)$ are all less than one for $y \ne 0$. Hence the only zero of $|\mathbf{I}-\mathbf{V}(\lambda)|$ on the line $Re(\lambda) = x$ is at $\lambda = x$, and so we can choose a $0 < \delta_2 < \delta_1$ sufficiently small such that the only zero of $|\mathbf{I}-\mathbf{V}(\lambda)|$ in the strip $x-\delta_2 \le Re(\lambda) \le x+\delta_2$ is at $\lambda = x$. This completes the proof.

□

Lemma 4.6 shows that, when $\lambda = \Delta_w < \Delta_v$ is the smallest positive root of $K_c(\lambda) = 1$, there is a strip centred on the line $Re(\lambda) = \Delta_w$ in which the only zero of $|\mathbf{I} - \mathbf{V}(\lambda)|$ is at $\lambda = \Delta_w$. Hence there are no zeros in the left hand side of this strip where $Re(\lambda) < \Delta_w$ and so in this reduced strip, from equations (4.23) and Lemma 4.5,

$$\mathbf{W}(\lambda) = -\frac{1}{|\mathbf{I} - \mathbf{V}(\lambda)|} Adj(\mathbf{I} - \mathbf{V}(\lambda))\mathbf{R}(\lambda) = -\sum_{j=1}^{s} \frac{1}{(\lambda - \Delta_w)^j}\mathbf{B}_j + \mathbf{h}(\lambda),$$

for some vectors of constants $\mathbf{B}_j$ for $j = 1, ..., s$ and vector of analytic functions $\mathbf{h}(\lambda)$. Note that $s = 1$ if $c > c_0$ and s is even when $c = c_0$.

A modified Tauberian theorem of Delange [D3], which is essentially a special case of Proposition 2 from Lui [L1], is now stated. The expression for $\mathbf{W}(\lambda)$ given above can then be used in Theorem 4.9 to tie down the behaviour of a wave solution $w_i(u)$ for $i = 1, .., n$ at speed $c \geq c_0$ as $u \to \infty$.

LEMMA 4.7. *Suppose that the real-valued function w is non-negative and monotone non-increasing on $\Re_+$, and there exists a Δ_w such that $W(\lambda) = \int_0^\infty e^{\lambda x} w(x)dx$ converges for $Re(\lambda) < \Delta_w$. Also for some positive integer s and constants $B_1, ..., B_s$ and some function $h : \Re \to \Re$,*

$$\lim_{x \uparrow \Delta_w} \left[W(x+iy) + \sum_{j=1}^{s} B_j (x + iy - \Delta_w)^{-j} \right] = h(y)$$

uniformly on compact subsets of $\Re$. Then

$$\lim_{x \to \infty} w(x) e^{\Delta_w x}/x^{s-1} = (-1)^{(s-1)} B_s/(s-1)!.$$

□

THEOREM 4.9. *Let $K_c(\lambda) = 1$ have its smallest positive root at $\lambda = \Delta_w < \Delta_v$ and let $\lambda = \Delta_w$ be a zero of multiplicity s (where $s = 1$ if $c > c_0$ and $s = 2r$ for some positive integer r if $c = c_0 > 0$). If $w_i(u)$, for $i = 1, ..., n$, is a non-trivial, monotone decreasing, differentiable solution to equations (4.6) with continuous bounded derivatives and with $\lim_{u\to\infty} w_i(u) = 0$, then $w_i(u)e^{\Delta_w u}/u^{s-1} \to \{\mathbf{A}\}_i$ as $u \to \infty$, where $\mathbf{A} > \mathbf{0}$ depends on c and is such that $\mathbf{V}(\Delta_w)\mathbf{A} = \mathbf{A}$.*

PROOF. For any positive speed $c \geq c_0$, consider any solution $w_i(u)$ $(i = 1, ..., n)$ to equations (4.6) with the conditions specified. In Theorem 4.4 we showed that $(\mathbf{I} - \mathbf{V}(\lambda))\mathbf{W}(\lambda) = -\mathbf{R}(\lambda)$, for $0 < Re(\lambda) < \Delta_w$. Also, since $\Delta_w < \Delta_v$, $\mathbf{R}(\lambda)$, $\mathbf{V}(\lambda)$ and $Adj(\mathbf{I} - \mathbf{V}(\lambda))$ have entries which exist and are analytic in a strip $|Re(\lambda) - \Delta_w| \leq \varepsilon$ for ε sufficiently small and positive. From Lemma 4.6, ε may be chosen so that the only zero of $|\mathbf{I} - \mathbf{V}(\lambda)|$ in the strip is at $\lambda = \Delta_w$.

Now $Adj(\mathbf{I} - \mathbf{V}(\Delta_w)) > \mathbf{0}$, $\mathbf{R}(\Delta_w) > \mathbf{0}$ and $|\mathbf{I} - \mathbf{V}(\lambda)|$ has a zero of multiplicity s at $\lambda = \Delta_w$. Hence $1/|\mathbf{I} - \mathbf{V}(\lambda)|$ has a pole of multiplicity s at $\lambda = \Delta_w$. Therefore we can write

$$-\frac{1}{|\mathbf{I} - \mathbf{V}(\lambda)|} Adj(\mathbf{I} - \mathbf{V}(\lambda))\mathbf{R}(\lambda) = -\sum_{j=1}^{s} \frac{1}{(\lambda - \Delta_w)^j}\mathbf{B}_j + \mathbf{h}(\lambda),$$

where $\mathbf{h}(\lambda)$ has entries which are analytic in the strip $|Re(\lambda) - \Delta_w| \leq \varepsilon$. Observe also that $\mathbf{B}_s = (1/C(\Delta_w))Adj(\mathbf{I} - \mathbf{V}(\Delta_w))\mathbf{R}(\Delta_w)$, where $C(\lambda) = \dfrac{1}{s!}\dfrac{d^s|\mathbf{I} - \mathbf{V}(\lambda)|}{d\lambda^s}$.

From Lemma 4.5 $C(\Delta_w)$ is positive if $s = 1$ (i.e. for $c > c_0$) and is negative if $s = 2r$ for some positive integer r (i.e. if $c = c_0 > 0$). So in all cases, if we define $\mathbf{A} = (-1)^{(s-1)}\mathbf{B}_s/(s-1)!$, then $\mathbf{A} > \mathbf{0}$. Also, since $(\mathbf{I} - \mathbf{V}(\Delta_w))\mathbf{A} = \mathbf{0}$, necessarily $\mathbf{A}$ is the right eigenvector corresponding to the eigenvalue $K_c(\Delta_w)$, so is unique up to a multiple.

But, from equations (4.19) for $\Delta_w - \varepsilon \leq Re(\lambda) < \Delta_w$,

$$\mathbf{W}(\lambda) = -\frac{1}{|\mathbf{I} - \mathbf{V}(\lambda)|}Adj(\mathbf{I} - \mathbf{V}(\lambda))\mathbf{R}(\lambda),$$

and therefore

$$\mathbf{W}(\lambda) + \sum_{j=1}^{s} \frac{1}{(\lambda - \Delta_w)^j}\mathbf{B}_j = \mathbf{h}(\lambda).$$

Since $\mathbf{h}(\lambda)$ is a vector of analytic functions, $\lim_{x \uparrow \Delta_w} \{\mathbf{h}(x + iy)\}_j = \{\mathbf{h}(\Delta_w + iy)\}_j$ uniformly in compact subsets of $\Re$. Then Lemma 4.7 can be used to obtain the result that, for all $i = 1, .., n$,

$$\lim_{u \to \infty} \frac{w_i(u)e^{\Delta_w u}}{u^{s-1}} = \frac{(-1)^{(s-1)}}{(s-1)!}\{\mathbf{B}_s\}_i = \{\mathbf{A}\}_i.$$

□

A Tauberian theorem has been used to establish the behaviour of $w_i(u)$ for u large. An alternative, but less direct approach, based on martingales was used for the case $n = 1$, model 1, by Barbour [B3] and was discussed for the host vector epidemic in Radcliffe, Rass and Stirling [R17]. This probabilistic approach may be generalised to cover multi-type models.

The uniqueness of the wave solution, modulo translation, can now be established at each speed $c \geq c_0$. We first comment on the effect of choosing a translate of the wave solution at speed c.

Suppose that $\lim_{u \to \infty} w_i(u)e^{\Delta_w u}/u^{s-1} = \{\mathbf{A}\}_i$, where $w_i(u)$, for $i = 1, ..., n$, is a solution to equations (4.6) for fixed c. Then the translate $v_i(u) = w_i(u + t)$, for $i = 1, ..., n$, is also a solution to (4.6), and

$$\lim_{u \to \infty} \frac{v_i(u)e^{\Delta_w u}}{u^{s-1}} = \lim_{u \to \infty} \frac{w_i(u+t)e^{\Delta_w u}}{u^{s-1}} = e^{-\Delta_w t} \lim_{u \to \infty} \frac{w_i(u)e^{\Delta_w u}}{u^{s-1}} = e^{-\Delta_w t}\{\mathbf{A}\}_i.$$

Hence using a translate has the effect of scaling the eigenvector $\mathbf{A}$ used in the limiting behaviour of the solution as $u \to \infty$. This implies that a specific eigenvector may be used by an appropriate choice of translate.

Note that Theorem 4.9 does not cover the unusual situation when $K_{c_0}(\Delta_v) = 1$, since then Δ_w is not strictly less than Δ_v. Hence Theorem 4.10 also has this restriction.

THEOREM 4.10. *If $K_c(\lambda) = 1$ has its smallest positive root at $\lambda = \Delta_w < \Delta_v$, then any non-trivial, monotone decreasing differentiable solution $w_i(u)$ $(i = 1, ..., n)$ of equations (4.6), with continuous bounded derivatives and with $\lim_{u\to\infty} w_i(u) = 0$, is unique modulo translation.*

PROOF. Suppose that $w_i(u)$ and $w_i^*(u)$ $(i = 1, ..., n)$ are two solutions at speed c, where one is not a translate of the other. By choosing a suitable translate we can ensure from Theorem 4.9 that, for $i = 1, ..., n$,

$$\lim_{u\to\infty} w_i^*(u)e^{\Delta_w u}/u^{s-1} = \lim_{u\to\infty} w_i(u)e^{\Delta_w u}/u^{s-1} = \{\mathbf{A}\}_i.$$

We show that this leads to a contradiction.

Define $g(x) = 1 - e^{-x}$. Then $\lim_{u\to\infty} g(w_i(u))/g(w_i^*(u)) = 1$. Also, from Theorem 4.1, $\lim_{u\to-\infty} g(w_i(u))/g(w_i^*(u)) = 1$. Since $w_i(u)$ and $w_i^*(u)$ are positive, continuous functions of u for all $i = 1, ..., n$ and all $u \in \Re$, it follows that $g(w_i(u))/g(w_i^*(u))$ is continuous for all i and all $u \in \Re$.

As the wave solutions are not identical, without loss of generality we may assume that $w_i(u) > w_i^*(u)$ for some i and u. Hence there exists an m and a B such that

$$\max_i \sup_u \frac{g(w_i(u))}{g(w_i^*(u))} = \frac{g(w_m(B))}{g(w_m^*(B))} > 1.$$

For any i and u, from equations (4.6) we have

$$\begin{aligned} w_i(u) &= \sum_{j=1}^{n} \int_{-\infty}^{\infty} g(w_j(r))v_{ij}(r-u)dr \\ &\leq \frac{g(w_m(B))}{g(w_m^*(B))} \sum_{j=1}^{n} \int_{-\infty}^{\infty} g(w_j^*(r))v_{ij}(r-u)dr \qquad (4.29) \\ &= \frac{g(w_m(B))}{g(w_m^*(B))} w_i^*(u). \end{aligned}$$

Define $h(x) = g(x)/x$. Since $h(x)$ is a strictly decreasing function of x for $x > 0$, and $w_m(B) > w_m^*(B)$, then $h(w_m(B)) < h(w_m^*(B))$. Thus from inequality (4.29), we obtain for all i and u,

$$\frac{w_i(u)}{w_i^*(u)} \leq \frac{g(w_m(B))}{g(w_m^*(B))} = \frac{h(w_m(B))}{h(w_m^*(B))}\frac{w_m(B)}{w_m^*(B)} < \frac{w_m(B)}{w_m^*(B)}.$$

This result clearly is false if $i = m$ and $u = B$. A contradiction has therefore been obtained. Hence any non-trivial monotone decreasing solution $w_i(u)$, for $i = 1, ..., n$, of equations (4.6) corresponding to a positive speed $c \geq c_0$ is unique modulo translation.

□

CHAPTER 5

The asymptotic speed of propagation

5.1 Some preliminaries

In Section 3.1 spatial models were set up to describe an epidemic which is triggered by the introduction of infectives from outside. It was shown in Section 3.2 that this leads to a study of the following equations

$$(5.1)\qquad w_i(\mathbf{s},t)=\sum_{j=1}^{n}\int_0^t\int_{\Re^N}\left(1-e^{-w_j(\mathbf{s}-\mathbf{r},t-\tau)}\right)\gamma_{ij}(\tau)p_{ij}(\mathbf{r})d\mathbf{r}d\tau+H_i(\mathbf{s},t),$$

for $i=1,...,n$, where $w_i(\mathbf{s},t)$ is non-negative and monotone increasing in t with $w_i(\mathbf{s},0)\equiv 0$ and $w_i(\mathbf{s},t)$ is jointly continuous in $\mathbf{s}$ and t and is partially differentiable with respect to t with the partial derivative jointly continuous and uniformly bounded over all $\mathbf{s}$ and over t in a finite interval. Note that the constraint on the partial derivative of $w_i(\mathbf{s},t)$ follows immediately from equations (3.2) using the same constraint on the partial derivative of $x_i(\mathbf{s},t)$ and the positivity of $x_i(\mathbf{s},t)$. Here $\gamma_{ij}(\tau)=\sigma_j\lambda_{ij}(\tau)$, $\gamma^*_{ij}(\tau)=\sigma\lambda^*_{ij}(\tau)$ and $H_i(\mathbf{s},t)=\sum_{j=1}^{k}\int_0^t\int_{\Re^N}\int_0^\infty\varepsilon_j(\mathbf{s}-\mathbf{r},\tau)p^*_{ij}(\mathbf{r})\gamma^*_{ij}(u+\tau)d\tau d\mathbf{r}du$. The contact distributions are assumed, as in Chapter 3, to have densities which are bounded continuous functions and to be radially symmetric. As in Chapters 2 and 3, the $\gamma_{ij}(t)$ and $\gamma^*_{ij}(t)$ are restricted to be bounded with continuous, bounded derivatives with $\gamma^*_{ij}=\int_0^\infty\gamma^*_{ij}(t)dt$ finite.

Define $\varepsilon_j(\mathbf{r})=\int_0^\infty\varepsilon_j(\mathbf{r},\tau)d\tau$ and $\varepsilon_j=\int_{\Re^N}\varepsilon_j(\mathbf{r})d\mathbf{r}$. As in Chapter 3 it is assumed that the infectives from outside have some infectious influence so that $H_i(\mathbf{s},t)>0$ for some i and t and $\mathbf{s}$ in some open set $F\in\Re^N$. Extra restrictions are imposed on the $\varepsilon_j(\mathbf{r})$ and $p^*_{ij}(\mathbf{r})$ in this chapter, which ensure that the continuing effect of the infectives from outside does not dominate the ultimate behaviour of the epidemic. Specifically the $\varepsilon_i(\mathbf{r})$ are taken to be not only bounded but to have finite support and the $p^*_{ij}(\mathbf{r})$ are assumed to have Laplace transforms existing within the region of convergence of the transforms for the $p_{ij}(\mathbf{r})$.

In Theorem 3.1 it was shown that there exists a unique continuous, monotone increasing solution to equations (5.1) with each $w_i(\mathbf{s},t)\equiv 0$, and that the solution satisfies the specified conditions on the partial derivatives.

We need to define the asymptotic speed c^* of propagation of infection for a model described by equations (5.1). The intuition behind the definition was explained in Chapter 1. The epidemic is said to spread at rate c^* if both the following occur. When you run faster than c^* you leave the epidemic behind, whereas if you run slower than c^* you will eventually be surrounded by the epidemic.

The asymptotic speed of propagation may be expressed either in terms of $w_i(\mathbf{s},t)$, which is the natural form to use when proving the theorems, or in terms of $v_i(\mathbf{s},t) = 1 - x_i(\mathbf{s},t) = 1 - e^{-w_i(\mathbf{s},t)}$. Here $x_i(\mathbf{s},t)$ and $v_i(\mathbf{s},t)$ respectively represent the proportion of type i individuals at position $\mathbf{s}$ who are still susceptible at time t and who have suffered the epidemic by time t. Note that $g(w) = 1 - e^{-w}$ is a strictly increasing function of w, so that results for $w_i(\mathbf{s},t)$ are easily converted to results for $v_i(\mathbf{s},t) = g(w_i(\mathbf{s},t))$ for all $i = 1, ..., n$.

The asymptotic speed of propagation, as defined by Aronson and Weinberger [A4,A5], is said to be c^* if, for any c_1 and c_2 with $0 < c_1 < c^* < c_2$,

1. the solution $w_i(\mathbf{s},t)$, and hence $v_i(\mathbf{s},t)$, tends uniformly to zero in the region $|\mathbf{s}| \geq c_2 t$ as t tends to infinity;
2. the solution $w_i(\mathbf{s},t)$, and hence $v_i(\mathbf{s},t)$, is bounded away from zero uniformly in the region $|\mathbf{s}| \leq c_1 t$ for t sufficiently large.

In this chapter we use this definition. When considering spatial spread in biological systems, it is possible to define the rate of spread in more than one way. In Chapter 7 we use a different definition for what we term the speed of first spread. For certain of the epidemics considered here, the asymptotic speed of propagation and the speed of first spread of infection are shown to be the same. The functions $x_i(\mathbf{s},t)$ and the associated functions $v_i(\mathbf{s},t)$ and $w_i(\mathbf{s},t)$ are monotone in t. In biological systems in which the functions can oscillate the definition of the asymptotic speed of propagation used in this chapter will not be appropriate. For a discussion see van den Bosch, Metz and Diekmann [V1].

The infection matrix is $\mathbf{\Gamma} = (\gamma_{ij})$, where $\gamma_{ij} = \int_0^\infty \gamma_{ij}(\tau)d\tau$. In the non-spatial model it has been shown that a necessary condition for a major epidemic is that $\rho(\mathbf{\Gamma}) > 1$. When $\rho(\mathbf{\Gamma}) \leq 1$, it is shown in this chapter that the $w_i(\mathbf{s},t)$, and hence the $v_i(\mathbf{s},t)$, tend uniformly to zero in the region $|\mathbf{s}| \geq ct$ as t tends to infinity for all i and for any $c > 0$. This is interpreted as zero speed of propagation.

Wave solutions were considered in Chapter 4. The condition $\rho(\mathbf{\Gamma}) > 1$ is necessary for waves to exist. In this chapter we assume that the contact distributions are radially symmetric. For waves to exist, radial symmetry is not required, however it is necessary that the marginal contact distributions in the specified direction are exponentially dominated in the tail. For radially symmetric contact distributions this marginal distribution is direction independent. Hence this corresponds to the following condition for each $p_{ij}(\mathbf{r})$. For some positive real λ, $P_{ij}(\lambda) = \int_{\Re^N} p_{ij}(\mathbf{r})e^{\lambda\{\mathbf{r}\}_1}d\mathbf{r}$ is finite, where $\{\mathbf{r}\}_1$ is the first component of $\mathbf{r}$. Under these conditions the existence of a critical speed c_0 was established. Wave solutions were shown to exist and be unique modulo translation at each positive speed $c \geq c_0$; no wave solutions being possible at speeds below c_0. For radially symmetric contact distributions the critical wave speed c_0 is shown in Lemma 5.1 of Section 5.3 to be positive, so that the critical speed in this case is the minimum wave speed.

We first consider a model described by equations (5.1) with $\rho(\mathbf{\Gamma}) > 1$ with each $p_{ij}(\mathbf{r})$ exponentially dominated in the tail. It is then shown that there is a finite speed of propagation. The main purpose of this chapter is to prove that this asymptotic speed of propagation is in fact c_0, the minimum speed for which waves exist. This is a central result of the monograph. When $\rho(\mathbf{\Gamma}) \leq 1$ the speed of propagation is shown to be zero.

In Section 5.6 the behaviour when the contact distributions are radially symmetric, but not all exponentially dominated in the tail, is briefly considered. Essentially if $\rho(\mathbf{\Gamma}) > 1$, i.e. the epidemic is major, then it will then spread too fast for wave solutions to exist and for it to eventually propagate at a finite speed. This may be considered as corresponding to an infinite speed of propagation. The case $\rho(\mathbf{\Gamma}) \leq 1$ again corresponds to zero speed of propagation. In this latter case, in order to prove the results some extra conditions are imposed on the contact distributions.

In Section 5.7 we use the results of this chapter to prove the pandemic theorem for all dimensions N. This is in contrast to Chapter 3 where, for $\mathbf{\Gamma}$ finite and $\rho(\mathbf{\Gamma}) > 1$, the pandemic theorem was only established for dimensions $N = 1$ and 2.

Finally in Section 5.8 a shape theorem is established. When $\rho(\mathbf{\Gamma}) > 1$ and the contact distribution are radially symmetric and exponentially dominated in the tail, it is shown that, as t tends to infinity, the region of infection spreads like a sphere of radius $c_0 t$. Specifically it is shown that for any $\delta > 0$ and t sufficiently large, almost no infection has occurred at points outside the open sphere of radius $(c_0+\delta)t$ centred at the origin, whilst within the concentric sphere of radius $(c_0-\delta)t$, almost all the infection which will occur has occurred already. Hence new infection at time t is mainly taking place for $\mathbf{s}$ in the region $(c_0 - \delta)t < |\mathbf{s}| < (c_0 + \delta)t$.

5.2 The single population case

As in previous chapters we first give a brief intuitive outline for the case $n = 1$ before giving the proofs for general n. Results for this case were obtained independently by Diekmann [D7] and Thieme [T2]. The approach presented here is based on that of the former author. We drop the subscripts and observe that $\rho(\mathbf{\Gamma}) = \gamma$. First consider the case when the contact distribution $p(\mathbf{s})$, and hence also $p^*(\mathbf{s})$, is exponentially dominated in the tail. Since the contact distribution is radially symmetric, when $\gamma > 1$ the minimum wave speed c_0 is positive. A simple proof when γ is finite was given in Chapter 4, Section 4.3 when discussing the characteristic equation.

In order to show that, when $\gamma > 1$, c_0 is the asymptotic speed of propagation, we need to show that it satisfies parts 1 and 2 of Aronson and Weinberger's definition of the asymptotic speed of propagation. When $\gamma \leq 1$, we need only consider part 1, since in this case we will show that the speed of propagation is zero.

Part 1 of the speed of propagation.

The proof of this part is relatively straightforward. This proves that, for a specific range of values of c^*, $w(\mathbf{s}, t)$ tends uniformly to zero in the region $|\mathbf{s}| \geq c^* t$ The precise statement of the result proved is given below.

1. If $\gamma \leq 1$, then for any $c^* > 0$, $\lim_{t\to\infty} \sup\{w(\mathbf{s}, t) : |\mathbf{s}| \geq c^* t\} = 0$ or equivalently $\lim_{t\to\infty} \sup\{v(\mathbf{s}, t) : |\mathbf{s}| \geq c^* t\} = 0$. For notational convenience we define c_0 to be zero in this case. Note that there are no wave solutions at any speed when $\gamma \leq 1$, so that c_0 does not then represent the infimum of the possible positive wave speeds.
2. If $\gamma > 1$, then for any $c^* > c_0$, $\lim_{t\to\infty} \sup\{w(\mathbf{s}, t) : |\mathbf{s}| \geq c^* t\} = 0$ or equivalently $\lim_{t\to\infty} \sup\{v(\mathbf{s}, t) : |\mathbf{s}| \geq c^* t\} = 0$. Here c_0 is the minimum possible positive wave speed.

In both cases, consider any $c^* > c_0$, and take c so that $c^* > c > c_0$. It was shown in Chapter 4 when considering the wave solutions that, for such a c, there exists a $\lambda \in (0, \Delta_v)$ with $K_c(\lambda) < 1$, where $K_c(\lambda) = \sigma P(\lambda)\Lambda(c\lambda)$ and $\Lambda(\theta) = \int_0^\infty e^{-\theta\tau}\lambda(\tau)d\tau$.

Consider this c and the corresponding λ. Note that, from the constraints on $p^*(\mathbf{r})$, the Laplace transform $P^*(\lambda) = \int_{\Re^N} p^*(\mathbf{r}))e^{\lambda\{\mathbf{r}\}_1}d\mathbf{r}$ is finite.

In Chapter 3 the epidemic equation was shown to admit a unique solution. Hence if we construct the solution in a different manner the same solution is obtained. Define $y^{(0)}(\mathbf{s},t) = H(\mathbf{s},t)e^{\lambda(\{\mathbf{s}\}_1 - ct)}$. Then for each integer $m \geq 0$ and all i define

$$y^{(m+1)}(\mathbf{s},t) = \int_0^t \int_{\Re^N} \left(1 - e^{-y^{(m)}(\mathbf{s}-\mathbf{r},t-\tau)e^{-\lambda(\{\mathbf{s}-\mathbf{r}\}_1 - c(t-\tau))}}\right) e^{\lambda(\{\mathbf{s}-\mathbf{r}\}_1 - c(t-\tau))} \times p(\mathbf{r})e^{\lambda\{\mathbf{r}\}_1}\gamma(\tau)e^{-\lambda c\tau}d\mathbf{r}d\tau + H(\mathbf{s},t)e^{\lambda(\{\mathbf{s}\}_1 - ct)}.$$

It is easy to establish Cauchy convergence of the sequence of non-negative functions $y^{(m)}(\mathbf{s},t)$ (for $m \geq 0$) to a limit which we denote by $y(\mathbf{s},t)$. Dominated convergence can be used to obtain the equation satisfied by $y(\mathbf{s},t)$. Then if we define $w(\mathbf{s},t) = y(\mathbf{s},t)e^{-\lambda(\{\mathbf{s}\}_1 - ct)}$, it can be shown that $w(\mathbf{s},t)$ satisfies equation (5.1) with $n = 1$ and the subscripts dropped. From Theorem 3.1 $w(\mathbf{s},t) = y(\mathbf{s},t)e^{-\lambda(\{\mathbf{s}\}_1 - ct)}$ is therefore the unique solution to this equation.

Since $\varepsilon(\mathbf{s})$ has bounded support, there exists a positive A such that this support lies in the closed ball of radius A centred on the origin. Therefore

$$y^{(0)}(\mathbf{s},t) \leq \int_0^\infty \int_0^\infty \int_{\Re^N} \varepsilon(\mathbf{s}-\mathbf{r},\tau)e^{\lambda\{\mathbf{s}-\mathbf{r}\}_1}\gamma^*(u)p^*(\mathbf{r})e^{\lambda\{\mathbf{r}\}_1}d\mathbf{r}d\tau du \leq e^{\lambda A}\gamma^* P^*(\lambda) \sup_{\mathbf{r}\in\Re^N} \varepsilon(\mathbf{r}).$$

Denote this upper bound by D and let $y^{(m)}$ denote the sup of $y^{(m)}(\mathbf{s},t)$ over all $\mathbf{s} \in \Re^N$ and $t \geq 0$. Then $y^{(0)} \leq D$. For all $m \geq 0$, since $1 - e^{-y} \leq y$ for all non-negative y, it is easily shown that $y^{(m+1)} \leq K_c(\lambda)y^{(m)} + D$ and hence that $y^{(m)} \leq D\sum_{j=0}^m (K_c(\lambda))^j \leq D/(1 - K_c(\lambda))$. We therefore obtain the uniform upper bound $D^* = D/(1 - K_c(\lambda))$ for $y^{(m)}(\mathbf{s},t)$ for all $m \geq 0$, and hence for $y(\mathbf{s},t)$, for all $\mathbf{s} \in \Re^N$ and $t \geq 0$.

The bound D^* is invariant to rotation of the co-ordinate axes since $p(\mathbf{s})$ and $p^*(\mathbf{s})$ are radially symmetric. Therefore $w^{(m)}(\mathbf{s},t) \leq D^*e^{\lambda(ct - |\mathbf{s}|)}$. Then, for all $|\mathbf{s}| \geq c^*t$, we obtain the bound

$$w(\mathbf{s},t) \leq D^*e^{\lambda(ct - |\mathbf{s}|)} \leq D^*e^{\lambda(ct - c^*t)} = D^*e^{-\lambda(c^* - c)t}.$$

Hence

$$\sup\{w(\mathbf{s},t) : |\mathbf{s}| \geq c^*t\} \leq D^*e^{-\lambda t(c^* - c)}.$$

Since $c < c^*$ the right hand side of this inequality tends to zero as $t \to \infty$. Hence $\lim_{t\to\infty} \sup\{w(\mathbf{s},t) : |\mathbf{s}| \geq c^*t\} = 0$.

When $\gamma \leq 1$ this result holds for any $c^* > 0$, so that the asymptotic speed of propagation is zero. When $\gamma > 1$, the result holds for all $c > c_0 > 0$. This establishes the result that, when $\gamma > 1$, c_0 satisfies part 1 of the definition of the asymptotic speed of propagation, and so specifies that the asymptotic speed of propagation is at most c_0.

Part 2 of the speed of propagation.

The proof that, when $\gamma > 1$, c_0 satisfies part 2 of the definition of the asymptotic speed of propagation is considerably more complex than the proof of part 1 and involves the construction of a subsolution and the use of a comparison lemma. The results are derived in Diekmann [D7]. The construction of a subsolution adapts previous work of Aronson and Weinberger [A3,A4,A5,W1].

Let B_R denote a ball of radius R centred on the origin. To establish part 2 of the definition of the speed of propagation it is not sufficient to show that, for any R, $\inf w(\mathbf{s}, t)$ is eventually bounded away from zero for $\mathbf{s} \in B_R$. We need to show that it is in fact eventually bounded away from zero for $\mathbf{s}$ in a ball whose radius grows at any rate c where $0 < c < c_0$. The result proved is that if $\gamma > 1$, then for any $c_1 \in (0, c_0)$, there exist a positive constant b and a constant T^* sufficiently large such that $\inf\{w(\mathbf{s}, t) : |\mathbf{s}| \leq c_1 t\} \geq b$ for all $t > T^*$. This shows that the asymptotic speed of propagation is at least c_0.

The approach used is to choose c such that $c_1 < c < c_0$ and to construct a non-negative function $\psi(\mathbf{s}, t)$, for $t \geq 0$ and $\mathbf{s} \in \Re^N$, which takes its maximum value σm everywhere in the closed ball of radius $(a + ct)$ centred on the origin. Here a is positive and σ is a scaling factor. For each t, the support of $\psi(\cdot, t)$ is the closed ball of radius $(a^* + ct)$ centred on the origin, where $a^* > a$.

Both balls have radii which grow asymptotically at speed $c > c_1$. The function $\psi(\mathbf{s}, t)$ is chosen so that it can be wedged under $w(\mathbf{s}, t + t_0)$ for t_0 sufficiently large, to give $w(\mathbf{s}, t + t_0) > \psi(\mathbf{s}, t)$ for $|\mathbf{s}| \leq a^* + ct$ and $t \geq 0$.

Once this result is established, the remainder of the proof is simple. For any $t \geq t_0$ it follows that $w(\mathbf{s}, t) \geq \sigma m$ for $|\mathbf{s}| \leq a + c(t - t_0)$, and hence that $w(\mathbf{s}, t) \geq \sigma m$ for $|\mathbf{s}| \leq c_1 t$ provided that $c_1 t \leq a + c(t - t_0)$ i.e. provided that $t \geq (ct_0 - a)/(c - c_1)$. We need only take $b = \sigma m$ and $T^* = \max(t_0, (ct_0 - a)/(c - c_1))$ to obtain the required result that $\inf\{w(\mathbf{s}, t) : |\mathbf{s}| \leq c_1 t\} \geq b$ for all $t \geq T^*$.

A brief explanation is now given to explain first how it is proved that $w(\mathbf{s}, t_0 + t) > \psi(\mathbf{s}, t)$ for $|\mathbf{s}| \leq a^* + ct$ and $t \geq 0$, and then how the function $\psi(\mathbf{s}, t)$ is constructed.

Consider the first result. It is simple to prove that, for a given $T > 0$, there exists a t_0 sufficiently large such that $w(\mathbf{s}, t_0 + t) > \psi(\mathbf{s}, t)$ for $|\mathbf{s}| \leq a^* + ct$ provided t is restricted to the finite range $0 \leq t \leq T$. Now

$$w(\mathbf{s}, t) = \int_0^\infty \int_{\Re^N} \left(1 - e^{-w(\mathbf{s}-\mathbf{r}, t-\tau)}\right) p(\mathbf{r}) \gamma(\tau) d\mathbf{r} d\tau + H(\mathbf{s}, t).$$

Since there is some effect of the initial infection and $H(\mathbf{s}, t)$ is a continuous function of $\mathbf{s}$ and t, there exists a $t_1 \geq 0$ and a closed ball B_{A_1} of radius $A_1 > 0$ such that $H(\mathbf{s}, t_1) > 0$ for $\mathbf{s} \in B_{A_1}$. Note that we may choose the origin for $\mathbf{s}$ so that this open ball is centred on the origin. Since $w(\mathbf{s}, t)$ is monotone increasing in t, this implies that $w(\mathbf{s}, t) > 0$ for $\mathbf{s} \in B_{A_1}$ and $t \geq t_1$. Since $p(\mathbf{s})$ is radially symmetric, there exists a positive A_2 such that the convolution $p * p(\mathbf{s}) > 0$ for $\mathbf{s} \in B_{A_2}$. Also, as $\gamma > 0$, there exists a $t_2 > 0$ such that $\int_0^{t_2} \gamma(t)dt > 0$. Then, using two steps of the above integral equation, it immediately follows that $w(\mathbf{s}, t) > 0$ for $\mathbf{s} \in B_{A_1+A_2}$ and $t \geq t_1 + 2t_2$. This argument may be repeated so that, for any positive integer r, $w(\mathbf{s}, t) > 0$ for $\mathbf{s} \in B_{A_1+rA_2}$ and $t \geq t_1 + 2rt_2$.

Choose r such that $A_1 + rA_2 > a^* + cT$ and then for this r take $t_0 = t_1 + 2rt_2$. Then $w(\mathbf{s}, t) > 0$ for $|\mathbf{s}| \leq a^* + cT$ and $t \geq t_0$. Since $w(\mathbf{s}, t_0)$ is a continuous function of $\mathbf{s}$ it achieves its infimum in a bounded region. Therefore, for any $|\mathbf{s}| \leq a^* + ct$ and $0 \leq t \leq T$, using the monotonicity of $w(\mathbf{s}, t)$ in t,

$$w(\mathbf{s}, t_0 + t) \geq w(\mathbf{s}, t_0) \geq \inf_{|\mathbf{s}| \leq a^* + cT} w(\mathbf{s}, t_0) > 0.$$

If we take σm to be less than this infimum, then we immediately obtain the desired result that $w(\mathbf{s}, t_0 + t) > \psi(\mathbf{s}, t)$ for $|\mathbf{s}| \leq a^* + ct$ for $0 \leq t \leq T$. So the function $\psi(\cdot, t)$ has been wedged under $w(\cdot, t_0 + t)$ for the restricted range $0 \leq t \leq T$.

The restriction on t is removed by using a comparison lemma. Define an operator E_T by

$$E_T(u(\mathbf{s}, t)) = \int_0^T \int_{|\mathbf{r}| \leq R} \left(1 - e^{-u(\mathbf{s}-\mathbf{r}, t-\tau)}\right) p(\mathbf{r})\gamma(\tau) d\mathbf{r} d\tau.$$

Then clearly $w(\mathbf{s}, t) \geq E_T(w(\mathbf{s}, t))$ for $\mathbf{s} \in \Re^N$ and $t \geq 0$. For a suitable choice of T and R, the function $\psi(\mathbf{s}, t)$ is constructed so that $\psi(\mathbf{s}, t) < E_T(\psi(\mathbf{s}, t))$ for $|\mathbf{s}| \leq a^* + ct$ and $t \geq T$. A comparison lemma, Lemma 5.3, can then be used to show that the result that $w(\mathbf{s}, t_0 + t) > \psi(\mathbf{s}, t)$ for $|\mathbf{s}| \leq a^* + ct$, which was established for the range $0 \leq t \leq T$, in fact holds for the unrestricted range $t \geq 0$.

Finally we explain how $\psi(\mathbf{s}, t)$ is constructed and shown to satisfy the condition that $E_T(\psi(\mathbf{s}, t)) > \psi(\mathbf{s}, t)$ for $|\mathbf{s}| \leq a^* + ct$ and $t \geq T$ for a suitable choice of T.

Since $\gamma > 1$ and $0 < c < c_0$, from the definition of c_0 given in Chapter 4 Section 4.3, $K_c(\lambda) = \int_0^\infty \int_{\Re^N} e^{\lambda(\{\mathbf{r}\}_1 - c\tau)} p(\mathbf{r})\gamma(\tau) d\mathbf{r} d\tau > 1$ for all $\lambda \in [0, \Delta_v]$. Then R and T can be chosen sufficiently large and $h < 1$ may be chosen to be sufficiently close to 1 such that

$$L(y) = h \int_0^T \int_{|\mathbf{r}| \leq R} e^{\lambda(\{\mathbf{r}\}_1 - c\tau)} p(\mathbf{r})\gamma(\tau) d\mathbf{r} d\tau > 1$$

for all real λ. Hence $L(\lambda)$ is the Laplace transform of the non-negative function

$$k(y) = h \int_0^T \tilde{p}(y + c\tau)\gamma(\tau) d\tau,$$

where the function k has compact support and $\tilde{p}(\{\mathbf{r}\}_1)$ is obtained by truncating $p(\mathbf{r})$ outside B_R and integrating out the truncated function over all but the first entry of $\mathbf{r}$.

Define

$$q(y) = \begin{cases} e^{-\alpha y} sin(\beta y), & \text{for } 0 \leq y \leq \pi/\beta, \\ 0, & \text{elsewhere.} \end{cases}$$

Since k has compact support and has Laplace transform $L(\lambda) > 1$ for all λ, it may be shown that there exist positive β, α and δ such that $q * k(y) > q(y - \delta)$ for all $0 \leq y - \delta \leq \pi/\beta$, (see Lemma 5.4).

For this β, α and δ, the function q is now used to construct $\psi(\mathbf{s},t)$. Now $q'(y) = 0$ has a unique solution for $y \in [0, \pi/\beta]$. Let ρ be this value of y and take $m = q(\rho)$. Now take $a \geq \rho$ and $a^* = a + \pi/\beta - \rho$. Define

$$\psi(\mathbf{s},t) = \begin{cases} \sigma m, & \text{for } |\mathbf{s}| \leq a + ct, \\ \sigma q(|\mathbf{s}| + \rho - a - ct), & \text{for } a + ct \leq |\mathbf{s}| \leq a^* + ct, \\ 0, & \text{for } |\mathbf{s}| > a^* + ct. \end{cases}$$

Note that this is equivalent to defining $\psi(\mathbf{s},t) = \sigma \max_{\eta \geq \rho - a - ct} q(|\mathbf{s}| + \eta)$ for all $\mathbf{s}$ and t.

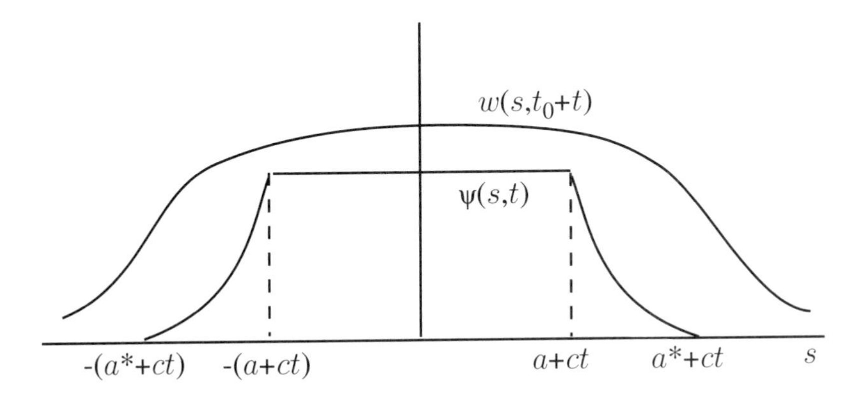

FIGURE 5.1. The functions $w(s, t_0 + t)$ and $\psi(s,t)$ for the case $N = 1$.

Then $\psi(\mathbf{s},t)$ is a continuous function of $\mathbf{s}$ and t which is monotone decreasing in $|\mathbf{s}|$ and takes its maximum value at all points of the set $|\mathbf{s}| \leq a + ct$ and $t \geq 0$. The support of $\psi(\cdot, t)$ is the closed ball B_{a^*+ct}. For a suitable choice of a, the inequality, $q * k(y) > q(y - \delta)$ for $0 \leq y - \delta \leq \pi/\beta$, may be used to show that $\psi(\mathbf{s},t)$ satisfies the inequality $E_T(\psi(\mathbf{s},t)) > \psi(\mathbf{s},t)$ for $|\mathbf{s}| \leq a^* + ct$ and $t \geq T$ provided σ is at most a specified value σ^*. A sketch of the proof is given below; a detailed proof for the general n-type model is given in Lemma 5.5.

Choose σ^* such that $1 - e^{-x} \geq hx$ for $0 \leq x \leq \sigma^* m$. Then for any $0 < \sigma \leq \sigma^*$,

$$E_T(\psi(\mathbf{s},t)) \geq h \int_0^T \int_{B_R} \psi(\mathbf{s} - \mathbf{r}, t - \tau) p(\mathbf{r}) \gamma(\tau) d\mathbf{r} d\tau.$$

The proof is split into two cases, $|\mathbf{s}| \leq a + c(t - T) - R$ and $a + c(t - T) - R < |\mathbf{s}| \leq a^* + ct$. The first case is simple to prove since $\psi(\mathbf{s},t) = \sigma m$ within this range. Also when $\mathbf{r} \in B_R$, $|\mathbf{s}| \leq a + c(t - T) - R$ and $0 \leq \tau \leq T \leq t$, then $|\mathbf{s} - \mathbf{r}| \leq |\mathbf{s}| + |\mathbf{r}| \leq a + c(t - \tau)$ and so $\psi(\mathbf{s} - \mathbf{r}, t - \tau) = \sigma m$. Hence, for all $|\mathbf{s}| \leq a + c(t - T) - R$ and $t \geq T$,

$$E_T(\psi(\mathbf{s},t)) \geq h \int_0^T \int_{B_R} \sigma m p(\mathbf{r}) \gamma(\tau) d\mathbf{r} d\tau = \sigma m L(0) > \sigma m = \psi(\mathbf{s},t).$$

The second case uses the alternative form of the definition of the function ψ. Since $\psi(\mathbf{s},t)$ is radially symmetric, we may take $\mathbf{s}$ to be in the direction of the first co-ordinate. For $\mathbf{r}=(r_i)\in\Re^N$ and $t\geq T$ with $|\mathbf{s}|$ and t in the specified ranges, it can then be shown, using a terminating Taylor expansion, that $|\mathbf{s}-\mathbf{r}|\leq|\mathbf{s}|-r_1+\delta$, provided a is chosen so that $a\geq\max(\rho,R+R^2/(2\delta))$. Defining $\eta^*=\eta-c\tau$ and $x=r_1-c\tau$ within the appropriate integral, we then obtain the following sequence of inequalities for all $a+c(t-T)-R<|\mathbf{s}|\leq a^*+ct$ and $t\geq T$.

$$\begin{aligned}
E_T(\psi(\mathbf{s},t)) &\geq h\int_0^T\int_{B^R}\psi(\mathbf{s}-\mathbf{r},t-\tau)p(\mathbf{r})\gamma(\tau)d\mathbf{r}d\tau\\
&= h\int_0^T\int_{B_R}\sigma\max_{\eta\geq\rho-a-c(t-\tau)}q(|\mathbf{s}-\mathbf{r}|+\eta)p(\mathbf{r})\gamma(\tau)d\mathbf{r}d\tau\\
&\geq h\int_0^T\int_{B^R}\sigma\max_{\eta\geq\rho-(a+c(t-\tau))}q(|\mathbf{s}|-r_1+\delta+\eta)p(\mathbf{r})\gamma(\tau)d\mathbf{r}d\tau\\
&= h\int_0^T\int_{\Re}\sigma\max_{\eta\geq\rho-(a+c(t-\tau))}q(|\mathbf{s}|-r_1+\delta+\eta)\tilde{p}(r_1)\gamma(\tau)dr_1d\tau\\
&= h\int_0^T\int_{\Re}\sigma\max_{\eta^*\geq\rho-(a+ct)}q(|\mathbf{s}|+\delta+\eta^*-x)\tilde{p}(x+c\tau)\gamma(\tau)dxd\tau\\
&\geq \max_{\eta^*\geq\rho-(a+ct)}\int_{\Re}\sigma q(|\mathbf{s}|+\eta^*+\delta-x)k(x)dx\\
&> \sigma\max_{\eta^*\geq\rho-(a+ct)}q(|\mathbf{s}|+\eta^*)=\psi(\mathbf{s},t).
\end{aligned}$$

Hence, for $a\geq\max(\rho,R+R^2/(2\delta))$ and any $0\leq\sigma\leq\sigma^*$, $E_T(\psi(\mathbf{s},t))>\psi(\mathbf{s},t)$ for all $|\mathbf{s}|\leq a^*+ct$ and $t\geq T$.

When the contact distribution is not exponentially dominated in the tail.

For $\gamma>1$ and any positive speed c, R and T may be chosen sufficiently large and $h<1$ may be chosen to be sufficiently close to 1 so that

$$h\int_0^T\int_{|\mathbf{r}|\leq R}e^{\lambda(\{\mathbf{r}\}_1-c\tau)}p(\mathbf{r})\gamma(\tau)d\mathbf{r}d\tau>1,$$

for all $\lambda\in\Re^N$. Hence in an identical fashion to the last subsection it can be shown that the speed of propagation is infinite. It is shown that, for any $c>0$ there exists a $b>0$ and T sufficiently large such that $w(\mathbf{s},t)>b$ for $|\mathbf{s}|\leq ct$ and $t\geq T$.

When $\gamma\leq 1$ the asymptotic speed of propagation can be shown to be zero. Additional conditions are required on the contact distributions. Specifically $\varepsilon(\mathbf{s},\tau)$, $p(\mathbf{s})$ and $p^*(\mathbf{s})$ are taken to be radially symmetric, monotone decreasing, continuous functions of $|\mathbf{s}|$. In addition, the contact distributions are assumed to be differentiable (except possibly at a finite number of points) with bounded derivatives. They are also taken to be convex functions of the radial distance $|\mathbf{r}|$ for $|\mathbf{r}|$ sufficiently large. This latter condition implies that each contact distribution is a convex function in the tails of the first co-ordinate $\{\mathbf{r}\}_1$ when the remaining co-ordinates are held fixed. These conditions are precisely the conditions used in Lemma 3.6 to prove that $w(\mathbf{s},t)$ and $H(\mathbf{s},t)$ are radially symmetric and monotone decreasing in $|\mathbf{s}|$. Note that these functions are also continuous and monotone increasing in t.

Define $a(\mathbf{s}) = \lim_{t\to\infty} H(\mathbf{s},t)$ and $w(\mathbf{s}) = \lim_{t\to\infty} w(\mathbf{s},t)$. Then taking limits in the equation for $w(\mathbf{s},t)$ we obtain

$$w(\mathbf{s}) = \gamma \int_{\Re^N} \left(1 - e^{-w(\mathbf{s}-\mathbf{r})}\right) p(\mathbf{r}) d\mathbf{r} + a(\mathbf{s}).$$

If we take $v(\mathbf{s}) = 1 - e^{-w(\mathbf{s})}$, this is just the final size equation (3.6) when $n = 1$, which was obtained in Theorem 3.2 part 1.

Now $a(\mathbf{s})$ and $w(\mathbf{s})$ are continuous and monotone decreasing in $|\mathbf{s}|$. Since the initial infection is confined to a bounded region, it can be shown that $a(\mathbf{s}) \to 0$ as $|\mathbf{s}| \to \infty$. If we define $w = \lim_{|\mathbf{s}|\to\infty} w(\mathbf{s})$, then w must satisfy the equation $w = \gamma\left(1 - e^{-w}\right)$. Since $\gamma \le 1$, the only solution to this equation is $w = 0$.

Consider any $c > 0$. Using the radial symmetry and monotonicity of $w(\mathbf{s},t)$ in $|\mathbf{s}|$ and t, $\sup_{|\mathbf{s}|\ge ct} w(\mathbf{s},t) \le w((ct\ \mathbf{0}')',t) \le w((ct\ \mathbf{0}')')$.
But $w((ct\ \mathbf{0}')') \downarrow w$, where $w = 0$, as $t \to \infty$. Hence $\lim_{t\to\infty} \sup_{|\mathbf{s}|\ge ct} w(\mathbf{s},t) = 0$ for any speed $c > 0$, and so the speed of propagation is zero.

The pandemic result.

The pandemic theorem of Chapter 3 shows that, when $\gamma > 1$, the epidemic spreads everywhere and there is a positive lower bound on the final size of the epidemic at all points. The proof required the contact distribution to be radially symmetric with the dimensionality $N = 1$ or $N = 2$. The results concerning the speed of propagation may be used to derive the results for all dimensions N.

Let $v(\mathbf{s}) = \lim_{t\to\infty} v(\mathbf{s},t)$ be the final size of the epidemic at position $\mathbf{s}$. This is the proportion at this position who eventually suffer the epidemic. The pandemic lemma, Lemma 3.3, shows that either $v(\mathbf{s}) \ge \eta(0) > 0$ or $v = \inf_{\mathbf{s}\in\Re^N} v(\mathbf{s}) = 0$. This lemma does not require symmetry or restrict the dimensionality of the contact distribution.

Part 2 of the results concerning the speed of propagation of the epidemic is used to show that, for any dimension N, $v > 0$ when the contact distribution is radially symmetric. If the contact distribution is exponentially dominated in the tail there is a finite positive speed of propagation c_0 otherwise the speed of propagation is infinite. In the former case we choose a positive value $c < c_0$ and in the latter case we choose any positive c. For such a c, there exists a $b > 0$ and $T > 0$ such that $w(\mathbf{s},t) > b$ for all $|\mathbf{s}| \le ct$ and all $t \ge T$. Since $v(\mathbf{s},t) = 1 - e^{-w(\mathbf{s},t)}$, it immediately follows that $v(\mathbf{s},t) > B$ for all $|\mathbf{s}| \le ct$ and all $t \ge T$, where $B = 1 - e^{-b}$.

Consider any $\mathbf{s}$ and choose $t^* \ge T$ such that $|\mathbf{s}| \le ct^*$. Now $v(\mathbf{s},t)$ is an increasing function of t, with $v(\mathbf{s}) = \lim_{t\to\infty} v(\mathbf{s},t)$. Therefore $v(\mathbf{s}) \ge v(\mathbf{s},t^*) \ge B$. Since this holds for any $\mathbf{s}$, $v = \inf_{\mathbf{s}\in\Re^N} v(\mathbf{s}) \ge B > 0$. The pandemic result, that $v(\mathbf{s}) \ge \eta(0)$ for all $\mathbf{s}$, then follows immediately from the pandemic lemma.

The asymptotic shape.

When $\gamma > 1$ and the contact distribution is radially symmetric and exponentially dominated in the tail, then there is a finite positive speed of propagation which is the minimum wave speed c_0. In this case it can be shown that the infection spreads out asymptotically like a sphere centred on the origin. For large t the sphere has radius $c_0 t$. Most of the infection taking place occurs in the region near the boundary of this sphere. Specifically it is shown that for any $\varepsilon > 0$ and $\delta > 0$ there exists a T sufficiently large such that for any $t \geq T$, the following results hold. For all $\mathbf{s}$ outside the open sphere of radius $(c_0 + \delta)t$ centred at the origin the proportion at position $\mathbf{s}$ who have already suffered the epidemic is less than ε. Also at all points in the closed sphere of radius $(c_0 - \delta)t$ centred at the origin the proportion at position $\mathbf{s}$ who have already suffered the epidemic is between $\eta(0) - \varepsilon$ and $\eta(b)$, where b depends on the initial infection and tends to zero as the amount of initial infection tends to zero. Most of the infection taking place at time t is occurring in the region between these concentric spheres.

The first result is easily derived from the results in part 1 of the asymptotic speed of propagation. Consider any $\varepsilon > 0$ and $\delta > 0$. Take $c = c_0 + \delta$ and $\varepsilon^* = -\log(1 - \varepsilon)$. Then there exists a T such that $w(\mathbf{s}, t) < \varepsilon^*$ for all $|\mathbf{s}| \geq ct$ and all $t \geq T$. Hence $v(\mathbf{s}, t) < \varepsilon$ for all $|\mathbf{s}| \geq (c_0 + \delta)t$ and all $t \geq T$.

Note that for all $\mathbf{s}$ and $t > 0$, from the monotonicity of $v(\mathbf{s}, t)$ in t and from Theorem 3.4, $v(\mathbf{s}, t) \leq v(\mathbf{s}) \leq \eta(b)$. Hence in order to prove the second result we need to strengthen the result obtained in part 2 of the speed of propagation. Define $\gamma(R, T^*) = \int_0^{T^*} \gamma(t)dt \int_{B_R} p(\mathbf{r})d\mathbf{r}$ and consider any $c > c_0$. We use an iterative procedure to show that, for any $\varepsilon^* > 0$ and R and T^* sufficiently large such that $\gamma(R, T^*) > 1$, there exists a T such that $w(\mathbf{s}, t) \geq \zeta(R, T^*) - \varepsilon^*$ for all $|\mathbf{s}| \leq ct$ and all $t \geq T$. Here $z = \zeta(R, T^*)$ is the unique positive solution to $z = \gamma(R, T^*)(1 - e^{-z})$. Note that $\eta(0)$ is just the limit as T and R tend to infinity of $1 - e^{-\zeta(R,T^*)}$.

Now consider any $\varepsilon > 0$ and $\delta > 0$. Take $c = c_0 - \delta$ and take R and T sufficiently large and ε^* sufficiently small so that $\eta(0) - \varepsilon \leq 1 - e^{-\zeta(R,T^*) - \varepsilon^*}$. Then there exists a T such that $w(\mathbf{s}, t) \geq \zeta(R, T^*) - \varepsilon^*$ for all $|\mathbf{s}| \leq (c_0 - \delta)t$ and all $t \geq T$. Therefore $v(\mathbf{s}, t) \geq \eta(0) - \varepsilon$, and hence $\eta(0) - \varepsilon \leq v(\mathbf{s}, t) \leq \eta(b)$ for all $|\mathbf{s}| \leq (c_0 - \delta)t$ and all $t \geq T$.

5.3 An upper bound on the speed of propagation

We now turn our attention to the n-type model. The results concerning the asymptotic speed of propagation appear in Radcliffe and Rass [R7] for the non-reducible case covered in this chapter. Results for the reducible case may be found in Radcliffe and Rass [R12]. We consider the case when all the contact distributions are exponentially dominated in the tail. Lemma 5.1 shows that the condition restricting the $p_{ij}(\mathbf{r})$ to be radial functions ensures that $c_0 > 0$.

In Chapter 4, $P_{ij}(\lambda)$ was taken to be the two-sided Laplace transform of the projection of $p_{ij}(\mathbf{r})$ in a specified direction. Since the contact distributions are radially symmetric this will be direction independent, so that $P_{ij}(\lambda) = \int_{\Re^N} e^{\lambda\{\mathbf{r}\}_1} p(\mathbf{r})d\mathbf{r}$. The corresponding transform for $p^*_{ij}(\mathbf{r})$ is denoted by $P^*_{ij}(\lambda)$ and we define $\Lambda_{ij}(\lambda) = \int_0^\infty e^{-\lambda\tau}\lambda_{ij}(\tau)d\tau$. Note that $K_c(\lambda) = \rho(\mathbf{V}(\lambda))$, where $\mathbf{V}(\lambda) = (V_{ij}(\lambda))$ and $V_{ij}(\lambda) = \sigma_j P_{ij}(\lambda)\Lambda_{ij}(\lambda)$.

LEMMA 5.1. *If* $\rho(\mathbf{\Gamma}) > 1$, *then* $c_0 > 0$.

PROOF. It is first shown that, for any Δ_v, there exists a $\lambda^* > 0$ and $c^* > 0$ such that $K_c(\lambda) > 1$ for $\lambda \geq \lambda^*$ and $0 < c \leq c^*$. A contradiction argument may then be used to show that there exists a $c_2 > 0$ such that $K_{c_2}(\lambda) > 1$ for all $\lambda \in [0, \Delta_v]$.

From Lemma 4.4 part 3, $K_c(\lambda)$ is a monotone decreasing function of c. Therefore $K_c(\lambda) > 1$ for all $\lambda \in [0, \Delta_v]$ and all c such that $0 < c \le c_2$. Hence this result implies that $c_0 > 0$.

When Δ_v is infinite, since $\mathbf{\Gamma}$ is non-reducible, there exists a distinct sequence $i_1, ..., i_{m-1}$, for some $1 \le m-1 \le n$, such that $\gamma_{i_j i_{j+1}} \ne 0$ for $j = 1, ..., (m-1)$, where $i_m \equiv i_1$. Hence, as in the proof of Theorem A.3 part 3, $K_c(\lambda) \ge \{\prod_{s=1}^{m-1} V_{i_s i_{s+1}}(\lambda)\}^{\frac{1}{m-1}}$. Define $\gamma(\tau)$ to be the convolution of all of the functions $\gamma_{i_j i_{j+1}}(\tau)$ for $j = 1, ..., (m-1)$. Also let $p(x)$ be a similar convolution of $\tilde{p}_{i_j i_{j+1}}(x)$, where $\tilde{p}_{i_j i_{j+1}}(\{\mathbf{r}\}_1) = \int_{\Re^{N-1}} p_{i_j i_{j+1}}(\mathbf{r}) d\{\mathbf{r}\}_2 d\{\mathbf{r}\}_3 ... d\{\mathbf{r}\}_N$. As the $p_{i_j i_{j+1}}(\mathbf{r})$ are radial functions, each $\tilde{p}_{i_j i_{j+1}}(x)$ and hence $p(x)$ is symmetric about $x = 0$. Also $\int_0^\infty \gamma(\tau) d\tau = \prod_{j=1}^{m-1} \gamma_{i_j i_{j+1}} \ne 0$. Hence there exist positive reals a, b, A, B, γ^* and p^* such that $\gamma(\tau) \ge \gamma^*$ for $\tau \in [a, b]$ and $p(x) \ge p^*$ for $x \in [A, B]$. Hence for λ real and positive,

$$\{K_c(\lambda)\}^{(m-1)} \ge p^* \gamma^* (b-a)(B-A) e^{\lambda(A-cb)}.$$

Thus for $0 < c^* < A/b$, $\lim_{\lambda \to \infty} K_{c^*}(\lambda) = \infty$. Take such a c^*. Then there exists a finite positive λ^* such that $K_{c^*}(\lambda) > 1$ for $\lambda \ge \lambda^*$. Since $K_c(\lambda)$ is a decreasing function of c, it immediately follows that $K_c(\lambda) > 1$ for $\lambda \ge \lambda^*$ and $0 < c \le c^*$.

When Δ_v is finite and $P_{ij}(\Delta_v)$ is infinite for some i, j (for which $\gamma_{ij} \ne 0$), then from Lemma 4.4 part 4 $\lim_{\lambda \uparrow \Delta_v} K_c(\lambda)$ is infinite for all $c > 0$. Hence we can take any $c^* > 0$ and choose $0 < \lambda^* < \Delta_v$ such that $K_{c^*}(\lambda) > 1$ for all $\lambda \ge \lambda^*$. If Δ_v is finite and $P_{ij}(\Delta_v)$ is finite for all i, j such that $\gamma_{ij} \ne 0$, then for each i, j $P_{ij}(\Delta_v) > P_{ij}(0) = 1$ since $p_{ij}(\mathbf{r})$ is a radial function. Hence $\lim_{c \downarrow 0} K_c(\Delta_v) \ge \rho(\mathbf{\Gamma}) > 1$. Thus we can choose $c^* > 0$ such that $K_{c^*}(\Delta_v) > 1$. Using continuity of $K_{c^*}(\lambda)$ we can then choose $0 < \lambda^* < \Delta_v$ so that $K_{c^*}(\lambda) > 1$ for all $\lambda \ge \lambda^*$. In both cases, since $K_c(\lambda)$ is monotone decreasing in c, $K_c(\lambda) > 1$ for all $\lambda \ge \lambda^*$ and $0 < c \le c^*$.

Hence, for any Δ_v, there exist a $c^* > 0$ and a $0 < \lambda^* < \Delta_v$ such that $K_c(\lambda) > 1$ for all $\lambda \ge \lambda^*$ and $0 < c \le c^*$.

Now suppose that for each $c > 0$ there exists a $\lambda \in [0, \Delta_v]$ such that $K_c(\lambda) < 1$. Then there exists a monotone decreasing sequence $\{c_n\}$ with $c_1 \le c^*$, and a sequence $\{\lambda_n\}$ such that $K_{c_n}(\lambda_n) < 1$ for $n = 1, 2, ...$, and $\lim_{n \to \infty} c_n = 0$. Since $c_n \le c^*$, necessarily $\lambda_n \le \lambda^*$. Hence there exists a subsequence $n_1, n_2 ...$, with $c_{n_j} > 0$ and $\lambda_{n_j} \to \lambda$ as $j \to \infty$, where $\lambda \in [0, \lambda^*]$. Hence $K_0(\lambda) \le 1$. But, for $\mathbf{\Gamma}$ finite, each $P_{ij}(\lambda) \sigma_j \Lambda_{ij}(0) = \gamma_{ij} P_{ij}(\lambda) \ge \gamma_{ij} P_{ij}(0) = \gamma_{ij}$. Thus $K_0(\lambda) \ge \rho(\mathbf{\Gamma}) > 1$ and we have obtained a contradiction. If $\mathbf{\Gamma}$ has an infinite entry then $\lim_{c \downarrow 0} K_c(\lambda) = \infty$ for any $\lambda \in [0, \lambda^*]$ so that a contradiction is also obtained. Therefore $c_0 > 0$.

□

Consider $K_c(\lambda)$. Since the contact distributions are radially symmetric, if $\gamma_{ij} \ne 0$, then $V'_{ij}(0) < 0$ for all $c > 0$. Therefore, from the proof of Theorem A.3 part 1,

$$K'_c(0) = \sum_i \sum_j V'_{ij}(0) \{Adj(\rho(\mathbf{\Gamma})\mathbf{I} - \mathbf{V}(0))\}_{ij} / trace(Adj(\rho(\mathbf{\Gamma})\mathbf{I} - \mathbf{V}(0))),$$

and hence, from Theorem A.2 part 8, $K'_c(0) < 0$.

If $\rho(\mathbf{\Gamma}) \le 1$, since $K_c(0) = \rho(\mathbf{\Gamma}) \le 1$ and $K'_c(0) < 0$, it immediately follows that for each $c > 0$ there exists a $\lambda^* \in (0, \Delta_v)$ such that $K_c(\lambda^*) < 1$.

When $\rho(\mathbf{\Gamma}) > 1$, since $c_0 > 0$ from Lemma 5.1, using Lemma 4.4 and the definition of c_0, there is a unique $\lambda_0 \in (0, \Delta_v]$ such that $K_{c_0}(\lambda_0) = 1$. Note that λ_0 may be equal to Δ_v if Δ_v is finite. Monotonicity of $K_c(\lambda)$ in c then implies that $K_c(\lambda_0) < 1$ for all $c > c_0$. For each $c > c_0$, continuity of $K_c(\lambda)$ in λ implies that $K_c(\lambda_0) < 1$ and hence that there exists a λ^* with $0 < \lambda^* < \lambda \le \Delta_v$ such that $K_c(\lambda^*) < 1$.

This suggests the formulation of Theorem 5.1, from which Corollaries 5.1 and 5.2, relating to the cases $\rho(\mathbf{\Gamma}) \le 1$ and $\rho(\mathbf{\Gamma}) > 1$ respectively, are immediate.

THEOREM 5.1. *For any* $c^* > 0$ *such that* $K_{c^*}(\lambda) < 1$ *for some* $\lambda \in (0, \Delta_v)$, $\lim_{t\to\infty} \sup\{w_i(\mathbf{s}, t) : |\mathbf{s}| \ge c^* t\} = 0$ *for* $i = 1, ..., n$.

PROOF. Consider any $c^* > 0$ and $\lambda \in (0, \Delta_v)$ such that $K_{c^*}(\lambda) < 1$. Since $K_c(\lambda)$ is a continuous function of c for each fixed λ, there exists a positive $c < c^*$ such that $K_c(\lambda) < 1$. Observe that if $\rho(\mathbf{\Gamma}) > 1$ then necessarily $c > c_0$.

Take such a λ and c. We now construct a solution to equations (5.1) in a different manner to Theorem 3.1 in order to obtain a uniform upper bound for each $w_i(\mathbf{s}, t)e^{\lambda(\{\mathbf{s}\}_1 - ct)}$. We already know from Theorem 3.1 that the solution is unique.

For each i, define $y_i^{(0)}(\mathbf{s}, t) = H_i(\mathbf{s}, t)e^{\lambda(\{\mathbf{s}\}_1 - ct)}$. Then, for all $i = 1, ..., n$, successively define the non-negative sequence of functions $y_i^{(m+1)}(\mathbf{s}, t)$ for $m \ge 0$ by

$$y_i^{(m+1)}(\mathbf{s}, t) = \sum_{j=1}^{n} \int_0^t \int_{\Re^N} e^{\lambda(\{\mathbf{s}-\mathbf{r}\}_1 - c(t-\tau))} \left(1 - e^{-y_j^{(m)}(\mathbf{s}-\mathbf{r}, t-\tau)e^{-\lambda(\{\mathbf{s}-\mathbf{r}\}_1 - c(t-\tau))}}\right)$$
$$\times\, p_{ij}(\mathbf{r})e^{\lambda\{\mathbf{r}\}_1}\gamma_{ij}(\tau)e^{-\lambda c\tau} d\mathbf{r} d\tau + H_i(\mathbf{s}, t)e^{\lambda(\{\mathbf{s}\}_1 - ct)}.$$

Define $y_i^{(m)} = \sup y_i^{(m)}(\mathbf{s}, t)$ and $u_i^{(m+1)} = \sup |y_i^{(m+1)}(\mathbf{s}, t) - y_i^{(m)}(\mathbf{s}, t)|$ for all i and $m \ge 0$, where each sup is taken over all $\mathbf{s} \in \Re^N$ and $t \ge 0$. Now $\varepsilon_i(\mathbf{s})$ has finite support for $i = 1, ..., n$. Hence there exists a positive constant A such that the support of each of these functions is contained in the closed ball of radius A centred on the origin. It then immediately follows that

$$y_i^{(0)}(\mathbf{s}, t) \le e^{\lambda\{\mathbf{s}\}_1} H_i(\mathbf{s}, t)$$
$$\le \sum_{j=1}^{k} \int_0^\infty \int_0^\infty \int_{\Re^N} \varepsilon_j(\mathbf{s} - \mathbf{r}, \tau)e^{\lambda\{\mathbf{s}-\mathbf{r}\}_1}\gamma_{ij}^*(u)p_{ij}^*(\mathbf{r})e^{\lambda\{\mathbf{r}\}_1} d\mathbf{r} d\tau du$$
$$= \sum_{j=1}^{k} \gamma_{ij}^* \int_{\Re^N} \varepsilon_j(\mathbf{s} - \mathbf{r})e^{\lambda\{\mathbf{s}-\mathbf{r}\}_1} p_{ij}^*(\mathbf{r})e^{\lambda\{\mathbf{r}\}_1} d\mathbf{r} \le D_i,$$

where $D_i = \sum_{j=1}^{n} P_{ij}^*(\lambda)\gamma_{ij}^* \sup_{\mathbf{s}\in\Re^N} \varepsilon_j(\mathbf{s})e^{\lambda A}$.

Since $1 - e^{-y} \le y$ for $y \ge 0$,

$$|y_i^{(1)}(\mathbf{s},t) - y_i^{(0)}(\mathbf{s},t)| \leq \sum_{j=1}^{n} \int_0^\infty \int_{\Re^N} y_j^{(0)}(\mathbf{s}-\mathbf{r},t-\tau) p_{ij}(\mathbf{r}) e^{\lambda\{\mathbf{r}\}_1} \gamma_{ij}(\tau) e^{-\lambda c\tau} d\mathbf{r} d\tau$$
$$\leq \sum_{j=1}^{n} D_j P_{ij}(\lambda)\sigma_j \Lambda_{ij}(c\lambda) = \sum_{j=1}^{n} D_j V_{ij}(\lambda).$$

Hence $u_i^{(1)} \leq \sum_{j=1}^n D_j V_{ij}(\lambda)$.

Also for $m \geq 0$, in similar manner,

$$y_i^{(m+1)}(\mathbf{s},t) \leq \sum_{j=1}^{n} \int_0^\infty \int_{\Re^N} y_j^{(m)}(\mathbf{s}-\mathbf{r},t-\tau) p_{ij}(\mathbf{r}) e^{\lambda\{\mathbf{r}\}_1} \gamma_{ij}(\tau) e^{-\lambda c\tau} d\mathbf{r} d\tau + D_i$$
$$\leq \sum_{j=1}^{n} y_j^{(m)} P_{ij}(\lambda)\sigma_j \Lambda_{ij}(c\lambda) + D_i = \sum_{j=1}^{n} y_j^{(m)} V_{ij}(\lambda) + D_i.$$

Therefore $y_i^{(m+1)} \leq \sum_{j=1}^n V_{ij}(\lambda) y_j^{(m)} + D_i$.

Finally, since $|e^{-x} - e^{-y}| \leq |x-y|$ for $x > 0$ and $y > 0$, for each $m \geq 0$ it also follows that

$$|y_i^{(m+1)}(\mathbf{s},t) - y_i^{(m)}(\mathbf{s},t)| \leq \sum_{j=1}^{n} \int_0^\infty \int_{\Re^N} |y_i^{(m)}(\mathbf{s}-\mathbf{r},t-\tau) - y_i^{(m-1)}(\mathbf{s}-\mathbf{r},t-\tau)|$$
$$\times p_{ij}(\mathbf{r}) e^{\lambda\{\mathbf{r}\}_1} \gamma_{ij}(\tau) e^{-\lambda c\tau} d\mathbf{r} d\tau$$
$$\leq \sum_{j=1}^{n} u_j^{(m)} V_{ij}(\lambda),$$

and hence $u_i^{(m+1)} \leq \sum_{j=1}^n V_{ij}(\lambda) u_j^{(m)}$.

Let $\mathbf{D} = (D_i)$, $\mathbf{y}^{(m)} = (y_i^{(m)})$ and $\mathbf{u}^{(m)} = (u_i^{(m)})$, and take $\mathbf{a}' = (a_i)' > \mathbf{0}'$ to be the left eigenvector of $\mathbf{V}(\lambda)$ corresponding to $K_c(\lambda)$.

We first establish Cauchy convergence for each $y_i^{(m)}(\mathbf{s},t)$. It has been shown that $\mathbf{u}^{(m+1)} \leq \mathbf{V}(\lambda)\mathbf{u}^{(m)}$ for all $m \geq 1$ and $\mathbf{u}^{(1)} \leq \mathbf{V}(\lambda)\mathbf{D}$. Hence for any $m \geq 0$, $\mathbf{a}'\mathbf{u}^{(m)} \leq (K_c(\lambda))^m \mathbf{a}'\mathbf{D}$. Therefore for any $i = 1, ..., n$, $m \geq 0$ and $r \geq 1$,

$$|y_i^{(m+r)}(\mathbf{s},t) - y_i^{(m)}(\mathbf{s},t)| \leq \frac{\mathbf{a}'\mathbf{D}(K_c(\lambda))^{m+1}}{a_i(1-K_c(\lambda))}.$$

Since $K_c(\lambda) < 1$, the right hand side of the inequality tends to zero as m tends to infinity. Therefore each $y_i^{(m)}(\mathbf{s},t)$ tends to a limit $y_i(\mathbf{s},t)$ as m tends to infinity. The joint continuity of these limit functions follows from the same property of the sequences $y_i^{(m)}(\mathbf{s},t)$, which is easily established as in Theorem 3.1. Also $y_i(\mathbf{s},0) \equiv 0$ since $y_i^{(m)}(\mathbf{s},0) \equiv 0$ for all $m \geq 0$. Using dominated convergence, the $y_i(\mathbf{s},t)$ satisfy the equations

$$y_i(\mathbf{s},t) = \sum_{j=1}^{n} \int_0^t \int_{\Re^N} e^{\lambda(\{\mathbf{s}-\mathbf{r}\}_1 - c(t-\tau))} \left(1 - e^{-y_j(\mathbf{s}-\mathbf{r},t-\tau)e^{-\lambda(\{\mathbf{s}-\mathbf{r}\}_1 - c(t-\tau))}}\right)$$
$$\times\, p_{ij}(\mathbf{r}) e^{\lambda\{\mathbf{r}\}_1} \gamma_{ij}(\tau) e^{-\lambda c\tau} d\mathbf{r} d\tau + H_i(\mathbf{s},t) e^{\lambda(\{\mathbf{s}\}_1 - ct)},$$

for $i = 1, ..., n$. Now define $w_i(\mathbf{s},t) = e^{-\lambda(\{\mathbf{s}\}_1 - ct)} y_i(\mathbf{s},t)$. Then $w_i(\mathbf{s},t)$, $i = 1, ..., n$, satisfy equations (5.1) and so from Theorem 3.1 is the unique solution to these equations. Note that differentiability, and hence monotonicity, of the continuous solution is easily established as in that theorem.

Bounds are now obtained for $y_i(\mathbf{s},t)$, and hence for $w_i(\mathbf{s},t)$. We have shown that $\mathbf{y}^{(0)} \le \mathbf{D}$ and $\mathbf{y}^{(m+1)} \le \mathbf{V}(\lambda)\mathbf{y}^{(m)} + \mathbf{D}$. Therefore $\mathbf{a}'\mathbf{y}^{(m)} \le \mathbf{a}'\mathbf{D} \sum_{j=0}^{m} (K_c(\lambda))^j$ for all $m \ge 0$. Hence $y_i(\mathbf{s},t) \le \mathbf{a}'\mathbf{D}/(a_i(1 - K_c(\lambda))$ for all $\mathbf{s} \in \Re^N$ and $t \ge 0$. Since the upper bound is unaffected by rotation of the co-ordinate axes, for all $\mathbf{s} \in \Re^N$, $t \ge 0$ and $i = 1, ..., n$,

$$w_i(\mathbf{s},t) \le \frac{\mathbf{a}'\mathbf{D}}{a_i(1 - K_c(\lambda))} e^{-\lambda(|\mathbf{s}| - ct)}.$$

Let $D_i^* = \mathbf{a}'\mathbf{D}/(a_i(1 - K_c(\lambda)))$. Then $\sup\{w_i(\mathbf{s},t) : |\mathbf{s}| \ge c^*t\} \le D_i^* e^{\lambda(ct - c^*t)} = D_i^* e^{-\lambda t(c^* - c)}$. The result then follows immediately since $c^* > c$.

□

The following two corollaries are an immediate consequence of Theorem 5.1.

COROLLARY 5.1. *If* $\rho(\mathbf{\Gamma}) \le 1$, *then for any* $c > 0$, $\lim_{t\to\infty} \sup\{w_i(\mathbf{s},t) : |\mathbf{s}| \ge ct\} = 0$, *and hence* $\lim_{t\to\infty} \sup\{v_i(\mathbf{s},t) : |\mathbf{s}| \ge ct\} = 0$, *for* $i = 1, ..., n$.

□

COROLLARY 5.2. *If* $\rho(\mathbf{\Gamma}) > 1$, *then for any* $c > c_0$, $\lim_{t\to\infty} \sup\{w_i(\mathbf{s},t) : |\mathbf{s}| \ge ct\} = 0$, *and hence* $\lim_{t\to\infty} \sup\{v_i(\mathbf{s},t) : |\mathbf{s}| \ge ct\} = 0$, *for* $i = 1, ..., n$.

□

Corollary 5.1 shows that if $\rho(\mathbf{\Gamma}) \le 1$, no matter how slowly you travel, the proportion of infectives ahead of you will tend to zero. This corresponds to an epidemic which is not severe. The speed of propagation may be considered to be zero.

When $\rho(\mathbf{\Gamma}) > 1$, Corollary 5.2 establishes that c_0 satisfies part 1 of the definition of the asymptotic speed of propagation.

5.4 The critical speed as a lower bound

In Theorem 5.2 we will prove that, when $\rho(\mathbf{\Gamma}) > 1$ and the contact distributions are each radially symmetric and exponentially dominated in the tail, then c_0 satisfies part 2 of the definition of the asymptotic speed of propagation of the epidemic. Specifically we show that for any $c_1 \in [0, c_0)$, there exist positive constants b_i and constants T_i sufficiently large such that $\min\{w_i(\mathbf{s},t) : |\mathbf{s}| \le c_1 t\} \ge b_i$ for all $t > T_i$ and $i = 1, ..., n$. Together with Corollary 5.2, this then shows that the critical wave speed c_0 is this speed of propagation.

Before proceeding to the proof of Theorem 5.2, it is necessary to prove certain lemmas and state a comparison principle, so that the theorem is proved at the end of this section. Additional notation is also introduced. The mth power of a matrix $\mathbf{V}(\lambda)$ is denoted by $\mathbf{V}^m(\lambda)$; the equivalent notation being used when the entries of the matrix involve several parameters.

We assume in this section that $\rho(\mathbf{\Gamma}) > 1$. Observe that if $c_1 < c < c_0$ then $K_c(\lambda) > 1$ for all $\lambda \in (0, \Delta_v)$. For any type i, we need to construct a operator E_T such that $w_i(\mathbf{s}, t) \geq E_T[w_i](\mathbf{s}, t)$ with an associated function $k(y)$ with Laplace transform which is everywhere greater than one. We can then use a subsolution and comparison lemma as for the case $n = 1$.

Using equations (5.1) we have

$$w_i(\mathbf{s}, t) \geq \sum_{j=1}^{n} \int_0^t \int_{\Re^N} \left(1 - e^{-w_j(\mathbf{s}-\mathbf{r}, t-\tau)}\right) \gamma_{ij}(\tau) p_{ij}(\mathbf{r}) d\mathbf{r} d\tau.$$

We may reuse this inequality for $w_j(\mathbf{s} - \mathbf{r}, t - \tau)$ in the integrand on the right hand side to obtain a further inequality. This may be repeated any number of times and in the final replacement of the inequality, the summation may be truncated to include the term $j = i$ only. The associated $k(y)$ then has Laplace transform $\{\mathbf{V}^m(\lambda)\}_{ii} = \sum_{j_1=1}^{n} \sum_{j_2=1}^{n} \cdots \sum_{j_{m-1}=1}^{n} \{\mathbf{V}(\lambda)\}_{ij_1} \{\mathbf{V}(\lambda)\}_{j_1 j_2} \cdots \{\mathbf{V}(\lambda)\}_{j_{m-1} i}$. For c such that $0 < c < c_0$ we show that m can be chosen such that $\{\mathbf{V}^m(\lambda)\}_{ii} > 1$ for all $\lambda \in [0, \Delta_v]$. In fact E is defined so that $k(y)$ has compact support with its Laplace transform arbitrarily close to $\{\mathbf{V}^m(\lambda)\}_{ii}$, and hence still able to be made greater than one for all λ. It is therefore appropriate to define this Laplace transform and prove a lemma concerning it, before defining the operator E_T.

We first make the following definitions. Let $\mathbf{V}(\lambda; R, T)$ be the matrix with ij^{th} element

$$\{\mathbf{V}(\lambda; R, T)\}_{ij} = \int_0^T \int_{B_R} \exp(\lambda[\{\mathbf{x}\}_1 - c\tau]) p_{ij}(\mathbf{x}) \gamma_{ij}(\tau) d\mathbf{x} d\tau.$$

Since $p_{ij}(\mathbf{x})$ is radially symmetric, $\{\mathbf{V}(\lambda; R, T)\}_{ij} \geq \{\mathbf{V}(|\lambda|; R, T)\}_{ij}$ for λ real and negative; the inequality being strict if $\gamma_{ij} \neq 0$.

Define $\{\mathbf{V}^s(\lambda)\}_{ij} = \lim_{R,T\to\infty} \{\mathbf{V}^s(\lambda; R, T)\}_{ij}$ for any positive integer s. Then $\mathbf{V}^s(\lambda)$ has the usual interpretation when $\mathbf{V}(\lambda)$ is finite; in addition it is defined when some of the elements of $\mathbf{V}(\lambda)$ are infinite. Note that we may choose R and T sufficiently large so that $\{\mathbf{V}(\lambda; R, T)\}_{ij} = 0$ if and only if $\gamma_{ij} = 0$. Hence the zero elements of $\{\mathbf{V}^s(\lambda)\}$ are in the same positions for all real finite λ.

Let $\hat{p}_{ij}(\mathbf{x})$ be $p_{ij}(\mathbf{x})$ truncated outside the region $\mathbf{x} \in B_{R/m}$ and

$$\tilde{p}_{i,j_1,\cdots,j_{m-1},i}(\{\mathbf{x}\}_1) = \int_{-\infty}^{\infty} \cdots \int_{-\infty}^{\infty} \hat{p}_{ij_1} * \cdots * \hat{p}_{j_{m-1}i}(\mathbf{x}) d\{\mathbf{x}\}_2 \cdots d\{\mathbf{x}\}_N.$$

Let $\hat{\gamma}_{ij}(t)$ be $\gamma_{ij}(t)$ truncated outside $[0, T/m]$ and let $\tilde{\gamma}_{i,j_1,\cdots,j_{m-1},i}(u)$ denote the convolution

$$\tilde{\gamma}_{i,j_1,\cdots,j_{m-1},i}(u) = \hat{\gamma}_{ij_1} * \hat{\gamma}_{j_1 j_2} * \cdots * \hat{\gamma}_{j_{m-1}i}(u).$$

Note that $\tilde{p}_{i,j_1,\cdots,j_{m-1},i}(r) = 0$ if $|r| > R$ and $\tilde{\gamma}_{i,j_1,\cdots,j_{m-1},i}(u) = 0$ if $u > T$.

For any positive c, R and T, define

$$\begin{aligned} L_c(\lambda; R, T, m) = & \int_{-\infty}^{\infty} \int_{0}^{\infty} e^{\lambda(v-cu)} \sum_{j_1=1}^{n} \sum_{j_2=1}^{n} \cdots \\ & \sum_{j_{m-1}=1}^{n} \tilde{p}_{i,j_1,\cdots,j_{m-1},i}(v) \tilde{\gamma}_{i,j_1,\cdots,j_{m-1},i}(u) du dv \\ = & \sum_{j_1=1}^{n} \sum_{j_2=1}^{n} \cdots \sum_{j_{m-1}=1}^{n} \{\mathbf{V}(\lambda; R/m, T/m)\}_{ij_1} \\ & \times \{\mathbf{V}(\lambda; R/m, T/m)\}_{j_1 j_2} \cdots \{\mathbf{V}(\lambda; R/m, T/m)\}_{j_{m-1} i} \\ = & \{\mathbf{V}^m(\lambda; R/m, T/m)\}_{ii}. \end{aligned}$$

When λ is real, then $\{\mathbf{V}(\lambda; R/m, T/m)\}_{ij} \geq \{\mathbf{V}(|\lambda|; R/m, T/m)\}_{ij}$ for all i, j. Hence $L_c(\lambda; R, T, m) \geq L_c(|\lambda|; R, T, m)$ for all real λ.

LEMMA 5.2. *For any positive $c < c_0$, there exists a positive integer m, a positive real $h < 1$ and positive reals R_0 and T_0 sufficiently large such that $hL_c(\lambda; R, T, m) > 1$ for $\lambda \in \Re$ and $R \geq R_0$ and $T \geq T_0$.*

PROOF. Since $L_c(\lambda; R, T, m) \geq L_c(|\lambda|; R, T, m)$ for all real λ, the result that $hL_c(\lambda; R, T, m) > 1$ need only be established for $\lambda \geq 0$.

We first show that, for any suffix i there exists a positive integer m such that $\{\mathbf{V}^m(\lambda)\}_{ii} > 1$ for all $\lambda \geq 0$; with $\{\mathbf{V}^m(\lambda)\}_{ii}$ infinite for $\lambda > \Delta_v$ when Δ_v is finite.

The result in Theorem A.2 part 11 is used. Let $\mathbf{A}$ be a finite non-reducible, non-negative square matrix, which is not the 1×1 zero matrix. There exist positive integers l and s so that, by choosing a suitable permutation to apply to both rows and columns,

$$\mathbf{A}^{ls} = \begin{pmatrix} \mathbf{B}_1 & \mathbf{0} & \dots & \mathbf{0} \\ \mathbf{0} & \mathbf{B}_2 & \mathbf{0} \dots & \\ \vdots & \vdots & \ddots & \vdots \\ \mathbf{0} & \mathbf{0} & \dots & \mathbf{B}_l \end{pmatrix},$$

where $\mathbf{B}_j$ is a positive square matrix with $\rho(\mathbf{B}_j) = (\rho(\mathbf{A}))^{ls}$ for $j = 1, ..., l$. The positions of the zero entries of $\mathbf{A}$ determine the appropriate permutation, the integer l, the smallest possible s and the sizes of $\mathbf{B}_1, ..., \mathbf{B}_l$. Take $\mathbf{A}$ with $\{\mathbf{A}\}_{ij} = 0$ if and only if $\gamma_{ij} = 0$. Find l and s. Then for R and T sufficiently large so that $\{\mathbf{V}(\lambda; R, T)\}_{ij} = 0$ if and only if $\gamma_{ij} = 0$, by relabelling the populations $1, ..., n$,

$$\mathbf{V}^{ls}(\lambda; R, T) = \begin{pmatrix} \mathbf{B}_1(\lambda; R, T) & \mathbf{0} & \dots & \mathbf{0} \\ \mathbf{0} & \mathbf{B}_2(\lambda; R, T) & \mathbf{0} \dots & \\ \vdots & \vdots & \ddots & \vdots \\ \mathbf{0} & \mathbf{0} & \dots & \mathbf{B}_l(\lambda; R, T) \end{pmatrix},$$

where $\mathbf{B}_j(\lambda; R, T) > 0$ and $\rho(\mathbf{B}_j(\lambda; R, T)) = (\rho(\mathbf{V}(\lambda; R, T)))^{ls}$.

Now if $\mathbf{V}(\lambda^*)$ has an infinite element, then $\rho(\mathbf{V}^{ls}(\lambda^*;R,T)) \to \infty$ as R and T tend to infinity. Hence, for all $j = 1,...,l$, $\rho(\mathbf{B}_j(\lambda^*;R,T)) \to \infty$ as R and T both tend to infinity. Therefore, as R and T tend to infinity, each positive matrix $\mathbf{B}_j(\lambda^*;R,T)$ must have an element which tends to infinity and so any power $r \geq 3$ of this matrix has all its entries tending to infinity. Hence if $\mathbf{B}_j^r(\lambda) = \lim_{R,T\to\infty} \mathbf{B}_j^r(\lambda;R,T)$, then

$$\mathbf{V}^{rls}(\lambda) = \begin{pmatrix} \mathbf{B}_1^r(\lambda) & \mathbf{0} & \dots & \mathbf{0} \\ \mathbf{0} & \mathbf{B}_2^r(\lambda) & \mathbf{0}\dots & \\ \vdots & \vdots & \ddots & \vdots \\ \mathbf{0} & \mathbf{0} & \dots & \mathbf{B}_l^r(\lambda) \end{pmatrix},$$

where $\mathbf{B}_j^r(\lambda)$ has all infinite elements for $r \geq 3, \lambda = \lambda^*$ and $j = 1,...,l$.

Note that if Δ_v is finite this implies that $\{\mathbf{V}^{rls}(\lambda)\}_{ii} = \infty$ for $r \geq 3$ and $\lambda > \Delta_v$. If Δ_v is finite with $K_c(\Delta_v)$ infinite the result also holds at $\lambda = \Delta_v$. When $K_c(0) = \rho(\mathbf{\Gamma}) = \infty$ the result is valid at $\lambda = 0$. When Δ_v is finite take $m^* = 3ls$.

If Δ_v is infinite, since $K_c(\lambda) > 1$ for all λ, necessarily $\lim_{\lambda\to\infty} K_c(\lambda) = \infty$ and so $\lim_{\lambda\to\infty} \rho(\mathbf{B}_j(\lambda)) = \infty$ for all $j = 1,...,l$. Hence for each j there exists a distinct sequence $i_1,...,i_t$ such that, if we define $i_{t+1} = i_1$, then $\lim_{\lambda\to\infty} \prod_{k=1}^t \{\mathbf{B}_j(\lambda)\}_{i_k i_{k+1}} = \infty$, (Theorem A.3 part 4). Then $\lim_{\lambda\to\infty}\{\mathbf{B}_j^t(\lambda)\}_{i_1 i_1} = \infty$, so that $\{\mathbf{B}_j^t(\lambda)\}_{i_1 i_1}$ is the Laplace transform of a function whose support contains an open interval of the positive real line. For any i, find j and k so that type i corresponds to the kth entry of $\mathbf{B}_j(\lambda)$. Now, for any $w \geq 2$,

$$\{\mathbf{B}_j^{wt}(\lambda)\}_{kk} \geq \left(\{\mathbf{B}_j^t(\lambda)\}_{i_1 i_1}\right)^{w-2} \{\mathbf{B}_j^t(\lambda)\}_{k i_1} \{\mathbf{B}_j^t(\lambda)\}_{i_1 k}.$$

The right hand side of this inequality is the Laplace transform of a function $c(x)$ which is a convolution of continuous functions. For w sufficiently large, the function $c(x)$ has support which contains an open interval of the positive real line. Choose such a $w \geq 3$, so that $\lim_{\lambda\to\infty}\{\mathbf{B}_j^{tw}(\lambda)\}_{kk} = \infty$. Take $m^* = lstw$. Then $\lim_{\lambda\to\infty} \{\mathbf{V}^{m^*}(\lambda)\}_{ii} = \infty$.

When $\mathbf{V}(0)$ is finite, take $\lambda_0 = 0$. If $\mathbf{V}(0)$ has an infinite element, since $m^* = rls$ with $r \geq 3$, $\{\mathbf{V}^{m^*}(0)\}_{ii} = \infty$ for any $i = 1,...,n$. Hence we may find a $\lambda_0 > 0$ such that $\{\mathbf{V}^{m^*}(\lambda)\}_{ii} > 1$ for $0 \leq \lambda \leq \lambda_0$. When Δ_v is finite with $\mathbf{V}(\Delta_v)$ finite, take $\lambda_0^* = \Delta_v$. In all other cases $\lim_{\lambda\to\Delta_v} \{\mathbf{V}^{m^*}(\lambda)\}_{ii} = \infty$, and we may find a $\lambda_0^* < \Delta_v$ such that $\{\mathbf{V}^{m^*}(\lambda)\}_{ii} > 1$ for $\lambda \geq \lambda_0^*$.

Thus in all cases $\{\mathbf{V}^{m^*}(\lambda)\}_{ii} > 1$, and hence $\{\mathbf{V}^{m^*q}(\lambda)\}_{ii} > 1$ for all positive integers q, for $0 \leq \lambda \leq \lambda_0$ and $\lambda \geq \lambda_0^*$. Note that $\{\mathbf{V}^{m^*}(\lambda)\}_{ii}$ is finite for $\lambda \in [\lambda_0, \lambda_0^*]$.

It is now sufficient to show that there exists a positive integer q such that $\{\mathbf{V}^{m^*q}(\lambda)\}_{ii} > 1$ for $\lambda \in [\lambda_0, \lambda_0^*]$. For $\lambda \in [\lambda_0, \lambda_0^*]$

$$\mathbf{V}^{m^*q}(\lambda) = \begin{pmatrix} \mathbf{B}_1^*(\lambda) & \mathbf{0} & \dots & \mathbf{0} \\ \mathbf{0} & \mathbf{B}_2^*(\lambda) & \mathbf{0}\dots & \\ \vdots & \vdots & \ddots & \vdots \\ \mathbf{0} & \mathbf{0} & \dots & \mathbf{B}_l^*(\lambda) \end{pmatrix},$$

where $\mathbf{B}_j^*(\lambda) = \mathbf{B}_j^3(\lambda)$ or $\mathbf{B}_j^{tw}(\lambda)$. Note that $\rho(\mathbf{B}_j^*(\lambda)) > 1$ for $\lambda \geq 0$.

Assume that there does not exist a q such that $(\mathbf{B}_j^*(\lambda))^q > 1$ for $\lambda \in [\lambda_0, \lambda_0^*]$. Therefore there exists a sequence $\{\lambda_r,\ r = 1, 2, ...\}$ with $\lambda_r \in [\lambda_0, \lambda_0^*]$ such that $\{(\mathbf{B}_j^*(\lambda))^r\}_{ii} \leq 1$. Hence there exists a subsequence $\{\lambda_{q_r},\ r = 1, 2, ...\}$ with $\{(\mathbf{B}_j^*(\lambda_{q_r}))^{q_r}\}_{ii} \leq 1$ such that $\lambda_{q_r} \to \bar{\lambda}$ as $r \to \infty$.

Now $\dfrac{1}{\rho((\mathbf{B}_j^*(\bar{\lambda}))^q)}(\mathbf{B}_j^*(\bar{\lambda}))^q$ tends to $E(\bar{\lambda})$ as $q \to \infty$, where $E(\bar{\lambda})$ is the idempotent corresponding to $\rho(\mathbf{B}_j^*(\bar{\lambda}))$ in the spectral expansion of $\mathbf{B}_j^*(\bar{\lambda})$. As the elements of $E(\bar{\lambda})$ are positive, we can find k and Q so that $(\mathbf{B}_j^*(\bar{\lambda}))^q \geq \rho((\mathbf{B}_j^*(\bar{\lambda}))^q)k\mathbf{1}\mathbf{1}'$ if $q \geq Q$.

Since the elements of $\mathbf{B}_j^*(\lambda)$ are continuous functions, and in addition $\mathbf{B}_j^*(\bar{\lambda}) > \mathbf{0}$ with $\rho(\mathbf{B}_j^*(\bar{\lambda})) > 1$, there exists a $\delta > 0$ such that

$$\mathbf{B}_j^*(\lambda) \geq \left(\frac{1}{\frac{1 + \rho(\mathbf{B}_j^*(\bar{\lambda}))}{2}} \right) \mathbf{B}_j^*(\bar{\lambda})$$

if $|\lambda - \bar{\lambda}| \leq \delta$. Thus

$$(\mathbf{B}_j^*(\lambda))^q \geq \left(\frac{\rho(\mathbf{B}_j^*(\bar{\lambda}))}{\frac{1 + \rho(\mathbf{B}_j^*(\bar{\lambda}))}{2}} \right)^q k\mathbf{1}\mathbf{1}'$$

if $|\lambda - \bar{\lambda}| \leq \delta$ and $q \geq Q$.

Choose Q so that the right hand side $\geq \mathbf{1}\mathbf{1}'$. Thus $\{(\mathbf{B}_j^*(\lambda))^q\}_{ii} \geq 1$ if $|\lambda - \bar{\lambda}| \leq \delta$ and $q \geq Q$. But for r sufficiently large both $|\lambda_{q_r} - \bar{\lambda}| \leq \delta$ and $q_r \geq Q$ and also $\{(\mathbf{B}_j^*(\lambda_r))^{q_r}\}_{ii} > 1$. This contradicts $\{(\mathbf{B}_j^*(\lambda_r))^{q_r}\}_{ii} \leq 1$. Hence by contradiction there exists a q such that $\{(\mathbf{B}_j^*(\lambda))^q\}_{ii} > 1$. Therefore for each i there exists a q such that $\{\mathbf{V}^{m^*q}(\lambda)\}_{ii} > 1$ for $\lambda \in [\lambda_0, \lambda_0^*]$, and hence for all $\lambda \geq 0$. Take $m = qm^*$. Note that $\{\mathbf{V}^m(\lambda)\}_{ii} = \infty$ for $\lambda > \Delta_v$ if Δ_v is finite.

Now $L_c(\lambda; R, T, m)$ is the Laplace transform of a non-negative function. In addition it is an increasing function of R and T, which tends to a limit $\{\mathbf{V}^m(\lambda)\}_{ii} > 1$ as R and T tend to infinity. We can clearly choose R and T sufficiently large so that $L_c(\lambda; R, T, m) \to \infty$ as $\lambda \to \infty$. Take R_1, T_1 and λ^* so that $L_c(\lambda; R, T, m) > 1$ for $\lambda \geq \lambda^*$, $R \geq R_1$ and $T \geq T_1$. Suppose no R_0 and T_0 exist such that $L_c(\lambda; R_0, T_0, m) > 1$ for $\lambda \in [0, \infty)$. Then there exist sequences $\{R_j\}$, $\{T_j\}$ and $\{\lambda_j\}$ such that $L_c(\lambda_j; R_j, T_j, m) \leq 1$ with $\lim_{j\to\infty} R_j = \infty$, $\lim_{j\to\infty} T_j = \infty$ and necessarily $\lambda_j \in [0, \lambda^*)$. Hence there exists a convergent subsequence. Then there is a $\bar{\lambda} \in [0, \lambda^*)$ with $\lim_{R\to\infty, T\to\infty} L_c(\bar{\lambda}; R, T, m) \leq 1$, which gives a contradiction. Thus there exists an m, R_0 and T_0 such that $L_c(\lambda; R_0, T_0, m) > 1$ for $\lambda \in [0, \infty)$, and hence for all real λ since $L_c(\lambda; R, T, m) \geq L_c(|\lambda|; R, T, m)$ for $\lambda \in \Re$.

Since also $\lim_{\lambda\to\infty} L_c(\lambda; R_0, T_0, m) = \infty$, the infimum of $L_c(\lambda; R_0, T_0, m)$ for $\lambda \in \Re$ is attained for some $\lambda \geq 0$. Therefore $\inf_{\lambda\in\Re} L_c(\lambda; R_0, T_0, m) > 1$. Take h such that$\{\inf_{\lambda\in\Re} L_c(\lambda; R_0, T_0, m)\}^{-1} \leq h < 1$. The lemma then follows immediately.

□

For a specific i, we define an operator E_T, operating on $\psi(\mathbf{s},t)$. Let $g(x) = 1 - e^{-x}$. Define

$$\left\{\mathbf{E}_T^{(0)}[\psi](\mathbf{s},t)\right\}_j = \int_0^t \int_{\Re^N} \hat{\gamma}_{ji}(\tau)\hat{p}_{ji}(\mathbf{x})g(\psi(\mathbf{s}-\mathbf{x},t-\tau))d\mathbf{x}d\tau.$$

Then successively for $1 \le k \le (m-2)$ we define

$$\left\{\mathbf{E}_T^{(k)}[\psi](\mathbf{s},t)\right\}_j = \sum_{l=1}^{n} \int_0^t \int_{\Re^N} \hat{\gamma}_{jl}(\tau)\hat{p}_{jl}(\mathbf{x})g\left(\left\{\mathbf{E}_T^{(k-1)}[\psi](\mathbf{s}-\mathbf{x},t-\tau))\right\}_l\right) d\mathbf{x}d\tau.$$

Finally we define

$$E_T[\psi](\mathbf{s},t) = \sum_{j=1}^{n} \int_0^t \int_{\Re^N} \hat{\gamma}_{ij}(\tau)\hat{p}_{ij}(\mathbf{x})g\left(\left\{\mathbf{E}_T^{(m-2)}[\psi](\mathbf{s}-\mathbf{x},t-\tau))\right\}_j\right) d\mathbf{x}d\tau.$$

E_T maps a function defined on $\Re^N \times \Re_+$ onto a function defined on $\Re^N \times [T,\infty)$. Note that $w_i(\mathbf{x},t) \ge E_T[w_i](\mathbf{x},t)$ for all $t \ge T$.

Having defined the operator E_T we are now in a position to state and prove a comparison principle, which is essentially Lemma 1 of Diekmann [D7]. Consider any type i. Choose an m, R_0, T_0 and h as in Lemma 5.2. Take R and T so that $R \ge R_0$ and $T \ge T_0$. Then $hL_c(\lambda; R, T, m) > 1$ for all real λ. Observe that m, R_0, T_0 and h and E_T depend on i. Also define $\phi \succ \zeta$ if ϕ and ζ are continuous functions defined in $\Re^N$ with $\phi(\mathbf{x}) \ge \zeta(\mathbf{x})$, the inequality being strict for $\mathbf{x} \in \text{supp}(\zeta)$, where $\text{supp}(\zeta)$ denotes the support of the function ζ. The comparison principle is now stated for $w_i(\mathbf{s},t)$.

LEMMA 5.3 (COMPARISON PRINCIPLE). *Suppose that $E_T[\psi](\cdot,t) \succ \psi(\cdot,t)$ for $t \ge T$, where $\psi : \Re^N \times \Re_+ \to \Re$ is a non-negative continuous function such that*

1. *for any $t_1 > 0$ there exists an $S = S(t_1) < \infty$ such that for any $t \in [0,t_1]$, $supp(\psi(\cdot,t)) \subset B_S$;*
2. *if $\{(\mathbf{s}_n,t_n)\}_{n=1}^{\infty} \subset \Re^N \times R_+$ is a sequence for which $\mathbf{s}_n \in supp(\psi(\cdot,t_n))$ and $lim_{n\to\infty}(\mathbf{s}_n,t_n) = (\mathbf{s},t)$, then necessarily $\mathbf{s} \in supp(\psi(\cdot,t))$.*

If there exists a $t_0 \ge 0$ such that $w_i(\cdot,t_0+t) \succ \psi(\cdot,t)$ for all $0 \le t \le T$, then $w_i(\cdot,t_0+t) \succ \psi(\cdot,t)$ for all $t \ge 0$.

PROOF. Let $\bar{t} = \sup\{t \ge T | w(\cdot,t_0+t) \succ \psi(\cdot,t)\}$ and suppose $\bar{t} < \infty$. Then there exists a sequence $\{(\mathbf{s}_n,t_n)\}_{n=1}^{\infty} \subset \Re^N \times R_+$ such that: (i) $\mathbf{s}_n \in \text{supp}(\psi(\cdot,t_n))$; (ii) $w_i(\mathbf{s}_n,t_0+t_n) \le \psi(\mathbf{s}_n,t_n)$ and (iii) $t_n \downarrow \bar{t}$ as $n \to \infty$. It follows from 1 that the sequence $\{(\mathbf{s}_n\}_{n=1}^{\infty}$ is contained in a compact subset of $\Re^N$, and hence contains a convergent subsequence. Hence by 2 and (ii), there exists $\bar{\mathbf{s}} \in \text{supp}(\psi(\cdot,\bar{t}))$ such that $w_i(\bar{\mathbf{s}},t_0+\bar{t}) \le \psi(\bar{\mathbf{s}},\bar{t})$.

Since $w_i(\mathbf{x},t_0+t) \ge \psi(\mathbf{x},t)$ for all $\mathbf{x}$ and all $0 \le t < \bar{t}$, the definition of E_T implies that $E_T[w_i](\mathbf{s},t_0+\bar{t}) \ge E_T[\psi](\mathbf{s},\bar{t})$. Therefore

$w_i(\bar{\mathbf{s}},t_0+\bar{t}) \ge E_T[w_i](\bar{\mathbf{s}},t_0+\bar{t}) \ge E_T[\psi](\bar{\mathbf{s}},\bar{t}) > \psi(\bar{\mathbf{s}},\bar{t})$.

Thus we have obtained a contradiction. So the assumption that $\bar{t} < \infty$ is not true and the lemma is proved.

□

Consider $L(\lambda) = hL_c(\lambda; R, T, m) = \int_{\Re} e^{\lambda y} k(y)dy$, where

$$k(y) = h \sum_{j_1=1}^{n} \sum_{j_2=1}^{n} \cdots \sum_{j_{m-1}=1}^{n} \int_0^\infty \tilde{p}_{i,j_1,\cdots,j_{m-1},i}(y + c\tau)\tilde{\gamma}_{i,j_1,\cdots,j_{m-1},i}(\tau)d\tau.$$

Note that $k(y)$ has compact support and $L(\lambda) > 1$ for all real λ. We now define a subsolution $\psi(\mathbf{s}, t)$. The subsolution ψ is chosen to satisfy the condition $E_T[\psi](\cdot, t) \succ \psi(\cdot, t)$ for $t \geq T$.

Define

$$q(y; \alpha, \beta) = \begin{cases} e^{-\alpha y} \sin(\beta y), & \text{for } 0 \leq y \leq \pi/\beta, \\ 0, & \text{for } y \in \Re \setminus [0, \pi/\beta]. \end{cases}$$

The choice of α and β used to define the subsolution $\psi(\mathbf{s}, t)$ is determined by the following lemma, which is proved in Diekmann [D7].

LEMMA 5.4. *Let $k \in L_1(\Re)$ be a non-negative function with compact support such that $L(\lambda) = \int_{-\infty}^{\infty} e^{\lambda y} k(y)dy > 1$ for all $\lambda \in \Re$. Then there exists a positive number β_0, a continuous function $\tilde{\alpha} = \tilde{\alpha}(\beta)$ and a positive function $\Delta(\beta)$ defined on $[0, \beta_0]$ such that, for any $\beta \in [0, \beta_0]$ and $\delta \in [0, \Delta(\beta)]$,*

$$\phi * k \succ \phi_\delta,$$

where $\phi(y) = q(y; \tilde{\alpha}(\beta), \beta)$ and $\phi_\delta(y) = \phi(y - \delta)$.

PROOF. Since $k(y)$ is non-negative, $L(\lambda)$ is a convex function such that $L(\lambda) \to \infty$ as $|\lambda| \to \infty$. Thus $L(\lambda)$ achieves its infimum. Suppose this occurs at $\lambda = \mu$. Then

$$L'(\mu) = \int_\infty^\infty y e^{\mu y} k(y)dy = 0.$$

Define $H(\alpha, \beta)$ by

$$H(\alpha, \beta) = \begin{cases} \dfrac{1}{\beta} \displaystyle\int_{-\infty}^{\infty} e^{\alpha y} \sin(\beta y) k(y)dy, & \text{for } \beta \neq 0, \\ \displaystyle\int_{-\infty}^{\infty} y e^{\alpha y} k(y)dy, & \text{for } \beta = 0. \end{cases}$$

(Note that $H(\alpha, 0) = \lim_{\beta \to 0} H(\alpha, \beta)$).

Now μ is chosen so that $H(\mu, 0) = \int_{-\infty}^{\infty} y e^{\mu y} k(y)dy = 0$. Also, for all real α, $\dfrac{\partial H(\alpha, 0)}{\partial \alpha} = \displaystyle\int_{-\infty}^{\infty} y^2 e^{\alpha y} k(y)dy > 0$. It then follows from the implicit function theorem that there exists $\beta_1 > 0$ and a continuous function $\tilde{\alpha}(\beta)$ with $\tilde{\alpha}(0) = \mu$ such that $H(\tilde{\alpha}(\beta), \beta) = 0$ for $0 \leq \beta \leq \beta_1$. Thus for $0 < \beta \leq \beta_1$

$$\int_{-\infty}^{\infty} e^{\tilde{\alpha}(\beta) y} \sin(\beta y) k(y)dy = 0.$$

Now $\int_{-\infty}^{\infty} e^{\alpha y} \cos(\beta y) k(y)dy > 1$ when $\alpha = \mu$ and $\beta = 0$. Therefore there exists β_2 such that

$$\int_{-\infty}^{\infty} e^{\alpha y}\cos(\beta y)k(y)dy > 1$$

for $\alpha = \tilde{\alpha}(\beta)$ and $0 \leq \beta \leq \beta_2$.

We now show that $\phi * k \succ \phi$ if β is sufficiently small and positive. First consider the case when y does not lie in the support of ϕ, so that $y \in \Re\backslash[0, \pi/\beta]$ for $\beta > 0$. Then $\phi(y) = 0$ and $\phi * k(y) = \int_0^{\pi/\beta} e^{-\tilde{\alpha}(\beta)\eta}\sin(\beta\eta)k(y-\eta)d\eta \geq 0$, since $k(u)$ is non-negative for all u. So in this case $\phi * k(y) \geq \phi(y)$ as required.

Next consider the case when $y = 0$ or $y = \pi/\beta$. For each of these values of y, $\phi(y) = 0$. Let $B > 0$ be such that $\text{supp}(k) \in [-B, B]$. Provided β is chosen so that $0 < \beta < \pi/B$, then $\int_0^{\pi/\beta} k(\eta)d\eta = \int_0^B k(\eta)d\eta > 0$ and $\int_0^{\pi/\beta} k(-\eta)d\eta = \int_0^B k(-\eta)d\eta > 0$. Therefore

$$\phi * k(0) = \int_0^{\pi/\beta} e^{-\tilde{\alpha}(\beta)\eta}\sin(\beta\eta)k(-\eta)d\eta > 0 = \phi(0)$$

and

$$\phi * k(\pi/\beta) = \int_0^{\pi/\beta} e^{-\tilde{\alpha}(\beta)(\pi/\beta-\eta)}\sin(\beta\eta)k(\eta)d\eta > 0 = \phi(\pi/\beta).$$

Hence $\phi * k(y) > \phi(y)$ for $y = 0$ and $y = \pi/\beta$ for any $0 < \beta < \pi/B$.

Finally consider the case when $y \in (0, \pi/\beta)$ for $\beta > 0$. The inequality

$$\phi * k(y) \geq \int_{-\infty}^{\infty} e^{-\tilde{\alpha}(\beta)\eta}\sin(\beta\eta)k(y-\eta)d\eta$$

is true if we can choose β so that for $\eta \in \Re\backslash[0, \pi/\beta]$ either $\sin(\beta\eta) \leq 0$ or $k(y-\eta) = 0$. Now $\text{supp}(k) \in [-B, B]$. If $y \in [0, \pi/\beta]$ and $|y - \eta| \leq B$, then $\eta \in [-B, B + \pi/\beta] \in [-\pi/\beta, 2\pi/\beta]$ provided $\beta \leq \pi/B$. Thus the above inequality is true provided $\beta \leq \pi/B$. Therefore, for $0 < \beta < \min(\beta_1, \beta_2, \pi/B)$,

$$\begin{aligned}
\phi * k(y) &= \int_0^{\pi/\beta} e^{-\tilde{\alpha}(\beta)\eta}\sin(\beta\eta)k(y-\eta)d\eta \\
&\geq \int_{-\infty}^{\infty} e^{-\tilde{\alpha}(\beta)\eta}\sin(\beta\eta)k(y-\eta)d\eta \\
&= \int_{-\infty}^{\infty} e^{-\tilde{\alpha}(\beta)(y-u)}\sin(\beta y - \beta u)k(u)du \\
&= e^{-\tilde{\alpha}(\beta)y}\sin(\beta y)\int_{-\infty}^{\infty} e^{\tilde{\alpha}(\beta)u}\cos(\beta u)k(u)du \\
&\quad - e^{-\tilde{\alpha}(\beta)y}\cos(\beta y)\int_{-\infty}^{\infty} e^{\tilde{\alpha}(\beta)u}\sin(\beta u)k(u)du \\
&\geq e^{-\tilde{\alpha}(\beta)y}\sin(\beta y) \\
&= \phi(y).
\end{aligned}$$

Observe that the last inequality holds, and is strict for $y \in (0, \pi/\beta)$ since $\int_{-\infty}^{\infty} e^{\tilde{\alpha}(\beta)u}\cos(\beta u)k(u)du > 1$ and $\int_{-\infty}^{\infty} e^{\tilde{\alpha}(\beta)u}\sin(\beta u)k(u)du = 0$ for $0 < \beta < \min(\beta_1, \beta_2)$. Hence $\phi * k(y) > \phi(y)$ for $0 < y < \pi/\beta$ provided $0 < \beta < \min(\beta_1, \beta_2, \pi/B)$.

Hence we have proved that $\phi * k(y) > \phi(y)$ for $y \in \mathrm{supp}(y)$ and $\phi * k(y) \geq \phi(y)$ otherwise. So $\phi * k \succ \phi$ for $\beta \in [0, \beta_0]$, where $\beta_0 = \min\{\beta_1, \beta_2, \pi/B]$.

Now $f(y) = \phi * k(y) - \phi(y)$ is a continuous function with $f(y) > 0$ for $y \in [0, \pi/\beta]$ and $\phi(y)$ is a continuous function with $\phi(y) = 0$ for $y \in \Re/[0, \pi/\beta]$. It follows from continuity considerations that for δ sufficiently small $\phi * k(y) - \phi(y - \delta) > 0$ for $y \in [0, \pi/\beta + \delta]$ and $\phi * k(y) - \phi(y-\delta) \geq 0$ for $y \in \Re/[0, \pi/\beta + \delta]$. Hence $\phi * k \succ \phi_\delta$. Here we can take $\delta \in [0, \Delta_\beta)$, where Δ_β is given by

$$\Delta_\beta = \inf_{0 \leq y \leq \pi/\beta} \{\sup\{\epsilon > 0 | \phi * k(y) - \phi(y - \epsilon) > 0\}\}.$$

□

For the specific type i and $0 < c < c_0$ obtain $k(y)$. Using this $k(y)$ and setting $\phi(y) = q(y; \alpha, \beta)$, from Lemma 5.4 we may choose β, and hence α and δ, so that $\phi * k \succ \phi_\delta$, where $\phi(y) = q(y; \alpha, \beta)$. The notation can now be simplified by defining $q(y) = q(y; \alpha, \beta)$ for these values of α and β. For $y \in [0, \pi/\beta]$, $q'(y) = 0$ has a unique solution $y = \rho$ for which the function is a maximum. Let $M = q(\rho)$.

The function $\psi(\mathbf{s}, t)$, for the specified i and c, is now defined in terms of q. For any $\sigma > 0$ and $D > 0$, define

$$\psi(\mathbf{s}, t) = \begin{cases} \sigma M, & \text{for } |\mathbf{s}| \leq D + ct + \rho, \\ \sigma q(|\mathbf{s}| - D - ct), & \text{for } D + ct + \rho \leq |\mathbf{s}| \leq D + ct + (\pi/\beta), \\ 0, & \text{for } |\mathbf{s}| \geq D + ct + (\pi/\beta). \end{cases}$$

This is easily seen to be equivalent to defining

$$\psi(\mathbf{s}, t) = \sigma \max_{\eta \geq -D - ct} q(|\mathbf{s}| + \eta).$$

Lemma 5.5 now shows that $\psi(\mathbf{s}, t)$ is a subsolution for E_T for a suitable choice of σ and D.

LEMMA 5.5. *There exists a $\sigma^* > 0$ and $D^* > 0$ such that, for any $0 < \sigma < \sigma^*$ and any $D > D^*$, the function ψ constructed for this choice of σ and D satisfies $E_T[\psi](\cdot, t) \succ \psi(\cdot, t)$ for all $t \geq T$.*

PROOF. Note that $\psi(\mathbf{s}, t) \leq \sigma M$ for all $\mathbf{s} \in \Re^N$ and $t \geq 0$. We can choose σ^* sufficiently small so that $(1 - e^{-x}) \geq h^{1/m} x$ for all x such that $0 \leq x \leq \sigma^* M \max\left(1, \{n \max_{jk} \int_0^{T/m} \gamma_{jk}(\tau) d\tau\}^{m-1}\right)$.

Define $D^* = R^2/(2\delta) - \rho + R$, and take any $D > D^*$ and $0 < \sigma < \sigma^*$ to construct $\psi(\mathbf{s}, t)$. Then

$$
\begin{aligned}
E_T[\psi](\mathbf{s},t) \geq h & \int_{\Re^N} \int_0^t \int_{\Re^N} \int_0^{t-\tau_{m-1}} \cdots \int_{\Re^N} \\
& \times \int_0^{t-\sum_{j=1}^{m-1}\tau_j} \psi\left(\mathbf{s} - \sum_{j=1}^{m-1}\mathbf{s}_j, t - \sum_{j=1}^{m-1}\tau_j\right) \sum_{j_1=1}^{n}\sum_{j_2=1}^{n} \cdots \sum_{j_{m-1}=1}^{n} \hat{\gamma}_{ij_1}(\tau_{m-1}) \\
& \times \hat{\gamma}_{j_1 j_2}(\tau_{m-2}) \cdots \hat{\gamma}_{j_{m-1}i}(\tau_0)\hat{p}_{ij_1}(\mathbf{s}_{m-1}) \cdots \hat{p}_{j_{m-1}i}(\mathbf{s}_0) d\mathbf{s}_0 d\tau_0 \cdots d\mathbf{s}_{m-1} d\tau_{m-1} \\
= h & \int_{\Re^N} \int_0^t \psi(\mathbf{s} - \mathbf{r}, t - \tau) \sum_{j_1=1}^{n}\sum_{j_2=1}^{n} \cdots \sum_{j_{m-1}=1}^{n} \hat{p}_{ij_1} * \cdots * \hat{p}_{j_{m-1}i}(\mathbf{r}) \\
& \times \hat{\gamma}_{ij_1} * \hat{\gamma}_{j_1 j_2} * \cdots * \hat{\gamma}_{j_{m-1}i}(\tau) d\mathbf{r} d\tau.
\end{aligned}
$$

We need to prove that, for any given $t \in [T, \infty]$, $E_T[\psi](\mathbf{s},t) > \psi(\mathbf{s},t)$ for all $\mathbf{s} \in \text{supp}(\psi(\cdot,t))$ and $E_T[\psi](\mathbf{s},t) \geq \psi(\mathbf{s},t) = 0$ otherwise.

For a given t we split the range of $|\mathbf{s}|$ into three intervals (1) $|\mathbf{s}| > D + ct + \pi/\beta$, (2) $|\mathbf{s}| \leq D + c(t-T) + \rho - R$ and (3) $D + c(t-T) + \rho - R < |\mathbf{s}| \leq D + ct + \pi/\beta$. Note that ranges (2) and (3) correspond to $\mathbf{s} \in \text{supp}(\psi(\cdot,t))$ whilst the range (1) is just the complement of the support of $\psi(\cdot,t)$.

Therefore for range (1) the result is trivial since $\psi(\mathbf{s},t) = 0$ for $|\mathbf{s}| > D+ct+\pi/\beta$. Now $\psi(\mathbf{s},t) \geq 0$ for all $\mathbf{s}$ and $t \geq 0$ and, from the definition of E_T, $E_T[\psi](\mathbf{s},t)$ is therefore non-negative. Hence $E_T[\psi](\mathbf{s},t) \geq \psi(\mathbf{s},t) = 0$ for $|\mathbf{s}| > D + ct + \pi/\beta$ as required.

Next consider the range $|\mathbf{s}| \leq D + c(t-T) + \rho - R$. Since $|\mathbf{s} - \mathbf{r}| \leq |\mathbf{s}| + |\mathbf{r}|$, the function $\psi(\mathbf{s}-\mathbf{r}, t-\tau) = \sigma M$ if $|\mathbf{s}| + |\mathbf{r}| \leq D + c(t-\tau) + \rho$, i.e. if $|\mathbf{s}| \leq D + c(t-\tau) + \rho - |\mathbf{r}|$. This is true for all $\mathbf{r} \in B_R$ and $\tau \in [0,T]$ provided $|\mathbf{s}| \leq D + c(t-T) + \rho - R$. Thus if $|\mathbf{s}| \leq D + c(t-T) + \rho - R$, then for $t \geq T$

$$
\begin{aligned}
E_T[\psi](\mathbf{s},t) \geq & h \int_{\Re^N} \int_0^t \sigma M \sum_{j_1=1}^{n}\sum_{j_2=1}^{n} \cdots \sum_{j_{m-1}=1}^{n} \hat{p}_{ij_1} * \cdots * \hat{p}_{j_{m-1}i}(\mathbf{r}) \\
& \times \hat{\gamma}_{ij_1} * \hat{\gamma}_{j_1 j_2} * \cdots * \hat{\gamma}_{j_{m-1}i}(\tau) d\{\mathbf{r}\} d\tau \\
& = \sigma M h L_c(0; R, T, m) > \sigma M \geq \psi(\mathbf{s},t).
\end{aligned}
$$

Thus in this case $E_T[\psi](\mathbf{s},t) > \psi(\mathbf{s},t)$.

Finally consider the case when $D + c(t-T) + \rho - R < |\mathbf{s}| \leq D + ct + \pi/\beta$. Now we have chosen β, α and δ when defining the function q and hence ψ so that

$$
\int_{\infty}^{\infty} q(y - u + \delta) k(u) du > q(y), \tag{5.2}
$$

for $y \in \text{supp}(q(\cdot))$ i.e. for $0 \leq y \leq \pi/\beta$.

Using a terminating Taylor expansion, and noting that $-2\mathbf{s}'\mathbf{x} + |\mathbf{x}|^2 \geq -|\mathbf{s}|^2$,

$$
\begin{aligned}
|\mathbf{s}-\mathbf{x}| &= ((\mathbf{s}-\mathbf{x})'(\mathbf{s}-\mathbf{x}))^{1/2} \\
&= (\mathbf{s}'\mathbf{s} - 2\mathbf{s}'\mathbf{x} + \mathbf{x}'\mathbf{x})^{1/2} \\
&= |\mathbf{s}|\left(1 + \left(-2\frac{\mathbf{s}'\mathbf{x}}{|\mathbf{s}|^2} + \frac{|\mathbf{x}|^2}{|\mathbf{s}|^2}\right)\right)^{1/2} \\
&\le |\mathbf{s}|\left(1 + \frac{1}{2}\left(-2\frac{\mathbf{s}'\mathbf{x}}{|\mathbf{s}|^2} + \frac{|\mathbf{x}|^2}{|\mathbf{s}|^2}\right)\right) \\
&= |\mathbf{s}| - \frac{\mathbf{s}'\mathbf{x}}{|\mathbf{s}|} + \frac{|\mathbf{x}|^2}{2|\mathbf{s}|}.
\end{aligned}
$$

Now in this case $|\mathbf{s}| > D + c(t-T) + \rho - R \ge D + \rho - R$. Thus if $|\mathbf{x}| < R$, then

$$
\begin{aligned}
|\mathbf{s}-\mathbf{x}| &\le |\mathbf{s}| - \frac{\mathbf{s}'\mathbf{x}}{|\mathbf{s}|} + \frac{R^2}{2(D+\rho-R)} \\
&\le |\mathbf{s}| - \frac{\mathbf{s}'\mathbf{x}}{|\mathbf{s}|} + \delta,
\end{aligned}
$$

since we have taken $D > D^* = \left(R^2/\{2\delta\}\right) - \rho + R$.

As $q(y)$ is a decreasing function of y and $\hat{p}_{ij_1} * \cdots * \hat{p}_{j_{m-1}i}(\mathbf{x})$ is a radial function, then if we rotate the co-ordinate axes so that the vector $\mathbf{s}$ is in the direction of the first co-ordinate and consider any $t \ge T$,

$$
\begin{aligned}
E_T[\psi](\mathbf{s},t) &\ge h\int_0^t \int_{\Re^N} \psi(\mathbf{s}-\mathbf{r}, t-\tau) \\
&\times \sum_{j_1=1}^n \sum_{j_2=1}^n \cdots \sum_{j_{m-1}=1}^n \hat{p}_{ij_1} * \cdots * \hat{p}_{j_{m-1}i}(\mathbf{r})\tilde{\gamma}_{ij_1\cdots j_{m-1}i}(\tau) d\mathbf{r} d\tau \\
&= h\sigma \int_0^t \int_{\Re^N} \max_{\eta \ge -D-c(t-\tau)} q(|\mathbf{s}-\mathbf{r}| + \eta) \\
&\times \sum_{j_1=1}^n \sum_{j_2=1}^n \cdots \sum_{j_{m-1}=1}^n \hat{p}_{ij_1} * \cdots * \hat{p}_{j_{m-1}i}(\mathbf{r})\tilde{\gamma}_{ij_1\cdots j_{m-1}i}(\tau) d\mathbf{r} d\tau \\
&\ge h\sigma \int_0^t \int_{\Re^N} \max_{\eta \ge -D-c(t-\tau)} q(|\mathbf{s}| - \{\mathbf{r}\}_1 + \delta + \eta) \\
&\times \sum_{j_1=1}^n \sum_{j_2=1}^n \cdots \sum_{j_{m-1}=1}^n \hat{p}_{ij_1} * \cdots * \hat{p}_{j_{m-1}i}(\mathbf{r})\tilde{\gamma}_{ij_1\cdots j_{m-1}i}(\tau) d\mathbf{r} d\tau \\
&= h\sigma \int_0^t \int_{-\infty}^{\infty} \max_{\eta \ge -D-c(t-\tau)} q(|\mathbf{s}| - x + \delta + \eta) \\
&\times \sum_{j_1=1}^n \sum_{j_2=1}^n \cdots \sum_{j_{m-1}=1}^n \tilde{p}_{i,j_1,\ldots,j_{m-1},i}(x)\tilde{\gamma}_{ij_1\cdots j_{m-1}i}(\tau) dx d\tau.
\end{aligned}
$$

Let $u = x - c\tau$ and $\eta^* = \eta - c\tau$. Then for $t \ge T$, since $\tilde{\gamma}_{ij_1\cdots j_{m-1}i}(\tau) = 0$ for $\tau > T$,

$$
\begin{aligned}
E_T[\psi](\mathbf{s},t) &\geq \sigma \int_{-\infty}^{\infty}\int_0^{\infty} \max_{\eta^*\geq -D-ct} q(|\mathbf{s}|-u+\delta+\eta^*) \\
&\quad \times h\sum_{j_1=1}^{n}\sum_{j_2=1}^{n}\cdots\sum_{j_{m-1}=1}^{n} \tilde{p}_{i,j_1,\ldots,j_{m-1},i}(u+c\tau)\tilde{\gamma}_{ij_1\cdots j_{m-1}i}(\tau)d\tau du \\
&= \sigma\int_{-\infty}^{\infty} \max_{\eta^*\geq -D-ct} q(|\mathbf{s}|-u+\delta+\eta^*)k(u)du \\
&\geq \sigma \max_{\eta^*\geq -D-ct}\int_{-\infty}^{\infty} q(|\mathbf{s}|-u+\delta+\eta^*)k(u)du \\
&> \sigma \max_{\eta^*\geq -D-ct} q(|\mathbf{s}|+\eta^*) \\
&= \psi(\mathbf{s},t).
\end{aligned}
$$

Here we have used the inequality (5.2) for the penultimate step, which is a consequence of Lemma 5.4. For this inequality to be strict, for any $\mathbf{s}$ satisfying $|\mathbf{s}| \leq D+ct+\pi/\beta$, we need $|\mathbf{s}|+\eta^*$ to lie in the support of q for some $\eta^* \geq -D-ct$. This is easily seen since $|\mathbf{s}| - D - ct \leq \pi/\beta$, and so the range of $|\mathbf{s}| + \eta^*$ for $\eta^* \geq -D - ct$ includes the non-empty set $[\max(0, |\mathbf{s}| - D - ct), \pi/\beta]$ which is contained in the support of q.

Thus in this final range $E_T[\psi](\mathbf{s},t) > \psi(\mathbf{s},t)$ and the proof of the theorem is complete.

□

We need to show that $w_i(\mathbf{s},t) > 0$ for $(\mathbf{s},t) \in B_R \times [t_0, t_0+T]$. Lemma 5.6 gives a simple proof of this result. Since $w_i(\mathbf{s},t)$ is monotone increasing in t, the infimum of $w_i(\mathbf{s},t)$ in this range is identical to $\inf w_i(\mathbf{s},t_0)$ over $\mathbf{s} \in B_R$. From Theorem 3.1 $x_i(\mathbf{s},t)$, and hence $w_i(\mathbf{s},t_0)$, is continuous in $\mathbf{s}$. This then implies that $\inf w_i(\mathbf{s},t_0) > 0$ for $(\mathbf{s},t) \in B_R \times [t_0, t_0+T]$. This enables us to complete the proof of Theorem 5.2.

LEMMA 5.6. *For any $R > 0$ there exists a $t_0 = t_0(R)$ such that $w_i(\mathbf{s},t) > 0$ for $(\mathbf{s},t) \in B_R \times [t_0,\infty)$.*

PROOF. The conditions imposed on the $H_i(\mathbf{s},t) > 0$ ensure that there exists an open set F in $\Re^N$ and a positive constant T, such that for some j, $H_j(\mathbf{s},t) > 0$ for $t \geq T$ and $\mathbf{s} \in F$. We may shift the origin so that, without loss of generality, $F \supset B_A$ for some positive A. Hence $w_j(\mathbf{s},t) > 0$ for $\mathbf{s} \in B_A$ and $t \geq T$.

Since $\mathbf{\Gamma}$ is non-reducible, there exists a sequence $j_1, \ldots, j_l$ with $j_1 = j$ and $j_l = j$ and $\gamma_{j_{s+1}j_s} \neq 0$ for $s = 1, \ldots, (l-1)$, Now $w_{j_{s+1}}(\mathbf{s},t) > 0$ for $t \geq T$ if

$$\int_0^T \int_{\Re^N} w_{j_s}(\mathbf{r},\tau)\gamma_{j_{s+1}j_s}(t-\tau)p_{j_{s+1}j_s}(\mathbf{s}-\mathbf{r})d\mathbf{r}d\tau > 0.$$

It is easily seen therefore that $w_j(\mathbf{s},t) > 0$ for $t \geq T$ if

$$\int_0^T \int_{\Re^N} w_j(\mathbf{r},\tau)\gamma_{j_l j_{l-1}} * \cdots * \gamma_{j_2 j_1}(t-\tau)p_{j_l j_{l-1}} * \cdots * p_{j_2 j_1}(\mathbf{s}-\mathbf{r})d\mathbf{r}d\tau > 0.$$

Now for some B, C, T_1, T_2, $p_{j_l j_{l-1}} * \cdots * p_{j_2 j_1}(\mathbf{x}) > 0$ for $B \leq |\mathbf{x}| \leq C$ and $\gamma_{j_l j_{l-1}} * \cdots * \gamma_{j_2 j_1}(\tau) > 0$ for $\tau \in [T_1, T_2]$, with $B < C$ and $T_1 < T_2$. Hence $w_j(\mathbf{s}, t) > 0$ for $\mathbf{s} \in B_A$ and/or $B - A \leq |\mathbf{s}| \leq A + C$ and $t \geq T + T_1$. Repeating this procedure, in two steps we obtain the result that $w_j(\mathbf{s}, t) > 0$ for $\mathbf{s} \in B_{A+C-B}$ and $t \geq T + 2T_1$.

If $j = i$, choose a non-negative integer r such that $R \leq A + r(C - B)$; then $w_i(\mathbf{s}, t) > 0$ for $\mathbf{s} \in B_R$ and $t \geq T + 2rT_1$. Hence in this case we take $t_0 = T + 2rT_1$.

If $j \neq i$, there exists a sequence $i_1, ...i_k$, with $i_k = i$, $i_1 = j$ and $\gamma_{i_{s+1} i_s} \neq 0$ for $s = 1, ...(k-1)$. Now $w_i(\mathbf{s}, t) > 0$ if $\int_0^t \int_{\Re^N} w_j(\mathbf{r}, \tau) \gamma_{i_k i_{k-1}} * \cdots * \gamma_{i_2 i_1}(t - \tau) p_{i_k i_{k-1}} * \cdots * p_{i_2 i_1}(\mathbf{s} - \mathbf{r}) d\mathbf{r} d\tau > 0$. There exist non-negative reals S_1, S_2, D and E such that $\gamma_{i_k i_{k-1}} * \cdots * \gamma_{i_2 i_1}(t) > 0$ for $t \in [S_1, S_2]$ and $p_{i_k i_{k-1}} * \cdots * p_{i_2 i_1}(\mathbf{x}) > 0$ for $D \leq |\mathbf{x}| \leq E$. Choose r such that $A + r(C - B) \geq \max(R, D)$. Then $w_j(\mathbf{s}, t) > 0$ for $\mathbf{s} \in B_R$ and $t \geq T + 2rT_1$ and hence $w_i(\mathbf{s}, t) > 0$ for $\mathbf{s} \in B_R$ and $t \geq T + 2rT_1 + S_1$. In this case we take $t_0 = T + 2rT_1 + S_1$.

Hence in either case $w_i(\mathbf{s}, t) > 0$ for $(\mathbf{s}, t) \in B_R \times [t_0, \infty)$. This completes the proof of the lemma.

□

When $\rho(\mathbf{\Gamma}) > 1$ and each of the contact distributions are radially symmetric and are also exponentially dominated in the tail, we are now able to use Lemmas 5.3, 5.5 and 5.6 to prove that the critical wave speed provides a lower bound for the asymptotic speed of propagation of the epidemic. This can then be used with the result from the last section to show that the asymptotic speed of propagation is the minimum wave speed.

THEOREM 5.2. *For* $\rho(\mathbf{\Gamma}) > 1$, *and any* $c_1 \in (0, c_0)$, *there exist positive constants* b_i *and constants* T_i *sufficiently large such that* $\min\{w_i(\mathbf{s}, t) : |\mathbf{s}| \leq c_1 t\} \geq b_i$, *and therefore* $\min\{v_i(\mathbf{s}, t) : |\mathbf{s}| \leq c_1 t\} \geq B_i = (1 - e^{-b_i})$, *for all* $t > T_i$ *and* $i = 1, ..., n$.

PROOF. Take any c such that $c_1 < c < c_0$ and any integer i with $1 \leq i \leq n$. Choose h, R, T and m so that $L(\lambda) = hL_c(\lambda; R, T, m) > 1$ for all real λ. From this construct the operator E_T and the subsolution $\psi(\mathbf{s}, t)$.

From Lemma 5.6, for any finite positive T, $\inf w_i(\mathbf{s}, t) > 0$ for $(\mathbf{s}, t) \in B_R \times [t_0, t_0 + T]$. Choose σ such that $0 < \sigma < \sigma^*$ and $\sigma M < \inf w_i(\mathbf{s}, t)$ where the inf is taken over $(\mathbf{s}, t) \in B_R \times [t_0, t_0 + T]$. Hence $w_i(\mathbf{s}, t_0 + t) \succ \psi(\mathbf{s}, t)$ for $0 \leq t \leq T$.

Now in Lemma 5.5 it was shown that $E_T(\psi(\cdot, t)) \succ \psi(\cdot, t)$ for $t \geq T$. Using the comparison principle, Lemma 5.3, we then obtain the result that $w_i(\cdot, t_0 + t) \succ \psi(\cdot, t)$ for all $t \geq 0$.

Hence $w_i(\mathbf{s}, t_0 + t) > \sigma M$ for $|s| \leq \rho + D + ct$ and $t \geq 0$. Thus $w_i(\mathbf{s}, t) > \sigma M$ for $|s| \leq \rho + D + c(t - t_0)$ and $t \geq 0$. Therefore $w_i(\mathbf{s}, t_0 + t) > \sigma M$ for $|s| \leq c_1 t$ provided $c_1 t \leq \rho + D + c(t - t_0)$ and $t \geq t_0$, i.e. if $t \geq \max(t_0, (ct_0 - \rho - D)/(c - c_1))$. Thus if we take $b_i = \sigma M > 0$ and $T_i = \max(t_0, (ct_0 - \rho - D)/(c - c_1))$ we obtain the result that $\min\{w_i(\mathbf{s}, t) : |\mathbf{s}| \leq c_1 t\} \geq b_i$ for all $t \geq T_i$. This proves that c_0 satisfies part 2 of the definition of the asymptotic speed of propagation.

Note that we can define the appropriate operator E_T for each value of i, and hence obtain the corresponding b_i and T_i so that Theorem 2 holds for all $i = 1, ..., n$.

□

5.5 The asymptotic speed of propagation

The results of Sections 5.3 and 5.4 are now combined to obtain the asymptotic speed of propagation of the epidemic. The results given here require the contact distributions to be radially symmetric and exponentially dominated in the tail.

Provided $\rho(\mathbf{\Gamma}) \leq 1$, so that the threshold value for a major epidemic is not exceeded, the asymptotic speed of propagation is zero.

When $\rho(\mathbf{\Gamma}) > 1$ the asymptotic speed of propagation in any direction is the minimum wave speed c_0, which is direction-independent and positive because of the radial symmetry of the contact distributions. This latter result was established in Lemma 5.1.

The minimum wave speed can therefore be given in the direction of the first co-ordinate axis. Thus

$$c_0 = \inf\{c > 0 : K_c(\lambda) < 1 \text{ for some } \lambda \in (0, \Delta_v)\},$$

where $K_c(\lambda) = \rho(\mathbf{V}(\lambda))$ and $\{\mathbf{V}(\lambda)\}_{ij} = \sigma_j \Lambda_{ij}(c\lambda) P_{ij}(\lambda)$. Here the $\Lambda_{ij}(\lambda)$ are the Laplace transforms of the infection rates and the $P_{ij}(\lambda)$ are the Laplace transforms of the projections of the contact distributions in the direction of the first co-ordinate axis.

In Chapter 7 the speed of first spread is obtained for certain epidemic models using the saddle point method and is also shown to be c_0, the minimum wave speed. However the saddle point method does not require radial symmetry of the contact distributions. This suggests that the minimum wave speed, which is direction-dependent in general, will still give the asymptotic speed of propagation when the contact distributions are not radially symmetric. Note that the definition of the asymptotic speed of propagation would need to be adjusted. We have only stated it for the radially symmetric case (see Aronson and Weinberger [A4,A5]).

THEOREM 5.3. *If $\rho(\mathbf{\Gamma}) \leq 1$, then for any $c > 0$, $\lim_{t\to\infty} \sup\{v_i(\mathbf{s}, t) : |\mathbf{s}| \geq ct\} = 0$, for $i = 1, ..., n$.*

When $\rho(\mathbf{\Gamma}) > 1$ then for any $c > c_0$, $\lim_{t\to\infty} \sup\{v_i(\mathbf{s}, t) : |\mathbf{s}| \geq ct\} = 0$, for $i = 1, ..., n$. Also for any $0 < c < c_0$, there exist positive constants b_i and constants T_i sufficiently large such that $\min\{v_i(\mathbf{s}, t) : |\mathbf{s}| \leq c_1 t\} \geq B_i = (1 - e^{-b_i})$, for all $t > T_i$ and $i = 1, ..., n$.

PROOF. The first result follows immediately from Corollary 5.1. The second result follows from Corollary 5.2 and Theorem 5.2.

□

Note that $v_i(\mathbf{s}, t) = 1 - x_i(\mathbf{s}, t)$ is the proportion of type i individuals at position $\mathbf{s}$ who have suffered the epidemic by time t.

5.6 Non-exponentially dominated contact distributions

In this section the behaviour of the epidemic is considered when at least one $P_{ij}(\lambda)$ is infinite for λ real and non-zero. The existence and uniqueness of $w_i(\mathbf{s}, t)$ has been established in Chapter 3, Theorem 3.1. We first treat the case when $\rho(\mathbf{\Gamma}) > 1$ and prove the analogue of Lemma 5.2.

LEMMA 5.7. *When $\rho(\mathbf{\Gamma}) > 1$, for any $c > 0$ there exists a positive integer m, a positive real $h < 1$ and positive reals R_0 and T_0 sufficiently large such that $hL_c(\lambda; R, T, m) > 1$ for $\lambda \in \Re$ and $R \geq R_0$ and $T \geq T_0$.*

PROOF. As in Lemma 5.2, there exist positive integers l and s so that, by relabelling the populations $1, ..., n$,

$$\mathbf{V}^{ls}(\lambda) = \begin{pmatrix} \mathbf{B}_1(\lambda) & \mathbf{0} & \dots & \mathbf{0} \\ \mathbf{0} & \mathbf{B}_2(\lambda) & \mathbf{0}\dots & \\ \vdots & \vdots & \ddots & \vdots \\ \mathbf{0} & \mathbf{0} & \dots & \mathbf{B}_l(\lambda) \end{pmatrix},$$

where $\mathbf{B}_j(\lambda) > 0$. Also $\{\mathbf{V}^{rls}(\lambda)\}_{ii} = \infty$ for $\lambda > 0$, $r \geq 3$ and all i. If $\mathbf{V}(0)$ has an infinite element, then $\{\mathbf{V}^{rls}(0)\}_{ii} = \infty$ for $r \geq 3$ and all i. Take $m = 3ls$.

When $\mathbf{V}(0)$ is finite, since $K_c(0) = \rho(\mathbf{\Gamma}) > 1$ and $\mathbf{B}_j(0) > 0$ for all j,

$$\lim_{r\to\infty} \mathbf{B}_j^r(0)/(\rho(\mathbf{\Gamma}))^{rls} = \mathbf{E}_j(0) > \mathbf{0};$$

where $\mathbf{E}_j(0)$ is the idempotent of $\mathbf{B}_j(0)$ corresponding to the eigenvalue $(\rho(\mathbf{\Gamma}))^{ls}$. Then for r sufficiently large $\{\mathbf{V}^{rls}(0)\}_{ii} > 1$ for all i. Take such an $r \geq 3$ and let $m = rls$.

Then in both cases $\{\mathbf{V}^m(\lambda)\}_{ii} > 1$ for all $\lambda \geq 0$ and all i, and $\{\mathbf{V}^m(\lambda)\}_{ii} = \infty$ for $\lambda > 0$. We may then proceed exactly as in the proof of Lemma 5.2, noting that $\lim_{R,T\to\infty} L_c(\lambda; R, T, m) = \{\mathbf{V}^m(\lambda)\}_{ii} > 1$ for $\lambda \geq 0$.

□

In an identical manner to Section 5.4 we may then use this lemma to define the operator E_T and hence prove the following theorem.

THEOREM 5.4. *When $\rho(\mathbf{\Gamma}) > 1$, for any $c > 0$ there exist positive constants b_i and T_i with T_i sufficiently large so that $\min\{w_i(\mathbf{s}, t) : |\mathbf{s}| \leq c_1 t\} \geq b_i$, and hence $\min\{v_i(\mathbf{s}, t) : |\mathbf{s}| \leq c_1 t\} \geq B_i$, for $t > T_i$ and $i = 1, ..., n$, where $B_i = 1 - e^{-b_i}$.*

□

Theorem 5.4 shows that when $\rho(\mathbf{\Gamma}) > 1$ and at least one $p_{ij}(\mathbf{r})$ is not exponentially dominated in the tail, no matter how fast you run the epidemic will always overtake you. The asymptotic speed of propagation may then be considered to be infinite.

In order to establish the result that the speed of propagation is zero if $\rho(\mathbf{\Gamma}) \leq 1$, we impose the same further conditions imposed in Section 3.7 when considering the behaviour at infinity.

Specifically each $p_{ij}(\mathbf{s})$ and $p_{ij}^*(\mathbf{s})$ is taken to be monotone decreasing in $|\mathbf{r}|$, differentiable (except possibly at a finite number of points) with bounded derivatives, and a convex function of the radial distance $|\mathbf{r}|$ for $|\mathbf{r}|$ sufficiently large. This latter condition implies that each contact distribution is a convex function in the tails of the first co-ordinate $\{\mathbf{r}\}_1$ when the remaining co-ordinates are held fixed. In addition each $\varepsilon_i(\mathbf{s}, \tau)$ is assumed to be a radially symmetric, monotone decreasing function of $|\mathbf{s}|$. Note that this last condition is easy to relax, this being discussed at the end of the section.

A direct proof is given in Theorem 5.5 based on Theorem 3.2 part 1 and Lemma 3.6. Theorem 3.5 may alternatively be used.

THEOREM 5.5. *If* $\rho(\mathbf{\Gamma}) \leq 1$, *then* $\lim_{t\to\infty} \sup\{v_i(\mathbf{s},t) : |\mathbf{s}| \geq ct\} = 0$, *and hence* $\lim_{t\to\infty} \sup\{w_i(\mathbf{s},t) : |\mathbf{s}| \geq ct\} = 0$, *for any* $c > 0$ *and* $i = 1, ..., n$.

PROOF. From Lemma 3.6 each $H_i(\mathbf{s},t)$ and $w_i(\mathbf{s},t)$, and hence $v_i(\mathbf{s},t)$, is a radial function of $\mathbf{s}$ and is monotone decreasing in $|\mathbf{s}|$, so that this is also true for the limits $a_i(\mathbf{s}) = \lim_{t\to\infty} H_i(\mathbf{s},t)$ and $v_i(\mathbf{s}) = \lim_{t\to\infty} v_i(\mathbf{s},t)$. Note that $v_i(\mathbf{s})$ is the final size for type i at position $\mathbf{s}$. Let $\tilde{p}_{ij}(\{\mathbf{r}\}_1)$ and $\tilde{p}^*_{ij}(\{\mathbf{r}\}_1)$ be the marginal density functions obtained by integrating $p_{ij}(\mathbf{r})$ and $p^*_{ij}(\mathbf{r})$ over all but the first co-ordinate of $\mathbf{r}$. Then these marginal density functions are symmetric about zero. From radial symmetry the co-ordinates may be rotated so that $\mathbf{s}$ is in the direction of the first co-ordinate. Hence $v_i(\mathbf{s}) = v_i((|\mathbf{s}|\ \mathbf{0}')')$, $a_i(\mathbf{s}) = a_i((|\mathbf{s}|\ \mathbf{0}')')$ and $\varepsilon_i(\mathbf{s},t) = \varepsilon_i((|\mathbf{s}|\ \mathbf{0}')',t)$.

From the final size equations (3.6) for the $v_i(\mathbf{s})$, derived in Theorem 3.2 part 1, we then obtain

$$(5.3)\quad -\log(1 - v_i((|\mathbf{s}|\ \mathbf{0}')')) = \sum_{j=1}^{n} \gamma_{ij} \int_{-\infty}^{\infty} \tilde{p}_{ij}(u) v_j(([|\mathbf{s}| - u]\ \mathbf{0}')')du + a_i((|\mathbf{s}|\ \mathbf{0}')'),$$

for $i = 1, ..., n$. Since each $a_i((|\mathbf{s}|\ \mathbf{0}')')$ and $v_i((|\mathbf{s}|\ \mathbf{0}')')$ is non-negative and monotone decreasing in $|\mathbf{s}|$ each tends to a limit as $|\mathbf{s}|$ tends to infinity. Denote the corresponding limits by v_i and a_i. We now prove that $a_i = 0$ for all i.

Since each $\varepsilon_i(\mathbf{s})$ has bounded support there exists a positive A such that each support is contained in the closed ball of radius A centred on the origin. Then for each $i = 1, ..., n$,

$$\begin{aligned} a_i((|\mathbf{s}|\ \mathbf{0}')') &= \sum_{j=1}^{k} \int_0^{\infty} \int_0^{\infty} \int_{-\infty}^{\infty} \varepsilon_j(([|\mathbf{s}| - u]\ \mathbf{0}')', \tau) \tilde{p}^*_{ij}(u) \gamma^*_{ij}(\tau + t) du d\tau dt \\ &\leq \sum_{j=1}^{k} \gamma^*_{ij} \int_{-\infty}^{\infty} \varepsilon_j(([|\mathbf{s}| - u]\ \mathbf{0}')') \tilde{p}^*_{ij}(u) du \\ &\leq \sum_{j=1}^{k} \gamma^*_{ij} \sup_{\mathbf{r}} \varepsilon_j(\mathbf{r}) \int_{|\mathbf{s}|-A}^{|\mathbf{s}|+A} \tilde{p}^*_{ij}(u) du. \end{aligned}$$

Since this last term tends to zero as $|\mathbf{s}|$ tends to infinity, $a_i = 0$ for all i.

Using dominated convergence in equations (5.3), since $a_i = 0$ for all i gives the equation

$$-\log(1 - v_i) = \sum_{j=1}^{n} \gamma_{ij} v_j, \quad (i = 1, ..., n).$$

From Theorem B.2 part 2, since $\rho(\mathbf{\Gamma}) \leq 1$, the only non-negative solution to this system of equations is $v_i = 0$ for all i.

Now, for any $c > 0$, $v_i(\mathbf{s},t)$ is radially symmetric and a monotone decreasing function of $|\mathbf{s}|$ and a monotone increasing function of t. Hence for all $i = 1, ..., n$,

$$\sup\{v_i(\mathbf{s},t) : |\mathbf{s}| \geq ct\} = v_i((ct\ \mathbf{0}')',t) \leq v_i((ct\ \mathbf{0}')'). \tag{5.4}$$

For each i, the right hand side of inequality (5.4) tends to $v_i = 0$ as $t \to \infty$. Therefore we obtain the result that for any $c > 0$ and $i = 1, ...n$, $\lim_{t\to\infty} \sup\{v_i(\mathbf{s},t) : |\mathbf{s}| \geq ct\} = 0$. This may trivially be re-written in terms of the $w_i(\mathbf{s},t)$.

□

Note that we do not need to restrict $\varepsilon_i(\mathbf{s},\tau)$ to be radially symmetric and monotone decreasing in $|\mathbf{s}|$ for all i, it is sufficient to have each function bounded above by a function $\varepsilon_i^*(\mathbf{s},\tau)$ which does satisfy these conditions and which also has bounded support. Let $w_i^*(\mathbf{s},t)$ satisfy equations (5.1) with $\varepsilon_j^*(\mathbf{s},\tau)$ replacing $\varepsilon_j(\mathbf{s},\tau)$. Then from the construction in Theorem 3.1, $w_i(\mathbf{s},t) \leq w_i^*(\mathbf{s},t)$ for all $\mathbf{s}$, t and i. Therefore, applying Theorem 5.5 to $w_i^*(\mathbf{s},t)$,

$$\lim_{t\to\infty} \sup\{w_i(\mathbf{s},t) : |\mathbf{s}| \geq ct\} \leq \lim_{t\to\infty} \sup\{w_i^*(\mathbf{s},t) : |\mathbf{s}| \geq ct\} = 0.$$

5.7 The pandemic theorem revisited

In Chapter 3 we proved the pandemic lemma for $\rho(\mathbf{\Gamma}) > 1$. This holds for any finite N. The extension to the pandemic theorem required the use of Essén's results, which restricted the proof to dimensions $N = 1$ and $N = 2$. The problem was to show, in the case where $\mathbf{\Gamma}$ is finite, $\rho(\mathbf{\Gamma}) > 1$ and $a_i = 0$ for all i, that $\inf_{\mathbf{s}} v_i(\mathbf{s}) > 0$. However, now that we have the results on the asymptotic speed of propagation, this follows immediately from part 2. This result requires the contact distributions to be radially symmetric. We can then deduce the pandemic theorem for any finite N. (Note that the assumptions used in this chapter imply that $a_i = 0$ for all i.)

Proof of Theorem 3.3 (for general N).

Proof. In order to extend Lemma 3.3 to Theorem 3.3 for general dimension N it is only necessary to show that $\inf_{\mathbf{s}} v_i(\mathbf{s}) > 0$. Consider any positive $c < c_0$ if the contact distributions are exponentially dominated in the tail, and any positive c otherwise. From Theorems 5.2 and 5.3, it follows that there exist T_i such that $w_i(\mathbf{s},t) \geq b_i$ for all $|\mathbf{s}| \leq ct$ and all $t \geq T_i$. Consider any $\mathbf{s}$ and take $t^* \geq T_i$ so that $|\mathbf{s}| \leq ct^*$. Hence $w_i(\mathbf{s},t^*) \geq b_i$. Now $w_i(\mathbf{s},t)$ is an increasing function of t. Hence $\lim_{t\to\infty} w(\mathbf{s},t) \geq b_i$ for all $\mathbf{s}$.

Hence $v_i(\mathbf{s}) = 1 - \lim_{t\to\infty} e^{-w_i(\mathbf{s},t)} \geq B_i > 0$, where $B_i = 1 - e^{-b_i}$ and hence $\inf_{\mathbf{s}} v_i(\mathbf{s}) \geq B_i > 0$. This completes the proof of the pandemic theorem.

□

5.8 The asymptotic shape

When $\rho(\mathbf{\Gamma}) > 1$ and the contact distributions are radially symmetric and exponentially dominated in the tail, the asymptotic speed of propagation for any type has been shown to be the minimum wave speed c_0, that speed being positive. In terms of $v_i(\mathbf{s},t)$, the proportion of type i individuals at position $\mathbf{s}$ who have suffered the epidemic by time t, the following results are given in Theorem 5.3. For each $c > c_0$, for any $\varepsilon > 0$ there exists a T sufficiently large such that $\sup_{|\mathbf{s}| \geq ct} v_i(\mathbf{s}, t) \leq \varepsilon$ for all i and all $t \geq T$. Also for each positive $c < c_0$ there exist positive constants B_i and T such that $\inf_{|\mathbf{s}| \leq ct} v_i(\mathbf{s}, t) \geq B_i$ for all i and all $t \geq T$.

The first result may be used directly and the second strengthened to give a result concerning the asymptotic shape of infection. Define $\{\mathbf{b}\}_i = \sup_{\mathbf{s} \in \Re^N} a_i(\mathbf{s})$, where

$$a_i(\mathbf{s}) = \sum_{j=1}^{k} \int_0^\infty \int_{\Re^N} \int_0^\infty \varepsilon_j(\mathbf{s} - \mathbf{r}, \tau) p_{ij}^*(\mathbf{r}) \gamma_{ij}^*(t + \tau) d\tau d\mathbf{r} dt.$$

When $\mathbf{\Gamma}$ is finite the shape theorem then says that for any $\varepsilon > 0$ and $\delta > 0$ there exists a T such that, for all $t \geq T$, the following two results hold;

$$v_i(\mathbf{s}, t) \leq \varepsilon \text{ for } |\mathbf{s}| \geq (c_0 + \delta)t$$

and

$$\eta_i(\mathbf{0}) - \varepsilon < v_i(\mathbf{s}, t) \leq \eta_i(\mathbf{b}) \text{ for } |\mathbf{s}| \leq (c_0 - \delta)t,$$

where $\eta_i(\mathbf{0})$ and $\eta_i(\mathbf{b})$ are the lower and upper bounds obtained for the final size for type i individuals in the Pandemic Theorem, Theorem 3.3, and in Theorem 3.4. An equivalent result holds when $\mathbf{\Gamma}$ has at least one infinite entry. Hence the region of infection grows like a sphere whose radius for large time t is $c_0 t$. Most of the new infection taking place at time t is close to the boundary of the sphere.

To prove the shape theorem the bounds b_i for $w_i(\mathbf{s}, t)$, and hence the bounds B_i for $v_i(\mathbf{s}, t)$, are first strengthened using the following lemma which is proved using a generalisation of the technique given by Diekmann [D7] for the one-type model, and is given in Radcliffe and Rass [R7] for the general n-type model.

LEMMA 5.8. *Let $\mathbf{B} = (b_{ij})$ be a non-negative non-reducible matrix of finite entries with $\rho(\mathbf{B}) > 1$, and let $\mathbf{z} = (z_i) = \boldsymbol{\phi}$ be the unique positive solution to*

$$z_i = \sum_{j=1}^{n} b_{ij} \left(1 - e^{-z_j}\right), \quad (i = 1, ..., n), \tag{5.5}$$

as given in Theorem B.1. Take $\mathbf{u} = (u_i)$ to be the positive right eigenvector of $\mathbf{B}$ corresponding to $\rho(\mathbf{B})$, which is scaled so that $\mathbf{u}$ is sufficiently close to $\mathbf{0}$ to make $\rho(\mathbf{B}) u_i > -\log(1 - u_i)$ for all $i = 1, ..., n$. Define $\{\mathbf{N}^{(0)}\}_i = -\log(1 - u_i)$ and successively, for $m \geq 0$, define

$$\{\mathbf{N}^{(m+1)}\}_i = \sum_{j=1}^{n} b_{ij} \left(1 - e^{-\{\mathbf{N}^{(m)}\}_j}\right). \tag{5.6}$$

Then for every $\boldsymbol{\varepsilon} > \mathbf{0}$ there exists an M such that $\mathbf{N}^{(m)} > \boldsymbol{\phi} - \boldsymbol{\varepsilon}$ for $m \geq M$.

PROOF. Induction on m is used to show that $\{\mathbf{N}^{(m)}\}$ is a strictly increasing sequence of vectors. From equations (5.6) with $m = 0$,

$$\begin{aligned}\{\mathbf{N}^{(1)}\}_i &= \sum_{j=1}^{n} b_{ij}\left(1 - e^{-\{\mathbf{N}^{(0)}\}_j}\right)\\ &= \sum_{j=1}^{n} b_{ij}u_j\\ &= \{\mathbf{Bu}\}_i = \rho(\mathbf{B})u_i > -\log(1-u_i)\\ &= \{\mathbf{N}^{(0)}\}_i.\end{aligned}$$

Hence $\mathbf{N}^{(1)} > \mathbf{N}^{(0)}$. Now assume that $\mathbf{N}^{(m+1)} > \mathbf{N}^{(m)}$ for all $0 \leq m \leq M$. From equations (5.6) and the inductive hypothesis with $m = M$,

$$\begin{aligned}\{\mathbf{N}^{(M+2)}\}_i &= \sum_{j=1}^{n} b_{ij}\left(1 - e^{-\{\mathbf{N}^{(M+1)}\}_j}\right)\\ &> \sum_{j=1}^{n} b_{ij}\left(1 - e^{-\{\mathbf{N}^{(M)}\}_j}\right)\\ &= \{\mathbf{N}^{(M+1)}\}_i.\end{aligned}$$

Hence $\mathbf{N}^{(M+2)} > \mathbf{N}^{(M+1)}$, and so by induction $\mathbf{N}^{(m+1)} > \mathbf{N}^{(m)}$ for all $m \geq 0$.

Now $\{\mathbf{N}^{(m+1)}\}_i \leq \sum_{j=1}^{n} b_{ij}$, so that the sequence of vectors is bounded above. Also the sequence is bounded below by $\mathbf{N}^{(0)} > \mathbf{0}$. Therefore it tends to a positive limit as m tends to infinity which satisfies equations (5.5). Hence $\lim_{m\to\infty} \mathbf{N}^{(m)} = \boldsymbol{\phi}$. The lemma is then immediate.

□

Lemma 5.8 is now used to sharpen the lower bound obtained in Theorem 5.2. This result is obtained for the case when $\rho(\mathbf{\Gamma}) > 1$ and the contact distributions are all exponentially dominated in the tail, which is precisely when there is a finite positive speed of propagation. Define

$$\{\mathbf{\Gamma}(R,T)\}_{ij} = \int_0^T \int_{|\mathbf{r}|\leq R} p_{ij}(\mathbf{r})\gamma_{ij}(\tau)d\mathbf{r}d\tau.$$

Hence for R and T sufficiently large, $\{\mathbf{\Gamma}(R,T)\}_{ij} \neq 0$ if $\gamma_{ij} \neq 0$ so that $\mathbf{\Gamma}(R,T)$ is non-reducible. For such values of R and T, from Theorem B.2 $\rho(\mathbf{\Gamma}(R,T))$ is a continuous, increasing function of R and T with $\lim_{R\to\infty,T\to\infty}\rho(\mathbf{\Gamma}(R,T)) = \rho(\mathbf{\Gamma}) > 1$. Hence for R and T sufficiently large, $\rho(\mathbf{\Gamma}(R,T))$ is non-reducible with $\rho(\mathbf{\Gamma}(R,T)) > 1$.

THEOREM 5.6. *When $\rho(\mathbf{\Gamma}) > 1$ consider any R and T sufficiently large that $\mathbf{\Gamma}(R,T)$ is non-reducible with $\rho(\mathbf{\Gamma}(R,T)) > 1$. Then for any $c \in (0, c_0)$ and any $\varepsilon > 0$, there exists a $t^* > T$ sufficiently large such that $\inf\{w_i(\mathbf{s},t) : |\mathbf{s}| \leq ct\} \geq \phi_i(R,T) - \varepsilon$ for all $t > t^*$ and all $i = 1,...,n$, where $z_i = \phi_i(R,T)$ is the unique positive solution to equations (5.5) with $\mathbf{B} = \mathbf{\Gamma}(R,T)$.*

PROOF. Consider any $c < c_0$ and choose $d_0 \in (c, c_0)$. From Theorem 5.2 there exist positive constants b_i and T_0 such that $\inf\{w_i(\mathbf{s},t) : |\mathbf{s}| \leq d_0 t\} \geq b_i$ for all $i = 1, ..., n$ and $t > T_0 \geq T$. Now in Lemma 5.8 take $\mathbf{B} = \mathbf{\Gamma}(R,T)$ and choose the scaling to the eigenvector $\mathbf{u}$ so that $-\log(1-u_i) \leq b_i$ for all i. Then $w_i(\mathbf{s},t) \geq \{\mathbf{N}_0\}_i$ for all $|\mathbf{s}| \leq d_0 t$, all i and all $t \geq T_0$.

Now for each i and all $t > T_0 \geq T$,

$$\begin{aligned} w_i(\mathbf{s},t) &\geq \sum_{j=1}^{n} \int_0^T \int_{\mathbf{r}\in B_R} \left(1 - e^{-w_j(\mathbf{s}-\mathbf{r},t-\tau)}\right) p_{ij}(\mathbf{r})\gamma_{ij}(\tau) d\mathbf{r} d\tau \\ &\geq \sum_{j=1}^{n} \{\mathbf{\Gamma}(R,T)\}_{ij} \left(1 - e^{-\{\mathbf{N}_0\}_i}\right) \\ &= \{\mathbf{N}_1\}_i, \end{aligned}$$

provided that $|\mathbf{s} - \mathbf{r}| \leq d_0(t - \tau)$ for all $\mathbf{r} \in B_R$, $\tau \leq T$ and $t - \tau \geq T_0$. This will certainly hold if $t \geq T_0 + T$ and $|\mathbf{s}| \leq -R - d_0 T + d_0 t$.

Now take $d_1 \in (c, d_0)$. Note that $|\mathbf{s}| \leq -R - d_0 T + d_0 t$ when $|\mathbf{s}| \leq d_1 t$ if we take t sufficiently large so that $d_1 t \leq -R - d_0 T + d_0 t$, i.e. provided $t \geq (R + d_0 T)/(d_0 - d_1)$. Then for all i, $w_i(\mathbf{s},t) \geq \{\mathbf{N}_1\}_i$ for all $|\mathbf{s}| \leq d_1 t$ and $t \geq T_1$ if we take $T_1 = \max(T + T_0, (R + d_0 T)/(d_0 - d_1))$.

By successively choosing $d_{m+1} \in (c, d_m)$ and defining $T_{m+1} = \max(T+T_m, (R+d_{m+1}T)/(d_m - d_{m+1}))$ using Lemma 5.8 we obtain the result that, for all i and any positive integer m,

$$\inf\{w_i(\mathbf{s},t) : |\mathbf{s}| \leq d_m t\} \geq \{\mathbf{N}^{(m)}\}_i \text{ for } t \geq T_m.$$

For any $\varepsilon > 0$, choose M in Lemma 5.8 so that $\mathbf{N}^{(M)} > \boldsymbol{\phi}(R,T) - \varepsilon\mathbf{1}$. Then we immediately obtain the required result that $\inf\{w_i(\mathbf{s},t) : |\mathbf{s}| \leq ct\} \geq \phi_i(R,T) - \varepsilon$ for all $t > T_M$ and all $i = 1, ..., n$.

□

An equivalent result may then easily be obtained for $v_i(\mathbf{s},t)$ by observing that if $w_i(\mathbf{s},t) \geq \phi_i(R,T) - \varepsilon$ then

$$v_i(\mathbf{s},t) \geq \left(1 - e^{-\phi_i(R,T)}\right) - e^{-\phi_i(R,T)}(e^\varepsilon - 1) = \eta_i(R,T) - (1 - \eta_i(R,T))(e^\varepsilon - 1),$$

where $y_i = \eta_i(R,T)$ for $i = 1, ..., n$ is the unique positive solution to

$$-\log(1 - y_i) = \sum_{j=1}^{n} \{\mathbf{\Gamma}(R,T)\}_{ij} y_j, \quad (i = 1, ..., n).$$

COROLLARY 5.3. *When $\rho(\mathbf{\Gamma}) > 1$ consider any R and T sufficiently large that $\mathbf{\Gamma}(R,T)$ is non-reducible with $\rho(\mathbf{\Gamma}(R,T)) > 1$. Then for any $c \in (0, c_0)$ and any $\varepsilon > 0$, there exists a $t^* > T$ sufficiently large such that* $\inf\{v_i(\mathbf{s},t) : |\mathbf{s}| \leq ct\} \geq$ *$\eta_i(R,T) - \varepsilon$ for all $t > t^*$ and all $i = 1, ..., n$.*

□

Corollaries 5.2 and 5.3 are then used to establish the following shape theorem when the contact distributions are radially symmetric and exponentially dominated in the tail and when $\rho(\mathbf{\Gamma}) > 1$, so that the speed of propagation c_0 is finite and positive.

THEOREM 5.7. *For every $\varepsilon^* > 0$ and $\delta^* > 0$ there exists a T^* such that $v_i(\mathbf{s},t) \leq \varepsilon^*$ for $|\mathbf{s}| \geq (c_0 + \delta^*)t$ for all $i = 1, ..., n$ provided $t \geq T^*$. A second result also holds which depends upon the structure of the infection matrix and is specified in the three cases below.*

1. *When $\mathbf{\Gamma}$ has all finite entries, then $\eta_i(\mathbf{0}) - \varepsilon^* \leq v_i(\mathbf{s},t) \leq \eta_i(\mathbf{b})$ for $|\mathbf{s}| \leq (c_0 - \delta^*)t$ and $i = 1, ..., n$ provided $t \geq T^*$, where $\eta_i(\mathbf{0})$ and $\eta_i(\mathbf{b})$ are the lower and upper bounds for the final size for type i obtained in Theorem 3.2 part 1 and Theorem 3.3 part 1.*
2. *When $\mathbf{\Gamma}$ has an infinite entry in every row, then $1 - \varepsilon^* \leq v_i(\mathbf{s},t) \leq 1$ for $|\mathbf{s}| \leq (c_0 - \delta^*)t$ and $i = 1, ..., n$ provided $t \geq T^*$.*
3. *The final case corresponds to a partitioning of $\mathbf{\Gamma}$ as in Theorem 3.2 part 3, re-ordering the types if necessary, so that*

$$\mathbf{\Gamma} = \begin{pmatrix} \mathbf{\Gamma}_{11} & \mathbf{\Gamma}_{12} \\ \mathbf{\Gamma}_{21} & \mathbf{\Gamma}_{22} \end{pmatrix},$$

where $\mathbf{\Gamma}_{11}$ is an $m \times m$ matrix, $(\mathbf{\Gamma}_{11}\mathbf{\Gamma}_{12})$ has no infinite element and $(\mathbf{\Gamma}_{21}\mathbf{\Gamma}_{22})$ has at least one infinite element in each row.

Then for $i = m+1, ..., n$, $1-\varepsilon^* \leq v_i(\mathbf{s},t) \leq 1$ for $|\mathbf{s}| \leq (c_0-\delta^*)t$ provided $t \geq T^*$. For $i = 1, ..., m$, $\eta_i^*(\mathbf{a}^*) - \varepsilon^* \leq v_i(\mathbf{s},t) \leq \eta_i^*(\mathbf{b}^*)$ for $|\mathbf{s}| \leq (c_0 - \delta^*)t$ provided $t \geq T^*$, where $\eta_i^*(\mathbf{a}^*)$ and $\eta_i^*(\mathbf{b}^*)$ are the lower and upper bounds for the final size obtained in Theorem 3.2 part 3 and Theorem 3.3 part 3.

PROOF. From Corollary 5.2 it immediately follows that, for any $c > c_0$ and $\varepsilon > 0$ there exists a T such that $\sup\{v_i(\mathbf{s},t) : |\mathbf{s}| \geq ct\} \leq \varepsilon$ for $t \geq T$ and $i = 1, ..., n$. Hence for all i, $v_i(\mathbf{s},t) \leq \varepsilon$ for $|\mathbf{s}| \geq ct$ and $t \geq T$. Take $c = c_0 + \delta^*$ and $\varepsilon = \varepsilon^*$ and find the corresponding value T_1 of T. Then $v_i(\mathbf{s},t) \leq \varepsilon^*$ for $|\mathbf{s}| \geq (c_0 + \delta^*)t$ provided $t \geq T_1$.

The cases are now treated separately.

1. From Theorem 3.3 part 1, $v_i(\mathbf{s},t) \leq \eta_i(\mathbf{b})$ for all $\mathbf{s}$ and t and hence for the restricted range $|\mathbf{s}| \leq (c_0 - \delta^*)t$ for any $t \geq 0$ and all i.

   Using Theorem B.3 part 1, $\eta_i(R,T) \uparrow \eta_i(\mathbf{0})$ as $R, T \to \infty$. Thus we can choose an R and T which satisfy the conditions of Corollary 5.3 and are sufficiently large so that $\eta_i(\mathbf{0}) - \varepsilon^*/2 \leq \eta_i(R,T) \leq \eta_i(\mathbf{0})$ for all i. Now take this R and T and take $\varepsilon = \varepsilon^*/2$ and $c = c_0 - \delta^*$. From Corollary 5.3 we can find the corresponding t^* so that, for all i, $v_i(\mathbf{s},t) \geq \eta_i(R,T) - \varepsilon^*/2$ for $|\mathbf{s}| \leq (c_0 - \delta^*)t$ provided $t \geq t^*$. Then, for all i,

$$v_i(\mathbf{s},t) \geq \eta_i(R,T) - \varepsilon^*/2 \geq \eta_i(\mathbf{0}) - \varepsilon^*$$

   for $|\mathbf{s}| \leq (c_0 - \delta^*)t$ provided $t \geq t^*$.

2. The result that $v_i(\mathbf{s},t) \leq 1$ trivially holds for all $\mathbf{s}$ and t and hence for the restricted range $|\mathbf{s}| \leq (c_0 - \delta^*)t$ for any $t \geq 0$ and all i.

   Using Theorem B.3 part 2, $\eta_i(R,T) \uparrow 1$ as $R, T \to \infty$. Thus we can choose an R and T which satisfy the conditions of Corollary 5.3 and are sufficiently large so that $1 - \varepsilon^*/2 \leq \eta_i(R,T) \leq 1$ for all i. Now take this R and T and take $\varepsilon = \varepsilon^*/2$ and $c = c_0 - \delta^*$. From Corollary 5.3 we can find the corresponding t^* so that, for all i, $v_i(\mathbf{s},t) \geq \eta_i(R,T) - \varepsilon^*/2$ for $|\mathbf{s}| \leq (c_0 - \delta^*)t$ provided $t \geq t^*$. Then, for all i,

$$v_i(\mathbf{s},t) \geq \eta_i(R,T) - \varepsilon^*/2 \geq 1 - \varepsilon^*$$

for $|\mathbf{s}| \leq (c_0 - \delta^*)t$ provided $t \geq t^*$.

3. The result that $v_i(\mathbf{s},t) \leq 1$ trivially holds for all i, $\mathbf{s}$ nd t. It therefore holds for $i = m+1, ..., n$, for the restricted range $|\mathbf{s}| \leq (c_0 - \delta^*)t$ for any $t \geq 0$. Also, from Theorem 3.3 part 3, $v_i(\mathbf{s},t) \leq \eta_i^*(\mathbf{b}^*)$ for $i = 1, ..., m$ and all $\mathbf{s}$ and t and hence for $i = 1, ..., m$ for the restricted range $|\mathbf{s}| \leq (c_0 - \delta^*)t$ for any $t \geq 0$.

   Using Theorem B.3 part 3, $\eta_i(R,T) \uparrow 1$ for $i = m+1, ..., n$ and $\eta_i(R,T) \uparrow \eta_i^*(\mathbf{a}^*)$ for $i = 1, ..., m$ as $R, T \to \infty$. Thus we can choose an R and T which satisfy the conditions of Corollary 5.3 and are sufficiently large so that $1 - \varepsilon^*/2 \leq \eta_i(R,T) \leq 1$ for $i = m+1, ..., n$ and $\eta_i(\mathbf{0}) - \varepsilon^*/2 \leq \eta_i(R,T) \leq \eta_i(\mathbf{0})$ for $i = 1, ..., m$. Now take this R and T and take $\varepsilon = \varepsilon^*/2$ and $c = c_0 - \delta^*$. From Corollary 5.3 we can find the corresponding t^* so that, for all i, $v_i(\mathbf{s},t) \geq \eta_i(R,T) - \varepsilon^*/2$ for $|\mathbf{s}| \leq (c_0 - \delta^*)t$ provided $t \geq t^*$. Then, for $i = m+1, ..., n$,

$$v_i(\mathbf{s},t) \geq \eta_i(R,T) - \varepsilon^*/2 \geq 1 - \varepsilon^*$$

   for $|\mathbf{s}| \leq (c_0 - \delta^*)t$ provided $t \geq t^*$. Also, for $i = 1, ..., m$,

$$v_i(\mathbf{s},t) \geq \eta_i(R,T) - \varepsilon^*/2 \geq \eta_i^*(\mathbf{a}^*) - \varepsilon^*$$

   for $|\mathbf{s}| \leq (c_0 - \delta^*)t$ provided $t \geq t^*$.

For each case, the theorem then follows by taking $T^* = \max(T_1, t^*)$.

□

Observe that $\eta_i(\mathbf{b})$ can be made arbitrarily close to $\eta_i(\mathbf{0})$ for all i for case 1, and $\eta_i^*(\mathbf{b}^*)$ can be made arbitrarily close to $\eta_i^*(\mathbf{a}^*)$ for $i = 1, ..., m$ for case 3, provided the amount of initial infection is made sufficiently small.

CHAPTER 6

An epidemic on sites

6.1 A one-type finite site spatial model

The non-spatial theory of Chapter 2 may be used to immediately obtain results for an epidemic on a finite number of sites. These results are presented first. Consideration is then given to the discrete space analogue of the continuous space models of Chapter 3 in which individuals are located at a symmetrical N-dimensional lattice of sites, which may be represented by the integer lattice.

A single type epidemic on N sites can be regarded as an N-type non-spatial epidemic by treating the sites as types. Consider the model with varying infectivity specified in Chapter 2. For the model on sites, σ_i denotes the size of the population at the ith site and $\lambda_{ij}(\tau)$ denotes the rate of infection of a susceptible individual at the ith site by an infectious individual at the jth site who was infected time τ ago. Also $x_i(t)$ represents the proportion of susceptible individuals at site i and time t. Then the $x_i(t)$ satisfy equations (2.10) with $n = N$.

Since $x_i(t)$ is monotone decreasing in t and is bounded, it tends to a limit as t tends to infinity. Let $v_i = 1 - \lim_{t\to\infty} x_i(t)$, which is the final size of the epidemic at site i. The results then follow immediately from Theorems 2.2 and 2.3.

Define $\mathbf{\Gamma} = (\gamma_{ij})$, as in Section 2.5, where $\gamma_{ij} = \sigma_j \int_0^\infty \lambda_{ij}(\tau)d\tau$. We assume that all sites are 'reachable' from all other sites, so that $\mathbf{\Gamma}$ is non-reducible. When $\rho(\mathbf{\Gamma}) \leq 1$ a major epidemic does not occur. The final size tends to zero at all sites if we let the amount of initial infection tend to zero. When $\rho(\mathbf{\Gamma}) > 1$, and the infectives introduced at time zero have some infectious influence, a major epidemic occurs affecting all sites. As the amount of initial infection tends to zero, the final sizes v_i, for $i = 1, ..., N$, tend to the unique positive values specified in Theorem 2.3 part 2. For any site i for which there exists a site j with γ_{ij} infinite, the final size $v_i = 1$ regardless of the amount of initial infection and we define $\eta_i = 1$. For any other site i, v_i tends to η_i as the amount of initial infection tends to zero, where $0 < \eta_i < 1$ and the η_i are the unique positive values satisfying $-\log(1 - \eta_i) = \sum_{j=1}^N \gamma_{ij}\eta_j$ for all such sites i.

Some simple results linking the spatial to the non-spatial results are obtained if some additional assumptions are made. The population size at each site is assumed to be constant, so that $\sigma_i = \sigma$ for all i. The infection rate $\lambda_{ij}(\tau)$, of a susceptible individual at site i by an infective individual at site j who was infected time τ ago, can be considered as being a product of two terms. We write $\lambda_{ij}(\tau) = \alpha(\tau)r_{ij}$, where $\alpha(\tau)$ is the infection rate per contact between a susceptible and an infected individual infected time τ ago, and r_{ij} is the contact rate between an individual at site i and one at site j. It is reasonable to assume that the contact rates between individuals at more extreme sites will increase to compensate for less between site contacts. We assume therefore that the total contact rate for an individual at site i, $\sum_{j=1}^{N} r_{ij}$, does not depend on i. Then for all i we define $\lambda(\tau) = \sum_{j=1}^{N} \lambda_{ij}(\tau) = \alpha(\tau) \sum_{j=1}^{N} r_{ij}$ and $\gamma = \sigma \int_0^\infty \lambda(\tau) d\tau$.

With these constraints imposed, we can prove the result that no major epidemic occurs if $\gamma \leq 1$. When $\gamma > 1$ then, as the amount of initial infection tends to zero, the final size tends to a common value η at all sites, where $\eta = 1$ if γ is infinite, otherwise η is the unique positive solution to $-\log(1-\eta) = \gamma\eta$. This corresponds to the results for the limit of the final size of a non-spatial epidemic with population size σ, infection rate $\lambda(\tau)$ for an individual infected time τ ago and $\gamma = \sigma \int_0^\infty \lambda(\tau) d\tau$. The proof is given below.

We first show that when γ is infinite then $\mathbf{\Gamma}$ has an infinite entry in every row and when γ is finite then $\mathbf{\Gamma}$ is a finite matrix. Now $\gamma_{ij} = \sigma \int_0^\infty \lambda_{ij}(\tau) d\tau$ and

$$\gamma = \sigma \int_0^\infty \lambda(\tau) d\tau = \sum_{j=1}^{N} \sigma \int_0^\infty \lambda_{ij}(\tau) d\tau = \sum_{j=1}^{N} \gamma_{ij}$$

for all $i = 1, ..., N$. The result is then immediate.

Therefore when γ is infinite $\rho(\mathbf{\Gamma})$ is also infinite. Also, from Theorem 2.3, the final size $v_i = 1$ at all sites regardless of the amount of initial infection.

We next show that when γ is finite then $\rho(\mathbf{\Gamma}) = \gamma$. Now when γ is finite, $\mathbf{\Gamma}$ has all finite entries and $\gamma\mathbf{1} = \mathbf{\Gamma}\mathbf{1}$. Hence γ is an eigenvalue of $\mathbf{\Gamma}$ with corresponding eigenvector $\mathbf{1}$. Perron Frobenius theory (see Theorem A.2 part 7) then implies that $\rho(\mathbf{\Gamma}) = \gamma$.

Hence the condition $\rho(\mathbf{\Gamma}) > 1$ for a major epidemic is equivalent to $\gamma > 1$. Therefore when $\gamma \leq 1$, and hence $\rho(\mathbf{\Gamma}) \leq 1$, from Theorem 2.3 the final size of the epidemic tends to zero at all sites as the amount of initial infection tends to zero.

When γ is finite with $\gamma > 1$, then $\mathbf{\Gamma}$ is finite with $\rho(\mathbf{\Gamma}) > 1$. Hence from Theorem 2.3 the final size of the epidemic at site i tends to $\eta_i > 0$ as the amount of initial infection tends to zero. Here $\{\mathbf{y}\}_i = y_i = \eta_i$, $i = 1, ..., N$, is the unique positive solution to the equations

$$-\log(1 - y_i) = \{\mathbf{\Gamma}\mathbf{y}\}_i \tag{6.1}$$

for $i = 1, ..., N$. Consider $\mathbf{y} = \eta\mathbf{1}$, where $-\log(1-\eta) = \gamma\eta$. Then, for any $i = 1, ..., N$,

$$\{\mathbf{\Gamma}\mathbf{y}\}_i = \eta\{\mathbf{\Gamma}\mathbf{1}\}_i = \eta\gamma = -\log(1-\eta) = -\log(1-y_i).$$

Hence $\mathbf{y} = \eta\mathbf{1}$ is a solution to equations (6.1), and so is the unique positive solution to these equations. Hence $\eta_i = \eta$ for all sites i. Therefore when γ is finite with $\gamma > 1$, the final size of the epidemic tends to $\eta > 0$ at all sites as the amount of initial infection tends to zero, where $-\log(1-\eta) = \gamma\eta$. This completes the proof.

6.2 The multi-type finite site spatial model

Now consider an n-type epidemic on N sites. Again we only consider the model with varying infectivity and assume that the epidemic is non-reducible so that an infective of any type j at any site r can infect a susceptible of any type i at any site s, possibly through a series of infections. Let $\lambda_{ij}(s,r,\tau)$ be the rate of infection of a type i susceptible individual at site s by a type j infective individual at site r who was infected time τ ago. Denote the corresponding rate by $\lambda^*_{ij}(s,r,\tau)$ when the infective individual was one of those introduced from outside at time zero. Let $\sigma_i(s)$ be the size of the ith population at site s and let $x_i(s,t)$ denote the proportion of type i individuals at site s who are still susceptible at time t. Define $\sigma(s) = \sum_{j=1}^{n} \sigma_j(s)$ and take $\sigma(r)\varepsilon_i(r,\tau)$ to be the number of individuals of type j who were introduced from outside at site r at time zero and had at that time been infected time τ ago.

This epidemic can also be treated as a non-spatial epidemic. In this case there will be nN 'types', which consist of all type and site combinations. Results for this finite site model are then immediately obtained from Theorem 2.3, as for the one-type model on sites. Let $v_i(s) = 1 - \lim_{t\to\infty} x_i(s,t)$, so that $v_i(s)$ is the final size of the epidemic for type i individuals at site s. Define $\gamma_{ij}(s,r) = \sigma_j \int_0^\infty \lambda_{ij}(s,r,\tau)d\tau$. Note that, from Theorem 2.3, if there exists a j and an r such that $\gamma_{ij}(s,r)$ is infinite, then $v_i(s) = 1$. For all other type site combinations, i and s, such that $\gamma_{ij}(s,r)$ is finite for all j,r the final size equations are

$$-\log(1 - v_i(s)) = \sum_{j=1}^{n} \sum_{r=1}^{N} v_j(r)\gamma_{ij}(s,r) + a_i(s).$$

If we define $\gamma^*_{ij}(s,r,\tau) = \sigma(r)\lambda^*_{ij}(s,r,\tau)$, then $a_i(s) = \sum_{j=1}^{k} \sum_{r=1}^{N} \int_0^\infty \int_0^\infty \gamma^*_{ij}(s,r,t+\tau)\varepsilon_j(r,\tau)d\tau dt$. It is assumed that $a_i(s) > 0$ for some i and s.

A way of ordering the type site combinations is to let type i and site s correspond to the combined 'type' $(i-1)N+s$, so that the ordering sets $v_{(i-1)N+s} = v_i(s)$ and $\gamma_{(i-1)N+s,(j-1)N+r} = \gamma_{ij}(s,r)$ for $i,j = 1,...n$ and $r,s = 1,...N$. Then in Theorem 2.3 n is replaced by Nn and $\{\mathbf{\Gamma}\}_{ij} = \gamma_{ij}$. When $\rho(\mathbf{\Gamma}) \leq 1$ then the final size $v_i(s)$ tends to zero for all types i and sites s as the amount of initial infection tends to zero. When $\rho(\mathbf{\Gamma}) > 1$, for any i,s such that there exists a pair j,r with $\gamma_{ij}(s,r)$ infinite, then $v_i(s) = 1$ regardless of the amount of initial infection and we define $\eta_i(s) = 1$. For all other type and site combinations i and s, $v_i(s)$ tends to $\eta_i(s) > 0$ as the amount of initial infection tends to zero, where $-\log(1 - \eta_i(s)) = \sum_{j=1}^{n} \sum_{r=1}^{N} \gamma_{ij}(s,r)\eta_j(r)$ for all such i,s.

We now impose equivalent constraints to those imposed for the one-type model. We assume that the population sizes for different types are constant over all sites and can then define $\sigma_i = \sigma_i(s)$, and hence $\sigma = \sigma(s)$, for all s. Then $\sigma = \sum_{j=1}^{n} \sigma_j$. In similar manner to Chapter 2 Section 2.1 we can write $\lambda_{ij}(s,r,\tau) = r_{ij}(s,r)\alpha_{ij}(\tau)$, where $r_{ij}(s,r)$ represents the contact rate between type i and type j individuals at sites s and r respectively and $\alpha_{ij}(\tau)$ represents the infection rate per contact between a type i susceptible and a type j infective who was infected time τ ago. Then we assume that the total contact rate between a type i individual at site s with type j individuals over all sites is independent of s, so that $\sum_{r=1}^{N} r_{ij}(s,r) = r_{ij}$. If we define $\lambda_{ij}(\tau) = \alpha_{ij}(\tau)r_{ij}$, then $\lambda_{ij}(\tau) = \sum_{r=1}^{N} \lambda_{ij}(s,r,\tau)$ for all sites s. Now define $\{\tilde{\mathbf{\Gamma}}\}_{ij} = \tilde{\gamma}_{ij}$, where $\tilde{\gamma}_{ij} = \sigma_j \int_0^\infty \lambda_{ij}(\tau)d\tau$. We then obtain the result that, for all s,

$$\tilde{\gamma}_{ij} = \sum_{r=1}^{N} \sigma_j \int_0^\infty \lambda_{ij}(s,r,\tau)d\tau = \sum_{r=1}^{N} \gamma_{ij}(s,r).$$

Define $\{\mathbf{\Gamma}_{ij}\}_{sr} = \gamma_{ij}(s,r)$ and $\{\mathbf{v}_i\}_s = v_i(s)$. Then using the ordering of the site and type combinations given previously, the non-reducible matrix $\mathbf{\Gamma}$ and vector $\mathbf{v}$ are partitioned so that

$$\mathbf{\Gamma} = \begin{pmatrix} \mathbf{\Gamma}_{11} & \dots & \mathbf{\Gamma}_{1n} \\ \vdots & \ddots & \vdots \\ \mathbf{\Gamma}_{n1} & \dots & \mathbf{\Gamma}_{nn} \end{pmatrix} \text{ and } \mathbf{v} = \begin{pmatrix} \mathbf{v}_1 \\ \vdots \\ \mathbf{v}_n \end{pmatrix}.$$

We can then prove the following theorem giving the limit of the final sizes of the epidemic as the amount of initial infection tends to zero.

THEOREM 6.1.

1. *If $\tilde{\mathbf{\Gamma}}$ has all finite entries then so does $\mathbf{\Gamma}$. In this case $\rho(\tilde{\mathbf{\Gamma}}) = \rho(\mathbf{\Gamma})$. Also if $\rho(\tilde{\mathbf{\Gamma}}) \leq 1$ then the final size $v_i(s)$, for all i and s, tends to zero as the amount of initial infection tends to zero.*
2. *When $\tilde{\mathbf{\Gamma}}$ has all finite entries and $\rho(\tilde{\mathbf{\Gamma}}) > 1$, then $v_i(s)$ tends to $\tilde{\eta}_i > 0$ for all s as the amount of initial infection tends to zero. Here $\{\mathbf{y}\}_i = y_i = \tilde{\eta}_i$, for $i = 1, ..., n$ is the unique positive solution to*

$$-\log(1 - y_i) = \{\tilde{\mathbf{\Gamma}}\mathbf{y}\}_i$$

   for $i = 1, ..., n$.
3. *If $\tilde{\mathbf{\Gamma}}$ has an infinite entry in every row, then $v_i(s) = 1$ for all $i = 1, ..., n$ and $s = 1, ..., N$ regardless of the amount of initial infection.*
4. *For the remaining case, we may rearrange the types so that $\tilde{\mathbf{\Gamma}}$ has an infinite entry in each of the last $n - k$ rows only. In this case $v_i(s) = 1$ for all s and all $i = k + 1, ..., n$. As the amount of initial infection tends to zero, for all s and $i = 1, ..., k$, $v_i(s)$ tends to $\tilde{\eta}_i > 0$. Here $\tilde{\eta}_i$, for $i = 1, ..., k$, satisfies the relation*

$$-\log(1 - \tilde{\eta}_i) = \sum_{j=1}^{k} \{\tilde{\mathbf{\Gamma}}\}_{ij}\tilde{\eta}_j + \sum_{j=k+1}^{n} \{\tilde{\mathbf{\Gamma}}\}_{ij},$$

   for $i = 1, ..., k$.

PROOF.

1. Since $\tilde{\gamma}_{ij} = \{\mathbf{\Gamma}_{ij}\mathbf{1}\}_s$ for all sites s, if $\mathbf{\Gamma}_{ij}$ has an infinite entry in any row then it has one in every row. This occurs precisely when $\tilde{\gamma}_{ij}$ is infinite. Hence $\tilde{\mathbf{\Gamma}}$ has all finite entries if and only if $\mathbf{\Gamma}$ has all finite entries.

   Consider the case when $\tilde{\mathbf{\Gamma}}$ has all finite entries. Let $\mathbf{u}$ be the right eigenvector of $\tilde{\mathbf{\Gamma}}$ corresponding to $\rho(\tilde{\mathbf{\Gamma}})$ and let $u_i = \{\mathbf{u}\}_i$. Define $\mathbf{w}$ to be the nN vector with $\mathbf{w}' = (u_1\mathbf{1}', u_2\mathbf{1}'..., u_n\mathbf{1}')$, where each of the component vectors are length N. Then

$$\mathbf{\Gamma w} = \begin{pmatrix} \mathbf{\Gamma}_{11} & \dots & \mathbf{\Gamma}_{1n} \\ \vdots & \ddots & \vdots \\ \mathbf{\Gamma}_{n1} & \dots & \mathbf{\Gamma}_{nn} \end{pmatrix} \begin{pmatrix} u_1 \mathbf{1} \\ \vdots \\ u_n \mathbf{1} \end{pmatrix} = \begin{pmatrix} \sum_{j=1}^n u_j \mathbf{\Gamma}_{1j} \mathbf{1} \\ \vdots \\ \sum_{j=1}^n u_j \mathbf{\Gamma}_{nj} \mathbf{1} \end{pmatrix}$$
$$= \begin{pmatrix} \sum_{j=1}^n \tilde{\gamma}_{1j} u_j \mathbf{1} \\ \vdots \\ \sum_{j=1}^n \tilde{\gamma}_{nj} u_j \mathbf{1} \end{pmatrix} = \begin{pmatrix} \rho(\tilde{\mathbf{\Gamma}}) u_1 \mathbf{1} \\ \vdots \\ \rho(\tilde{\mathbf{\Gamma}}) u_n \mathbf{1} \end{pmatrix} = \rho(\tilde{\mathbf{\Gamma}}) \mathbf{w}.$$

Hence $\mathbf{w} > \mathbf{0}$ is the eigenvector of $\mathbf{\Gamma}$ corresponding to the eigenvalue $\rho(\tilde{\mathbf{\Gamma}})$, therefore from Perron-Frobenius theory $\rho(\mathbf{\Gamma}) = \rho(\tilde{\mathbf{\Gamma}})$ (see Theorem A.2 part 7).

From Theorem 2.3, when $\rho(\mathbf{\Gamma}) \leq 1$ then $\mathbf{v}$ tends to $\mathbf{0}$, and hence $v_i(s)$ tends to 0 for all i, s, as the amount the amount of initial infection tends to zero.

2. If $\tilde{\mathbf{\Gamma}}$ is finite with $\rho(\tilde{\mathbf{\Gamma}}) > 1$ then, from part 1, $\mathbf{\Gamma}$ is finite with $\rho(\mathbf{\Gamma}) > 1$. Hence from Theorem 2.3 $\mathbf{v}$ tends to a limit $\boldsymbol{\eta}$ as the amount of initial infection tends to zero. Partition $\boldsymbol{\eta}' = (\boldsymbol{\eta}_1', ..., \boldsymbol{\eta}_n')$ to correspond to $\mathbf{v} = (\mathbf{v}_1', ..., \mathbf{v}_n')$. Define $\{\boldsymbol{\phi}_i\}_r = -\log(1 - \{\boldsymbol{\eta}_i\}_r)$. Then $\boldsymbol{\eta}$ is the unique positive solution to

$$\begin{pmatrix} \boldsymbol{\phi}_1 \\ \vdots \\ \boldsymbol{\phi}_n \end{pmatrix} = \begin{pmatrix} \mathbf{\Gamma}_{11} & \dots & \mathbf{\Gamma}_{1n} \\ \vdots & \ddots & \vdots \\ \mathbf{\Gamma}_{n1} & \dots & \mathbf{\Gamma}_{nn} \end{pmatrix} \begin{pmatrix} \boldsymbol{\eta}_1 \\ \vdots \\ \boldsymbol{\eta}_n \end{pmatrix}. \tag{6.2}$$

In order to show that the final size $v_i(s)$ tends to $\tilde{\eta}_i > 0$ for all i, s, where $-\log(1 - \tilde{\eta}_i) = \sum_{j=1}^n \tilde{\gamma}_{ij} \tilde{\eta}_j$ for all i, we need only show that $\boldsymbol{\eta}_i = \tilde{\eta}_i \mathbf{1}$ for $i = 1, ..., n$ is a solution to equation (6.2).

Now in this case

$$\begin{pmatrix} \mathbf{\Gamma}_{11} & \dots & \mathbf{\Gamma}_{1n} \\ \vdots & \ddots & \vdots \\ \mathbf{\Gamma}_{n1} & \dots & \mathbf{\Gamma}_{nn} \end{pmatrix} \begin{pmatrix} \boldsymbol{\eta}_1 \\ \vdots \\ \boldsymbol{\eta}_n \end{pmatrix} = \begin{pmatrix} \mathbf{\Gamma}_{11} & \dots & \mathbf{\Gamma}_{1n} \\ \vdots & \ddots & \vdots \\ \mathbf{\Gamma}_{n1} & \dots & \mathbf{\Gamma}_{nn} \end{pmatrix} \begin{pmatrix} \tilde{\eta}_1 \mathbf{1} \\ \vdots \\ \tilde{\eta}_n \mathbf{1} \end{pmatrix}$$
$$= \begin{pmatrix} \sum_{j=1}^n \tilde{\eta}_j \mathbf{\Gamma}_{1j} \mathbf{1} \\ \vdots \\ \sum_{j=1}^n \tilde{\eta}_j \mathbf{\Gamma}_{nj} \mathbf{1} \end{pmatrix} = \begin{pmatrix} \sum_{j=1}^n \tilde{\eta}_j \tilde{\gamma}_{1j} \mathbf{1} \\ \vdots \\ \sum_{j=1}^n \tilde{\eta}_j \tilde{\gamma}_{nj} \mathbf{1} \end{pmatrix}$$
$$= \begin{pmatrix} -\log(1 - \tilde{\eta}_1) \mathbf{1} \\ \vdots \\ -\log(1 - \tilde{\eta}_n) \mathbf{1} \end{pmatrix} = \begin{pmatrix} \boldsymbol{\phi}_1 \\ \vdots \\ \boldsymbol{\phi}_n \end{pmatrix}.$$

So $\boldsymbol{\eta}_i = \tilde{\eta}_i \mathbf{1}$, for $i = 1, ..., n$, is a solution to equation (6.2), and hence is the unique positive solution. Hence the final size $v_i(s)$ tends to $\tilde{\eta}_i$, for all i, s, as the amount of initial infection tends to zero.

3. If $\tilde{\mathbf{\Gamma}}$ has an infinite entry in each row, then for each i there exists a j such that $\tilde{\gamma}_{ij}$ is infinite. Hence from the proof of part 1, $\mathbf{\Gamma}_{ij}$ has an infinite entry

in each row and therefore so has $\mathbf{\Gamma}$. Then from Theorem 2.3, $\mathbf{v} = \mathbf{1}$, and so $v_i(s) = 1$ for all i, s, regardless of the amount of initial infection.

4. Rearranging the n types if necessary, $\tilde{\mathbf{\Gamma}}$ has an infinite entry in row i for each $i = k+1, ..., n$, and all finite entries in the first k rows. Hence, as in the proof of part 3, $\mathbf{\Gamma}$ has an infinite entry in each of the last $N(n-k)$ rows and all finite entries in the first Nk rows.

As in the proof of part 3, it immediately follows that $\mathbf{v}_i = \mathbf{1}$ for $i = k+1, ..., n$. Hence the final size $v_i(s) = 1$ for all s and all $i = k+1, ..., n$ regardless of the amount of initial infection.

Also from Theorem 2.3, $\mathbf{v}_i$ tends to $\boldsymbol{\eta}_i$ for $i = 1, ..., k$ as the amount of initial infection tends to zero. Here $\boldsymbol{\eta}_i$ for $i = 1, ..., k$ is the unique positive solution to

$$\begin{pmatrix} \boldsymbol{\phi}_1 \\ \vdots \\ \boldsymbol{\phi}_k \end{pmatrix} = \begin{pmatrix} \mathbf{\Gamma}_{11} & \cdots & \mathbf{\Gamma}_{1k} \\ \vdots & \ddots & \vdots \\ \mathbf{\Gamma}_{k1} & \cdots & \mathbf{\Gamma}_{kk} \end{pmatrix} \begin{pmatrix} \boldsymbol{\eta}_1 \\ \vdots \\ \boldsymbol{\eta}_k \end{pmatrix} + \begin{pmatrix} \mathbf{\Gamma}_{1,(k+1)} & \cdots & \mathbf{\Gamma}_{1,n} \\ \vdots & \ddots & \vdots \\ \mathbf{\Gamma}_{k,(k+1)} & \cdots & \mathbf{\Gamma}_{k,n} \end{pmatrix} \begin{pmatrix} \mathbf{1} \\ \vdots \\ \mathbf{1} \end{pmatrix}, \tag{6.3}$$

where $\{\boldsymbol{\phi}_i\}_s = -\log(1 - \{\boldsymbol{\eta}_i\}_s$.

As in part 2, it is only necessary to show that $\boldsymbol{\eta}_i = \tilde{\eta}_i \mathbf{1} > \mathbf{0}$, for $i = 1, ..., k$, is a solution to equation (6.3). Here $\tilde{\eta}_i$, for $i = 1, ..., k$, is the unique positive solution to $-\log(1 - \tilde{\eta}_i) = \sum_{j=1}^{k} \tilde{\gamma}_{ij} \tilde{\eta}_j + \sum_{j=k+1}^{n} \tilde{\gamma}_{ij}$ for $i = 1, ..., k$.

Now for this value of the $\boldsymbol{\eta}_i$,

$$\begin{aligned}
&\begin{pmatrix} \mathbf{\Gamma}_{11} & \cdots & \mathbf{\Gamma}_{1k} \\ \vdots & \ddots & \vdots \\ \mathbf{\Gamma}_{k1} & \cdots & \mathbf{\Gamma}_{kk} \end{pmatrix} \begin{pmatrix} \boldsymbol{\eta}_1 \\ \vdots \\ \boldsymbol{\eta}_k \end{pmatrix} + \begin{pmatrix} \mathbf{\Gamma}_{1,(k+1)} & \cdots & \mathbf{\Gamma}_{1,n} \\ \vdots & \ddots & \vdots \\ \mathbf{\Gamma}_{k,(k+1)} & \cdots & \mathbf{\Gamma}_{k,n} \end{pmatrix} \begin{pmatrix} \mathbf{1} \\ \vdots \\ \mathbf{1} \end{pmatrix} \\
&= \begin{pmatrix} \mathbf{\Gamma}_{11} & \cdots & \mathbf{\Gamma}_{1k} \\ \vdots & \ddots & \vdots \\ \mathbf{\Gamma}_{k1} & \cdots & \mathbf{\Gamma}_{kk} \end{pmatrix} \begin{pmatrix} \tilde{\eta}_1 \mathbf{1} \\ \vdots \\ \tilde{\eta}_k \mathbf{1} \end{pmatrix} + \begin{pmatrix} \mathbf{\Gamma}_{1,(k+1)} & \cdots & \mathbf{\Gamma}_{1,n} \\ \vdots & \ddots & \vdots \\ \mathbf{\Gamma}_{k,(k+1)} & \cdots & \mathbf{\Gamma}_{k,n} \end{pmatrix} \begin{pmatrix} \mathbf{1} \\ \vdots \\ \mathbf{1} \end{pmatrix} \\
&= \begin{pmatrix} \sum_{j=1}^{k} \tilde{\eta}_j \mathbf{\Gamma}_{1j} \mathbf{1} + \sum_{j=k+1}^{n} \mathbf{\Gamma}_{1j} \mathbf{1} \\ \vdots \\ \sum_{j=1}^{k} \tilde{\eta}_j \mathbf{\Gamma}_{kj} \mathbf{1} + \sum_{j=k+1}^{n} \mathbf{\Gamma}_{kj} \mathbf{1} \end{pmatrix} = \begin{pmatrix} \left[\sum_{j=1}^{k} \tilde{\eta}_j \tilde{\gamma}_{1j} + \sum_{j=k+1}^{n} \tilde{\gamma}_{1j}\right] \mathbf{1} \\ \vdots \\ \left[\sum_{j=1}^{k} \tilde{\eta}_j \tilde{\gamma}_{kj} + \sum_{j=k+1}^{n} \tilde{\gamma}_{kj}\right] \mathbf{1} \end{pmatrix} \\
&= \begin{pmatrix} -\log(1 - \tilde{\eta}_1) \mathbf{1} \\ \vdots \\ -\log(1 - \tilde{\eta}_k) \mathbf{1} \end{pmatrix} = \begin{pmatrix} \boldsymbol{\phi}_1 \\ \vdots \\ \boldsymbol{\phi}_k \end{pmatrix}.
\end{aligned}$$

Hence $\boldsymbol{\eta}_i = \tilde{\eta}_i \mathbf{1} > \mathbf{0}$, for $i = 1, ..., k$, is a solution to equation (6.3). Therefore $v_i(s)$ tends to $\tilde{\eta}_i$, for all s and $i = 1, ..., k$, as the amount of initial infection tends to zero. This completes the proof of part 4 and hence of the theorem. □

Note that the limit of the final size for a type i individual at site s, which is the same for all sites s, is the limiting final size for a non-spatial epidemic. This non-spatial epidemic has size σ_i for type i individuals. The infection rate of a type i susceptible by a type j infective who was infected time τ ago is $\lambda_{ij}(\tau) = \alpha_{ij}(\tau)r_{ij}$ so that $\tilde{\gamma}_{ij} = \sigma_j \int_0^\infty \lambda_{ij}(\tau)d\tau$ and $\tilde{\mathbf{\Gamma}} = (\tilde{\gamma}_{ij})$ is the infection matrix.

6.3 The infinite sites spatial model

The discrete space analogue of the continuous space model of Chapter 3 is now considered. Individuals are located on a symmetrical N-dimensional lattice of sites, which may be represented by the integer lattice Z^N. The material in the remainder of this chapter is contained in Rass and Radcliffe [R19]. The final size equations are obtained. An infinite matrix formulation of the final size equations enables us to give a simple proof of the pandemic theorem for all N when a particular interpretation of the infection rates is given and symmetry and 'reachability' conditions are imposed on the contact distributions. A 'reachability' constraint is clearly needed since otherwise infection could miss some sites entirely, so that the lower bound on the spatial final size would necessarily be zero. A general proof of the pandemic theorem may also be obtained, which does not require the constraint on the infection rates. The proof is quite complex and is not included.

The upper bound on the spatial final size obtained in Chapter 3 Theorem 3.4 for the continuous space model is valid for the discrete space model also. Hence the limits of the spatial final size as the amount of initial infection tends to zero may also be obtained for the discrete space model of an epidemic on sites.

This model is now set up. Consider n types of individuals, but now type i has uniform size σ_i at all points of the N-dimensional lattice of integers $\mathbf{Z}^N$. Define $\sigma = \sum_{i=1}^n \sigma_i$. The proportion of susceptibles in population i at position s and time t is denoted by $x_i(\mathbf{s}, t)$. Let $I_i(\mathbf{s}, t, \tau)d\tau$ be the proportion of individuals in population i at site $\mathbf{s}$ at time t who were infected in the time interval $(t-\tau-d\tau, t-\tau)$. Let $\lambda_{ij}(\tau)$ be the rate of infection of susceptible individuals of type i by infectious individuals of type j who were infected time τ ago. Take $p_{ij}(\mathbf{r})$ to be the corresponding contact distribution representing the vector displacement $\mathbf{r}$ over which infection occurs.

Infected individuals of k types are introduced from outside at time $t = 0$; the number of such individuals of type j at site $\mathbf{s}$ who were infected in the time interval $(-\tau - d\tau, -\tau)$ being $\sigma\varepsilon_j(\mathbf{s}, \tau)d\tau$. The rate of infection by such individuals, of susceptibles from population i is $\lambda_{ij}^*(\tau)$ and the corresponding contact distribution is $p_{ij}^*(\mathbf{r})$. Let $\sigma\epsilon_i(\mathbf{s}) = \int_0^\infty \sigma\epsilon_i(\mathbf{s}, \tau)d\tau$ be the number of these type i infectives at site $\mathbf{s}$, where $\epsilon_i(\mathbf{s})$ is assumed to be uniformly bounded for each i.

The model is described by the equations

$$
\begin{aligned}
&\frac{\partial x_i(\mathbf{s},t)}{\partial t} = -x_i(\mathbf{s},t)\left(\sum_{j=1}^n \sum_{\mathbf{r}\in Z^N} \int_0^t I_j(\mathbf{s}-\mathbf{r},t,\tau)p_{ij}(\mathbf{r})\gamma_{ij}(\tau)d\tau + h_i(\mathbf{s},t)\right),\\
(6.4)\quad &x_i(\mathbf{s},t) = 1 - \int_0^t I_i(\mathbf{s},t,\tau)d\tau,\\
&I_i(\mathbf{s},t,\tau) = I_i(\mathbf{s},t-\tau,0),\ (i = 1,...,n),
\end{aligned}
$$

where $\gamma_{ij}(t) = \sigma_j\lambda_{ij}(t)$, $\gamma_{ij}^*(t) = \sigma\lambda_{ij}^*(t)$ and

(6.5)
$$h_i(\mathbf{s},t) = \sum_{j=1}^{k} \sum_{\mathbf{r} \in Z^N} p^*_{ij}(\mathbf{r}) \int_0^{\infty} \varepsilon_j(\mathbf{s}-\mathbf{r},\tau)\gamma^*_{ij}(t+\tau)d\tau.$$

The initial conditions are $x_i(\mathbf{s},0) \equiv 1$, for $i = 1, ..., n$. It is assumed that $h_i(\mathbf{s},t) > 0$ for some i and $\mathbf{s}$ and for t in some open interval of $\Re_+$.

The conditions on the infection rates are the same as in the non-spatial case. As in Section 2.4, the $\gamma_{ij}(t)$ and $\gamma^*_{ij}(t)$ are restricted to be bounded with continuous, bounded derivatives. Since the epidemic is only triggered by infection from outside the integrated infection rates, $\gamma^*_{ij} = \int_0^{\infty} \gamma^*_{ij}(t)dt$, are taken to be finite.

Symmetry conditions are imposed on the contact distributions. It is assumed that each $p_{ij}(\mathbf{r})$ is symmetric about zero for each of its entries. Hence, for all i, j and $\mathbf{r}$, we have the condition that $p_{ij}(\mathbf{r}) = p_{ij}(\mathbf{r}^*)$, where $\{\mathbf{r}^*\}_k = |\{\mathbf{r}\}_k|$ for all k.

Note that we could model the infection rate for one type i susceptible by one type j infective individual vector distance $\mathbf{r}$ away who was infected time τ ago by $s_i c_{ij}(\mathbf{r})d_j(\tau)$, where $d_j(\tau)$ is the infection rate per contact for a type j individual infected time τ ago and $c_{ij}(\mathbf{r})$ represents the contact rate between two type i and j individuals with $\mathbf{r}$ specifying the vector difference in their positions. Define $c_{ij} = \sum_{Z^N} c_{ij}(\mathbf{r})$. Then $\lambda_{ij}(\tau) = s_i c_{ij} d_j(\tau)$ and $p_{ij}(\mathbf{r}) = c_{ij}(\mathbf{r})/c_{ij}$.

Sensible constraints to impose are therefore the symmetry conditions $c_{ij}(\mathbf{r}) = c_{ji}(\mathbf{r})$, and hence $c_{ij} = c_{ji}$, and $p_{ij}(\mathbf{r}) = p_{ji}(\mathbf{r})$ for all i, j and $\mathbf{r}$. These are the additional constraints used for the proof in Section 6.6 of the multi-type pandemic theorem based on an infinite matrix formulation of the final size equations.

6.4 The final size equations

From the description of the model and equations (6.5), $x_i(\mathbf{s},t)$ is partially differentiable with respect to t and $I_i(\mathbf{s},t,\tau)$ (and hence $I_i(\mathbf{s},\tau,0)$) is a continuous function of τ. Also $x_i(\mathbf{s},t) = 1 - \int_0^t I_i(\mathbf{s},u,0)du$, so that differentiating we obtain $(\partial/\partial t)x_i(\mathbf{s},t) = -I_i(\mathbf{s},t,0) \leq 0$, so that the partial derivative is a continuous function of t and $x_i(\mathbf{s},t)$ is monotone decreasing in t.

Substituting for $I_i(\mathbf{s},t)$ in equations (6.5) gives the equivalent system of equations for the $x_i(\mathbf{s},t)$,

(6.6)
$$\frac{\partial x_i(\mathbf{s},t)}{\partial t} = -x_i(\mathbf{s},t)\left(\sum_{j=1}^{n} \sum_{\mathbf{r} \in Z^N} p_{ij}(\mathbf{r}) \int_0^t \frac{\partial x_i(\mathbf{s}-\mathbf{r},t-\tau)}{\partial t}\gamma_{ij}(\tau)d\tau + h_i(\mathbf{s},t)\right),$$

for $i = 1, ..., n$, where each $x_i(\mathbf{s},t)$ is non-negative, monotone decreasing in t and partially differentiable with respect to t, with its derivative continuous in t, and with $x_i(\mathbf{s},0) \equiv 1$. The conditions on the infection rates and on the infection from outside give the uniform bound $h_i(\mathbf{s},t) \leq h_i$, where $h_i = \sum_{j=1}^{k} \sup_\tau \gamma^*_{ij}(\tau) \sup_{\mathbf{r}} \varepsilon_j(\mathbf{r})$. Also

$$\left|\frac{\partial x_i(\mathbf{s},t)}{\partial t}\right| \leq x_i(\mathbf{s},t)\left(\sum_{j=1}^{n}\sup_{\tau}\gamma_{ij}(\tau)\sum_{\mathbf{r}\in Z^N}p_{ij}(\mathbf{r})\int_0^t I_j(\mathbf{s}-\mathbf{r},u,0)du + h_i\right)$$
$$\leq C_i x_i(\mathbf{s},t),$$

where $C_i = \sum_{j=1}^{n}\sup_{\tau}\gamma_{ij}(\tau) + h_i$. It is then simple to show, as in Section 3.1, that $x_i(\mathbf{s},t) > 0$ for all $\mathbf{s} \in Z^N$ and $t \geq 0$ and that the partial derivative of $\log x_i(\mathbf{s},t)$ is uniformly bounded.

If equations (6.6) are now integrated from 0 to t a system of integral equations is obtained, namely

$$(6.7) \quad -\log x_i(\mathbf{s},t) = \sum_{j=1}^{n}\sum_{\mathbf{r}\in Z^N}\int_0^t (1 - x_j(\mathbf{s}-\mathbf{r},t-\tau))p_{ij}(\mathbf{r})\gamma_{ij}(\tau)d\tau + H_i(\mathbf{s},t)$$

for $i = 1, ..., n$, where $H_i(\mathbf{s},t) = \int_0^t h_i(\mathbf{s},w)dw$.

There is a one-to-one correspondence between the solutions of equations (6.6) and (6.7), hence we can base the spatial analysis on equations (6.7). The results are summarised in Lemma 6.1, which may be proved in a similar fashion to Lemma 3.2. Theorem 6.2 then shows that equations (6.7) admit a unique solution of a specific form. The proof follows in similar manner to the proof of Theorem 3.1. Both proofs are therefore omitted.

LEMMA 6.1. *There exists a one to one correspondence between the positive, monotone decreasing (in t) solutions $x_i(\mathbf{s},t)$, with $x_i(\mathbf{s},0) = 1$, $(i = 1, ..., n)$, of equations (6.6) and (6.7) which are partially differentiable with respect to t with the partial derivatives continuous in t and uniformly bounded for all $\mathbf{s}$ and for t in a finite interval. Any such solution has $(\partial/\partial t)(-\log x_i(\mathbf{s},t))$ uniformly bounded for all $\mathbf{s}$ and $t \geq 0$.*

□

Note that it is simple to show from equations (6.6) that the partial derivative of $-\log x_i(\mathbf{s},t)$ with respect to t is uniformly bounded for all $\mathbf{s}$ and $t \geq 0$ by $\sum_{j=1}^{n}\sup_{\tau}\lambda_{ij}(\tau) + h_i$.

THEOREM 6.2 (EXISTENCE AND UNIQUENESS OF SOLUTIONS FOR THE SITES MODEL). *There exists a positive, monotone decreasing and continuous (in t) solution $x_i(\mathbf{s},t)$ to equations (6.7) with $x_i(\mathbf{s},0) = 1$, $(i = 1, ..., n)$, which is unique. The solution is partially differentiable with respect to t and the partial derivative is continuous in t and uniformly bounded for all $\mathbf{s}$ and for t in a finite interval.*

□

Consider $x_i(\mathbf{s},t)$, for each type i and for any site $\mathbf{s}$. It is a monotone decreasing function of t and is bounded below by zero, hence it tends to a limit which gives the proportion of type i individuals at site $\mathbf{s}$ who are unaffected by the epidemic. The corresponding proportion who eventually suffer the epidemic (termed the final size), which is denoted by $v_i(\mathbf{s})$, is then given by $v_i(\mathbf{s}) = \lim_{t\to\infty}(1 - x_i(\mathbf{s},t))$.

Now $H_i(\mathbf{s},t)$ is monotone increasing and is also bounded above since

$$
\begin{aligned}
H_i(\mathbf{s},t) &= \sum_{j=1}^{k} \sum_{\mathbf{r}\in Z^N} \int_0^\infty \int_\tau^{t+\tau} \epsilon_j(\mathbf{s}-\mathbf{r},\tau) p_{ij}^*(\mathbf{r}) \gamma_{ij}^*(v)\, dv\, d\tau \\
&\le \sum_{j=1}^{k} \left(\int_0^\infty \gamma_{ij}^*(v)\,dv \right) \sum_{\mathbf{r}\in Z^N} \epsilon_j(\mathbf{s}-\mathbf{r}) p_{ij}^*(\mathbf{r}) \\
&\le \sum_{j=1}^{k} \left(\int_0^\infty \gamma_{ij}^*(v)\,dv \right) \sup_{\mathbf{r}\in Z^N} \epsilon_j(\mathbf{r}).
\end{aligned}
$$

Hence $H_i(\mathbf{s},t)$ tends to a non-negative limit, which we denote by $a_i(\mathbf{s})$, as $t \to \infty$. Since $h_i(\mathbf{s},t) > 0$ for some i and $\mathbf{s}$ and some t in an open interval of $\Re_+$, $a_i(\mathbf{s}) > 0$ for some i and $\mathbf{s}$.

A relation is now obtained for the spatial final size. The discrete case follows in an almost identical manner to the continuous case. Note that in this theorem $*$ denotes convolution. The results are given in Theorem 6.3. Let $\gamma_{ij} = \int_0^\infty \gamma_{ij}(\tau)d\tau$, and define the infection matrix $\mathbf{\Gamma}$ by $\{\mathbf{\Gamma}\}_{ij} = \gamma_{ij}$.

THEOREM 6.3 (THE SPATIAL FINAL SIZE EQUATIONS).

1. *If $\rho(\mathbf{\Gamma})$ is finite, then the $v_i(\mathbf{s})$, satisfy the following equations*

$$
-\log(1 - v_i(\mathbf{s})) = \sum_{j=1}^{n} \gamma_{ij} p_{ij} * v_j(\mathbf{s}) + a_i(\mathbf{s}), \ (i = 1, ..., n, \ \mathbf{s} \in Z^N). \tag{6.8}
$$

2. *If $\mathbf{\Gamma}$ has an infinite element in every row, then $v_i(\mathbf{s}) = 1$ for all $\mathbf{s}$ and all $i = 1, ..., n$.*
3. *The remaining case corresponds to a partitioning of $\mathbf{\Gamma}$ (by permutation of the indices), as in Theorem 2.2 part 3, into*

$$
\mathbf{\Gamma} = \begin{pmatrix} \mathbf{\Gamma}_{11} & \mathbf{\Gamma}_{12} \\ \mathbf{\Gamma}_{21} & \mathbf{\Gamma}_{22} \end{pmatrix},
$$

where $(\mathbf{\Gamma}_{11}\mathbf{\Gamma}_{12})$ contains m rows and has no infinite element, and $(\mathbf{\Gamma}_{21}\mathbf{\Gamma}_{22})$ has at least one infinite element in each row.

Let $\mathbf{v}(\mathbf{s})$ and $\mathbf{a}(\mathbf{s})$ be the n-vectors which have $\{\mathbf{v}(\mathbf{s})\}_i = v_i(\mathbf{s})$ and $\{\mathbf{a}(\mathbf{s})\}_i = a_i(\mathbf{s})$. The corresponding partitioning of these vectors is given by $\mathbf{v}'(\mathbf{s}) = (\mathbf{v}_1'(\mathbf{s}), \mathbf{v}_2'(\mathbf{s}))$ and $\mathbf{a}'(\mathbf{s}) = (\mathbf{a}_1'(\mathbf{s}), \mathbf{a}_2'(\mathbf{s}))$.

Then $\mathbf{v}_2(\mathbf{s}) = \mathbf{1}$, and the entries of $\mathbf{v}_1(\mathbf{s})$ satisfy the equations

$$
-\log(1 - v_i(\mathbf{s})) = \sum_{j=1}^{m} \{\mathbf{\Gamma}_{11}\}_{ij} p_{ij} * v_j(\mathbf{s}) + \{\mathbf{\Gamma}_{12}\mathbf{1} + \mathbf{a}_1(\mathbf{s})\}_i, \ (i = 1, ..., m).
$$

□

In Sections 6.5 to 6.8 we restrict attention to a non-reducible epidemic, as has been done throughout the monograph. This restricts the infection matrix $\mathbf{\Gamma}$ to be non-reducible.

6.5 The pandemic theorem for the one-type case

A positive lower bound on the spatial final size can be obtained for the case when $\rho(\mathbf{\Gamma}) > 1$. We first consider the case when $n = 1$; dropping the subscripts, as they are unnecessary. Note that then $\gamma > 1$. If γ is infinite, from Theorem 6.3 part 2, we immediately have $v(\mathbf{s}) \equiv 1$. Only the case when γ is finite therefore needs to be considered.

It is first shown that $\inf_{\mathbf{s}\in Z^N} v(\mathbf{s}) > 0$. This can then be strengthened to give the pandemic result that $v(\mathbf{s}) \geq \eta(0)$ for all $\mathbf{s} \in Z^N$, where $z = \eta(0)$ is the unique positive solution to $-\log(1-z) = \gamma z$.

For simplicity, first consider sites on a line, i.e. the case $N = 1$. The final size equation (6.8) can be written in doubly infinite matrix form as

$$\begin{pmatrix} \vdots \\ \vdots \\ -\log(1-v(-1)) \\ -\log(1-v(0)) \\ -\log(1-v(1)) \\ \vdots \\ \vdots \end{pmatrix} = \gamma \begin{pmatrix} \ddots & \vdots & \vdots & \vdots & \ddots \\ \ddots & \vdots & \vdots & \vdots & \ddots \\ \cdots & p(0) & p(1) & p(2) & \cdots \\ \cdots & p(1) & p(0) & p(1) & \cdots \\ \cdots & p(2) & p(1) & p(0) & \cdots \\ \ddots & \vdots & \vdots & \vdots & \ddots \\ \ddots & \vdots & \vdots & \vdots & \ddots \end{pmatrix} \begin{pmatrix} \vdots \\ \vdots \\ v(-1) \\ v(0) \\ v(1) \\ \vdots \\ \vdots \end{pmatrix} + \begin{pmatrix} \vdots \\ \vdots \\ a(-1) \\ a(0) \\ a(1) \\ \vdots \\ \vdots \end{pmatrix}.$$

We assume that the contact distribution $p(r)$ is symmetric about 0 and that the contact distribution has finite mean, i.e. $\sum_{r=-\infty}^{\infty} |r|p(r) < \infty$.

Consider any k adjacent sites, $r+1,\ r+2,\ \ldots\ r+k$ for any integer r. Then the subset of the final size equations corresponding to these sites gives the vector inequality

$$\begin{pmatrix} -\log(1-v(r+1)) \\ -\log(1-v(r+2) \\ \vdots \\ -\log(1-v(r+k)) \end{pmatrix} \geq \gamma \begin{pmatrix} p(0) & p(1) & \cdots & p(k-1) \\ p(1) & p(0) & \cdots & p(k-2) \\ \vdots & \vdots & \ddots & \vdots \\ p(k-1) & p(k-2) & \cdots & p(0) \end{pmatrix} \begin{pmatrix} v(r+1) \\ v(r+2) \\ \vdots \\ v(r+k) \end{pmatrix} + \begin{pmatrix} a(r+1) \\ a(r+2) \\ \vdots \\ a(r+k) \end{pmatrix}.$$

Note that the matrix,

$$\mathbf{P}^{(k)} = \begin{pmatrix} p(0) & p(1) & \cdots & p(k-1) \\ p(1) & p(0) & \cdots & p(k-2) \\ \vdots & \vdots & \ddots & \vdots \\ p(k-1) & p(k-2) & \cdots & p(0) \end{pmatrix},$$

does not depend upon the site r chosen. A 'reachability' condition is imposed. This says that there exists a monotone increasing sequence $\{k_i\}$, with $k_i \to \infty$ as $i \to \infty$, for which each $\mathbf{P}^{(k_i)}$ is non-reducible.

LEMMA 6.2. *Let $\{k_i\}$ be a monotone increasing sequence with $k_i \to \infty$ as $i \to \infty$, for which each $k_i \times k_i$ matrix $\mathbf{P}^{(k_i)}$ is non-reducible. Then $\lim_{i\to\infty} \rho(\mathbf{P}^{(k_i)}) = 1$, where $\rho(\mathbf{P}^{(k_i)})$ is the Perron-Frobenius eigenvalue of $\mathbf{P}^{(k_i)}$.*

PROOF. It is easily seen that $\rho(\mathbf{P}^{(k_i)}) \le 1$. First observe that the maximum eigenvalue of a non-reducible non-negative matrix is strictly monotone increasing in its entries. We can increase each diagonal entry of $\mathbf{P}^{(k_i)}$ to make the matrix doubly stochastic (hence with maximum eigenvalue 1 and corresponding right and left eigenvectors the unit vector). The result is immediate.

Since the matrix $\mathbf{P}^{(k_i)}$ is real and symmetric, it has a spectral decomposition $\mathbf{P}^{(k_i)} = \sum_{j=1}^{k_i} \lambda_j \mathbf{E}_j$, where the λ_j are the (real) eigenvalues of $\mathbf{P}^{(k_i)}$ with corresponding idempotents $\mathbf{E}_j = \mathbf{u}_j\mathbf{u}_j{}'$ and $\mathbf{u}_1, ..., \mathbf{u}_{k_i}$ are an orthonormal set of (real) eigenvectors. Then

$$\begin{aligned}
\frac{1}{k_i}\mathbf{1}'\mathbf{P}^{(k_i)}\mathbf{1} &= \frac{1}{k_i}\sum_{j=1}^{k_i} \lambda_j (\mathbf{1}'\mathbf{u}_j)^2 \le \frac{1}{k_i}\rho\left(\mathbf{P}^{(k_i)}\right)\sum_{j=1}^{k_i}(\mathbf{1}'\mathbf{u}_j)^2 \\
&= \frac{1}{k_i}\rho\left(\mathbf{P}^{(k_i)}\right)\mathbf{1}'\left(\sum_{j=1}^{k_i}\mathbf{E}_j\right)\mathbf{1} \\
&= \frac{1}{k_i}\rho\left(\mathbf{P}^{(k_i)}\right)\mathbf{1}'\mathbf{1} = \rho\left(\mathbf{P}^{(k_i)}\right).
\end{aligned}$$

Also

$$\frac{1}{k_i}\mathbf{1}'\mathbf{P}^{(k_i)}\mathbf{1} = \frac{1}{k_i}\sum_{u=-(k_i-1)}^{(k_i-1)} (k_i - |u|)p(u).$$

Hence

$$\rho(\mathbf{P}^{(k_i)}) \ge \sum_{u=-(k_i-1)}^{(k_i-1)} p(u) - \frac{1}{k_i}\sum_{u=-(k_i-1)}^{(k_i-1)} |u|p(u).$$

The right hand side tends to 1 as $i \to \infty$ since $\int_{-\infty}^{\infty} |u|p(u)du$ is finite. Hence the limit of the left hand side is bounded below by 1. Therefore $\lim_{i\to\infty} \rho(\mathbf{P}^{(k_i)}) = 1$. □

A similar result is now established for general dimension N. This is then used to show that $v(\mathbf{s}) > 0$ for all $\mathbf{s} \in Z^N$. It is assumed that the contact distribution is symmetric about 0 in each of its entries. For all $1 \le r \le N$ and all sequences $1 \le i_1 < i_2 < ... < i_r \le N$, if we write the N-vector $\mathbf{u} = (u_i)$, then it is also assumed that $\sum_{u_1=-\infty}^{\infty} ... \sum_{u_N=-\infty}^{\infty} \prod_{j=1}^{r} |u_{i_j}|p(\mathbf{u}) < \infty$.

Consider the set of k^N sites,

$$A_k(\mathbf{r}) = \{\mathbf{z} \in Z^N : \{\mathbf{r}\}_i + 1 \le \{\mathbf{z}\}_i \le \{\mathbf{r}\}_i + k \text{ for } i = 1, ..., N\}.$$

Let $\mathbf{u}^*$, $\mathbf{v}^*$ and $\mathbf{a}^*$ be the vectors of length k^N which have entries corresponding to $-\log(1-v(\mathbf{z}))$, $v(\mathbf{z})$ and $a(\mathbf{z})$ respectively for $\mathbf{z} \in A_k(\mathbf{r})$. Then from the corresponding subset of the final size equations (6.8) we obtain

$$\mathbf{u}^* \geq \gamma \mathbf{P}^{(k)} \mathbf{v}^* + \mathbf{a}^*, \tag{6.9}$$

where $\mathbf{P}^{(k)}$ is a $k^N \times k^N$ matrix which does not depend upon $\mathbf{r}$. Its ij^{th} entry, if $\{\mathbf{v}^*\}_i = v(\mathbf{s})$ and $\{\mathbf{v}^*\}_j = v(\mathbf{t})$, is $p(\mathbf{s}-\mathbf{t})$. The same 'reachability' constraint is imposed on $\mathbf{P}^{(k)}$ as for the case $N = 1$. Let the sequence be denoted by $\{k_i\}$ as before.

LEMMA 6.3. *Let $\{k_i\}$ be a monotone increasing sequence with $k_i \to \infty$ as $i \to \infty$, for which each $k_i^N \times k_i^N$ matrix $\mathbf{P}^{(k_i)}$ is non-reducible. Then* $\lim_{i\to\infty} \rho(\mathbf{P}^{(k_i)}) = 1$.

PROOF. Again $\mathbf{P}^{(k_i)}$ is symmetric, and can be made doubly stochastic by increasing its diagonal entries, so that $\rho(\mathbf{P}^{(k_i)}) \leq 1$. Also, as in Lemma 6.2,

$$\begin{aligned}
\rho(\mathbf{P}^{(k_i)}) &\geq \frac{1}{k_i^N} \mathbf{1}'\mathbf{P}^{k_i}\mathbf{1} \\
&= \frac{1}{k_i^N} \sum_{u_N=-(k_i-1)}^{k_i-1} \cdots \sum_{u_1=-(k_i-1)}^{k_i-1} (k_i - |u_1|)(k_i - |u_2|)...(k_i - |u_N|)p(\mathbf{u}) \\
&= \sum_{u_N=-(k_i-1)}^{k_i-1} \cdots \sum_{u_1=-(k_i-1)}^{k_i-1} p(\mathbf{u}) \\
&- \frac{1}{k_i}\left(\sum_{u_N=-(k_i-1)}^{k_i-1} \cdots \sum_{u_1=-(k_i-1)}^{k_i-1} |u_1| p(\mathbf{u}) + \ldots \right. \\
&+ \left. \sum_{u_N=-(k_i-1)}^{k_i-1} \cdots \sum_{u_1=-(k_i-1)}^{k_i-1} |u_N| p(\mathbf{u}) \right) \\
&+ \ldots.. + (-1)^N \frac{1}{k_i^N} \sum_{u_N=-(k_i-1)}^{k_i-1} \cdots \sum_{u_1=-(k_i-1)}^{k_i-1} |u_1 \times u_2 ... \times u_N| p(\mathbf{u}).
\end{aligned}$$

Because of the finite expectation constraint, the right hand side of this inequality tends to 1 as $k_i \to \infty$. We therefore obtain the result that $\lim_{k_i\to\infty} \rho(\mathbf{P}^{(k_i)}) = 1$. □

Thus, for general N, we can choose $k > 1$ and sufficiently large so that the matrix $\mathbf{P}^{(k)}$ is non-reducible and $\gamma\rho(\mathbf{P}^{(k)}) > 1$. This result is now used to prove the pandemic theorem for $n = 1$.

THEOREM 6.4 (THE PANDEMIC THEOREM FOR THE ONE TYPE CASE). *If γ is infinite, then $\inf_{\mathbf{s}\in Z^N} v(\mathbf{s}) = 1$. When γ is finite and $\gamma > 1$, then $v(\mathbf{s}) \geq \eta(0)$ for all $\mathbf{s} \in Z^N$, where $z = \eta(0)$ is the unique positive solution to $-\log(1-z) = \gamma z$.*

PROOF. If γ is infinite then the result follows from Theorem 6.3 part 2.

Now consider the case when γ is finite with $\gamma > 1$. Define $v = \inf_{\mathbf{s} \in Z^N} v(\mathbf{s})$. The sites may be labelled so that there is some initial infection introduced at site $\mathbf{0}$, and hence $a(\mathbf{0}) > 0$. First choose k^* sufficiently large so that the matrix $\mathbf{P}^{(k^*)}$ is non-reducible with $\gamma\rho(\mathbf{P}^{(k^*)}) > 1$. Let $\mathbf{z} > \mathbf{0}$ be the unique positive solution to

$$-\log(1 - \{\mathbf{z}\}_i) = \gamma\{\mathbf{P}^{(k^*)}\mathbf{z}\}_i, \quad (i = 1, ..., {k^*}^N).$$

Define $\theta = \min_{1 \le i \le {k^*}^N} \{\mathbf{z}\}_i$. We then show that, for any site $\mathbf{s}$, $v(\mathbf{s}) \ge \theta$ and hence $v \ge \theta$.

Consider any site $\mathbf{s}$. Take $\mathbf{r}$ and k so that $\mathbf{P}^{(k)}$ is non-reducible, $\gamma\rho(\mathbf{P}^{(k)}) > 1$ and the vector of final sizes $\mathbf{v}^*$ in equation (6.9) includes the final size at both site $\mathbf{0}$ and site $\mathbf{s}$. Since $a(\mathbf{0}) > 0$, from Corollary B.1 part 1 we obtain the result that $v(\mathbf{s}) > 0$. Now consider inequality (6.9) with $k = k^*$ and $\mathbf{r}$ so that $\mathbf{v}^*$ includes the final size at site $\mathbf{s}$. Since $v(\mathbf{s}) > 0$, from Corollary B.1 part 3 we obtain the result that $\mathbf{v}^* \ge \mathbf{z} \ge \theta\mathbf{1}$. Hence $v(\mathbf{s}) \ge \theta$.

The result has been obtained for any site. Hence $v \ge \theta > 0$. We now show that the lower bound θ can be replaced by $\eta(0)$. Consider the final size equations (6.8) with $n = 1$. Then it immediately follows that $-\log(1 - v) \ge \gamma v$. From Corollary B.1, either $v \ge \eta(0)$, or $v = 0$. But we have just excluded this latter case. Hence $v(\mathbf{s}) \ge v \ge \eta(0)$ for all $\mathbf{s} \in Z^N$, which completes the proof of the theorem.

□

6.6 A matrix approach for the multi-type pandemic theorem

Consider the general n-type epidemic. The pandemic lemma is now quoted without proof, since it is identical to the continuous case.

LEMMA 6.4 (THE PANDEMIC LEMMA).

1. *If Γ has all finite entries and if $\rho(\Gamma) > 1$, then either $v_i(\mathbf{s}) \ge \eta_i(\mathbf{0})$ for all $\mathbf{s}$ and all i, or $\inf_{\mathbf{s}} v_i(\mathbf{s}) = 0$ for all $i = 1, ..., n$. Here $z_i = \eta_i(\mathbf{0})$ is the unique positive solution to*

$$-\log(1 - z_i) = \sum_{j=1}^{n} \gamma_{ij} z_j, \quad (i = 1, ..., n).$$

2. *If Γ has an infinite element in every row, then $v_i(\mathbf{s}) \equiv 1$ for all $i = 1, ..., n$.*
3. *The remaining case corresponds to a partitioning of Γ and $\mathbf{v}(\mathbf{s})$ as in Theorem 6.3, part 3. Then $\mathbf{v}_2(\mathbf{s}) \equiv \mathbf{1}$ and $\mathbf{v}_1(\mathbf{s}) \ge \eta^*(\mathbf{a}^*)$; where $\{\mathbf{a}^*\}_i = \sum_{j=m+1}^{n} \gamma_{ij}$ for $i = 1, ..., m$ and $\mathbf{z} = \eta^*(\mathbf{a}^*)$ is the unique positive solution to*

$$-\log(1 - \{\mathbf{z}\}_i) = \sum_{j=1}^{m} \gamma_{ij}\{\mathbf{z}\}_j + \{\mathbf{a}^*\}_i, \quad (i = 1, ..., m).$$

□

The pandemic result will follow immediately from this lemma provided we can show, for the case when $\mathbf{\Gamma}$ is finite with $\rho(\mathbf{\Gamma}) > 1$, that $\inf_{\mathbf{s} \in Z^N} v_i(\mathbf{s}) > 0$. A fairly simple proof is possible provided restrictions are made on the infection rates and contact distributions, as described in Section 6.3.

Take $\lambda_{ij}(\tau) = s_i c_{ij} d_j(\tau)$, where $c_{ij} = c_{ji}$ all i, j. Hence $\gamma_{ij} = \sigma_j s_i c_{ij} d_j$, where $d_j = \int_0^\infty d_j(\tau)d\tau$. Also let $p_{ij}(\mathbf{r})$ satisfy symmetry conditions; namely that $p_{ij}(\mathbf{r}) = p_{ji}(\mathbf{r})$ and $p_{ij}(\mathbf{r}) = p_{ij}(\mathbf{r}^*)$, where $\{\mathbf{r}^*\}_i = |\{\mathbf{r}^*\}_i|$.

Let $\mathbf{u} = (u_i)$ be an N-vector. The finite expectation conditions imposed are that, for all $1 \le r \le N$ and all sequences $1 \le t_1 < t_2 < ... < t_r \le N$, and for all i and j, then

$$\sum_{u_1=-\infty}^{\infty} \cdots \sum_{u_N=-\infty}^{\infty} \prod_{s=1}^{r} |u_{t_s}| p_{ij}(\mathbf{u}) < \infty.$$

Take the matrix $\mathbf{P}_{ij}^{(k)}$ to be identical to the matrix $\mathbf{P}^{(k)}$ defined in Section 6.5, except that the contact distribution $p(\mathbf{r})$ is replaced by $p_{ij}(\mathbf{r})$.

As in Section 6.5, consider k^N adjacent sites $A_k(\mathbf{r})$ for any $\mathbf{r} \in Z^N$. However now we consider the subset of the final size equation for the final sizes $v_i(\mathbf{s})$ for all types $i = 1, ..., n$ at the sites $\mathbf{s} \in A_k(\mathbf{r})$. Let $\mathbf{v}^*$ be the vector of these final sizes, where the order lists the final size for all sites successively for each type $i = 1, ..., n$. The vectors $\mathbf{u}^*$ and $\mathbf{a}^*$ are the corresponding ordered entries for $-\log(1 - v_i(\mathbf{s}))$ and $a_i(\mathbf{s})$. Then

$$\mathbf{u}^* \ge \mathbf{C}_k \mathbf{v}^* + \mathbf{a}^*, \tag{6.10}$$

where $\mathbf{C}_k$ does not depend upon $\mathbf{r}$ and

$$\mathbf{C}_k = \begin{pmatrix} \gamma_{11}\mathbf{P}_{11}^{(k)} & \gamma_{12}\mathbf{P}_{12}^{(k)} & \cdots & \gamma_{1n}\mathbf{P}_{1n}^{(k)} \\ \gamma_{21}\mathbf{P}_{21}^{(k)} & \gamma_{22}\mathbf{P}_{22}^{(k)} & \cdots & \gamma_{2n}\mathbf{P}_{2n}^{(k)} \\ \vdots & \vdots & \ddots & \vdots \\ \gamma_{n1}\mathbf{P}_{n1}^{(k)} & \gamma_{n2}\mathbf{P}_{n2}^{(k)} & \cdots & \gamma_{nn}\mathbf{P}_{nn}^{(k)} \end{pmatrix}.$$

The 'reachability' constraint is similar to Section 6.5, but is now given in terms of the $\mathbf{C}_k$. There exists a monotone increasing sequence $\{k_s\}$, with $k_s \to \infty$ as $s \to \infty$, for which each $nk_s^N \times nk_s^N$ matrix $\mathbf{C}_{k_s}$ is non-reducible for all s.

LEMMA 6.5. $\lim_{s\to\infty} \rho(\mathbf{C}_{k_s}) = \rho(\Gamma)$.

PROOF. First note that the diagonal entries of the $\mathbf{P}_{ij}^{(k_s)}$ can be increased to make them doubly stochastic. This increases some entries of $\mathbf{C}_{k_s}$. The matrix formed is $\mathbf{C}^*$. Let $\mathbf{u}$ be the right eigenvector of $\mathbf{\Gamma}$ corresponding to $\rho(\mathbf{\Gamma})$, so that $\mathbf{\Gamma}\mathbf{u} = \rho(\mathbf{\Gamma})\mathbf{u}$. Take $\mathbf{a}$ to be an $n \times k_s^N$ vector. This vector has successive k_s^N entries all equal; the ith common value being $\{\mathbf{u}\}_i$. Hence $\mathbf{C}^*\mathbf{a} = \rho(\mathbf{\Gamma})\mathbf{a}$, so that $\mathbf{C}^*$ has Perron-Frobenius eigenvalue $\rho(\Gamma)$. Hence $\rho(\mathbf{C}_{k_s}) \le \rho(\Gamma)$.

Next observe that for any non-reducible, non-negative matrix $\mathbf{B}$ and any vector $\mathbf{b} > \mathbf{0}$, the Perron-Frobenius eigenvalues of $\mathbf{B}$ and $diag(\mathbf{b})\mathbf{B}(diag(\mathbf{b}))^{-1}$ are identical.

Let $\mathbf{b}_{k_s}$ and $\mathbf{c}_{k_s}$ be vectors of length $n \times k_s^N$. Both vectors have each of the n successive entries of k_s^N elements identical, the jth common value being $\sqrt{\sigma_j d_j / s_j}$

for vector $\mathbf{b}_{k_s}$ and $\sqrt{s_j/(\sigma_j d_j)}$ for vector $\mathbf{c}_{k_s}$. Define $\mathbf{D}_{k_s} = diag(\mathbf{b}_{k_s})\mathbf{C}_{k_s} diag(\mathbf{c}_{k_s})$. Then $diag(\mathbf{c}_{k_s}) = (diag(\mathbf{b}_{k_s}))^{-1}$ and therefore $\rho(\mathbf{C}_{k_s}) = \rho(\mathbf{D}_{k_s})$.

Define $w_{ij} = c_{ij}\sqrt{s_i d_i \sigma_i s_j d_j \sigma_j}$. Then

$$\mathbf{D}_{k_s} = \begin{pmatrix} w_{11}\mathbf{P}_{11}^{(k_s)} & w_{12}\mathbf{P}_{12}^{(k_s)} & \dots & w_{1n}\mathbf{P}_{1n}^{(k_s)} \\ w_{21}\mathbf{P}_{21}^{(k_s)} & w_{22}\mathbf{P}_{22}^{(k_s)} & \dots & w_{2n}\mathbf{P}_{2n}^{(k_s)} \\ \vdots & \vdots & \ddots & \vdots \\ w_{n1}\mathbf{P}_{n1}^{(k_s)} & w_{n2}\mathbf{P}_{n2}^{(k_s)} & \dots & w_{nn}\mathbf{P}_{nn}^{(k_s)} \end{pmatrix}.$$

Note that $\mathbf{D}_{k_s}$ is a non-reducible non-negative symmetric matrix. The symmetry occurs since $w_{ij} = w_{ji}$ and each $\mathbf{P}_{ij}^{(k_s)}$ is symmetric with $\mathbf{P}_{ij}^{(k_s)} = \mathbf{P}_{ji}^{(k_s)}$.

Define $\{\mathbf{W}\}_{ij} = w_{ij}$. Then $\mathbf{W}$ is a non-reducible, non-negative symmetric matrix with $\rho(\mathbf{W}) = \rho(\Gamma)$. Now let $\mathbf{f}$ be the common right and left eigenvector of $\mathbf{W}$ corresponding to $\rho(\mathbf{W})$, which has been normalised so that $\mathbf{f}'\mathbf{f} = 1$. Take $\mathbf{f}^*$ to be the vector of length $n \times k_s^N$ with each of the n successive entries of k_s^N elements identical, the jth common value being $\{\mathbf{f}\}_j$.

Then using an orthonormal set of real eigenvectors of the real symmetric matrix $\mathbf{D}_{k_s}$, proceeding as in the proof of Lemma 6.2 we obtain

$$\frac{1}{k_s^N}\mathbf{f}^{*\prime}\mathbf{D}_{k_s}\mathbf{f}^* \leq \rho(\mathbf{D}_{k_s}). \tag{6.11}$$

The left hand side of this inequality, (6.11), is equal to

$$\frac{1}{k_s^N}\sum_i\sum_j \{f\}_i w_{ij}\{f\}_j \mathbf{1}'\mathbf{P}_{ij}^{(k_s)}\mathbf{1}.$$

When considering the case $n = 1$ in Lemma 6.3, we showed that $\lim_{k_s\to\infty}\frac{1}{k_s^N}\mathbf{1}'\mathbf{P}^{(k_s)}\mathbf{1} = 1$. The same result therefore holds if $\mathbf{P}^{(k_s)}$ is replaced by $\mathbf{P}_{ij}^{(k_s)}$. Hence the left hand side of inequality (6.11) tends to $\sum_i\sum_j\{f\}_i w_{ij}\{f\}_j = \rho(\mathbf{W}) = \rho(\Gamma)$ as $s \to \infty$. Therefore $\lim_{s\to\infty}\rho(\mathbf{D}_{k_s}) \geq \rho(\Gamma)$.

Also $\rho(\mathbf{D}_{k_s}) = \rho(\mathbf{C}_{k_s}) \leq \rho(\Gamma)$. Hence $\lim_{s\to\infty}\rho(\mathbf{D}_{k_s}) = \rho(\Gamma)$ and therefore also $\lim_{s\to\infty}\rho(\mathbf{C}_{k_s}) = \rho(\Gamma)$. This completes the proof of the lemma.

□

THEOREM 6.5. $\inf_{\mathbf{s}\in Z^N} v_i(\mathbf{s}) > 0$ *for all* $i = 1, ..., n$.

PROOF. The sites may be labelled so that there is some initial infection introduced at site $\mathbf{0}$, and hence $a_i(\mathbf{0}) > 0$ for some i. First, from Lemma 6.5, choose a k^* sufficiently large so that the matrix $\mathbf{C}_{k^*}$ is non-reducible with $\rho(\mathbf{C}_{k^*}) > 1$. Let $\mathbf{z}$ be the unique positive solution to

$$-\log(1 - \{\mathbf{z}\}_i) = \{\mathbf{C}_{k^*}\mathbf{z}\}_i, \quad (i = 1, ..., nk^{*N}).$$

Define $\eta = \min_{1\leq i\leq nk^{*N}}\{\mathbf{z}\}_i$. Then we will show that, for any site $\mathbf{s}$, $v_i(\mathbf{s}) \geq \eta$ for all $i = 1, ..., n$.

Consider any site $\mathbf{s}$. Take $\mathbf{r}$ and k so that $\mathbf{C}_k$ is non-reducible, $\rho(\mathbf{C}_k) > 1$ and the vector of final sizes $\mathbf{v}^*$ in inequality (6.10) includes the final size for all types at both site $\mathbf{0}$ and site $\mathbf{s}$. Since $a_i(\mathbf{0}) > 0$ for some i, from Corollary B.1 we obtain the result that $v_i(\mathbf{s}) > 0$ for all $i = 1, ..., n$.

Now consider inequality (6.10) with $k = k^*$ and $\mathbf{r}$ so that $\mathbf{v}^*$ includes the final sizes for all types at site $\mathbf{s}$. Since $v_i(\mathbf{s}) > 0$ for all i, from Corollary B.1 we obtain the result that $\mathbf{v}^* \geq \mathbf{z} \geq \eta \mathbf{1}$. Hence $v_i(\mathbf{s}) \geq \eta$ for all i.

The result has been obtained for any site. Hence $\inf_{\mathbf{s} \in Z^N} v_i(\mathbf{s}) \geq \eta > 0$ for all $i = 1, ..., n$, which completes the proof of the theorem.

□

Now that we have shown that $\inf_{\mathbf{s} \in Z^N} v_i(\mathbf{s}) > 0$, the proof of the pandemic theorem for the discrete case is easily established.

THEOREM 6.6 (THE PANDEMIC THEOREM).

1. *If Γ has all finite entries and $\rho(\Gamma) > 1$, then $v_i(\mathbf{s}) \geq \eta_i(\mathbf{0})$ for all $\mathbf{s}$ and all i, where $z_i = \eta_i(\mathbf{0})$ is the unique positive solution to*

$$-\log(1 - z_i) = \sum_{j=1}^{n} \gamma_{ij} z_j, \quad (i = 1, ..., n).$$

2. *If Γ has an infinite element in every row, then $v_i(\mathbf{s}) \equiv 1$ for all $i = 1, ..., n$.*
3. *The remaining case corresponds to a partitioning of Γ and $\mathbf{v}(\mathbf{s})$ as in Theorem 6.3, part 3. Then $\mathbf{v}_2(\mathbf{s}) \equiv \mathbf{1}$ and $\mathbf{v}_1(\mathbf{s}) \geq \eta^*(\mathbf{a}^*)$; where $\{\mathbf{a}^*\}_i = \sum_{j=m+1}^{n} \gamma_{ij}$ for $i = 1, ..., m$ and $\mathbf{z} = \eta^*(\mathbf{a}^*)$ is the unique positive solution to*

$$-\log(1 - \{\mathbf{z}\}_i) = \sum_{j=1}^{m} \gamma_{ij} \{\mathbf{z}\}_j + \{\mathbf{a}^*\}_i, \quad (i = 1, ..., m).$$

PROOF. Part 1 follows from the pandemic lemma, Lemma 6.4, and Theorem 6.5. Parts 2 and 3 follow immediately from the pandemic lemma parts 2 and 3.

□

6.7 The limit of the spatial final size

An upper bound can be obtained for the spatial final size. This was derived for the continuous space model in Chapter 3, Theorem 3.4. The proof remains valid for the discrete space model, as do the proofs of Corollaries 3.1 and 3.2 giving the limit of the spatial final size, provided we can show that $a_i(\mathbf{s}) \to 0$ for all i and $\mathbf{s}$ when $\varepsilon_j(\mathbf{r}) \to 0$ for all j and $\mathbf{r}$. This is easily seen since, from the proof of Theorem 6.2,

$$a_i(\mathbf{s}) = \lim_{t \to \infty} H_i(\mathbf{s}, t) \leq \sum_{j=1}^{k} \sup_{\mathbf{r} \in Z^N} \varepsilon(\mathbf{r}) \int_0^\infty \gamma_{ij}^*(\tau) d\tau.$$

The results are summarised here.

COROLLARY 6.1. *If $\mathbf{\Gamma}$ has all finite entries and $\rho(\mathbf{\Gamma}) \le 1$, then*

$$0 \le v_i(\mathbf{s}) \le \eta_i(\mathbf{b}),$$

where $y_i = \eta_i(\mathbf{b})$ is the unique positive solution to

$$-\log(1 - y_i) = \sum_{j=1}^{n} \gamma_{ij} y_j + \{\mathbf{b}\}_i, \ (i = 1, ..., n).$$

Hence when the amount of initial infection tends to zero, i.e. $\varepsilon_i \to 0$ for all $i = 1, ..., k$, the spatial size $v_i(\mathbf{s}) \to 0$ for all $i = 1, ..., n$ and all $\mathbf{s} \in Z^N$.

□

COROLLARY 6.2.

1. *If $\rho(\mathbf{\Gamma}) > 1$ and $\mathbf{\Gamma}$ has all finite entries, then $\eta_i(\mathbf{0}) \le v_i(\mathbf{s}) \le \eta_i(\mathbf{b})$ for all $\mathbf{s}$ and all i, where $\{\mathbf{b}\}_i = \sup_{\mathbf{s} \in Z^N} a_i(\mathbf{s})$ and $y_i = \eta_i(\mathbf{b})$ is the unique positive solution to*

$$-\log(1 - y_i) = \sum_{j=1}^{n} \gamma_{ij} y_j + \{\mathbf{b}\}_i, \ (i = 1, ..., n). \tag{6.12}$$

   Hence when the amount of initial infection tends to zero, i.e. $\varepsilon_i \to 0$ for all $i = 1, ..., k$, the spatial size $v_i(\mathbf{s}) \to \eta_i(\mathbf{0}) > 0$ for all $i = 1, ..., n$ and all $\mathbf{s} \in Z^N$.
2. *If $\mathbf{\Gamma}$ has an infinite element in every row, then $v_i(\mathbf{s}) = 1$ for all $\mathbf{s}$ and all $i = 1, ..., n$. This result still holds if we let the amount of initial infection tend to zero.*
3. *The remaining case corresponds to a partitioning of $\mathbf{\Gamma}$ and $\mathbf{v}(\mathbf{s})$ as in Theorem 2.2, part 3. Again take $\{\mathbf{b}\}_i = \sup_{\mathbf{s} \in Z^N} a_i(\mathbf{s})$. Then $\mathbf{v}_2(\mathbf{s}) \equiv \mathbf{1}$ and $\boldsymbol{\eta}^*(\mathbf{a}^*) \le \mathbf{v}_1(\mathbf{s}) \le \boldsymbol{\eta}^*(\mathbf{b}^*)$, where $\{\mathbf{a}^*\}_i = \sum_{j=m+1}^{n} \gamma_{ij}$ and $\{\mathbf{b}^*\}_i = \sum_{j=m+1}^{n} \gamma_{ij} + \{\mathbf{b}\}_i$ for $i = 1, ..., m$. Here $\mathbf{y} = \boldsymbol{\eta}^*(\mathbf{b}^*)$ is the unique positive solution to*

$$-\log(1 - \{\mathbf{y}\}_i) = \sum_{j=1}^{m} \gamma_{ij} \{\mathbf{y}\}_j + \{\mathbf{b}^*\}_i, \ (i = 1, ..., m). \tag{6.13}$$

   As the amount of initial infection tends to zero, the vectors of final sizes $\mathbf{v}_2(\mathbf{s}) \equiv \mathbf{1}$ and $\mathbf{v}_1(\mathbf{s}) \to \boldsymbol{\eta}^*(\mathbf{a}^*)$ for all $\mathbf{s} \in Z^N$.

□

These results parallel those for the continuous space models. No major epidemic occurs when $\rho(\mathbf{\Gamma})$ fails to exceed the threshold value of 1. When $\rho(\mathbf{\Gamma}) > 1$ a major epidemic occurs regardless of the amount of initial infection. The spatial final size is constrained to lie between two uniform positive bounds. As the amount of infection triggering the epidemic tends to zero, the final size everywhere tends to the lower bound, which is the corresponding limit in the non-spatial case.

CHAPTER 7

The saddle point method

7.1 Introduction

Saddle point techniques are a widely used mathematical tool providing a powerful means of examining the speed of spread in the forward front of systems of equations used to describe a wide variety of spatial models in biology. The specific saddle point method described in this chapter is in continuous time, where the space may be continuous or discrete. It is used to obtain the speed of first spread of the epidemic models of $S \to L \to I \to R$ type, which are special cases of the spatial models with varying infectivity described in Chapters 3 and 6.

In an epidemic a spatial system is considered where there are n types or populations, with all individuals being susceptible. Infectious individuals are introduced into the system in a bounded region, and the first spread of infection is modelled. This may be considered as a special case of a multi-type spatial system when the system is initially in a stable state. The system is then perturbed and the effect of this perturbation may be studied as it first spreads. Consider the spread in a specific direction, and model the system in a region corresponding to the forward front in that direction far from region of the original perturbation. In this region, the system of equations is often effectively linear. The saddle point method can be used to obtain the speed of spread of the forward front of the linear system (for which the results are exact), and hence the corresponding speed of first spread of the perturbation to the original system (where an approximation is involved).

The saddle point method is applied to two linear systems of equations, one in continuous space and the second in discrete space. The continuous space model is described by equations (7.1). Note that $p_{ij}(\mathbf{r})$ is a contact density function, so integrates to one.

$$(7.1)\qquad \frac{\partial y_i(\mathbf{s},t)}{\partial t} = \sum_{j=1}^{n} \alpha_{ij} \int_{\Re^N} p_{ij}(\mathbf{r}) y_j(\mathbf{s}-\mathbf{r},t) d\mathbf{r} + \sum_{j=1}^{n} \beta_{ij} y_j(\mathbf{s},t) - \mu_i y_i(\mathbf{s},t),$$

for $i = 1, ..., n$.

In the analogous discrete space model, individuals are at points of the N-dimensional integer lattice Z^N and $p_{ij}(\mathbf{r})$ represents a joint probability. The system is described by equations (7.2).

$$(7.2)\qquad \frac{\partial y_i(\mathbf{s},t)}{\partial t} = \sum_{j=1}^{n} \alpha_{ij} \sum_{\mathbf{r}\in Z^N} p_{ij}(\mathbf{r}) y_j(\mathbf{s}-\mathbf{r},t) + \sum_{j=1}^{n} \beta_{ij} y_j(\mathbf{s},t) - \mu_i y_i(\mathbf{s},t),$$

for $i = 1, ..., n$.

In both cases α_{ij}, β_{ij}, μ_i and $y_i(\mathbf{s}, 0)$ are non-negative for all i, j and $\mathbf{s}$. Hence necessarily $y_i(\mathbf{s}, t) \geq 0$ for all i and $\mathbf{s}$. Also the matrix with ij^{th} entry $\alpha_{ij} + \beta_{ij}$ is assumed to be non-reducible with at least one α_{ij} non-zero (as otherwise no spread could occur). In addition for some i, $y_i(\mathbf{s}, 0)$ is assumed to be positive for $\mathbf{s}$ in an open region of $\Re^N$ for the continuous case and at some site $\mathbf{s}$ for the discrete case.

The saddle point method is used to obtain the speed of spread of $y_i(\mathbf{s}, t)$ in a direction specified by the vector of direction cosines $\boldsymbol{\xi}$. Constraints are imposed on the $y_i(\mathbf{s}, 0)$ and the $p_{ij}(\mathbf{r})$ which are given in terms of the Laplace transforms. These are specified in Sections 7.2 and 7.4. The saddle point method is first explained in Section 7.2 in the simple case when there is a single equation. The rigorous development is later given for general n in Section 7.4 for the continuous space case. The analogous result for the discrete space model is also given, but without detailed proof.

In Section 7.3 it is shown how the results for the simple linear system can be used to obtain the speed of first spread in direction $\boldsymbol{\xi}$ for certain spatial models of one-type epidemics. The equation for the proportion of infectious individuals at position $\mathbf{s}$ at time t in the $S \to I$ and $S \to I \to R$ epidemic models in $\Re^N$ are approximately the same in the forward front, far from the initial focus of infection, as the linear system with $n = 1$ and $\beta_{11} = 0$, for a suitable choice of α_{11} and μ_1. Hence the results for the simple linear system can be used to obtain the speed of first spread of infection for both the $S \to I$ and $S \to I \to R$ epidemics. A simple adjustment can be used to show that, for the epidemic with removals, the speed of spread of the proportion of removed individuals is identical (as expected) with that of infectious individuals.

The minimum wave speed was obtained in Chapter 4 for these continuous space epidemic models, since they are just special cases of the model with varying infectivity. The speed of spread obtained by the saddle point method is shown to be the minimum wave speed obtained in Chapter 4. Note that the results concerning waves in that chapter did not require radial symmetry of the contact distribution, so that this result holds when there is no radial symmetry. In the radially symmetric case in Chapter 5 the asymptotic speed of propagation was also shown to be the minimum wave speed, so that the saddle point method yields the same result as the speed of propagation obtained by rigorous analytical method for the non-linear equation.

The saddle point method also yields an explicit form for the speed, whereas the method of Chapter 5 does not. Some simple explicit results are given in Section 7.3 for particular contact distributions.

Now consider the $S \to L \to I \to R$ epidemic model, the non-spatial form of which was described in Chapter 2. The approximate equations in the forward front do not lead to a single linear equation for the proportion of latent, infectious or removed individuals. Both of the epidemic models (with or without removal) which allow for a latent stage yield linked pairs of equations in the forward front for the proportions of latent and infectious individuals, so that we need to consider a pair of equations and hence a linear system with $n = 2$. For the model which allows for removal, a third differential equation links removed and latent individuals. One type models with a latent stage are therefore treated in sections 7.5 and 7.6 with the multi-type epidemic models.

In Section 7.5 the results for the general n-equation continuous space linear system are used to obtain the speed of spread of infection in the forward front in direction $\boldsymbol{\xi}$ for the corresponding n-type epidemic models. As for the one-type epidemic the method provides an explicit expression for this speed and does not require the contact distributions to be radially symmetric.

The saddle-point results for the linear discrete space model also enable results to be obtained for the models of Chapter 6. Specifically, the speed of spread of the forward front of infection in the direction of one of the co-ordinates of the N-dimensional lattice is obtained for the $S \to L \to I \to R$ and subset epidemic models. As for the continuous space model, the method does not yield results for the general model with varying infectivity.

In Section 7.6 the link between the speed of spread of the forward front of an epidemic, the minimum wave speed of Chapter 4 and the asymptotic speed of propagation obtained in Chapter 5 is investigated for multi-type continuous space models. Similar results are obtained as were derived for one-type models, these results being obtained in Radciffe and Rass [R11]. The speed of spread obtained by applying the saddle-point method to the approximate equations in the forward front is shown to be equal to the minimum wave speed, this result not being dependent on radial symmetry of the contact distributions. Hence in the radially symmetric case the asymptotic speed of propagation, which is equal to the the minimum wave speed, is identical to the speed of spread of the forward front obtained in this chapter.

In Chapter 5, where radial symmetry was assumed, the epidemic was shown to spread out from the initial focus of infection like an N-dimensional sphere, which has radius effectively $c_0 t$ when t is large. In the non-radially symmetric case the minimum wave speed c_0 was shown in Chapter 4 to depend upon the direction cosines $\boldsymbol{\xi}$, so that c_0 may be denoted by $c_0(\boldsymbol{\xi})$. Let

$$A = \{\mathbf{s} \in \Re^N : \mathbf{s}'\boldsymbol{\xi} = c_0(\boldsymbol{\xi}) \text{ for some vector of direction cosines } \boldsymbol{\xi}\}.$$

Then the saddle-point results suggest that the epidemic spreads out from the initial focus of infection like a shape determined by the N-dimensional set A. For large t the shape of the infection has boundaries tA, where $tA = \{\mathbf{s} \in \Re^N : (1/t)\mathbf{s} \in A\}$.

Application of the saddle point method to the $S \to I \to S$ epidemic and contact birth and branching processes is deferred to Chapters 8 and 9, where models of these processes are considered in detail.

7.2 The single equation case

Continuous space.

First consider the continuous space model described by equation (7.1) when $n = 1$, where the subscripts are dropped since they are unnecessary. A motivation for the definition of the speed of first spread is given which can be used for general N.

When $N = 1$ if $y(s,t)$ is a monotone decreasing function of s for each value of t there is a natural definition for the speed of first spread of $y(s,t)$. Consider η

small and positive, and for fixed t choose $s = s(t)$ so that $y(s(t), t) = \eta$. Then the speed of first spread of $y(s, t)$ in the forward tail is the limit of $s(t)/t$ as t tends to infinity. For the contact branching processes discussed in Chapter 9, $N = 1$ and $y(s, t)$ is monotone in s so that this simple approach can be used.

For many applications however $y(s, t)$ will not be monotone in s so that there is no unique value $s(t)$ such that $y(s(t), t) = \eta$. In addition there is no obvious extension to general N. A different approach is therefore adopted. For fixed t a value $s = s(t)$ is chosen so that $\int_{s(t)}^{\infty} y(s, t)dt = \eta$. Then the limit of $s(t)/t$ gives the speed of first spread of the integrated forward tail of $y(s, t)$ as t tends to infinity.

This can be generalised to processes in $\Re^N$ for general N. For a direction specified by the vector of direction cosines $\boldsymbol{\xi}$ and for fixed t, consider the integrated value of $y(\mathbf{r}, t)$ beyond the hyperplane $\boldsymbol{\xi}'\mathbf{r} = s$. If this integrated value is kept at a small constant value η as t changes, the value of s can be determined for each t. This value is denoted by $s(t)$. Then the speed of spread c of the forward region of the process in the specified direction is given by $c = \lim_{t\to\infty} s(t)/t$.

Define $L(\theta, t) = \int_{\Re^N} e^{\theta\boldsymbol{\xi}'\mathbf{r}} y(\mathbf{r}, t)d\mathbf{r}$ and $P(\theta) = \int_{\Re^N} e^{\theta\boldsymbol{\xi}'\mathbf{r}} p(\mathbf{r})d\mathbf{r}$. Then $L(\theta, t)$ and $P(\theta)$ are the Laplace transforms, with respect to u, of the results of integrating $y(\mathbf{r}, t)$ and $p(\mathbf{r})$ over the hyperplane $\boldsymbol{\xi}'\mathbf{r} = u$.

It is assumed that there exists an S such that $y(\mathbf{r}, 0) = 0$ for $\mathbf{r}$ such that $\boldsymbol{\xi}'\mathbf{r} > S$ with $L(\theta, 0)$ analytic and bounded for $Re(\theta)$ in a closed interval in $\Re_+$. This will certainly hold if $y(\mathbf{s}, 0)$ is a continuous function with bounded support. It also holds if $N = 1$ and $y(s, 0) = 1$ for $s < 0$ and $y(s, 0) = 0$ for $s \geq 0$, which is relevant when using the saddle point results for the branching process models considered in Chapter 9.

The Laplace transform $P(\theta)$ is taken to be analytic in the strip $0 < Re(\theta) < \Delta$, where the abscissa of convergence $\Delta > 0$ (and may be infinite). Also for any $0 < \lambda_1 < \lambda_2 < \Delta$ it is assumed that there exists a function k, with $\int_{-\infty}^{\infty} k(y)dy$ finite, such that $|P(\lambda + iy)| \leq k(y)$ for all $\lambda_1 \leq \lambda \leq \lambda_2$. This holds for most standard density functions.

We assume that $s(t)/t$ tends to a positive limit c as $t \to \infty$. Then for t sufficiently large necessarily $s(t) > S$, and so $\int_{\boldsymbol{\xi}'\mathbf{r}\geq s(t)} y(\mathbf{r}, 0)d\mathbf{r} = 0$. The saddle point method is now used to find c. Take the Laplace transform of the equation (7.1) with respect to $\mathbf{s}$ and replace the corresponding vector of parameters of the Laplace transform by the vector $\theta\boldsymbol{\xi}$. We then obtain

$$\frac{\partial L(\theta, t)}{\partial t} = (\alpha P(\theta) + \beta - \mu)L(\theta, t).$$

Therefore

$$L(\theta, t) = e^{(\alpha P(\theta)+\beta-\mu)t} L(\theta, 0). \tag{7.3}$$

Now define $L^*(\theta, t) = L(\theta, t) - e^{(\beta-\mu)t} L(\theta, 0)$. Then for any $0 < \lambda < \Delta$,

$$
\begin{aligned}
|L^*(\lambda + iy)| &\leq e^{(\beta-\mu)t}|L(\lambda,0)| \left| e^{\alpha t P(\lambda+iy)} - 1 \right| \\
&\leq e^{(\beta-\mu)t}|L(\lambda,0)|\alpha t|P(\lambda+iy)| \left| \sum_{j=1}^{\infty} \frac{(\alpha t P(\lambda))^{j-1}}{(j-1)!} \right| \\
&\leq e^{(\beta-\mu)t}|L(\lambda,0)|\alpha t|P(\lambda+iy)| \left(e^{\alpha t P(\lambda)} - 1 \right).
\end{aligned}
$$

Hence $L^*(\theta, t)$ is absolutely integrable over the line $Re(\theta) = \lambda$ for any $0 < \lambda < \Delta$, since $P(\theta)$ is absolutely integrable and $|L(\theta, 0)|$ is bounded. Note that $e^{(\beta-\mu)t}L(\theta, 0)$ is the Laplace transform of $y(\mathbf{r}, 0)$ integrated over the hyperplane $\boldsymbol{\xi}'\mathbf{r} = s$ and scaled by $e^{(\beta-\mu)t}$.

The Laplace transform is then inverted, so that the difference between the integral of $y(\mathbf{r}, t)$ and the integral of $e^{(\beta-\mu)t}y(\mathbf{r}, 0)$, each integrated over the hyperplane $\boldsymbol{\xi}'\mathbf{r} = u$, is given by the expression $(1/(2\pi i)) \int_{\lambda(t)-i\infty}^{\lambda(t)+i\infty} e^{-u\theta} L^*(\theta, t) d\theta$ for a suitable choice of $\lambda(t)$. Note that the integral of $e^{(\beta-\mu)t}y(\mathbf{r}, 0)$ over the hyperplane $\boldsymbol{\xi}'\mathbf{r} = u$ is zero if $u > S$.

Since $s(t)$ is determined from the result that $\int_{\boldsymbol{\xi}'\mathbf{r} \geq s(t)} y(\mathbf{r}, t) d\mathbf{r} = \eta$, where $\eta > 0$ and is small, we need a further integration over $u > s(t)$. Hence, by interchanging the order of integration, provided $\lambda(t) > 0$ and t is sufficiently large so that $s(t) > S$, we obtain

$$
\begin{aligned}
\eta &= \int_{\boldsymbol{\xi}'\mathbf{r} \geq s(t)} y(\mathbf{r}, t) d\mathbf{r} \\
&= \int_{s(t)}^{\infty} \frac{1}{2\pi i} \int_{\lambda(t)-i\infty}^{\lambda(t)+i\infty} e^{-u\theta} L^*(\theta, t) d\theta du \\
&= \frac{1}{2\pi i} \int_{\lambda(t)-i\infty}^{\lambda(t)+i\infty} \frac{1}{\theta} e^{-s(t)\theta} L^*(\theta, t) d\theta \\
&= \frac{1}{2\pi i} \int_{\lambda(t)-i\infty}^{\lambda(t)+i\infty} \frac{1}{\theta} e^{-s(t)\theta} \left(e^{(\alpha P(\theta)+\beta-\mu)t} - e^{(\beta-\mu)t} \right) L(\theta, 0) d\theta \\
&= \frac{1}{2\pi} \int_{-\infty}^{\infty} \frac{1}{\lambda(t) + iy} e^{-s(t)(\lambda(t)+iy)+(\beta-\mu)t} \left(e^{\alpha P(\lambda(t)+iy)t} - 1 \right) L(\lambda(t) + iy, 0) dy.
\end{aligned}
$$

Consider the two exponential terms in the integrand. The combined exponent is $g(\lambda(t) + iy)$, where $g(\theta) = (\alpha P(\theta) + \beta - \mu)t - \theta s(t)$ is an analytic function of θ for $0 < Re(\theta) < \Delta$ and is a convex function of θ for θ real. We choose $\theta = \lambda(t)$ to be the minimum of this convex function. Since $Re(g(\theta)) < g(Re(\theta))$ when θ has a non-zero imaginary component, $\theta = \lambda(t)$ is then the saddle point of $Re(g(\theta))$. The integral can then be rewritten to give

$$
\begin{aligned}
&\sqrt{t}\eta e^{-g(\lambda(t))} \\
&= \frac{\sqrt{t}}{2\pi} \int_{-\infty}^{\infty} \frac{1}{\lambda(t) + iy} e^{-\alpha t P(\lambda(t)) - iys(t)} \left(e^{\alpha t P(\lambda(t)+iy)} - 1 \right) L(\lambda(t) + iy, 0) dy.
\end{aligned}
$$

Now $Re(P(\lambda(t)+iy)) < P(\lambda(t))$ and $|L(\lambda(t)+iy,0)| < L(\lambda(t),0)$ for $y \neq 0$. Also $P(\lambda(t)+iy) \to 0$ as $|y| \to \infty$. The conditions on the $p(\mathbf{r})$ and $y(\mathbf{s},0)$ then imply that the dominant part of the integral for t large comes from the finite integral from $-A$ to A for a suitable choice of A. If this finite range is now split into two components,

$$\frac{\sqrt{t}}{2\pi}\int_{-A}^{A} \frac{1}{\lambda(t)+iy} e^{-\alpha t(P(\lambda(t))-P(\lambda(t)+iy))-iys(t)} L(\lambda(t)+iy,0)dy$$

and

$$\frac{\sqrt{t}}{2\pi}\int_{-A}^{A} \frac{1}{\lambda(t)+iy} e^{-\alpha t P(\lambda(t))-iys(t)} L(\lambda(t)+iy,0)dy,$$

then it is easily seen that the contribution from the latter term is negligible for t large.

The first integral may be rewritten as

$$\frac{\sqrt{t}}{2\pi}\int_{-A}^{A} \frac{1}{\lambda(t)+iy} e^{g(\lambda(t)+iy)-g(\lambda(t))} L(\lambda(t)+iy,0)dy$$

The dominant part of the exponential term in the integrand, and hence of the integral, when t is large comes from the region of the line very close to the saddle point on the real axis corresponding to $y=0$.

Since $\theta = \lambda(t)$ is the saddle point of $g(\theta)$, $g'(\lambda(t)) = 0$. Therefore, for y small, $g(\lambda(t)+iy) - g(\lambda(t)) \approxeq -\dfrac{1}{2}y^2\alpha P''(\lambda(t))t$. Hence, for t large there is a small positive δ such that,

$$\sqrt{t}\eta e^{-g(\lambda(t))} \approxeq \frac{\sqrt{t}L(\lambda(t),0)}{2\pi\lambda(t)} \int_{-\delta}^{\delta} e^{-\frac{1}{2}y^2\alpha t P''(\lambda(t))}dy.$$

Now compare the integrand with the density function of a normal random variable with mean zero and variance $1/(t\alpha P''(\lambda(t)))$ and observe that when t is large the variance is small. Then for t sufficiently large the integral of this normal density function over the restricted range $(-\delta,\delta)$ is effectively one, and hence the integral above is approximately $\sqrt{(2\pi)/(\alpha t P''(\lambda(t)))}$. Therefore

$$\sqrt{t}\eta e^{-g(\lambda(t))} \approxeq \frac{L(\lambda(t),0)}{\lambda(t)\sqrt{2\pi\alpha P''(\lambda(t))}}. \tag{7.4}$$

From the result that $g'(\lambda(t)) = 0$, because $g'(\lambda(t)) = \alpha P'(\lambda(t))t - s(t)$ we obtain the result that $s(t)/t = \alpha P'(\lambda(t))$. Now $P''(\lambda) > 0$ for all θ and $\lim_{t\to\infty} s(t)/t = c$ as $t \to \infty$. Hence $\lim_{t\to\infty}\lambda(t) = \lambda_0$, where λ_0 is the unique solution to $c = \alpha P'(\lambda_0)$. Note that in order to use the saddle point method we need $\lambda_0 > 0$.

Taking logs in equations (7.4), since $g(\lambda(t)) = (\alpha P(\lambda(t)) + \beta - \mu)t - \lambda(t)s(t)$, we obtain

$$\frac{s(t)}{t} \approxeq \frac{\alpha P(\lambda(t)) + \beta - \mu}{\lambda(t)} + \frac{\log(L(\lambda(t),0)/(\eta\lambda(t))) - \frac{1}{2}\log(2\pi\alpha P''(\lambda(t))) - \frac{1}{2}\log(t)}{t\lambda(t)}.$$

Taking the limit as $t \to \infty$ we obtain $c = (\alpha P(\lambda_0) + \beta - \mu)/\lambda_0$.

Therefore there are two relations linking c and λ_0, $c = \alpha P'(\lambda_0)$ and $c = f(\lambda_0)$, where $f(\lambda) = (\alpha P(\lambda) + \beta - \mu)/\lambda$. Consider the function $f(\lambda)$ for $\lambda > 0$. Differentiating we obtain $f'(\lambda) = (\alpha P'(\lambda) - f(\lambda))/\lambda$. This is zero when $\lambda = \lambda_0$. Also $f''(\lambda) = (\alpha P''(\lambda) - 2f'(\lambda))/\lambda$, which is positive for all $\lambda > 0$ such that $f'(\lambda) = 0$. Hence there is a unique minimum of the function $f(\lambda)$ at $\lambda = \lambda_0$. Therefore the speed of spread of the forward front for equation (7.1) with $n = 1$ is given by

$$c = \min_{\lambda \in (0,\Delta)} \frac{\alpha P(\lambda) + \beta - \mu}{\lambda}. \tag{7.5}$$

In general we can only obtain the speed of spread c by the saddle point method provided c is positive and lies in the range $(\alpha P'(0), \alpha P'(\Delta))$, where Δ is the abscissa of convergence of $P(\theta)$.

When $\alpha + \beta > \mu$, then $\lim_{\lambda \downarrow 0} f(\lambda) = \infty$. If $P(\Delta)$ is infinite, then $f(\lambda)$ has a minimum for $\lambda > 0$. Equation (7.5) then always gives the speed of spread of the forward front in the specified direction, provided $\alpha P(\lambda) + \beta > \mu$ for all $\lambda > 0$. This latter condition necessarily occurs when the mean of the projected contact distribution is zero. In this case $P'(0) = 0$ and hence for any $\lambda > 0$, $P'(\lambda) > 0$ and therefore $\alpha P(\lambda) + \beta > \alpha P(\lambda) + \beta = \alpha + \beta > \mu$.

However, when $\alpha + \beta < \mu$, $\lim_{\lambda \downarrow 0} f(\lambda) = -\infty$. Also when $\alpha + \beta = \mu$, $\lim_{\lambda \downarrow 0} f(\lambda) = \alpha P'(0)$. In this latter case, since any stationary point of $f(\lambda)$ is a minimum, in order to show that the infimum of $f(\lambda)$ occurs at $\lambda = 0$ we need only show that $\lim_{\lambda \downarrow 0} f'(\lambda) > 0$. A little manipulation shows that this limit is just $\alpha P''(0)/2$, which is positive, so the result is immediate. Hence in both cases $f(\lambda)$ has no minimum for $\lambda \in (0, \Delta)$. Therefore we obtain the speed of spread only when $\alpha + \beta > \mu$.

Note that if $N = 1$ and $y(s,t)$ is monotone decreasing in s, then it can easily be seen that the same value of c is obtained if we use the simpler approach and define $s(t)$ so that $y(s(t), t) = \eta$. Again $c = \lim_{t \to \infty} s(t)/t$. With this definition of $s(t)$, for t sufficiently large so that $y(s(t), 0) = 0$,

$$\eta = y(s(t), t) = \frac{1}{2\pi i} \int_{\lambda(t) - i\infty}^{\lambda(t) + i\infty} e^{-s(t)\theta} L^*(\theta, t) d\theta.$$

The method proceeds in an almost identical manner so that equation (7.4) is obtained without the term $\lambda(t)$ in the denominator of the right hand side. The same limit of $s(t)/t$ is therefore obtained.

Discrete space.

For the discrete space model we consider the speed of spread of the forward front in the direction of one of the co-ordinates of the lattice (which without loss of generality may be taken to be the first co-ordinate). This corresponds to using direction cosines $\boldsymbol{\xi}' = (1, 0, ..., 0)$.

For η small and positive, we define $s(t)$ to be the smallest integer such that $\sum_{\mathbf{u}:\{\mathbf{u}\}_1 \geq s(t)} y(\mathbf{u}, t) \leq \eta$. Notice that we cannot choose $s(t)$ so that the summation is equal to η. Define $\eta(t)$ to be the exact value of the summation. The speed of spread of the forward front is then $c = \lim_{t\to\infty} s(t)/t$. The value of c can again be obtained using the saddle point method.

Laplace transforms are defined as for the continuous case, except that integrals are replaced by summations. It is not necessary to have uniform bounds on the Laplace transforms which are integrable, we now only require the results that $P(\theta)$ exists for $\theta \in (0, \Delta)$, where the abscissa of convergence Δ is positive. The identical equation to (7.3) is obtained, however here $L(\theta, t)$ is the Laplace transform with respect to u of $\sum_{\mathbf{s}:\{\mathbf{s}\}_1=u} y(\mathbf{s}, t)$. We take $y(\mathbf{s}, 0)$ to be zero for $\{\mathbf{s}\}_1 > S$ for some integer S, so that $L(\theta, 0)$ exists for all real $Re(\theta) > 0$.

Inverting the Laplace transform, for a suitable choice of $\lambda(t)$ gives

$$\begin{aligned}\sum_{\mathbf{s}:\{\mathbf{s}\}_1=u} y(\mathbf{s}, t) &= \frac{1}{2\pi}\int_{-\pi}^{\pi} e^{-(\lambda(t)+iy)u} L(\lambda(t)+iy, t)dy \\ &= \frac{1}{2\pi}\int_{-\pi}^{\pi} e^{-(\lambda(t)+iy)u+(\alpha P(\lambda(t)+iy)+\beta-\mu)t} L(\lambda(t)+iy, 0)dy.\end{aligned}$$

Hence

$$\begin{aligned}\eta(t) &= \sum_{\mathbf{s}:\{\mathbf{s}\}_1\geq s(t)} y(\mathbf{s}, t) \\ &= \frac{1}{2\pi}\int_{-\pi}^{\pi} \frac{1}{(1-e^{-(\lambda(t)+iy)})} e^{-(\lambda(t)+iy)s(t)+(\alpha P(\lambda(t)+iy)+\beta-\mu)t} L(\lambda(t)+iy, 0)dy.\end{aligned}$$

You can then proceed almost identically to the continuous space case. The same speed of first spread c, as given in equation (7.5), is therefore obtained. In the expression for c, $P(\theta)$ is the Laplace transform of the marginal contact distribution in the appropriate co-ordinate direction on the lattice Z^N.

7.3 The speed of spread for certain one-type epidemic models

Consider the one-type continuous space analogue of the one-type $S \to I \to R$ non-spatial epidemic described in Chapter 2. Since $n = 1$ the subscripts may be dropped. We do not consider infection spread by individuals from outside, but assume that there is some initial infection in the population at time zero. The equations for the proportions $x(\mathbf{s}, t)$, $y(\mathbf{s}, t)$ and $z(\mathbf{s}, t)$ of susceptible, infectious and removed individuals are given below. For the $S \to I$ model, for which there are no removals, the last equation is omitted and $\mu = 0$.

$$\begin{aligned}\frac{\partial x(\mathbf{s}, t)}{\partial t} &= -\sigma\lambda x(\mathbf{s}, t)\int_{\Re^N} p(\mathbf{r})y(\mathbf{s}-\mathbf{r}, t)d\mathbf{r}, \\ \frac{\partial y(\mathbf{s}, t)}{\partial t} &= \sigma\lambda x(\mathbf{s}, t)\int_{\Re^N} p(\mathbf{r})y(\mathbf{s}-\mathbf{r}, t)d\mathbf{r} - \mu y(\mathbf{s}, t), \\ \frac{\partial z(\mathbf{s}, t)}{\partial t} &= \mu y(\mathbf{s}, t).\end{aligned}$$

If there is some initial infection which is confined to a bounded region of $\Re^N$, then at time t in the forward region in direction $\boldsymbol{\xi}$, far from the initial focus of infection, almost all individuals will be susceptible so that $x(\mathbf{s}, t)$ will be approximately equal to one. The equation for $y(\mathbf{s},t)$ in this forward region for the $S \to I \to R$ epidemic then becomes approximately

$$\frac{\partial y(\mathbf{s},t)}{\partial t} = \sigma\lambda \int_{\Re^N} p(\mathbf{r})y(\mathbf{s}-\mathbf{r},t)d\mathbf{r} - \mu y(\mathbf{s},t).$$

This is precisely the same as equation (7.1) with $n=1$ provided we take $\alpha = \sigma\lambda$ and $\beta = 0$. Since the equation for the non-linear epidemic model is approximately the same in the forward front as the linear equation this suggests that the speed of spread of the forward region is the same. The speed c of spread is therefore

$$c = \inf_{0<\theta<\Delta} \frac{\sigma\lambda P(\theta) - \mu}{\theta}.$$

This speed, in direction $\boldsymbol{\xi}$, was obtained for the linear system provided $\sigma\lambda P(\theta) > \mu$ for all real $\theta \in [0,\Delta]$ and the infimum occurred in $(0,\Delta)$ and was positive. Note that when $\mu = 0$, so that the model is for the $S \to I$ epidemic, the first condition must hold.

Now consider the minimum wave speed obtained in Chapter 4 for the model with varying infectivity, which has infection rate $\lambda(t)$, where t is the time since infection. Then $K_c(\theta) = \sigma P(\theta)\Lambda(c\theta)$, where $P(\theta)$ is the Laplace transform of the projection of the contact distribution in direction $\boldsymbol{\xi}$, and $\Lambda(\theta) = \int_0^\infty e^{-\theta t}\lambda(t)dt$.

Waves only exist provided $P(\theta)$ exists for some positive θ and $\gamma = \sigma\Lambda(0) > 1$. In this case, the minimum wave speed is given by

$$c_0 = \inf\{c > 0 \ : K_c(\theta) = 1 \text{ for some } \theta \in (0,\Delta)\ \}.$$

It was shown in Chapter 2 for the non-spatial models that the $S \to I \to R$ model may be regarded as a special case of the model with varying infectivity if we take $\lambda(t) = \lambda e^{-\mu t}$. This result is also true for spatial models. The corresponding result for the $S \to I$ epidemic is then obtained by setting $\mu = 0$ so that $\lambda(t) = \lambda$. Therefore for the $S \to I \to R$ epidemic $\gamma = \sigma\lambda/\mu$, which is infinite when $\mu = 0$. Also $K_c(\theta) = \sigma\lambda P(\theta)/(c\theta + \mu)$. Hence for a specific $c > 0$, $K_c(\theta) = 1$ if $c = (\sigma\lambda P(\theta) - \mu)/\theta$ for some $\theta \in (0,\Delta)$. Hence

$$c_0 = \max\left(0, \inf_{\theta\in(0,\Delta)} \frac{\sigma\lambda P(\theta) - \mu}{\theta}\right).$$

Therefore the speed of spread obtained by the saddle point method and the minimum wave speed are the same for all cases for which the speed of spread has been obtained. In the case when the contact distribution is radially symmetric, then the asymptotic speed of propagation obtained in Chapter 5 is equal to the minimum wave speed and so also equals the speed of spread obtained by the saddle point method for the $S \to I \to R$ and $S \to I$ models.

For the $S \to I \to R$ epidemic it is possible also to formally consider the speed of spread in direction $\boldsymbol{\xi}$ for the proportion of removed individuals at position $\mathbf{s}$ by time t (although it is intuitively obvious that it must also be c_0). The equation in the forward tail for $z(\mathbf{s},t)$ is

$$\frac{\partial z(\mathbf{s},t)}{\partial t} = \mu y(\mathbf{s},t).$$

Let $M(\theta)$ and $L(\theta)$ be the Laplace transforms of $z(\mathbf{r},t)$ and $y(\mathbf{r},t)$, each integrated over the hyperplane $\boldsymbol{\xi}'\mathbf{r} = u$. Then

$$\frac{\partial M(\theta,t)}{\partial t} = \mu L(\theta,t) = \mu e^{(\sigma\lambda P(\theta)-\mu)t} L(\theta,0).$$

Hence

$$M(\theta,t) = M(\theta,0) + \frac{\mu}{\sigma\lambda P(\theta)-\mu}\left(e^{(\sigma\lambda P(\theta)-\mu)t} - 1\right) L(\theta,0).$$

If we define $s(t)$ so that $\eta = \int_{\boldsymbol{\xi}'\mathbf{r}\geq s(t)} z(\mathbf{r},t)d\mathbf{r}$, since there will be no removals at time zero far from the initial focus of infection, for t sufficiently large

$$\eta = \frac{1}{2\pi i}\int_{\lambda(t)-i\infty}^{\lambda(t)+i\infty} \frac{\mu}{\theta(\sigma\lambda P(\theta)-\mu)} e^{-s(t)\theta} L^*(\theta,t)d\theta,$$

where $L^*(\theta,t) = \left(e^{(\sigma\lambda P(\theta)-\mu)t} - 1\right) L(\theta,0)$. The saddle point method then proceeds in an almost identical fashion to the method described in Section 7.2, with the same speed of spread being obtained, since the only extra term in the integral above is $\mu/(\sigma\lambda P(\theta)-\mu)$. Note that when the saddle point method can be applied $\sigma\lambda P(\theta) > \mu$ for all $0 \leq Re(\theta) \leq \Delta$. So the proportions of infected and removed individuals spread at the same rate, which is precisely the result which is expected.

Some simple results for the speed of spread can be obtained for the $S \to I$ epidemic for specific contact distributions. Note that $c = \sigma\lambda P(\theta_0)/\theta_0$, where $\theta_0 > 0$ satisfies the relation $\theta_0 P'(\theta_0) = P(\theta_0)$.

For a double exponential with $p(r) = (a/2)e^{-a|r|}$, and hence mean zero and variance $2/a^2$, $P(\theta) = a^2/(a^2-\theta^2)$. Therefore $\theta_0^2 = a^2/3$ and hence the speed of spread $c = \sqrt{2/a^2}\sigma\lambda(3\sqrt{3})/(2\sqrt{2})$. This expression for the minimum wave speed is given in Mollison [M7].

When the contact distribution is normal with zero mean and standard deviation a, then $P(\theta) = e^{\theta^2 a^2/2}$. Then $\theta_0^2 = 1/a^2$ and so $c = a\sigma\lambda\sqrt{e}$. A slightly more complicated result is also easily obtained when the mean is non-zero. In this case the speed of spread is

$$c = a\sigma\lambda\frac{e^{\alpha(\sqrt{1+\alpha^2}-\alpha)+1/2}}{\sqrt{1+\alpha^2}-\alpha},$$

where a is the standard deviation, b is the mean and $\alpha = b/(2a)$.

A fairly simple result may also be obtained for the $S \to I \to R$ epidemic when the contact distribution is double exponential with mean zero and variance $2/a^2$. Now $c = \sigma\lambda(P(\theta_0) - 1/\gamma)/\theta_0$, where $\theta_0 > 0$ satisfies the relation $\theta_0 P'(\theta_0) = P(\theta_0) - 1/\gamma$, where $\gamma = \sigma\lambda/\mu$. Then, using the equivalent result for c that $c = \sigma\lambda P'(\theta_0)$,

$$c = \sqrt{2/a^2}\sigma\lambda\frac{4\times\sqrt{\sqrt{9\gamma^2-8\gamma}-3\gamma+2}}{(3\gamma-\sqrt{9\gamma^2-8\gamma})^2}.$$

Now consider the continuous space model of the $S \to L \to I \to R$ epidemic. Let $x(\mathbf{s},t)$, $l(\mathbf{s},t)$, $y(\mathbf{s},t)$ and $z(\mathbf{s},t)$ represent the proportion of individuals at position $\mathbf{s}$ who are susceptible, latent, infectious and removed respectively at time t. The equations they satisfy are then given below.

$$\frac{\partial x(\mathbf{s},t)}{\partial t} = -x(\mathbf{s},t)\lambda\sigma \int_{\Re^N} p(\mathbf{r})y(\mathbf{s}-\mathbf{r},t)d\mathbf{r},$$
$$\frac{\partial l(\mathbf{s},t)}{\partial t} = x(\mathbf{s},t)\lambda\sigma \int_{\Re^N} p(\mathbf{r})y(\mathbf{s}-\mathbf{r},t)d\mathbf{r} - \alpha l(\mathbf{s},t),$$
$$\frac{\partial y(\mathbf{s},t)}{\partial t} = \alpha l(\mathbf{s},t) - \mu y(\mathbf{s},t),$$
$$\frac{\partial z(\mathbf{s},t)}{\partial t} = \mu y(\mathbf{s},t).$$

Let some infection, either latent or infectious, be present at time 0, but be confined to a compact region. Then at time t, in the forward region far from the initial focus of infection, almost all individuals are susceptible so that $x(\mathbf{s},t)$ is approximately one. The equations for $l(\mathbf{s},t)$, $y(\mathbf{s},t)$ and $z(\mathbf{s},t)$ are then approximated in the forward region by

$$\frac{\partial l(\mathbf{s},t)}{\partial t} = \lambda\sigma \int_{\Re^N} p(\mathbf{r})y(\mathbf{s}-\mathbf{r},t)d\mathbf{r} - \alpha l(\mathbf{s},t),$$
$$\frac{\partial y(\mathbf{s},t)}{\partial t} = \alpha l(\mathbf{s},t) - \mu y(\mathbf{s},t),$$
$$\frac{\partial z(\mathbf{s},t)}{\partial t} = \mu y(\mathbf{s},t).$$

There is no single linear equation to consider. If we consider the first two equations, then they are equivalent to equations (7.1) with $n = 2$ for a suitable choice of parameters by taking $y_1(\mathbf{s},t) = l(\mathbf{s},t)$ and $y_2(\mathbf{s},t) = y(\mathbf{s},t)$. Hence this requires the general linear system, for which the saddle point results are obtained in Section 7.4. It is therefore considered along with n-type models in Sections 7.5 and 7.6.

Note that the results for the speed of spread of infection for the discrete models of Chapter 6 are identical to those for the corresponding continuous space models.

7.4 The saddle point method in general

Continuous space.

Consider equations (7.1) describing a continuous space linear model. As in section 7.2 we will consider the speed of spread of $y_i(\mathbf{s},t)$ in a direction with direction cosines $\boldsymbol{\xi}$. As in the single equation case, $s(t)$ is defined so that the integrated value of $y_i(\mathbf{r},t)$ for $\boldsymbol{\xi}'\mathbf{r} \geq s(t)$ is kept at a constant value η as t changes. The speed of spread of the forward front of $y_i(\mathbf{s},t)$ in the specified direction is then $c = \lim_{t\to\infty} s(t)/t$. We assume that c exists and is positive, and use the saddle point method to determine its value. The method presented is adapted from Radcliffe and Rass [R5].

Define $\mu = \max_i(\mu_i)$. Take the Laplace transform of equations (7.1) with respect to $\mathbf{s}$. Replace the corresponding vector of parameters of the Laplace transform by the vector $\theta\boldsymbol{\xi}$. We then obtain

$$\frac{\partial \mathbf{L}(\theta,t)}{\partial t} = (\mathbf{K}(\theta) - \mu\mathbf{I})\mathbf{L}(\theta,t). \tag{7.6}$$

Here $\mathbf{L}(\theta, t)$ has i^{th} entry $L_i(\theta, t) = \int_{\Re^N} e^{\theta \boldsymbol{\xi}' \mathbf{r}} y_i(\mathbf{r}, t) d\mathbf{r}$ and $\mathbf{K}(\theta)$ has ij^{th} entry $\alpha_{ij} P_{ij}(\theta) + \beta_{ij} + \delta_{ij}(\mu - \mu_i)$, where $P_{ij}(\theta) = \int_{\Re^N} e^{\theta \boldsymbol{\xi}' \mathbf{r}} p_{ij}(\mathbf{r}) d\mathbf{r}$ and δ_{ij} is one if $i = j$ and is zero otherwise. Note that $L_i(\theta, t)$ and $P_{ij}(\theta)$ are the Laplace transforms, with respect to u, of the results of integrating $y_i(\mathbf{r}, t)$ and $p_{ij}(\mathbf{r})$ over the hyperplane $\boldsymbol{\xi}' \mathbf{r} = u$.

Integrating equation (7.6) gives

$$\mathbf{L}(\theta, t) = e^{t(\mathbf{K}(\theta) - \mu \mathbf{I})} \mathbf{L}(\theta, 0). \tag{7.7}$$

Each $y_i(\mathbf{r}, 0)$ is assumed to be zero for $\boldsymbol{\xi}' \mathbf{r} > S$ for some positive S and $L_i(\theta, 0)$ is taken to be analytic for all $Re(\theta) > 0$ and bounded for $Re(\theta)$ in a closed interval in $\Re_+$. The Laplace transforms $P_{ij}(\theta)$, $i, j = 1, ..., n$, are taken to be analytic in some region of the right hand half of the complex plane, and the minimum of the corresponding abscissae of convergence is denoted by $\Delta > 0$. We assume that, for any $0 < \lambda_1 < \lambda_2 < \Delta$, there exist functions k_{ij} such that $|P_{ij}(\lambda + iy)| \leq k_{ij}(y)$ for all i, j and $\lambda \in [\lambda_1, \lambda_2]$ where $\int_{-\infty}^{\infty} k_{ij}(y) dy$ are finite for all i, j.

Define $\mathbf{Q}(\theta)$ and $\mathbf{B}$ to be the matrices with ij^{th} entries $\alpha_{ij} P_{ij}(\theta)$ and β_{ij} respectively. In addition let $\mathbf{M} = \mathbf{B} + \mu \mathbf{I} - \operatorname{diag}(\mu_1, ..., \mu_n)$. Then $\mathbf{K}(\theta) = \mathbf{Q}(\theta) + \mathbf{M}$ is finite, non-negative and non-reducible for θ real and in the range $[0, \Delta)$.

Let $\{\mathbf{L}^*(\theta, t)\}_i = L_i^*(\theta, t)$ where we define

$$\mathbf{L}^*(\theta, t) = \mathbf{L}(\theta, t) - e^{t(\mathbf{M} - \mu \mathbf{I})} \mathbf{L}(\theta, 0) = e^{-\mu t} \left(e^{t \mathbf{K}(\theta)} - e^{t \mathbf{M}} \right) \mathbf{L}(\theta, 0). \tag{7.8}$$

The following lemma then justifies the inversion of equation (7.8). It also provides results required in Lemma 7.2 so that the proof is made more general.

LEMMA 7.1. *For any real λ such that $0 < \lambda < \Delta$, $L_i^*(\theta, t)$ is absolutely integrable over the line $Re(\theta) = \lambda$.*

PROOF. To establish the result, we need only show that $\int_{|y| \geq \alpha} |L_i^*(\lambda + iy, t)| dy$ is finite for some α.

From the conditions on the $P_{ij}(\theta)$, for any $0 < \lambda_1 \leq \lambda_2 < \Delta$ there exist integrable functions k_{ij} such that $|P_{ij}(\lambda + iy)| \leq k_{ij}(y)$ for all y and all $\lambda \in [\lambda_1, \lambda_2]$. Define $k(y) = \max_{i,j} k_{ij}(y)$. Then $k(y)$ is also integrable and hence we can choose an $\alpha > 0$ such that $0 < \sup_{|y| \geq \alpha} k(y) < 1$. Let $d = \sup_{|y| \geq \alpha} k(y)$ and define $\mathbf{S}$ to be the matrix with ij^{th} entry α_{ij}. Then if $\theta = \lambda + iy$,

$$\begin{aligned}
e^{\mu t} |\mathbf{L}^*(\theta, t)| &= \left| \left(\sum_{s=0}^{\infty} \frac{t^s (\mathbf{Q}(\theta) + \mathbf{M})^s}{s!} - \sum_{s=0}^{\infty} \frac{t^s (\mathbf{M})^s}{s!} \right) \mathbf{L}(\theta, 0) \right| \\
&= \left| \left(\sum_{s=1}^{\infty} \frac{t^s (\mathbf{Q}(\theta) + \mathbf{M})^s}{s!} - \sum_{s=1}^{\infty} \frac{t^s (\mathbf{M})^s}{s!} \right) \mathbf{L}(\theta, 0) \right| \\
&\leq \frac{k(y)}{d} \sum_{s=1}^{\infty} \frac{t^s}{s!} (d\mathbf{S} + \mathbf{M})^s |\mathbf{L}(\theta, 0)| \\
&= \frac{k(y)}{d} \left(e^{t(d\mathbf{S} + \mathbf{M})} - \mathbf{I} \right) |\mathbf{L}(\theta, 0)|.
\end{aligned}$$

Since $|\mathbf{L}(\theta, 0)|$ has bounded entries and $k(y)$ is integrable, then for fixed t and $\lambda \in [\lambda_1, \lambda_2]$ the entries of $|\mathbf{L}^*(\theta, t)|$ are absolutely integrable over the line $Re(\theta) = \lambda$. The result follows by taking $\lambda_1 = \lambda_2 = \lambda$.

□

From Lemma 7.1, for a suitable choice of $\lambda(t) \in (0, \Delta)$, we can invert the Laplace transform in equation (7.8) to obtain the result that

$$\int_{\boldsymbol{\xi}'\mathbf{r}=s} y_i(\mathbf{r}, t)d\mathbf{r} = \frac{1}{2\pi i}\int_{\lambda(t)-i\infty}^{\lambda(t)+i\infty} e^{-s\theta}L_i^*(\theta, t)d\theta + e^{-\mu t}\sum_{j=1}^{n}\{e^{t\mathbf{M}}\}_{ij}\int_{\boldsymbol{\xi}'\mathbf{r}=s} y_j(\mathbf{r}, 0)d\mathbf{r}.$$

Observe that $y_i(\mathbf{r}, 0) = 0$ for $\boldsymbol{\xi}'\mathbf{r} > S$, and we have assumed that $c > 0$; hence there exists a T_1^* such that the second integral will be zero for $s \geq s(t)$ when $t \geq T_1^*$. Therefore the latter term is zero for $t \geq T_1^*$.

Define $\rho(\theta) = \rho(\mathbf{K}(\theta))$ to be the Perron-Frobenius root of $\mathbf{K}(\theta)$ when θ is real. The definition may be extended to complex values of θ close to the real axes (see Theorem A.3 part 1). The line of integration of $L_i^*(\theta, t)$ is chosen to pass through the saddle point of $Re(\rho(\theta)t - \theta s(t))$ on the real line. By Theorem A.3 part 6, $\rho(\theta)$ is a convex function of θ for θ real. Provided $\rho'(0) < s(t)/t < \rho'(\Delta)$, there is therefore a unique value $\lambda(t)$ such that $\rho'(\lambda(t)) = s(t)/t$. Then $\lambda(t)$ is the required saddle point. As $t \to \infty$, $s(t)/t \to c$. Since $\rho(\lambda)$ is a convex function of the real variable λ, then $\lambda(t) \to \lambda_0$ as $t \to \infty$, where $\rho'(\lambda_0) = c$.

Now

$$\eta = \int_{\boldsymbol{\xi}'r \geq s(t)} y_i(\mathbf{r}, t)d\mathbf{r} = \frac{1}{2\pi i}\int_{s(t)}^{\infty}\int_{\lambda(t)-i\infty}^{\lambda(t)+i\infty} e^{-\theta x}L_i^*(\theta, t)d\theta dx.$$

From Lemma 7.1 this is absolutely integrable for $\lambda(t) > 0$, so that the order of integration can be changed. When the expectations of the projected contact distributions are zero, $\lambda_0 > 0$ when $c > 0$. In general we will need to assume that $0 < \lambda_0 < \Delta$. Then there exists a $T_2^* > T_1^*$ and $0 < \lambda_1 < \lambda_0 < \lambda_2 < \Delta$ such that $\lambda_1 \leq \lambda(t) \leq \lambda_2$ for $t \geq T_2^*$. Hence for $t \geq T_2^*$,

$$\eta = \frac{1}{2\pi i}\int_{\lambda(t)-i\infty}^{\lambda(t)+i\infty} \frac{1}{\theta}e^{-\theta s(t)}L_i^*(\theta, t)d\theta. \tag{7.9}$$

Let $h(t) = \rho(\lambda(t))t - \lambda(t)s(t)$. An approximation is obtained for $\sqrt{t}\eta e^{\mu t - h(\lambda(t))}$ when t is large in Lemma 7.2. This is then used in Theorem 7.1 to obtain the speed of spread.

It is assumed that $\mathbf{K}(\lambda_0)$ has distinct eigenvalues, $\mu_1(\lambda_0), \mu_2(\lambda_0) ..., \mu_n(\lambda_0)$, where $\rho(\lambda_0) = \mu_1(\lambda_0)$. Note that this will always hold for $n \leq 2$. Then there exists an open ball centred on $\theta = \lambda_0$ for which $\mathbf{K}(\theta)$ has distinct eigenvalues $\mu_j(\theta)$ for $j = 1, ..., n$ which are continuous functions of θ with $Re(\mu_j(\theta)) < \mu_1(\theta)$ for $j \neq 1$ (see Lemma A.1 and Theorem A.3 part 1). Then $\rho(\theta) = \mu_1(\theta)$ is the eigenvalue with largest real part.

Hence there exists a $\delta^* > 0$ and $T^* > T_2^*$ with the eigenvalues of $\mathbf{K}(\lambda(t) + iy)$ distinct and each is a continuous functions of y and t for $|y| < \delta^*$ and $t > T^*$.

For this range $\rho(\theta)$ is the eigenvalue with largest real part. Define $\mathbf{E}(\theta)$ to be the corresponding idempotent.

LEMMA 7.2. *Let $\rho(\lambda) > \mu$ for all $\lambda \in [0, \Delta]$ and assume that $c = \lim_{t\to\infty} s(t)/t$ exists, is positive and lies in the range $(\rho'(0), \rho'(\Delta))$. Then given any $\varepsilon > 0$, there exists a $0 < \delta < \delta^*$ and a corresponding $T > T^*$ such that*

$$\left| \sqrt{t}\eta e^{\mu t - h(t)} - \frac{\sqrt{t}\{\mathbf{E}(\lambda(t))\mathbf{L}(\lambda(t),0)\}_i}{2\pi\lambda(t)} \int_{|y|<\delta} \left(e^{t(\rho(\lambda(t)+iy)-\rho(\lambda(t)))-iys(t)} \right) dy \right| < \varepsilon,$$

for all $t > T$.

PROOF. For any $t > T^*$ and $0 < \delta < \delta^* < \alpha$, taking $\theta = \lambda(t) + iy$ in each integral, for any $0 < \delta < \delta^* < \alpha$ we may use equation (7.9) and write

$$\sqrt{t}\eta e^{\mu t - h(t)} = \frac{\sqrt{t}\{\mathbf{E}(\lambda(t))\mathbf{L}(\lambda(t),0)\}_i}{2\pi\lambda(t)} \int_{|y|<\delta} \left(e^{t(\rho(\theta t)-\rho(\lambda(t)))-iys(t)} \right) dy + \sum_{j=1}^{5} I_j,$$

where

$$I_1 = \frac{\sqrt{t}e^{-h(t)}}{2\pi} \int_{|y|>\alpha} \frac{e^{-\theta s(t)}}{\theta} \left\{ \left(e^{t\mathbf{K}(\theta)} - e^{t\mathbf{M}} \right) \mathbf{L}(\theta,0) \right\}_i dy,$$

$$I_2 = -\frac{\sqrt{t}e^{-h(t)}}{2\pi} \int_{|y|\le\alpha} \frac{e^{-\theta s(t)}}{\theta} \left\{ e^{t\mathbf{M}} \mathbf{L}(\theta,0) \right\}_i dy,$$

$$I_3 = \frac{\sqrt{t}e^{-h(t)}}{2\pi} \int_{\delta<|y|<\alpha} \frac{e^{-\theta s(t)}}{\theta} \left\{ e^{t\mathbf{K}(\theta)} \mathbf{L}(\theta,0) \right\}_i dy,$$

$$I_4 = \frac{\sqrt{t}e^{-h(t)}}{2\pi} \int_{|y|<\delta} \frac{e^{-\theta s(t)}}{\theta} \left\{ \left(e^{t\mathbf{K}(\theta)} - \mathbf{E}(\theta)e^{\rho(\theta)t} \right) \mathbf{L}(\theta,0) \right\}_i dy,$$

$$I_5 = \frac{\sqrt{t}e^{-h(t)}}{2\pi} \int_{|y|<\delta} e^{-\theta s(t)} \left(\frac{1}{\theta}\{\mathbf{E}(\theta)\mathbf{L}(\theta,0)\}_i - \frac{1}{\lambda(t)}\{\mathbf{E}(\lambda(t))\mathbf{L}(\lambda(t),0)\}_i \right) e^{\rho(\theta)t} dy,$$

with $\theta = \lambda(t) + iy$ in each integral. Note that since $\delta < \delta^*$ and $t > T^*$ that, in I_4 and I_5, $\rho(\theta)$ is the eigenvalue of $\mathbf{K}(\theta)$ with largest real part and $\mathbf{E}(\theta)$ is the corresponding idempotent (see Appendix A).

Therefore in order to prove the lemma we need only show that, given any $\varepsilon > 0$ there exists a $0 < \delta < \delta^* < \alpha$ and $T > T^*$ such that $|I_j| < \varepsilon/5$ for $j = 1, ..., 5$ for all $t > T$. The proof is split into five parts corresponding to the five components; however it is convenient not to consider them in consecutive order since I_3 is dependent on the δ selected.

1. First consider I_1. Now for $t > T^*$, $\lambda(t)$ is contained in the closed interval $[\lambda_1, \lambda_2]$ with $0 < \lambda_1 < \lambda_2 < \Delta$. Then

$$|I_1| \le \frac{\sqrt{t}}{2\pi\lambda_1} \int_{|y|>\alpha} e^{-\rho(\lambda(t))} \left\{ \left| \left(e^{t(\mathbf{Q}(\theta)+\mathbf{M})} - e^{t\mathbf{M}} \right) \mathbf{L}(\theta,0) \right| \right\}_i dy.$$

Consider the proof of Lemma 7.1, which gives a uniform bound on the term within the modulus in the integral above. Since the function k is integrable, we can choose α sufficiently large such that

$$d = \sup_{|y| \geq \alpha} k(y) < \min_{i,j} \inf_{\lambda \in [\lambda_1, \lambda_2]} P_{ij}(\lambda).$$

Hence

$$|I_1| \leq \frac{\sqrt{t}}{2\pi\lambda_1 d} \int_{|y|>\alpha} e^{-\rho(\lambda(t))} \left\{ e^{t(d\mathbf{S}+\mathbf{M})} \mathbf{L}(\lambda(t), 0) \right\}_i k(y) dy$$

for all $t \geq T^*$.

Now d has been chosen so that $d\mathbf{S} + \mathbf{M} < \mathbf{K}$, where the matrix $\mathbf{K}$ has $\{\mathbf{K}\}_{ij} = \sigma_{ij} \inf_{\lambda \in [\lambda_1, \lambda_2]} P_{ij}(\lambda) + \{\mathbf{M}\}_{ij}$. Hence from Theorem A.2 part 1 $\rho(d\mathbf{S} + \mathbf{M}) < \rho(\mathbf{K}) \leq \inf_{\lambda \in [\lambda_1, \lambda_2]} \rho(K(\lambda))$.

Using the spectral expansion of $d\mathbf{S}+\mathbf{M}$ and the boundedness of $\mathbf{L}(\theta, 0)$, there exists a $C > 0$ and a non-negative integer $c \leq n$, which can be non-zero if not all of the eigenvalues are distinct, such that

$$\{e^{t(d\mathbf{S}+\mathbf{M})} \mathbf{L}(\lambda(t), 0)\}_i \leq C t^c e^{t\rho(d\mathbf{S}+\mathbf{M})}$$

for $|y| > \alpha$ and $t \geq T^*$. Therefore

$$|I_1| \leq t^{c+0.5} e^{-t(\rho(\mathbf{K}) - \rho(d\mathbf{S}+\mathbf{M}))} \frac{C}{2\pi\lambda_1 d} \int_{|y|>\alpha} k(y) dy.$$

The right hand side tends to zero as t tends to infinity. Hence, for the α chosen, there exists a $T_1 > T^*$ such that $|I_1| < \varepsilon/5$ provided $t \geq T_1$.

2. Next consider I_2, with the selected α.

$$|I_2| \leq \frac{\sqrt{t}\alpha}{\pi\lambda_1} e^{-t\rho(\lambda(t))} \left\{ e^{t\mathbf{M}} \mathbf{L}(\lambda, 0) \right\}_i.$$

The same upper bound on $\{e^{t\mathbf{M}} \mathbf{L}(\lambda(t), 0)\}_i$ may be used as was used in the last part for $\{e^{t(d\mathbf{S}+\mathbf{M})} \mathbf{L}(\lambda(t), 0)\}_i$, since $d > 0$. Hence

$$|I_2| \leq t^{c+0.5} e^{-t(\rho(\mathbf{K}) - \rho(d\mathbf{S}+\mathbf{M}))} \frac{C\alpha}{\pi\lambda_1}.$$

The right hand side tends to zero as t tends to infinity. Hence there exists a $T_2 > T_1$ such that $|I_2| < \varepsilon$ for $t > T_2$.

3. We now consider I_5 out of sequence, since this decides the choice of $0 < \delta < \delta^*$.

$$|I_5| \leq \frac{\sqrt{t}}{2\pi} \int_{|y|<\delta} e^{t(Re(\rho(\lambda(t)+iy)) - \rho(\lambda(t)))}$$
$$\times \left| \frac{1}{\theta} \{\mathbf{E}(\theta)\mathbf{L}(\theta, 0)\}_i - \frac{1}{\lambda(t)} \{\mathbf{E}(\lambda(t))\mathbf{L}(\lambda(t), 0)\}_i \right| dy.$$

From Theorem A.3 parts 1 and 2 $\rho(\lambda(t)+iy)$ is analytic and the idempotent $\mathbf{E}(\lambda(t)+iy)$ has analytic entries for $|y|<\delta^*$ and $t>T^*$. Also, from Theorem A.3 part 6, $\rho(\lambda)$ is a strictly convex function of λ for $0<\lambda<\Delta$, so that $\rho''(\lambda)>0$, and its minimum over the closed interval $[\lambda_1,\lambda_2]$ is also positive.

Using uniform continuity of $\frac{1}{\theta}\{\mathbf{E}(\theta)\mathbf{L}(\theta,0)\}_i$ in a closed region, for any $\varepsilon_1>0$ there exists a $0<\delta_1<\delta^*$ and $T_5>T^*$ such that

$$\left|\frac{1}{\lambda(t)+iy}\{\mathbf{E}(\lambda(t)+iy)\mathbf{L}(\lambda(t)+iy,0)\}_i-\frac{1}{\lambda(t)}\{\mathbf{E}(\lambda(t))\mathbf{L}(\lambda(t),0)\}_i\right|<\varepsilon_1$$

for $|y|<\delta_1$ and $T>T_5$. Hence, for $0<\delta<\delta_1$,

$$|I_5|\le\frac{\sqrt{t}\varepsilon_1}{2\pi}\int_{|y|<\delta}e^{t(Re(\rho(\lambda(t)+iy))-\rho(\lambda(t)))}dy.$$

Using the analyticity of $\rho(\lambda+iy)$ and uniform continuity in a bounded region for the partial derivatives of the real and complex parts, there exists a positive A such that

$$|Re(\rho(\lambda(t)+iy)-\rho(\lambda(t))-\frac{y^2}{2}\rho''(\lambda(t)|\le\frac{A|y|^3}{3}.$$

for $|y|<\delta^*$ and $t>T^*$. Let $a=\inf_{\lambda\in[\lambda_1,\lambda_2]}\rho''(\lambda)$, which is positive. Now take $0<\delta_2<\delta^*$ such that $\delta_2<\frac{3a}{2A}$. Then

$$-\frac{y^2}{2}\rho''(\lambda(t))+\frac{A|y|^3}{3!}\le-\frac{y^2}{4}\rho''(\lambda(t)).$$

Hence for $t>T_5$ and $\delta<\min(\delta_1,\delta_2)$, using the result that a normal density function integrates to one gives

$$|I_5|\le\frac{\sqrt{t}\varepsilon_1}{2\pi}\int_{|y|<\delta}e^{t(-\frac{y^2}{4}\rho''(\lambda(t)))}dy\le\frac{\varepsilon_1}{\sqrt{\pi a}}.$$

Then if we take $\varepsilon_1<\sqrt{\pi a}\varepsilon/5$ and find the corresponding δ_1 and T_5, then $|I_5|\le\varepsilon/5$ for all $t>T_5$, provided we choose $\delta<\min(\delta_1,\delta_2)$.

4. Next consider I_4 for the δ selected. Since $\delta<\delta^*$, provided $t>T^*$ the eigenvalues of $\mathbf{K}(\lambda(t)+iy)$ are distinct and are continuous functions of t and y for all $|y|\le\delta$ and $t>T^*$.

   For $t\ge T^*$ and $|y|\le\delta$ the spectral expansion of $\mathbf{K}(\theta)$ is then given by $\mathbf{K}(\theta)=\sum_{j=1}^n\mu_j(\theta)\mathbf{E}_j(\theta)$ for $\theta=\lambda(t)+iy$, where $\mu_j(\theta)$ is the jth eigenvalue of $\mathbf{K}(\theta)$ with corresponding idempotent $\mathbf{E}_j(\theta)$. Note that $\mu_1=\rho(\theta)$ and $\mathbf{E}_1(\theta)=\mathbf{E}(\theta)$.

   From Theorem A.3 part 5, the idempotent $\mathbf{E}_j(\lambda+iy)$ has entries which are continuous functions of λ and y for $\lambda\in[\lambda_1,\lambda_2]$ and $|y|\le\delta<\delta^*$. Therefore there exist D_j such that $|\{\mathbf{E}_j(\lambda(t)+iy)\}_{rs}|\le D_j$ for all r,s, all $|y|\le\delta$ and all $t>T^*$.

   Now $Re(\rho(\lambda(t)+iy))>\max_{j>1}Re(\mu_j(\lambda(t)+iy))$ in this region. Continuity in a closed region then implies that there exists an $\eta>0$ such that

$$\sup_{|y|\le\delta,\ t\ge T^*} (Re(\rho(\lambda(t)+iy)) - \max_{j>1} Re(\mu_j(\lambda(t)+iy))) > \eta.$$

For $\theta = \lambda(t) + iy$ such that $t > T^*$ and $|y| \le \delta$,

$$e^{t\mathbf{K}(\theta)} - \mathbf{E}(\theta)e^{\rho(\theta)t} = \sum_{j=2}^{n} \mathbf{E}_j(\theta)e^{\mu_j(\theta)t}.$$

Also $\mathbf{L}(\theta, 0)$ is bounded so that there exists an L with $\mathbf{L}(\lambda(t)+iy) \le L\mathbf{1}$. Then, for $t > T^*$,

$$\begin{aligned} |I_4| &\le \frac{\sqrt{t}}{2\pi\lambda_1}\sum_{j=2}^{n}\int_{|y|\le\delta} e^{-t(Re(\rho(\lambda(t)+iy))-Re(\mu_j(\lambda(t)+iy)))} \\ &\quad\times \{|\mathbf{E}_j(\lambda(t)+iy)||L(\lambda(t)+iy,0)|\}_i dy \\ &\le \frac{n\delta L\sqrt{t}e^{-\eta t}}{\pi\lambda_1}\sum_{j=2}^{n} D_j. \end{aligned}$$

The right hand side of the inequality can be made arbitrarily small by taking t sufficiently large. Hence there exists a $T_4 > T^*$ such that $|I_4| < \varepsilon/5$ for all $t > T_4$.

5. Finally consider I_3 for the selected α and δ and consider any $t > T^*$. Then

$$\begin{aligned} |I_3| &\le \frac{\sqrt{t}}{2\pi\lambda_1}\int_{\delta\le|y|\le\alpha} e^{-t\rho(\lambda(t))}\left\{\left|e^{t\mathbf{K}(\lambda(t)+iy)}L(\lambda(t)+iy,0)\right|\right\}_i dy \\ &\le \frac{\sqrt{t}}{2\pi\lambda_1}\int_{\delta\le|y|\le\alpha} e^{-t\rho(\lambda(t))}\left\{e^{t|\mathbf{K}(\lambda(t)+iy)|}|L(\lambda(t)+iy,0)|\right\}_i dy. \end{aligned}$$

For a specific i, j such that $\alpha_{ij} \ne 0$, consider $|K_{ij}(\lambda+iy)|$ and a monotone decreasing positive sequence $\{a_r\}$ with $\lim_{r\downarrow\infty} a_r = 0$. Define

$$K^*_{ij}(a_r) = \sup_{\delta\le|y|\le\alpha, |\lambda-\lambda_0|\le a_r} |K_{ij}(\lambda+iy)|.$$

By uniform continuity in each of two bounded regions there exist λ_r and y_r such that $|K_{ij}(\lambda_r+iy_r)| = K^*_{ij}(a_r)$. Now $\lambda_r \to \lambda_0$ as $a_r \downarrow 0$, so there exists a subsequence of the a_r for which $\lambda_r \to \lambda_0$ and $y_r \to y_0$ for some y_0 such that $\delta \le |y_0| \le \alpha$.

If we also define $K^0_{ij}(a_r) = \inf_{|\lambda-\lambda_0|\le a_r} K_{ij}(\lambda)$. Then $K^0_{ij}(a_r) \to K_{ij}(\lambda_0)$ as $a_r \downarrow 0$. Therefore

$$\lim_{r\to\infty}(K^*_{ij}(a_r) - K^0_{ij}(a_r)) = |K_{ij}(\lambda_0+iy_0)| - K_{ij}(\lambda_0) < 0.$$

Hence there exists an $A_{ij} = a_r$ for r sufficiently large such that $K^*_{ij}(A_{ij}) < K^0_{ij}(A_{ij})$. Take $A = \min_{i,j:\alpha_{ij}\ne0} A_{ij}$. Since $\lambda(t) \to \lambda_0$ as $t \to \infty$, we can therefore choose a $T_0 > T^*$ such that $|\lambda(t) - \lambda_0| \le A$ for $t \ge T_0$.

Now let $\mathbf{K}^*(A)$ and $\mathbf{K}^0(A)$ be the matrices which have ij^{th} entries $K^*_{ij}(A)$ and $K^0_{ij}(A)$ respectively. Then, for all $t \geq T_0$ and $\delta \leq |y| \leq \alpha$, $|\mathbf{K}(\lambda(t)+iy)| \leq \mathbf{K}^*(A) \lneqq \mathbf{K}^0(A)$ and hence, from Theorem A.2 part 1, $\rho(\mathbf{K}^*(A)) < \rho(\mathbf{K}^0(A)) \leq \inf_{t\geq T_0} \rho(\lambda(t))$.

Therefore, for $t \geq T_0$,

$$|I_3| \leq \frac{\sqrt{t}2\alpha}{2\pi\lambda_1} e^{-t\rho(\mathbf{K}^0(A))} \left\{ e^{t\mathbf{K}^*(A)} \mathbf{L}(\lambda(t),0) \right\}_i .$$

Using the uniform bound on $\mathbf{L}(\lambda(t),0)$ and the spectral expansion of $\mathbf{K}^*(A)$ there exist a $C > 0$ and a non-negative integer c such that

$$\left\{ e^{t\mathbf{K}^*(A)} \mathbf{L}(\lambda(t),0) \right\}_i \leq Ct^c e^{t\rho(\mathbf{K}^*(A))}.$$

Therefore, for $t \geq T_0$,

$$|I_3| \leq \frac{\alpha C t^{c+0.5}}{\pi\lambda_1} e^{-t(\rho(\mathbf{K}^0(A))-\rho(\mathbf{K}^*(A)))}.$$

Since the right hand side tends to zero as $t \to \infty$, there exists a $T_3 > T_0$ such that $|I_3| < \varepsilon/5$ for $t \geq T_3$.

The lemma then follows by taking $T = \max_{1\leq j\leq 5} T_j$.

□

Lemma 7.2 is now used to derive the speed of spread of the linear system.

THEOREM 7.1. *If $\rho(\lambda) > \mu$ for all $\lambda \in [0,\Delta]$ and if $c = \lim_{t\to\infty} s(t)/t$ exists is positive and lies in the range $(\rho'(0), \rho'(\Delta))$, then the speed of spread is given by*

$$c = \inf_{0<\lambda<\Delta} \frac{\rho(\lambda)-\mu}{\lambda}. \tag{7.10}$$

PROOF. From the definition of $\lambda(t)$, note that $s(t) = t\rho'(\lambda(t))$. Then from Lemma 7.2, for any $\varepsilon > 0$ there exists a $0 < \delta < \delta^*$ and a $T > T^*$ such that, for any $t > T$, $\sqrt{t}\eta e^{\mu t - h(t)}$ is bounded above by $B(t)+\varepsilon$ and is bounded below by $B(t)-\varepsilon$, where

$$B(t) = Re\left(\frac{\sqrt{t}\{\mathbf{E}(\lambda(t))\mathbf{L}(\lambda(t),0)\}_i}{2\pi\lambda(t)} \int_{|y|<\delta} e^{t(\rho(\lambda(t)+iy)-\rho(\lambda(t))-iy\rho'(\lambda(t)))} dy \right).$$

In this theorem we modify these bounds and then use the modified bounds to obtain bounds for $(s(t)/t) - ((\rho(\lambda(t))-\mu)/\lambda(t))$. The limits of these latter bounds are both shown to be zero thus establishing the result that $c = (\rho(\lambda_0)-\mu)/\lambda_0$. The remainder of the proof is then easily established.

First consider an upper bound on $B(t)$. For $t \geq T^*$ and $|y| \leq \delta^*$, using uniform continuity in a bounded region of the third partial derivatives of the real and imaginary parts of $\rho(\theta)$ gives the result that

$$\rho(\lambda(t)+iy) = \rho(\lambda(t)) + iy\rho'(\lambda(t)) - \frac{y^2}{2}\rho''(\lambda(t)) + \frac{y^3}{3!}z(t,y),$$

where $|Re(z(t,y))| \leq A$ and $|Im(z(t,y))| \leq B$ for some positive constants A and B. Note that δ was chosen in the proof of Lemma 7.2 part 3 to be less than $3a/(2A)$ where $a = \inf_{\lambda\in[\lambda_1,\lambda_2]} \rho''(\lambda)$. Then, using the T and δ from Lemma 7.2, for $t \geq T$

$$\begin{aligned}
B(t) &\leq \frac{\sqrt{t}\{\mathbf{E}(\lambda(t))\mathbf{L}(\lambda(t),0)\}_i}{2\pi\lambda(t)} \int_{|y|<\delta} \left| e^{t(\rho(\lambda(t)+iy)-\rho(\lambda(t))-iy\rho'(\lambda(t))} \right| dy \\
&\leq \frac{\sqrt{t}\{\mathbf{E}(\lambda(t))\mathbf{L}(\lambda(t),0)\}_i}{2\pi\lambda(t)} \int_{|y|<\delta} e^{t(-(y^2/2)\rho''(\lambda(t))+(y^2/3!)\delta A)} dy \\
&\leq \frac{\sqrt{t}\{\mathbf{E}(\lambda(t))\mathbf{L}(\lambda(t),0)\}_i}{2\pi\lambda(t)} \int_{|y|<\delta} e^{-ty^2\rho''(\lambda(t))/4} dy \\
&\leq \frac{\{\mathbf{E}(\lambda(t))\mathbf{L}(\lambda(t),0)\}_i}{2\pi\lambda(t)} \int_{|y|<\sqrt{t}\delta} e^{-y^2\rho''(\lambda(t))/4} dy \\
&\leq \frac{\{\mathbf{E}(\lambda(t))\mathbf{L}(\lambda(t),0)\}_i}{\sqrt{\pi\rho''(\lambda(t))}\lambda(t)} \int_{|y|<\delta\sqrt{t\rho''(\lambda(t))/2}} \frac{1}{\sqrt{2\pi}} e^{-y^2/2} dy \\
&\leq \frac{\{\mathbf{E}(\lambda(t))\mathbf{L}(\lambda(t),0)\}_i}{\sqrt{\pi\rho''(\lambda(t))}\lambda(t)} \int_{-\infty}^{\infty} \frac{1}{\sqrt{2\pi}} e^{-y^2/2} dy \\
&= \frac{\{\mathbf{E}(\lambda(t))\mathbf{L}(\lambda(t),0)\}_i}{\sqrt{\pi\rho''(\lambda(t))}\lambda(t)}.
\end{aligned}$$

Now consider obtaining a lower bound on $B(t)$, which requires more thought. For any $t > T$, using the expansion for $\rho(\lambda(t))$, we may write

$$\begin{aligned}
B(t) &= \frac{\sqrt{t}\{\mathbf{E}(\lambda(t))\mathbf{L}(\lambda(t),0)\}_i}{2\pi\lambda(t)} \int_{|y|<\delta} e^{t(-(y^2/2)\rho''(\lambda(t))+(y^3/3!)Re(z(t,y))} \\
&\quad \times \cos(t(y^3/3!)Im(z(t,y)))dy \\
&= \frac{\{\mathbf{E}(\lambda(t))\mathbf{L}(\lambda(t),0)\}_i}{2\pi\lambda(t)} \int_{-\delta\sqrt{t}}^{\delta\sqrt{t}} e^{-(u^2/2)\rho''(\lambda(t))+(u^3/(3!\sqrt{t}))Re(z(t,u/\sqrt{t}))} \\
&\quad \times \cos((u^3/(3!\sqrt{t}))Im(z(t,u/\sqrt{t})))du.
\end{aligned}$$

To obtain the lower bound, first consider any $k > 0$. Define $\tau_1 = \max(T, k^2/(a\delta)^2)$, so that $\delta\sqrt{t} > k/a$ if $t > \tau_1$. Provided $t > \tau_1$ we can then split the range of the integral into two parts, $|u| \leq k/a$ and $k/a < |u| \leq \delta\sqrt{t}$.

In the first integral, for all u, $\cos((u^3/(3!\sqrt{t}))Im(z(t,u/\sqrt{t}))) \geq 1/\sqrt{2}$ provided $(k/a)^3B/(3!\sqrt{t}) \leq \pi/4$, i.e. provided $t \geq \tau = \max(\tau_1,\tau_2)$ where $\tau_2 = (4B^2k^6)/(9\pi^2a^6)$. In the second integral we simply use the result that cos is bounded below by -1. Additionally, in both of the integrals the result is used that $-A < Re(z(t,y)) < A$ for $t > T^*$ and $|y| < \delta$. Therefore for $t > \tau$, using the usual notation that Φ is the distribution function for the unit normal distribution,

$$\begin{aligned}
B(t) &\geq \frac{\{\mathbf{E}(\lambda(t)\mathbf{L}(\lambda(t),0)\}_i}{2\pi\lambda(t)} \int_{-k/a}^{k/a} e^{-(u^2/2)\rho''(\lambda(t))-(u^2A\delta/3!)} \frac{1}{\sqrt{2}} du \\
&- \frac{\{\mathbf{E}(\lambda(t)\mathbf{L}(\lambda(t),0)\}_i}{2\pi\lambda(t)} \int_{k/a<|u|\leq\delta\sqrt{t}} e^{-(u^2/2)\rho''(\lambda(t))+(u^2A\delta/3!)} du \\
&\geq \frac{\{\mathbf{E}(\lambda(t)\mathbf{L}(\lambda(t),0)\}_i}{2\pi\lambda(t)} \int_{-k/a}^{k/a} e^{-u^2\rho''(\lambda(t))} \frac{1}{\sqrt{2}} du \\
&- \frac{\{\mathbf{E}(\lambda(t)\mathbf{L}(\lambda(t),0)\}_i}{2\pi\lambda(t)} \int_{k/a<|u|} e^{-(u^2/4)\rho''(\lambda(t))} du \\
&= \frac{\{\mathbf{E}(\lambda(t)\mathbf{L}(\lambda(t),0)\}_i}{\lambda(t)2\sqrt{2\pi\rho''(\lambda(t))}} \left(2\Phi\left(\frac{k\sqrt{2\rho''(\lambda(t))}}{a} \right) - 1 \right) \\
&- \frac{\{\mathbf{E}(\lambda(t)\mathbf{L}(\lambda(t),0)\}_i}{\lambda(t)\sqrt{\pi\rho''(\lambda(t))}} 2 \left(1 - \Phi\left(\frac{k\sqrt{\rho''(\lambda(t))/2}}{a} \right) \right) \\
&\geq \frac{\{\mathbf{E}(\lambda(t)\mathbf{L}(\lambda(t),0)\}_i}{\lambda(t)2\sqrt{2\pi\rho''(\lambda(t))}} ((2\Phi(k\sqrt{2/a}) - 1) - 4\sqrt{2}(1 - \Phi(k/\sqrt{2a}))).
\end{aligned}$$

Since $((2\Phi(k\sqrt{2/a})-1)-4\sqrt{2}(1-\Phi(k/\sqrt{2a})))$ tends to 1 as k tends to infinity, we can choose k sufficiently large to make it at least $1/2$. For this k we can then find the corresponding τ. Hence for $t > \tau$,

$$B(t) \geq \frac{\{\mathbf{E}(\lambda(t)\mathbf{L}(\lambda(t),0)\}_i}{4\lambda(t)\sqrt{2\pi\rho''(\lambda(t))}}.$$

Therefore for $t > \tau$, we have obtained the result that

$$\frac{\{\mathbf{E}(\lambda(t)\mathbf{L}(\lambda(t),0)\}_i}{4\lambda(t)\sqrt{2\pi\rho''(\lambda(t))}} - \varepsilon \leq \sqrt{t}\eta e^{\mu t - h(t)} \leq \frac{\{\mathbf{E}(\lambda(t))\mathbf{L}(\lambda(t),0)\}_i}{\lambda(t)\sqrt{\pi\rho''(\lambda(t))}} + \varepsilon. \tag{7.11}$$

If we choose $\varepsilon < \dfrac{\{\mathbf{E}(\lambda_0)\mathbf{L}(\lambda_0,0)\}_i}{8\lambda_0\sqrt{2\pi\rho''(\lambda_0)}}$ and hence find the corresponding values of T, δ, k and τ, then there exists a $T_0 > \tau$ sufficiently large that the left and right hand sides of the above inequality for $\sqrt{t}\eta e^{\mu t - h(t)}$ are positive for $t > T_0$ and tend to finite positive limits as $t \to \infty$.

For $t > T_0$, scaling inequality (7.11) then taking logs and dividing by $t\lambda(t)$ gives the inequality

$$\begin{aligned}
&\frac{1}{\lambda(t)t}\left[\log\left(\frac{\{\mathbf{E}(\lambda(t)\mathbf{L}(\lambda(t),0)\}_i}{4\lambda(t)\sqrt{2\pi\rho''(\lambda(t))}} - \varepsilon \right) - \log(\eta\sqrt{t}) \right] \\
&\leq \frac{s(t)}{t} - \frac{\rho(\lambda(t)) - \mu}{\lambda(t)} \\
&\leq \frac{1}{\lambda(t)t}\left[\log\left(\frac{\{\mathbf{E}(\lambda(t)\mathbf{L}(\lambda(t),0)\}_i}{\lambda(t)\sqrt{\pi\rho''(\lambda(t))}} + \varepsilon \right) - \log(\eta\sqrt{t}) \right].
\end{aligned}$$

As $t \to \infty$, $[(s(t)/t) - (\rho(\lambda(t)) - \mu)/\lambda(t))] \to [c - (\rho(\lambda_0) - \mu)/\lambda_0]$ and both bounds tend to zero. Hence we obtain the result that $c = (\rho(\lambda_0) - \mu)/\lambda_0$.

Therefore there are two relations linking c and λ_0, $c = \rho'(\lambda_0)$ and $c = f(\lambda_0)$, where $f(\lambda) = (\rho(\lambda) - \mu)/\lambda$. Consider the function $f(\lambda)$ for $\lambda > 0$. Differentiating we obtain $f'(\lambda) = (\rho'(\lambda) - f(\lambda))/\lambda$. This is zero when $\lambda = \lambda_0$. Also $f''(\lambda) = (\rho''(\lambda) - 2f'(\lambda))/\lambda$, which is positive for all $\lambda > 0$ such that $f'(\lambda) = 0$. Hence there is a unique minimum of the function $f(\lambda)$ at $\lambda = \lambda_0$. Therefore the speed of spread of the forward front for equation (7.1) is given by

$$c = \inf_{0<\lambda<\Delta} \frac{\rho(\lambda) - \mu}{\lambda},$$

which completes the proof of the theorem.

□

Note that if $|\rho(\theta)\mathbf{I} - \mathbf{K}(\theta)| = 0$ is differentiated we obtain,

$$\rho'(\theta) = \frac{\sum_i \sum_j \mathbf{K}'_{ij}(\theta)\{Adj(\rho(\theta)\mathbf{I} - \mathbf{K}(\theta))\}_{ij}}{trace(Adj(\rho(\theta)\mathbf{I} - \mathbf{K}(\theta)))}.$$

Then if the projected contact distributions have mean zero, so that $K'_{ij}(0) = 0$ for all i, j, it immediately follows that $\rho'(0) = 0$.

When the projected contact distributions have mean zero and are infinite at their abscissae of convergence then $\rho'(0) = 0$, and hence $\rho(\lambda) > \rho(0)$ for $\lambda > 0$, and $\rho'(\Delta) = \infty$. Provided $\rho(0) > \mu$ the speed of spread will then be positive and given by equation (7.10).

In general when $\rho(0) > \mu$, then $\lim_{\lambda\downarrow 0} f(\lambda) = \infty$. If the $P_{ij}(\lambda)$ are infinite at their abscissae of convergence, then $f(\lambda)$ has a minimum for $\lambda > 0$. Equation (7.10) then gives the speed of spread of the forward front in the specified direction provided the value of c obtained is positive. A non-positive value is obtained precisely when $\rho(\lambda) \leq \mu$ for some $\lambda \in (0, \Delta)$, so that the condition that $\rho(\lambda) > \mu$ for all $\lambda \in (0, \Delta)$ does not hold.

If $\rho(0) > \mu$ and $\rho(\Delta)$ is finite it is also possible that the conditions of Theorem 7.1 are not met. If Δ is infinite then $\lim_{\lambda\to\Delta} \rho(\lambda) = 0$, and hence the infimum cannot be positive. When Δ is finite with $\rho'(\Delta) \leq 0$, then $\rho'(\lambda) \leq 0$ for all $0 \leq \lambda \leq \Delta$ and hence there cannot be a positive infimum within the range $(0, \Delta)$.

When $\rho(0) \leq \mu$ the conditions of Theorem 7.1 are not met. Note that when $\rho(0) < \mu$, $\lim_{\lambda\downarrow 0} f(\lambda) = -\infty$, so $f(\lambda)$ has no minimum. Also when $\rho(0) = \mu$ $\lim_{\lambda\downarrow 0} f'(\lambda) = \rho''(0)/2 > 0$. Again $f(\lambda)$ has no minimum for $\lambda > 0$.

Hence if $f(\lambda)$ does have a minimum for $\lambda \in (0, \Delta)$, and if this minimum is positive, then equation (7.10) gives the speed of spread. In all other cases the saddle point method does not give the speed of propagation.

Discrete space.

As in section 7.2 for the one type case, for the discrete space model we consider the speed of spread of the forward front in the direction of the first coordinate. For η small and positive, define $s(t)$ to be the smallest integer such that $\sum_{\mathbf{u}:\{\mathbf{u}\}_1 \geq s(t)} y_i(\mathbf{u}, t) \leq \eta$. Define $\eta(t)$ to be the exact value of the summation.

The speed of spread of the forward front for type i is then $c = \lim_{t\to\infty} s(t)/t$. The value of c can again be obtained using the saddle point method.

Laplace transforms are defined as for the continuous case, except that integrals are replaced by summations. An identical equation is obtained for the Laplace transform of $y_i(\mathbf{s}, t)$.

Inverting the Laplace transform, for a suitable choice of $\lambda(t)$ gives

$$\sum_{\mathbf{u}:\{\mathbf{u}\}_1=v} y_i(\mathbf{u}, t) = \frac{1}{2\pi}\int_{-\pi}^{\pi} e^{-(\lambda(t)+iy)v}\{\mathbf{L}(\lambda(t)+iy, t)\}_i dy$$

$$= \frac{1}{2\pi}\int_{-\pi}^{\pi} e^{-(\lambda(t)+iy)v} e^{(\mathbf{K}(\lambda(t)+iy)-\mu\mathbf{I})t}\mathbf{L}(\lambda(t)+iy, 0)\}_i dy.$$

Hence

$$\eta(t) = \sum_{\mathbf{u}:\{\mathbf{u}\}_1\geq s(t)} y_i(\mathbf{u}, t)$$

$$= \frac{1}{2\pi}\int_{-\pi}^{\pi} \frac{1}{1-e^{-(\lambda(t)+iy)}} e^{-(\lambda(t)+iy)s(t)} \left\{e^{(\mathbf{K}(\lambda(t)+iy)-\mu\mathbf{I})t}\mathbf{L}(\lambda(t)+iy, 0)\right\}_i dy.$$

As for continuous space, $\theta = \lambda(t)$ is taken to be the saddle point of $g(\theta) = Re((\rho(\theta)-\mu)t-\theta s(t))$ where $\rho(\theta)$ is the eigenvalue of $\mathbf{K}(\theta)$ with largest real part with corresponding idempotent $\mathbf{E}(\theta)$. Again $\mathbf{K}(\theta)$ is taken to have distinct eigenvalues when $\theta = \lambda_0 = \lim_{t\to\infty}\lambda(t)$.

Both Lemma 7.2 and Theorem 7.1 still hold. The main adjustment to the proof of Lemma 7.2 is that $\alpha = \pi$ and the components I_1 and I_2 are not required. The result for the speed of spread is then identical to the continuous case, with the speed c being given by equation (7.10).

7.5 The speed of spread of the forward front for epidemic models

The saddle point method is now applied to the model for the $S \to L \to I \to R$ epidemic, and sub-models which do not have latent and/or removed stages. These results were obtained for the non-reducible case in Radcliffe and Rass [R5] and for the reducible case in Radcliffe and Rass [R9]. As in previous chapters, attention is restricted to the non-reducible case.

The $S \to I \to R$ and $S \to I$ epidemic models.
The continuous space model of the $S \to I \to R$ epidemic is the obvious generalisation of the one-type model specified in Section 7.3. If there is some initial infection in a bounded region at time 0, then at time t in the forward region, far from this initial focus of infection, almost all individuals will be susceptible so that $x_i(\mathbf{s}, t)$ will be approximately equal to one. The equations for $y_i(\mathbf{s}, t)$ in this forward region then become approximately

$$\frac{\partial y_i(\mathbf{s}, t)}{\partial t} = \sum_{j=1}^{n}\lambda_{ij}\sigma_j\int_{\Re^N} p_{ij}(\mathbf{r})y_j(\mathbf{s}-\mathbf{r}, t)d\mathbf{r} - \mu_i y_i(\mathbf{s}, t),\ (i = 1, ..., n).$$

This is precisely the same as equations (7.1) provided we take $\alpha_{ij} = \lambda_{ij}\sigma_j$ and $\beta_{ij} = 0$. Since the equations for the non-linear system are approximately the same in the forward front as the equations for the linear system this suggests that the speed of spread of the forward region is the same. This speed was obtained, in direction $\boldsymbol{\xi}$, for the linear system provided $\rho(\mathbf{K}(\lambda)) > \mu$ for all $\lambda \in [0, \Delta]$. In the expression for c obtained it was also necessary that the infimum occurred in the range $(0, \Delta)$ and was positive.

Now $\mathbf{K}(\theta) = \mathbf{Q}(\theta) + \mu\mathbf{I} - \text{diag}(\boldsymbol{\mu})$, where $\mu = \max_i \mu_i$ and $\text{diag}(\boldsymbol{\mu})$ is the diagonal matrix with ii^{th} entry μ_i. Since $\alpha_{ij} = \lambda_{ij}\sigma_j$, the matrix $\mathbf{Q}(\theta)$ has ij^{th} entry $\lambda_{ij}\sigma_j P_{ij}(\theta)$, where $P_{ij}(\theta) = \int_{-\infty}^{\infty} e^{\theta\boldsymbol{\xi}'\mathbf{r}} p_{ij}(\mathbf{r}) d\mathbf{r}$. Then the speed of spread for the forward front of infection for the $\mathbf{S} \to I \to R$ epidemic in direction $\boldsymbol{\xi}$ is c, where

$$c = \inf_{0<\theta<\Delta} \frac{\rho(\mathbf{K}(\theta)) - \mu}{\theta},$$

provided the infimum is achieved for $\theta \in (0, \Delta)$ and the value of c obtained is positive.

As in the one-type case it is possible also to look at the equations involving removed individuals. The appropriate equations are

$$\frac{\partial z_i(\mathbf{s}, t)}{\partial t} = \mu_i y_i(\mathbf{s}, t), \quad (i = 1, ..., n).$$

If we take $\mathbf{M}(\theta, t)$ and $\mathbf{L}(\theta, t)$ to have i^{th} entries which are the Laplace transforms with respect to u of $z_i(\mathbf{r}, t)$ and $y_i(\mathbf{r}, t)$, each integrated over the hyperplane $\boldsymbol{\xi}'\mathbf{r} = u$, then solving the resulting equation gives

$$\mathbf{M}(\theta, t) = \mathbf{M}(\theta, 0) + \text{diag}(\boldsymbol{\mu})(\mathbf{K}(\theta) - \mu\mathbf{I})^{-1}\left(e^{t(\mathbf{K}(\theta) - \mu\mathbf{I})} - \mathbf{I}\right)\mathbf{L}(\theta, 0).$$

The saddle point method may then be used, with a slight adaptation, since there will be an extra term in the integrand when inverting the Laplace transform. The same speed of spread is obtained for removals as for infection.

Note that for the $S \to I$ epidemic, $\mu_i = 0$ for all $i = 1, ..., n$. Hence the speed of spread of the forward region is just

$$c = \inf_{0<\theta<\Delta} \frac{\rho(\mathbf{Q}(\theta))}{\theta},$$

which is necessarily positive.

If we consider the discrete space model of the $S \to I \to R$ epidemic, then the approximate equations in the forward region are

$$\frac{\partial y_i(\mathbf{s}, t)}{\partial t} = \sum_{j=1}^{n} \lambda_{ij}\sigma_j \sum_{\mathbf{r}\in Z^N} p_{ij}(\mathbf{r}) y_j(\mathbf{s} - \mathbf{r}, t) - \mu_i y_i(\mathbf{s}, t), \quad (i = 1, ..., n).$$

The same expression is obtained for the speed of spread as for the continuous case. However note that we consider the speed in the direction of a specific co-ordinate and hence $P_{ij}(\theta)$ is the Laplace transform of the marginal distribution in that co-ordinate direction obtained from the joint probability $p_{ij}(\mathbf{r})$.

The $S \to L \to I \to R$ and $S \to L \to I$ epidemic models.

Now consider the continuous space models of the $S \to L \to I \to R$ epidemic. Let $x_i(\mathbf{s},t)$, $l_i(\mathbf{s},t)$, $y_i(\mathbf{s},t)$ and $z_i(\mathbf{s},t)$ represent the proportion of individuals of type i at position $\mathbf{s}$ who are susceptible, latent, infectious and removed respectively at time t. The equations they satisfy are the obvious generalisation of the equations for the one-type model specified in Section 7.3.

If some infection, either latent or infectious, is present in a bounded region at time 0, then at time t in the forward region far from the initial focus of infection, almost all individuals are susceptible so that $x_i(\mathbf{s},t)$ is approximately one for all types i. The equations for $l_i(\mathbf{s},t)$ and $y_i(\mathbf{s},t)$ are then approximately

$$\frac{\partial l_i(\mathbf{s},t)}{\partial t} = \sum_{j=1}^{n} \lambda_{ij}\sigma_j \int_{\Re^N} p_{ij}(\mathbf{r}) y_j(\mathbf{s}-\mathbf{r},t) d\mathbf{r} - \alpha_i l_i(\mathbf{s},t),$$

$$\frac{\partial y_i(\mathbf{s},t)}{\partial t} = \alpha_i l_i(\mathbf{s},t) - \mu_i y_i(\mathbf{s},t), \ (i=1,...,n).$$

There are now $2n$ equations. If we compare these to equations (7.1) and add a $*$ to all terms in equation (7.1) to distinguish them from the terms in the above equations, then we require the following identification. For $i = 1,...,n$, take $y_i^*(\mathbf{s},t) = l_i(\mathbf{s},t)$, $y_{n+i}^*(\mathbf{s},t) = y_i(\mathbf{s},t)$, $\beta_{n+i,i}^* = \alpha_i$, $\mu_i^* = \alpha_i$ and $\mu_{n+i}^* = \mu_i$. Also for $i = 1,...,n$ and $j = 1,...,n$ take $\alpha_{i,n+j}^* = \sigma_j\lambda_{ij}$, $P_{i,n+j}^*(\mathbf{s}) = P_{ij}(\mathbf{s})$. All other β_{ij}^* and α_{ij}^* are taken to be zero. Then the speed of propagation for the proportion of latent or infectious individuals of each type is

$$c = \inf_{0<\theta<\Delta} \frac{\rho(\mathbf{K}(\theta)) - \mu}{\theta},$$

where $\mu = \max_i(\max(\alpha_i, \mu_i))$, $\boldsymbol{\alpha} = (\alpha_i)$, $\boldsymbol{\mu} = (\mu_i)$ and the matrix $\mathbf{K}(\theta)$ is given by

$$\mathbf{K}(\theta) = \begin{pmatrix} \mu\mathbf{I} - \operatorname{diag}(\boldsymbol{\alpha}) & (\sigma_j\lambda_{ij}P_{ij}(\theta)) \\ \operatorname{diag}(\boldsymbol{\alpha}) & \mu\mathbf{I} - \operatorname{diag}(\boldsymbol{\mu}) \end{pmatrix}. \tag{7.12}$$

The result holds provided the infimum is achieved in the range and is positive. Again the same result is obtained if we consider the speed of spread for the proportion of removed individuals.

For the $S \to L \to I$ epidemic, $\mu_i = 0$ for all $i = 1,...,n$ so that $\mu = \max_i(\alpha_i)$ and the matrix $\mathbf{K}(\theta)$ becomes

$$\mathbf{K}(\theta) = \begin{pmatrix} \mu\mathbf{I} - \operatorname{diag}(\boldsymbol{\alpha}) & (\sigma_j\lambda_{ij}P_{ij}(\theta)) \\ \operatorname{diag}(\boldsymbol{\alpha}) & \mu\mathbf{I} \end{pmatrix}.$$

Again the same expressions are obtained for the speed of spread if the epidemic is on the lattice Z^N.

7.6 The link with exact results

In Chapter 5 we used rigorous analytic methods to obtain the speed of propagation of infection in certain continuous space epidemic models. The models were formulated in terms of integro-differential equations involving general infection rates which depend on the time since infection. The spatial $S \to L \to I \to R$,

$S \to L \to I$, $S \to I \to R$ and $S \to I$ epidemics may be regarded as special cases of the model with varying infectivity, with specific forms of the infection rate. For the $S \to I \to R$ epidemic $\lambda_{ij}(t) = \lambda e^{-\mu_j t}$ and for the $S \to L \to I \to R$ model,

$$\lambda_{ij}(\tau) = \frac{\alpha_j \lambda_{ij}}{(\alpha_j - \mu_j)}(e^{-\mu_j \tau} - e^{-\alpha_j \tau}).$$

The form of $\lambda_{ij}(\tau)$ for the other two models can be obtained directly from these by taking $\mu_i = 0$ for all i since there is no removal.

When the contact distributions are radially symmetric, with the $P_{ij}(\theta)$ existing for an open region about $\theta = 0$, the exact methods used in Chapters 4 and 5 for the general model give the asymptotic speed of propagation of these epidemics as the minimum wave speed. Note that Δ_v in Chapters 4 and 5 is just the minimum, over i, j, of the abscissae of convergence of the $P_{ij}(\theta)$, so that $\Delta_v = \Delta$, where Δ is the abscissa of convergence for $\rho(\mathbf{K}(\theta))$ specified in Sections 7.4 and 7.5.

Provided $\rho(\mathbf{\Gamma}) > 1$ the asymptotic speed of propagation, which was obtained for the radially symmetric case, is given by the minimum wave speed c_0. The value of c_0 was shown in Lemma 5.1 to be positive. When $\rho(\mathbf{\Gamma}) \leq 1$ the speed of propagation was shown to be zero. Here $\mathbf{\Gamma}$ is the infection matrix.

The value of c_0 obtained for the minimum wave speed in a specific direction does not require the contact distributions to be radially symmetric. In this section we show that, whether or not the contact distributions are radially symmetric, the speed of first spread (obtained subject to the conditions of Theorem 7.1) is equal to the minimum wave speed in the specified direction. Hence, when $\rho(\mathbf{\Gamma}) > 1$ and the contact distributions are radially symmetric with Laplace transforms existing in some open strip of the complex plane, the speed of first spread is equal to the asymptotic speed of propagation of the epidemic; that speed being positive.

The $S \to I \to R$ and $S \to I$ epidemic models.

First consider the $S \to I \to R$ epidemic. In Section 7.5 it was shown that the speed of spread $c = \inf_{0<\theta<\Delta}(\rho(\mathbf{K}(\theta)) - \mu)/\theta$, where $\mathbf{K}(\theta) = \mathbf{Q}(\theta) + \mu\mathbf{I} - \text{diag}(\mu)$. Here $\mathbf{Q}(\theta)$ has ij^{th} entry $\sigma_j \lambda_{ij} P_{ij}(\theta)$, $\boldsymbol{\mu}$ is the vector with i^{th} entry μ_i and $\mu = \max_i \mu_i$. Note that the value of c is only obtained if $\rho(\mathbf{K}(0)) > \mu$ and if $c > 0$.

In Chapter 4, when waves exist, the minimum wave speed was shown to be

$$c_0 = \inf\{c > 0 : K_c(\theta) = 1 \text{ for some } \theta \in (0, \Delta_v)\}.$$

Here $K_c(\theta) = \rho(\mathbf{V}(\theta))$, where $\mathbf{V}(\theta)$ has ij^{th} entry $\sigma_j \Lambda_{ij}(c\theta) P_{ij}(\theta)$. For the $S \to I \to R$ epidemic

$$\Lambda_{ij}(c\theta) = \int_0^\infty e^{-c\theta t} \lambda_{ij} e^{-\mu_j t} dt = \frac{\lambda_{ij}}{c\theta + \mu_j}.$$

Hence $\mathbf{V}(\theta) = \mathbf{Q}(\theta)(c\theta\mathbf{I} + \text{diag}(\boldsymbol{\mu}))^{-1}$. Note that no wave solution is possible at any speed unless the $P_{ij}(\theta)$ exist in some open region about $\theta = 0$ and $\rho(\mathbf{\Gamma}) > 1$, where $\mathbf{\Gamma} = \mathbf{V}(0) = \mathbf{Q}(0)(\text{diag}(\boldsymbol{\mu}))^{-1}$.

Two lemmas are now established. The first shows that the condition $\rho(\mathbf{\Gamma}) > 1$ for the existence of wave solutions is just the condition $\rho(\mathbf{K}(0)) > \mu$ required to obtain the speed of spread. The second shows that $c_0 = c$ when both are positive. When there is radial symmetry and $\rho(\mathbf{\Gamma}) > 1$ it was shown in Lemma 5.1 that

$c_0 > 0$ and that the speed of propagation was c_0. Hence in this case the speed of spread and the asymptotic speed of propagation are identical for the $S \to I \to R$ epidemic.

LEMMA 7.3. *For the* $S \to I \to R$ *epidemic with* $\mu > 0$, $\rho(\mathbf{K}(0)) > \mu$ *if and only if* $\rho(\mathbf{\Gamma}) > 1$.

PROOF. Let $\mathbf{u}' > \mathbf{0}'$ be the left eigenvector of $\mathbf{\Gamma}$ corresponding to $\rho(\mathbf{\Gamma})$ (see Theorem A.1), so that

$$\rho(\mathbf{\Gamma})\mathbf{u}' = \mathbf{u}'\mathbf{\Gamma} = \mathbf{u}'\mathbf{Q}(0)(\mathrm{diag}(\boldsymbol{\mu}))^{-1}.$$

Therefore $\rho(\mathbf{\Gamma})\mathbf{u}'\mathrm{diag}(\boldsymbol{\mu}) = \mathbf{u}'\mathbf{Q}(0)$ and hence $(\rho(\mathbf{\Gamma})-1)\mathbf{u}'\mathrm{diag}(\boldsymbol{\mu}) = \mathbf{u}'(\mathbf{K}(0) - \mu\mathbf{I})$.

Now take $\mathbf{v} > \mathbf{0}$ to be the right eigenvector of $\mathbf{K}(0)$ corresponding to $\rho(\mathbf{K}(0))$. Then

$$(\rho(\mathbf{\Gamma}) - 1)\mathbf{u}'\mathrm{diag}(\boldsymbol{\mu})\mathbf{v} = \mathbf{u}'(\mathbf{K}(0) - \mu\mathbf{I})\mathbf{v} = (\rho(\mathbf{K}(0)) - \mu)\mathbf{u}'\mathbf{v}.$$

Since both $\mathbf{u}'\mathrm{diag}(\boldsymbol{\mu})\mathbf{v}$ and $\mathbf{u}'\mathbf{v}$ are positive, it therefore immediately follows that $\rho(\mathbf{K}(0)) > \mu$ if and only if $\rho(\Gamma) > 1$.

□

LEMMA 7.4. *For the* $S \to I \to R$ *epidemic, provided* $\rho(\mathbf{\Gamma}) > 1$ *and* $c > 0$, $c_0 = c$.

PROOF. Take any $c^* > 0$ with $c^* \geq c_0$ then there exists a $\theta^* \in (0, \Delta)$ such that $K_{c^*}(\theta^*) = 1$. For this value of c^*, $\mathbf{V}(\theta^*) = \mathbf{Q}(\theta^*)(c^*\theta\mathbf{I} + \mathrm{diag}(\boldsymbol{\mu}))^{-1}$. Take $\mathbf{u}' > \mathbf{0}'$ to be the left eigenvector of $\mathbf{V}(\theta^*)$ corresponding to $K_{c^*}(\theta^*) = \rho(\mathbf{V}(\theta^*)) = 1$. Then

$$\mathbf{u}' = \mathbf{u}'\mathbf{V}(\theta^*) = \mathbf{u}'\mathbf{Q}(\theta^*)(c^*\theta^*\mathbf{I} + \mathrm{diag}(\boldsymbol{\mu}))^{-1}.$$

Hence $\mathbf{u}'(c^*\theta^*\mathbf{I} + \mathrm{diag}(\boldsymbol{\mu})) = \mathbf{u}'\mathbf{Q}(\theta^*)$. Then $\mathbf{u}'(c^*\theta^*\mathbf{I} + \mu\mathbf{I}) = \mathbf{u}'\mathbf{K}(\theta^*)$ and so $\mathbf{u}'\mathbf{K}(\theta^*) = (c^*\theta^* + \mu)\mathbf{u}'$. From Theorem A.2 part 7 it therefore follows that $\rho(\mathbf{K}(\theta^*)) = c^*\theta^* + \mu$ and hence

$$c^* = \frac{\rho(\mathbf{K}(\theta^*)) - \mu}{\theta^*} \geq \inf_{0<\theta<\Delta} \frac{\rho(\mathbf{K}(\theta)) - \mu}{\theta} = c.$$

This holds for any positive $c^* \geq c_0$ so that $c_0 \geq c$.

Now take any $c^* \geq c$. From the definition of c, there exists a $\theta^* \in (0, \Delta)$ such that $c^* = (\rho(\mathbf{K}(\theta^*)) - \mu)/\theta^*$. Then $\rho(\mathbf{K}(\theta^*)) = c^*\theta^* + \mu$. Take $\mathbf{u}' > \mathbf{0}'$ to be the left eigenvector of $\mathbf{K}(\theta^*)$ corresponding to $\rho(\mathbf{K}(\theta^*))$, so that

$$(c^*\theta^* + \mu)\mathbf{u}' = \mathbf{u}'(\mathbf{Q}(\theta^*) + \mu\mathbf{I} - \mathrm{diag}(\boldsymbol{\mu})).$$

Therefore $\mathbf{u}'(c^*\theta^*\mathbf{I} + \mathrm{diag}\boldsymbol{\mu}) = \mathbf{u}'\mathbf{Q}(\theta^*)$ and so

$$\mathbf{u}' = \mathbf{u}'\mathbf{Q}(\theta^*)(c^*\theta^*\mathbf{I} + \mathrm{diag}\boldsymbol{\mu})^{-1} = \mathbf{u}'\mathbf{V}(\theta^*).$$

Hence from Theorem A.2 part 7, $K_{c^*}(\theta^*) = \rho(\mathbf{V}(\theta^*)) = 1$. Therefore $c^* \geq c_0$. Since this holds for any $c^* \geq c$ this implies that $c \geq c_0$ and hence $c_0 = c$.

□

Note that for the $S \to I$ epidemic, $\mu_i = 0$ for all i. In this case $\mathbf{\Gamma}$ is infinite, and so $\rho(\mathbf{\Gamma})$ is also infinite, and $\mathbf{K}(0) = \mathbf{Q}(0)$ so that $\rho(\mathbf{K}(0)) > 0 = \mu$. Lemma 7.3 is therefore unnecessary since the conditions $\rho(\mathbf{\Gamma}) > 1$ and $\rho(\mathbf{K}(0)) > \mu$ always hold. The proof of Lemma 7.4 is still valid when $\mu_i = 0$ for all i.

The $S \to L \to I \to R$ and $S \to L \to I$ epidemic models.
Now consider the $S \to L \to I \to R$ epidemic. Similar results may be proved in this case, which are also valid for the $S \to L \to I$ epidemic. Again it is shown that when the speed of spread $c > 0$, so the saddle point method can be used, then $c = c_0$ where c_0 is the minimum wave speed. In the radially symmetric case therefore this shows that the speed of spread and the asymptotic speed of propagation are identical.

Note that for this epidemic,

$$\Lambda_{ij}(c\theta) = \int_0^\infty e^{-c\theta t} \frac{\alpha_j \lambda_{ij}}{(\alpha_j - \mu_j)} (e^{-\mu_j t} - e^{-\alpha_j t}) dt = \frac{\alpha_j \lambda_{ij}}{(c\theta + \mu_j)(c\theta + \alpha_j)}.$$

Hence

$$\mathbf{V}(\theta) = \mathbf{Q}(\theta)\text{diag}(\boldsymbol{\alpha})(c\theta\mathbf{I} + \text{diag}\boldsymbol{\alpha})^{-1}(c\theta\mathbf{I} + \text{diag}\boldsymbol{\mu})^{-1},$$

where $\boldsymbol{\alpha}$ has i^{th} entry α_i and the Perron Frobenius root of $\mathbf{V}(\theta)$ is denoted by $K_c(\theta)$. Then $\mathbf{V}(0) = \mathbf{\Gamma} = \mathbf{Q}(0)(\text{diag}\boldsymbol{\mu})^{-1}$. The minimum wave speed is again given by

$$c_0 = \inf\{c > 0 : K_c(\theta) = 1 \text{ for some } \theta \in (0, \Delta_v)\}.$$

The matrix $\mathbf{K}(\theta)$ is given by equation (7.12). The speed of spread is just $c = \inf_{0<\theta<\Delta}(\rho(\mathbf{K}(\theta)) - \mu)/\theta$, where now $\mu = \max_i(\max(\alpha_i, \mu_i))$.

LEMMA 7.5. *$\rho(\mathbf{K}(0)) > \mu$ if and only if $\rho(\Gamma) > 1$.*

PROOF. Let $\rho(\mathbf{K}(0)) > \mu$ where $\mathbf{K}(\theta)$ is defined by equation (7.12) and is non-reducible. Let $\rho(\mathbf{K}(0)) = \mu + h$, where $h > 0$. Then from Theorem A.1 there exist a $\mathbf{u} > 0$ and $\mathbf{v} > 0$ such that

$$\mathbf{u}'(\mu\mathbf{I} - \text{diag}(\boldsymbol{\alpha})) + \mathbf{v}'\text{diag}(\boldsymbol{\alpha}) = (\mu + h)\mathbf{u}'$$

and

$$\mathbf{u}'\mathbf{Q}(0) + \mathbf{v}'(\mu\mathbf{I} - \text{diag}(\boldsymbol{\mu})) = (\mu + h)\mathbf{v}'.$$

Hence

$$\mathbf{u}'\mathbf{Q}(0) = \mathbf{u}'(\text{diag}(\boldsymbol{\alpha}) + h\mathbf{I})(\text{diag}(\boldsymbol{\alpha}))^{-1}(\text{diag}(\boldsymbol{\mu}) + h\mathbf{I}),$$

and so

$$\mathbf{u}'\mathbf{Q}(0)\text{diag}(\boldsymbol{\alpha})(\text{diag}(\boldsymbol{\alpha}) + h\mathbf{I})^{-1}(\text{diag}(\boldsymbol{\mu}) + h\mathbf{I})^{-1} = \mathbf{u}'.$$

Therefore, from Theorem A.2 part 7, $\rho(\mathbf{A}(h)) = 1$, where $\mathbf{A}(h)$ is the matrix with ij^{th} entry $\alpha_j\sigma_j\lambda_{ij}/((h + \mu_j)(h + \alpha_j))$.

Now each of the entries of $\mathbf{A}(h)$ is monotone decreasing in h, and from Theorem A.2 part 7 and Lemma A.1, $\rho(\mathbf{A}(h))$ is a strictly monotone increasing and continuous function of its entries. Hence, since $\mathbf{\Gamma} = \mathbf{A}(0)$, we have $\rho(\mathbf{\Gamma}) = \rho(\mathbf{A}(0)) > \rho(\mathbf{A}(h)) = 1$.

Now let $\rho(\mathbf{\Gamma}) = \rho(\mathbf{A}(0)) > 1$. Since $\lim_{h\to\infty} \mathbf{A}(h) = \mathbf{0}$, from Lemma A.1 $\lim_{h\to\infty} \rho(\mathbf{A}(h)) = 0$. The monotonicity and continuity of $\mathbf{A}(h)$ in h then implies that there exists an $h > 0$ such that $\rho(\mathbf{A}(h)) = 1$. Then there exists a $\mathbf{u} > 0$ such that

$$\mathbf{u}' = \mathbf{u}'\mathbf{A}(h) = \mathbf{u}'\mathbf{Q}(0)\text{diag}(\boldsymbol{\alpha})(\text{diag}(\boldsymbol{\alpha}) + h\mathbf{I})^{-1}(\text{diag}(\boldsymbol{\mu}) + h\mathbf{I})^{-1}.$$

Define $\mathbf{v}' = \mathbf{u}'(\text{diag}(\boldsymbol{\alpha}) + h\mathbf{I})(\text{diag}(\boldsymbol{\alpha}))^{-1}$. We can then reverse our steps and obtain $\rho(\mathbf{K}(0)) = \mu + h > \mu$.

□

The following lemma shows that, for the $S \to L \to I \to R$ epidemic, the speed of spread obtained by the saddle point method is the minimum wave speed.

LEMMA 7.6. *If $\rho(\mathbf{\Gamma}) > 1$ and $c > 0$, then the speed of spread $c = c_0$, the minimum wave speed.*

PROOF. Consider any $c^* > c$. Then there exists a θ^* such that $c^* = (\rho(\mathbf{K}(\theta^*)) - \mu)/\theta^*$ and hence $\rho(\mathbf{K}(\theta^*)) = c^*\theta^* + \mu$. Therefore there exist a $\mathbf{u} > 0$ and a $\mathbf{v} > 0$ such that

$$\mathbf{u}'(\mu\mathbf{I} - \text{diag}(\boldsymbol{\alpha})) + \mathbf{v}'\text{diag}(\boldsymbol{\alpha}) = (c^*\theta^* + \mu)\mathbf{u}'$$

and

$$\mathbf{u}'\mathbf{Q}(\theta^*) + \mathbf{v}'(\mu\mathbf{I} - \text{diag}(\boldsymbol{\mu})) = (c^*\theta^* + \mu)\mathbf{v}'.$$

Hence

$$\mathbf{u}'\mathbf{Q}(\theta^*) = \mathbf{u}'(\text{diag}(\boldsymbol{\alpha}) + c^*\theta^*\mathbf{I})(\text{diag}(\boldsymbol{\alpha}))^{-1}(\text{diag}(\boldsymbol{\mu}) + c^*\theta^*\mathbf{I}).$$

Therefore

$$\mathbf{u}'\mathbf{Q}(\theta^*)(\text{diag}(\boldsymbol{\alpha}) + c^*\theta^*\mathbf{I})^{-1}(\text{diag}(\boldsymbol{\alpha}))(\text{diag}(\boldsymbol{\mu}) + c^*\theta^*\mathbf{I})^{-1} = \mathbf{u}'.$$

But $\mathbf{Q}(\theta^*)(\text{diag}(\boldsymbol{\alpha}) + c^*\theta^*\mathbf{I})^{-1}(\text{diag}(\boldsymbol{\alpha}))(\text{diag}(\boldsymbol{\mu}) + c^*\theta^*\mathbf{I})^{-1}$ is just $\mathbf{V}(\theta^*)$ when $c = c^*$. Hence $K_{c^*}(\theta^*) = \rho(\mathbf{V}(\theta^*)) = 1$. Thus $c^* \geq c_0$ and hence $c \geq c_0$.

Now consider any $c^* \geq c_0$. There exists a θ^* such that $K_{c^*}(\theta^*) = 1$. Hence there exists a $\mathbf{u} > 0$ with $\mathbf{u}'\mathbf{V}(\theta^*) = \mathbf{u}'$ with c replaced by c^*. Define $\mathbf{v}' = \mathbf{u}'(\text{diag}(\boldsymbol{\alpha}) + c^*\theta^*\mathbf{I})(\text{diag}(\boldsymbol{\alpha}))^{-1}$. We can then reverse the steps and obtain $(\rho(\mathbf{K}(\theta^*) - \mu)/\theta^* = c^*$. Hence $c^* \geq c$ and therefore $c_0 \geq c$. We thus obtain the result that $c = c_0$.

□

When $\rho(\mathbf{\Gamma}) > 1$ it was shown in Lemma 5.1, for a more general epidemic with varying infectivity, that $c_0 > 0$ when the contact distributions are radially symmetric. For the epidemics to which we have applied the saddle point method, a simple proof that c_0 is positive is given which only requires the mean of each contact distribution in the specified direction to be zero. Let $\rho(\mathbf{K}(\theta))$ be denoted by $\rho(\theta)$.

LEMMA 7.7. *If $\rho(0) > \mu$, the means of the contact distributions in the direction considered are all zero and $f(\theta) = (\rho(\theta) - \mu)/\theta$ achieves its infimum for some $\theta \in (0, \Delta)$, then the speed of first spread c is positive.*

PROOF. The continuity and convexity of $\rho(\theta)$ were established in Theorem A.3 parts 1 and 6. Consider $\inf_{\theta\in(0,\Delta)} f(\theta)$, where $f(\theta) = \dfrac{\rho(\theta) - \mu}{\theta}$ and the infimum is achieved for some $\theta \in (0, \Delta)$. Now $f'(\theta) = (-f(\theta) + \rho'(\theta))/\theta = 0$ if $\rho'(\theta) = f(\theta)$. Also $f''(\theta) = \rho''(\theta)/\theta > 0$ when $\rho'(\theta) = f(\theta)$. Hence at the minimum, $\theta_0 \in (0, \Delta)$, $c_0 = f(\theta_0) = \rho'(\theta_0)$.

Differentiating $|\rho(\theta)\mathbf{I} - \mathbf{K}(\theta)| = 0$ we obtain,

$$\rho'(\theta) = \frac{\sum_i \sum_j \mathbf{K}'_{ij}(\theta)\{Adj(\rho(\theta)\mathbf{I} - \mathbf{K}(\theta))\}_{ij}}{trace(Adj(\rho(\theta)\mathbf{I} - \mathbf{K}(\theta)))}.$$

Using the strict convexity of $P_{ij}(\theta)$ and the result that $P'_{ij}(0) = 0$, we have $\mathbf{K}'_{ij}(\theta_0) \geq 0$ with strict inequality for some i and j. Also, from Theorem A.2 part 8, $Adj(\rho(\theta)\mathbf{I} - \mathbf{K}(\theta)) > 0$ for all $\theta \in [0, \Delta_v)$. Hence $\rho'(\theta_0) > 0$, so that $c_0 > 0$.

□

When $\rho(0) > \mu$ and the means of the projected contact distributions are zero the additional restriction in Lemma 7.7, required to prove that c_0 is positive, holds for most contact distributions. This is now briefly discussed.

When $\rho(0) > \mu$ necessarily $\lim_{\theta\downarrow 0} f(\theta)$ is infinite. If $\lim_{\theta\uparrow\Delta} f(\theta)$ is also infinite then the infimum of $f(\theta)$ must occur in the range $(0, \Delta)$.

When Δ is finite provided there exists an i and j with $\lambda_{ij} \neq 0$ and $P_{ij}(\Delta)$ infinite then, from Theorem A.3 part 3, $\rho(\Delta)$ is infinite. Hence $f(\Delta)$ is infinite. When $P_{ij}(\Delta)$ is finite for all i and j, then $f(\Delta)$ is finite. It is then possible to have $\lim_{\theta\uparrow\Delta} f'(\theta) \leq 0$, in which case the infimum of $f(\theta)$ occurs at $\theta = \Delta$. This is a case when c_0 has not been shown to be positive and also the saddle point method cannot be used, so that the speed of first spread has not been obtained.

When Δ is infinite, since the means of the projected contact distributions (in the direction with direction cosines $\boldsymbol{\xi}$) are all zero, the projected distributions are not one-sided. Then it can be shown that, for any such contact distributions, $f(\theta)$ must tend to infinity as θ tends to $\Delta = \infty$. First observe that, if $\tilde{p}_{ij}(u)$ is the integral of $p_{ij}(\mathbf{r})$ over the hyperplane $\boldsymbol{\xi}'\mathbf{r} = u$, then for any non-negative integer s,

$$\lim_{\theta\to\infty} \frac{P_{ij}(\theta)}{\theta^s} = \lim_{\theta\to\infty} \frac{P_{ij}^{(s)}(\theta)}{s!} = \lim_{\theta\to\infty} \frac{1}{s!} \int_{-\infty}^{\infty} u^s e^{\theta u} \tilde{p}_{ij}(u)du = \infty. \tag{7.13}$$

Consider the $S \to I \to R$ epidemic. Then $\rho(\theta) = \rho((\sigma_j \lambda_{ij} P_{ij}(\theta)))$. From the relation (7.13) with $s = 0$ and Theorem A.3 part 4, $\rho(\theta)$ tends to infinity as θ tends to infinity. Hence, using the relation (7.13) with $s = 1$ and observing that $\rho(a\mathbf{A}) = a\rho(\mathbf{A})$ for $a > 0$ and $\mathbf{A}$ a non-negative matrix,

$$\lim_{\theta\to\infty} f(\theta) = \lim_{\theta\to\infty} \frac{\rho(\theta)}{\theta} = \lim_{\theta\to\infty} \rho\left((\sigma_j \lambda_{ij} P_{ij}(\theta)/\theta)\right) = \infty.$$

Now consider the $S \to L \to I \to R$ epidemic. Then $\mathbf{K}(\theta)$ is given by equation (7.12) with $\boldsymbol{\alpha} > \mathbf{0}$. Define

$$\mathbf{K}^*(\theta) = \begin{pmatrix} \mathbf{0} & (\sigma_j \lambda_{ij} P_{ij}(\theta)) \\ \operatorname{diag}(\boldsymbol{\alpha}) & \mathbf{0} \end{pmatrix}.$$

Then $\mathbf{K}(\theta) \geq \mathbf{K}^*(\theta)$. Now $\rho\left((\mathbf{K}^*(\theta))^2\right) = \left(\rho(\mathbf{K}^*(\theta))\right)^2$ and $(\mathbf{K}^*(\theta))^2$ has maximum eigenvalue the maximum of $\rho((\alpha_i \sigma_j \lambda_{ij} P_{ij}(\theta)))$ and $\rho((\alpha_j \sigma_j \lambda_{ij} P_{ij}(\theta)))$.

Hence, from Theorem A.2 part 1, for any non-negative integer s

$$\frac{\rho(\theta)}{\theta^s} \geq \rho\left(\frac{1}{\theta^s}\mathbf{K}^*(\theta)\right) = \sqrt{\max[\rho((\alpha_i \sigma_j \lambda_{ij} P_{ij}(\theta)\theta^{-s})), \rho((\alpha_j \sigma_j \lambda_{ij} P_{ij}(\theta)\theta^{-s}))]}.$$

Since the right hand side tends to infinity as θ tends to infinity we therefore obtain the result that $\lim_{\theta\to\infty}(\rho(\theta)/\theta^s) = \infty$ for all non-negative integers s. The case $s = 0$ immediately shows that $\lim_{\theta\to\infty}\rho(\theta) = \infty$. This result, together with the case $s = 1$, then shows that $\lim_{\theta\to\infty} f(\theta) = \lim_{\theta\to\infty}\rho(\theta)/\theta = \infty$.

Thus when $\rho(0) > \mu$ and $\Delta = \infty$, the speed of spread of these epidemic models will be positive for any contact distributions whose projections have mean zero.

CHAPTER 8

Epidemics with return to the susceptible state

8.1 Introduction

So far in this book we have considered epidemics in which an individual once infected cannot return to the susceptible state. In these models an individual usually either dies or becomes immune and may be considered as removed. In many diseases an individual can recover and return to the susceptible state and later become reinfected. The $S \to I \to S$ epidemic is a simple model for a closed system in which individuals are allowed to return to the susceptible state. In this chapter we consider a general model which encompasses both $S \to I$ and $S \to I \to S$ epidemics. The $S \to I$ epidemic, in which individuals once infected remain infectious forever, was introduced in Chapter 2.

We first set up non-spatial models of a closed system and also an open system with a stable population size in which the birth and immigration rates into the system are balanced by the death and emigration rates from the system. One and two type epidemics are considered in Section 8.4. The threshold condition on the model parameters for the existence of an endemic equilibrium is obtained. A simple graphical method is used to show the global stability of the endemic equilibrium when the threshold value is exceeded. When no endemic equilibrium exists then the epidemic is shown to die out.

The equilibria are obtained for the n-type models in Section 8.5. Section 8.6 then examines the global stability of the equilibria. In Theorem 8.4, although we use what are essentially the same functions as were used by Lajmanovich and Yorke [L1] when proving global convergence for the multi-type $S \to I \to S$ epidemic, proofs are accomplished using simple analysis without the need to appeal to Liapunov theory. The proof has also been modified to encompass the more general models.

Results for an n-type epidemic on N sites are obtained by regarding it essentially as a non-spatial nN-type epidemic in a similar manner to Chapter 6, Sections 6.1 and 6.2. In particular, when the population sizes and total infection rates are site independent some neat global results are obtained linking the spatial to the non-spatial n-type results. In this case the spatial final size at all sites is shown to converge to the non-spatial global endemic equilibrium.

Spatial versions of the $S \to I \to S$ model in both $\Re^N$ and Z^N are set up. The general results of the saddle point method of Chapter 7 are applied to obtain the speed of first spread of infection from an initial focus of infection within a compact region.

For a discussion of more general models of $S \to I \to S$ type we refer the reader to Capasso [C1].

8.2 The non-spatial $S \to I \to S$ model

Consider n populations, each consisting of susceptible and infectious individuals. The rate of infection of a susceptible individual in population i by an infectious individual in population j is λ_{ij}. Infectious individuals in population i return to the susceptible state (i.e. recover) at rate $\beta_i \geq 0$. Note that $\beta_i > 0$, $i = 1, ..., n$, corresponds to the $S \to I \to S$ epidemic and $\beta_i = 0$, $i = 1, ..., n$, corresponds to the $S \to I$ epidemic.

Denote the numbers of susceptible and infectious individuals in the ith population at time t by $S_i(t)$ and $I_i(t)$ respectively. Then the model is described by the following system of equations,

$$(8.1) \qquad \begin{aligned} S_i'(t) &= -\sum_{j=1}^{n} \lambda_{ij} S_i(t) I_j(t) + \beta_i I_i(t), \\ I_i'(t) &= \sum_{j=1}^{n} \lambda_{ij} S_i(t) I_j(t) - \beta_i I_i(t), \ (i = 1, ..., n). \end{aligned}$$

Let the population size of the ith population be σ_i. Thus $\sigma_i = S_i(t) + I_i(t)$. Equations (8.1) can be re-written as

$$I_i'(t) = \sum_{j=1}^{n} \lambda_{ij} I_j(t)(\sigma_i - I_i(t)) - \beta_i I_i(t), \ (i = 1, ..., n),$$

where $0 \leq I_i(t) \leq \sigma_i$. Denote the proportion of individuals in population i who are infectious at time t by $y_i(t)$, i.e. $y_i(t) = I_i(t)/\sigma_i$. Then we obtain the system of equations,

$$(8.2) \qquad y_i'(t) = (1 - y_i(t)) \sum_{j=1}^{n} \sigma_j \lambda_{ij} y_j(t) - \beta_i y_i(t), \ (i = 1, ..., n).$$

8.3 The open $S \to I \to S$ model

In this section an open version of the $S \to I \to S$ model of Section 8.2 is considered. Individuals enter the population by birth and/or immigration and leave by death or emigration. The equations for this model turn out to be identical to those of the $S \to I \to S$ model but with different parameters. Thus we can analyse both models simultaneously.

Consider n populations, each consisting of susceptible and infectious individuals. The parameters are specified as in Section 8.2, with the additional parameters v_i, u_i and α_i representing, for population i, the rate at which susceptibles are born, the rate at which they immigrate into the population and the combined death/emigration rate respectively. The extended model is described by the following system of equations,

$$
\begin{aligned}
S_i'(t) &= v_i(S_i(t)+I_i(t)) + u_i - \alpha_i S_i(t) + \beta_i I_i(t) - \sum_{j=1}^{n} \lambda_{ij} S_i(t) I_j(t), \\
I_i'(t) &= \sum_{j=1}^{n} \lambda_{ij} S_i(t) I_j(t) - \alpha_i I_i(t) - \beta_i I_i(t), \ (i=1,...,n).
\end{aligned} \tag{8.3}
$$

We consider the case when the size σ_i, of the ith population remains constant. Thus $\sigma_i = S_i(t) + I_i(t)$ and hence $(v_i\sigma_i + u_i) = \alpha_i\sigma_i$. Note that the model of Section 8.2 (i.e. a disease in a closed population with no births, deaths, emigration or immigration) corresponds to the special case $v_i = u_i = \alpha_i = 0$. Equations (8.3) can be re-written as

$$
I_i'(t) = \sum_{j=1}^{n} \lambda_{ij} I_j(t)(\sigma_i - I_i(t)) - (\alpha_i + \beta_i) I_i(t), \ (i=1,...,n),
$$

where $0 \le I_i(t) \le \sigma_i$. Let $\mu_i = (\alpha_i + \beta_i)$ and denote the proportion of individuals in population i who are infectious at time t by $y_i(t)$, i.e. $y_i(t) = I_i(t)/\sigma_i$. Then we obtain the system of equations,

$$
y_i'(t) = (1 - y_i(t)) \sum_{j=1}^{n} \sigma_j \lambda_{ij} y_j(t) - \mu_i y_i(t), \ (i=1,...,n). \tag{8.4}
$$

In the closed system of Section 8.2 $\alpha_i = 0$, so that $\mu_i = \beta_i$, $i = 1, ..., n$. Thus equations (8.4) describe both closed and open systems where $\mu_i = \beta_i$ in a closed system and $\mu_i = (\alpha_i + \beta_i)$ in an open system. In general $\mu_i = 0$ models the situation when the duration of the epidemic is short relative to the life cycle of the individuals and no return to the susceptible state is possible; otherwise positive values of the μ_i will be required. A model with some of the μ_i zero and some positive is appropriate when the populations under consideration have quite varied life cycle lengths.

Let $\mathbf{\Lambda} = (\sigma_j \lambda_{ij})$ be non-reducible (and not the zero matrix when $n = 1$) so that susceptible individuals of all types can become infected.

8.4 One and two type epidemics

In this section we give an intuitive description of the cases when $n = 1$ or $n = 2$.

The one type epidemic.
When $n = 1$, dropping the subscripts since they are unnecessary, equation (8.4) becomes

$$
y'(t) = (1 - y(t))\sigma\lambda y(t) - \mu y(t). \tag{8.5}
$$

We consider the cases (i) $\mu > 0$ and (ii) $\mu = 0$ separately.

Case (i) $\mu > 0$.
Let $\rho = \sigma\lambda/\mu$ and $f(y) = y/(1-y) - \rho y$. The equilibrium values are the non-negative solutions of the equation $f(y) = 0$. Now $f(y)$ is a convex function and has

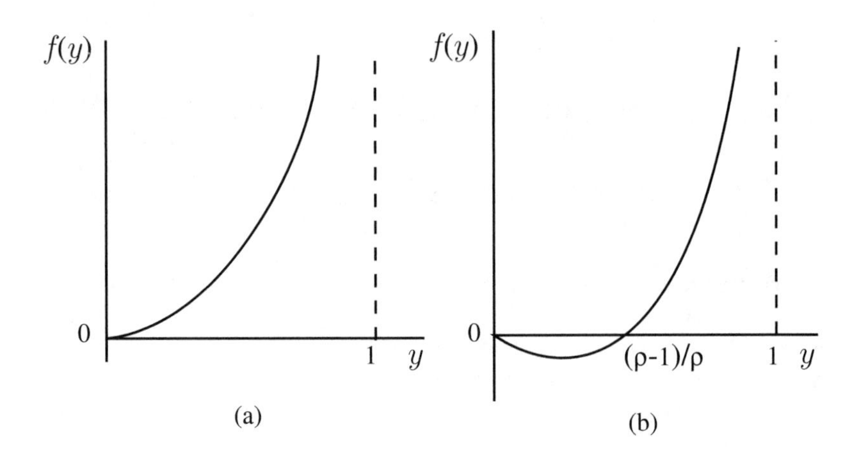

FIGURE 8.1. The one-type epidemic. Case (i) with $\mu > 0$. Here (a) corresponds to $\rho \leq 1$ and (b) corresponds to $\rho > 1$.

the form shown in Figure 8.1, with (a) and (b) corresponding to $\rho \leq 1$ and $\rho > 1$ respectively. When $\rho \leq 1$ there is only one equilibrium value, namely $y = 0$. When $\rho > 1$, there are two non-negative solutions $y = 0$ and $y = (\rho - 1)/\rho$, the latter being referred to as the endemic level.

Equation (8.5) can be rewritten as

$$y'(t) = \mu\rho y(t)(-y(t) + (\rho - 1)/\rho). \tag{8.6}$$

This has an explicit solution. Let $y(0) = y_0$. If $y_0 = 0$, then $y(t) = 0$ for all $t \geq 0$. If $y_0 > 0$ then

$$y(t) = \begin{cases} \dfrac{\rho - 1}{\rho\left[1 + \left(\frac{\rho-1}{\rho y_0} - 1\right) e^{-\mu(\rho-1)t}\right]}, & \text{for } \rho \neq 1, \\ \dfrac{1}{\frac{1}{y_0} + \mu t}, & \text{for } \rho = 1. \end{cases}$$

When $\rho > 1$ and $y_0 > 0$, $y(t)$ tends to the endemic level $(\rho - 1)/\rho$ as $t \to \infty$. If $\rho \leq 1$ and $y_0 > 0$, then $y(t) \to 0$ as $t \to \infty$.

The behaviour of $y(t)$ may also be deduced from the differential equation (8.6). In the case where $\rho \leq 1$, when $y(t) > 0$ then $y'(t) < 0$ and hence $y(t) \to 0$ as $t \to \infty$. In the case where $\rho > 1$, $y'(t) > 0$ for $0 < y(t) < (\rho - 1)/\rho$ and $y'(t) < 0$ for $y(t) > (\rho - 1)/\rho$. So that when $\rho > 1$, provided $y_0 > 0$, $y(t) \to (\rho - 1)/\rho$ as $t \to \infty$.

Case (ii) $\mu = 0$.
In this case

$$y'(t) = (1 - y(t))\sigma\lambda y(t). \tag{8.7}$$

The equilibrium values are the non-negative solutions of the equation $(1 - y)\sigma\lambda y = 0$. This has solutions $y = 0$ and $y = 1$. If $y(0) = y_0 > 0$, then equation (8.7) has the solution

$$y(t) = \frac{y_0}{y_0 + (1 - y_0)e^{-\sigma\lambda t}}.$$

It immediately follows that $y(t) \uparrow 1$ as $t \to \infty$ provided there is some initial infection. The same result may also easily be seen from the differential equation since $y'(t) > 0$ for $0 < y(t) < 1$. Thus if $y_0 \neq 0$, $y(t) \uparrow 1$ as $t \to \infty$.

The host-vector epidemic.
The host-vector epidemic is described by the equations

$$\begin{aligned} y_1'(t) &= (1 - y_1(t))\sigma_2\lambda_{12}y_2(t) - \mu_1 y_1(t), \\ y_2'(t) &= (1 - y_2(t))\sigma_1\lambda_{21}y_1(t) - \mu_2 y_2(t). \end{aligned} \tag{8.8}$$

There are three cases (i) $\mu_1 > 0$ and $\mu_2 > 0$, (ii) $\mu_1 = 0$ and $\mu_2 = 0$ and (iii) one of μ_1 and μ_2 is zero and the other is positive.

Case (i) $\mu_1 > 0$ and $\mu_2 > 0$.

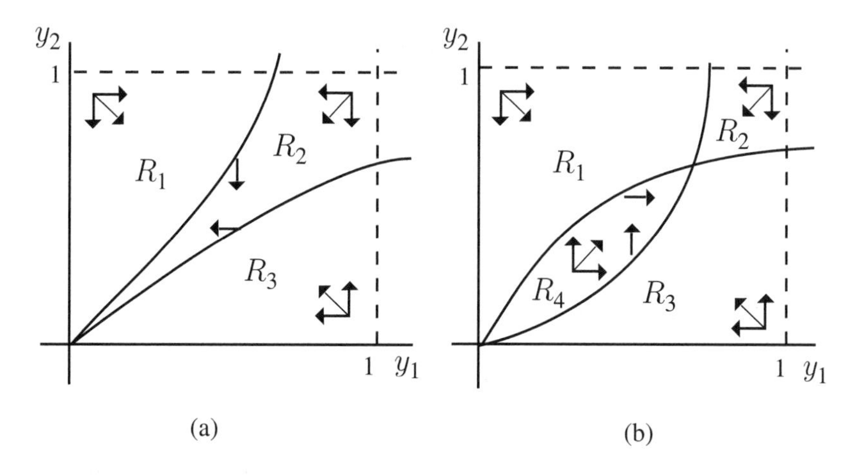

FIGURE 8.2. The host-vector epidemic. Case (i) with $\mu_1 > 0$ and $\mu_2 > 0$. Here (a) corresponds to $\rho \leq 1$ and (b) corresponds to $\rho > 1$.

The equilibrium values are the non-negative solutions of

$$y_2 = \frac{y_1}{\gamma_{12}(1 - y_1)},$$
$$y_1 = \frac{y_2}{\gamma_{21}(1 - y_2)},$$

where $\gamma_{ij} = \sigma_j\lambda_{ij}/\mu_i$. It is easily shown that this pair of equations has two solutions. The first is $y_1 = y_2 = 0$ and the second is $y_1 = \dfrac{\gamma_{12}\gamma_{21} - 1}{\gamma_{12}\gamma_{21} + \gamma_{21}}$ and $y_2 = \dfrac{\gamma_{12}\gamma_{21} - 1}{\gamma_{12}\gamma_{21} + \gamma_{12}}$. The second solution will be non-negative and not the zero solution precisely when $\gamma_{12}\gamma_{21} > 1$.

The first equilibrium equation has y_2 a convex function of y_1 with asymptote $y_1 = 1$. The second has y_1 a convex function of y_2 with asymptote $y_2 = 1$. These curves are shown in Figure 8.2. Let ρ be positive with $\rho^2 = \gamma_{12}\gamma_{21}$. In Figure 8.2, case (a) corresponds to $\rho \leq 1$ and case (b) corresponds to $\rho > 1$. For both cases (a) and (b) the curves intersect at (0,0). For case (b) the curves also intersect at the unique positive point

$$y_1 = \frac{\gamma_{12}\gamma_{21} - 1}{\gamma_{12}\gamma_{21} + \gamma_{21}} \text{ and } y_2 = \frac{\gamma_{12}\gamma_{21} - 1}{\gamma_{12}\gamma_{21} + \gamma_{12}}.$$

Now consider equations (8.8). A simple graphical analysis may be used to see the behaviour of $y_1(t)$ and $y_2(t)$. The derivative of $y_1(t)$ is positive in the interior of the regions R_1 and R_4 and negative in the interior of the regions R_2 and R_3. The derivative of $y_2(t)$ is positive in the interior of the regions R_3 and R_4 and negative in the interior of the regions R_1 and R_2. Hence in Figure 8.2 (a) all trajectories converge to the point (0,0), and in Figure 8.2 (b) they converge to the unique positive solution of the equations.

Case (ii) $\mu_1 = \mu_2 = 0$.

The equations are

$$\begin{aligned} y_1'(t) &= (1 - y_1(t))\sigma_2\lambda_{12}y_2(t), \\ y_2'(t) &= (1 - y_2(t))\sigma_1\lambda_{21}y_1(t). \end{aligned}$$

The equilibrium values are the solutions of

$$\begin{aligned} (1 - y_1)\sigma_2\lambda_{12}y_2 &= 0, \\ (1 - y_2)\sigma_1\lambda_{21}y_1 &= 0. \end{aligned}$$

There are two solutions $y_1 = 0,\ y_2 = 0$ and $y_1 = 1,\ y_2 = 1$.

Now $y_1'(t) > 0$ if $0 < y_1(t) < 1$ and $y_2(t) > 0$. Also $y_2'(t) > 0$ if $0 < y_2(t) < 1$ and $y_1(t) > 0$. Thus if $y_1(0) > 0$ and $y_2(0) > 0$, $y_1(t) \uparrow 1$ and $y_2(t) \uparrow 1$.

It is easily seen that if $y_1(0) = 0$ and $y_2(0) > 0$, since $y_1'(0) > 0$ then $y_1(t) > 0$ for $t > 0$. Similarly if $y_2(0) = 0$ and $y_1(0) > 0$, since $y_2'(0) > 0$ then $y_2(t) > 0$ for $t > 0$. Hence in these cases also $y_1(t) \uparrow 1$ and $y_2(t) \uparrow 1$.

Therefore when $\mu_1 = \mu_2 = 0$, provided there is some infection present at time $t = 0$, then $y_1(t) \uparrow 1$ and $y_2(t) \uparrow 1$, so that everyone in both populations eventually become infected.

Case (iii) By relabelling the types if necessary, this case corresponds to $\mu_1 = 0$ and $\mu_2 > 0$.

The equations are

$$\begin{aligned} y_1'(t) &= (1 - y_1(t))\sigma_2\lambda_{12}y_2(t), \\ y_2'(t) &= (1 - y_2(t))\sigma_1\lambda_{21}y_1(t) - \mu_2 y_2(t). \end{aligned} \tag{8.9}$$

The equilibrium values are then the solutions of

$$\begin{aligned} (1 - y_1)\sigma_2\lambda_{12}y_2 &= 0, \\ y_1 &= \frac{\mu_2 y_2}{\sigma_1\lambda_{21}(1 - y_2)}. \end{aligned}$$

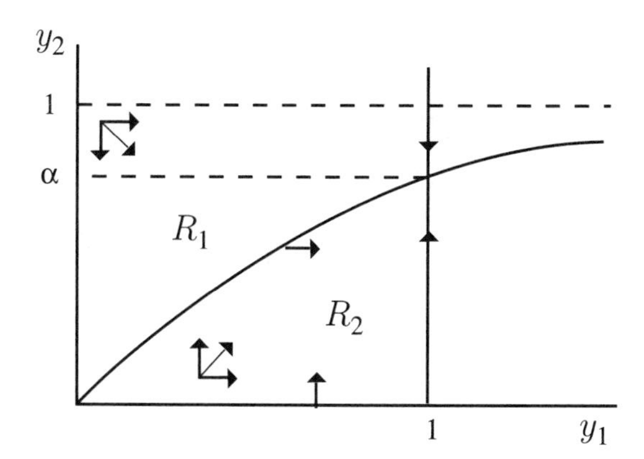

FIGURE 8.3. The host-vector epidemic. Case (iii) with $\mu_1 = 0$ and $\mu_2 > 0$.

Hence the equilibria are $(0, 0)$ and $(1, \alpha)$, where $\alpha = \dfrac{\sigma_1\lambda_{21}}{\sigma_1\lambda_{21} + \mu_2}$. The first equilibrium equation corresponds to the two lines $y_2 = 0$ and $y_1 = 1$, and the second equilibrium equation is a curve with asymptote $y_2 = 1$. These are shown in Figure 8.3.

As in the previous case, provided there is some initial infection, it is easily seen from equations (8.9) that $y_1(t) > 0$ for $t > 0$. In the interior of region R_2, the derivatives of both $y_1(t)$ and $y_2(t)$ are positive. In the interior of R_1 the derivative of $y_1(t)$ is positive and that of $y_2(t)$ is negative. A simple graphical analysis then shows that $(y_1(t), y_2(t))$ tends to the equilibrium point $(1, \alpha)$.

The general 2-type epidemic.

Consider the case $n = 2$ where $\mathbf{\Lambda} = (\sigma_j\lambda_{ij})$ is a non-reducible, non-negative matrix. The equations in this case are

$$\begin{aligned} y_1'(t) &= (1 - y_1(t))(\sigma_1\lambda_{11}y_1(t) + \sigma_2\lambda_{12}y_2(t)) - \mu_1 y_1(t), \\ y_2'(t) &= (1 - y_2(t))(\sigma_1\lambda_{21}y_1(t) + \sigma_2\lambda_{22}y_2(t)) - \mu_2 y_2(t). \end{aligned} \tag{8.10}$$

As for the host-vector model there are three cases (i) $\mu_1 > 0$ and $\mu_2 > 0$, (ii) $\mu_1 = 0$ and $\mu_2 = 0$ and (iii) one of μ_1 and μ_2 is zero and the other is positive.

Case (i) $\mu_1 > 0$ and $\mu_2 > 0$.

In this case the equilibrium values satisfy

$$\begin{aligned} y_2 &= \frac{1}{\sigma_2\lambda_{12}}\left(\frac{\mu_1 y_1}{(1 - y_1)} - \sigma_1\lambda_{11}y_1\right), \\ y_1 &= \frac{1}{\sigma_1\lambda_{21}}\left(\frac{\mu_2 y_2}{(1 - y_2)} - \sigma_2\lambda_{22}y_2\right). \end{aligned}$$

These curves always intersect at (0,0). The first curve has asymptote $y_1 = 1$ and the second has asymptote $y_2 = 1$.

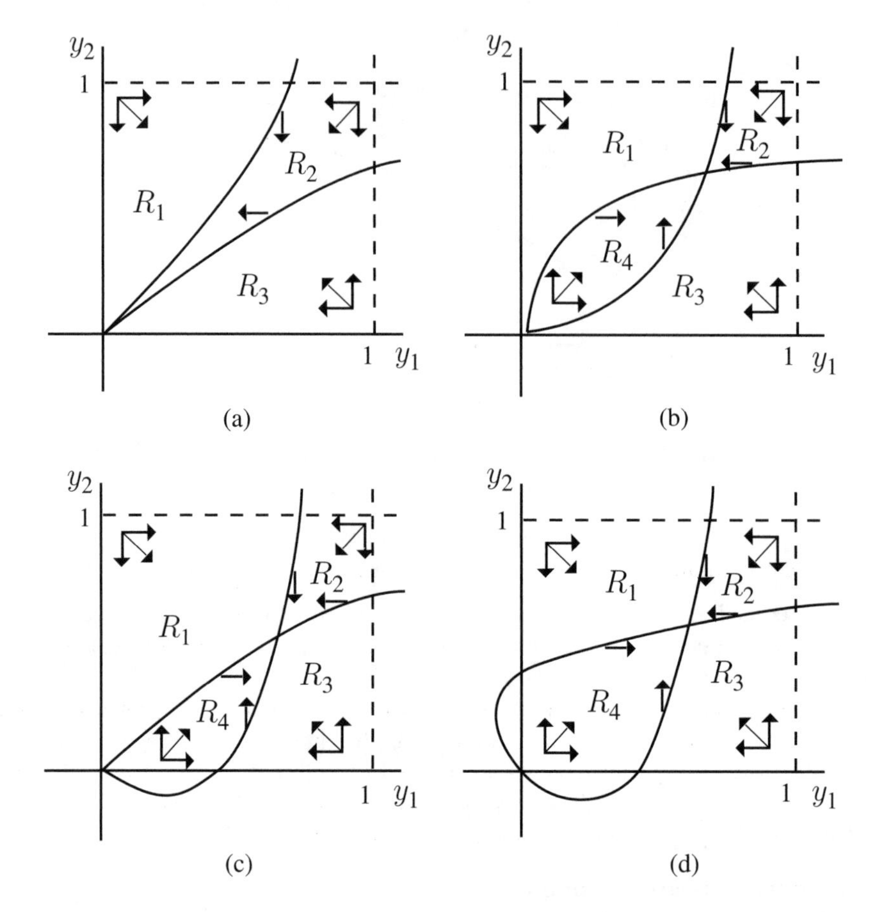

FIGURE 8.4. The two-type epidemic. Case (i) with $\mu_1 > 0$ and $\mu_2 > 0$. Case (a) corresponds to $\rho(\mathbf{\Gamma}) \leq 1$ and the other cases correspond to $\rho(\mathbf{\Gamma}) > 1$. Case (b) has $\sigma_1\lambda_{11} \leq \mu_1$ and $\sigma_2\lambda_{22} \leq \mu_2$. Case (c) has $\sigma_1\lambda_{11} \leq \mu_1$ and $\sigma_2\lambda_{22} > \mu_2$. Case (d) has $\sigma_1\lambda_{11} > \mu_1$ and $\sigma_2\lambda_{22} > \mu_2$.

On the first curve $\dfrac{dy_2}{dy_1} = \dfrac{1}{\sigma_2\lambda_{12}}\left(\dfrac{\mu_1}{(1-y_1)^2} - \sigma_1\lambda_{11}\right)$. This equals $\dfrac{\mu_1 - \sigma_1\lambda_{11}}{\sigma_2\lambda_{12}}$ at $(0,0)$, and so is negative when $\sigma_1\lambda_{11} > \mu_1$. On the second curve the derivative $\dfrac{dy_2}{dy_1}$ at $(0,0)$ is $\dfrac{\sigma_1\lambda_{21}}{\mu_2 - \sigma_2\lambda_{22}}$, which is negative when $\sigma_2\lambda_{22} > \mu_2$.

Thus there are four cases, which are illustrated in Figure 8.4. Case (a) corresponds to both derivatives positive at $(0,0)$ with

$$\frac{\mu_1 - \sigma_1\lambda_{11}}{\sigma_2\lambda_{12}} \geq \frac{\sigma_1\lambda_{21}}{\mu_2 - \sigma_2\lambda_{22}}.$$

In this case $(0,0)$ is the only point of intersection of the curves. Case (b) corresponds to both derivatives positive at $(0,0)$ with

$$\frac{\mu_1 - \sigma_1\lambda_{11}}{\sigma_2\lambda_{12}} < \frac{\sigma_1\lambda_{21}}{\mu_2 - \sigma_2\lambda_{22}}.$$

Cases (c) and (d) respectively correspond to exactly one derivative negative at $(0,0)$ and both derivatives negative at $(0,0)$. We can clearly relabel the types in case (c) so that $\sigma_1\lambda_{11} > \mu_1$, and hence the first derivative is negative at $(0,0)$. In all of cases (b), (c) and (d) the curves intersect at $(0,0)$ and at a unique point in the positive quadrant, the unique positive endemic solution.

Now consider equations (8.10). The derivative of $y_1(t)$ is positive in the interior of regions R_1 and R_4 and negative in the interior of regions R_2 and R_3. The derivative of $y_2(t)$ is positive in the interior of regions R_3 and R_4 and negative in the interior of regions R_1 and R_2. Define $\mathbf{\Gamma}$ to be the 2×2 matrix with $\{\mathbf{\Gamma}\}_{ij} = \gamma_{ij} = \dfrac{\sigma_j\lambda_{ij}}{\mu_i}$.

Case (a) corresponds to $\rho(\mathbf{\Gamma}) \leq 1$; in which case a simple graphical analysis shows that all trajectories converge to $(0,0)$. Cases (b), (c) and (d) all correspond to $\rho(\mathbf{\Gamma}) > 1$. In each case a simple graphical analysis shows that all trajectories converge to the unique positive endemic solution.

Case (ii) $\mu_1 = \mu_2 = 0$.
The equilibrium values satisfy

$$(1 - y_1)(\sigma_1\lambda_{11}y_1 + \sigma_2\lambda_{12}y_2) = 0,$$
$$(1 - y_2)(\sigma_1\lambda_{21}y_1 + \sigma_2\lambda_{22}y_2) = 0.$$

and are $(0,0)$ and $(1,1)$. As for the host vector epidemic, provided there is some initial infection $y_1(t) > 0$ and $y_2(t) > 0$ for $t > 0$. Hence when $t > 0$, if we consider equations (8.10) with $\mu_1 = \mu_2 = 0$, the derivative of $y_i(t)$ is positive for $y_i(t) < 1$ for each $i = 1,2$. Hence $y_1(t) \uparrow 1$ and $y_2(t) \uparrow 1$ as $t \to \infty$.

Case (iii) This case is, by relabelling the types if necessary, $\mu_1 = 0$ and $\mu_2 > 0$.

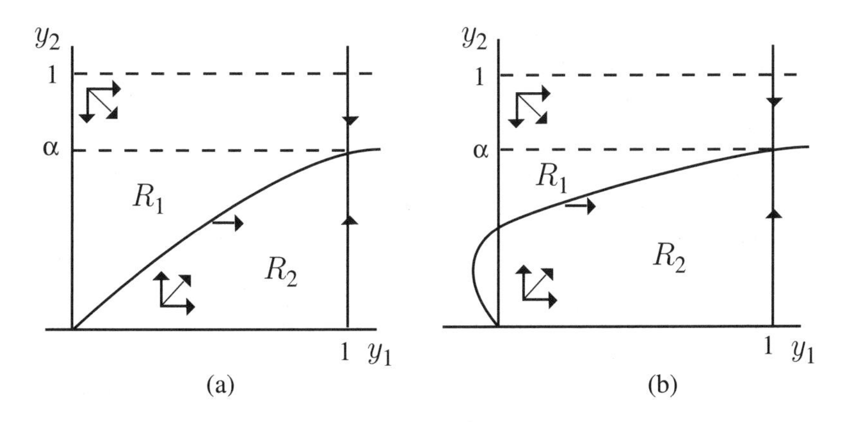

FIGURE 8.5. The two-type epidemic. Case (iii) with $\mu_1 = 0$ and $\mu_2 > 0$. Case (a) corresponds to $\sigma_2\lambda_{22} \leq \mu_2$. Case (b) corresponds to $\sigma_2\lambda_{22} > \mu_2$.

The equilibrium values are the solutions of

$$(1-y_1)(\sigma_1\lambda_{11}y_1+\sigma_2\lambda_{12}y_2)=0,$$
$$y_1=\frac{1}{\sigma_1\lambda_{21}}\left(\frac{\mu_2 y_2}{(1-y_2)}-\sigma_2\lambda_{22}y_2\right).$$

There are therefore two solutions, $(0,0)$ and $(1,\alpha)$ where α is the unique positive solution of the equation

$$\frac{y_2}{1-y_2}=\gamma_{22}y_2+\gamma_{21},$$

where $\gamma_{2j}=\sigma_j\lambda_{2j}/\mu_2$ for $j=1,2$ and $\gamma_{21}>0$ since $\mathbf{\Lambda}$ is non-reducible. Hence

$$\alpha=\begin{cases}\dfrac{\sqrt{(1+\gamma_{21}-\gamma_{22})^2+4\gamma_{21}\gamma_{22}}-(1+\gamma_{21}-\gamma_{22})}{2\gamma_{22}}, & \text{for } \gamma_{22}\neq 0,\\ \dfrac{\gamma_{21}}{1+\gamma_{21}}, & \text{for } \gamma_{22}=0.\end{cases}$$

Note that it is easily checked that $0<\alpha<1$.

The first equilibrium equation corresponds to the line $y_1=1$ and the point $y_1=y_2=0$. The second equilibrium equation is a curve, with y_1 a convex function of y_2, which has asymptote $y_2=1$. These are shown in Figure 8.5. Cases (a) and (b) correspond to $\dfrac{\sigma_2\lambda_{22}}{\mu_2}\le 1$ and $\dfrac{\sigma_2\lambda_{22}}{\mu_2}>1$ respectively.

Consider equations (8.10) with $\mu_1=0$. The derivative of $y_1(t)$ is positive in both regions R_1 and R_2. The derivative of $y_2(t)$ is negative in the interior of region R_1 and positive in the interior of region R_2. A simple graphical analysis then shows that all trajectories converge to the endemic solution $(1,\alpha)$.

8.5 The equilibrium solutions of the multi-type epidemic

Let $\mathbf{\Lambda}$ be the matrix with $\{\mathbf{\Lambda}\}_{ij}=\sigma_j\lambda_{ij}$ and define $\{\boldsymbol{\mu}\}_i=\mu_i$. The matrix $\mathbf{\Lambda}$ is assumed to be non-reducible. Consider the system of equations (8.4), where $\boldsymbol{\mu}\ge\mathbf{0}$ with initial conditions that $0\le y_i(0)\le 1$ for all i. Clearly $0\le y_i(t)\le 1$ for all i and $t\ge 0$. It is simple to prove the existence and uniqueness of a solution to equations (8.4); the result is therefore merely stated.

THEOREM 8.1. *There exists a unique, differentiable solution $y_i(t)$, $(i=1,...,n)$, to equations (8.4) with initial conditions $0\le y_i(0)\le 1$ for all i. The solution has each $y_i'(t)$ continuous and uniformly bounded for $t\ge 0$ and all i.*

□

The possible equilibrium solutions $y_i(t)=y_i$ to equations (8.4) satisfy the equations

$$(1-y_i)\sum_{j=1}^{n}\sigma_j\lambda_{ij}y_j=\mu_i y_i,$$

for $i=1,...,n$. Define $\{\mathbf{y}\}_i=y_i$. This may be then be rewritten as

$$\text{diag}(\mathbf{1}-\mathbf{y})\boldsymbol{\Lambda}\mathbf{y} = \text{diag}(\boldsymbol{\mu})\mathbf{y}. \tag{8.11}$$

A complete description of the possible equilibrium solutions $\mathbf{y}$ to equation (8.11) is easily obtained from Theorem B.2 with $f(y) = y/(1-y)$, and is given in Theorem 8.2.

THEOREM 8.2. *The equilibrium* $\mathbf{y} = \mathbf{0}$ *always occurs. This is the only possible equilibrium solution when* $\boldsymbol{\mu} > \mathbf{0}$ *and* $\rho(\boldsymbol{\Gamma}) \leq 1$, *where* $\boldsymbol{\Gamma} = (diag(\boldsymbol{\mu}))^{-1}\boldsymbol{\Lambda}$. *In this case we define* $\boldsymbol{\zeta} = \mathbf{0}$.

In all other cases there is a unique non-zero equilibrium $\mathbf{y} = \boldsymbol{\zeta} > \mathbf{0}$. *This is specified as follows:*

1. *When* $\boldsymbol{\mu} > \mathbf{0}$ *and* $\rho(\boldsymbol{\Gamma}) > 1$, *then* $\boldsymbol{\zeta} = \boldsymbol{\eta}(\boldsymbol{\Gamma}, \mathbf{0})$. *Here* $\mathbf{y} = (y_i) = \boldsymbol{\eta}(\boldsymbol{\Gamma}, \mathbf{0})$ *is the unique positive solution to*

$$f(y_i) = \frac{y_i}{1-y_i} = \{\boldsymbol{\Gamma}\mathbf{y}\}_i, \ (i = 1, ..., n).$$

2. *When* $\boldsymbol{\mu} = \mathbf{0}$, *then* $\boldsymbol{\zeta} = \mathbf{1}$.
3. *The remaining case corresponds to a partitioning of* $\boldsymbol{\Lambda}$, $\boldsymbol{\mu}$, *and* $\boldsymbol{\zeta}$ *(by permutation of the types) into*

$$\boldsymbol{\Lambda} = \begin{pmatrix} \boldsymbol{\Lambda}_{11} & \boldsymbol{\Lambda}_{12} \\ \boldsymbol{\Lambda}_{21} & \boldsymbol{\Lambda}_{22} \end{pmatrix}, \ \boldsymbol{\mu} = \begin{pmatrix} \boldsymbol{\mu}_1 \\ \boldsymbol{\mu}_2 \end{pmatrix} \ \textit{and} \ \boldsymbol{\zeta} = \begin{pmatrix} \boldsymbol{\zeta}_1 \\ \boldsymbol{\zeta}_2 \end{pmatrix},$$

where $\boldsymbol{\mu}_1$ *is the zero vector of length* m *for some* $0 < m < n$ *and* $\boldsymbol{\mu}_2 > \mathbf{0}$. *Define* $\boldsymbol{\Gamma}_{22} = (diag(\boldsymbol{\mu}_2))^{-1}\boldsymbol{\Lambda}_{22}$ *and* $\boldsymbol{\Gamma}_{21} = (diag(\boldsymbol{\mu}_2))^{-1}\boldsymbol{\Lambda}_{21}$.

In this case $\boldsymbol{\zeta}_1 = \mathbf{1}$ *and* $\boldsymbol{\zeta}_2 = \boldsymbol{\eta}_2(\boldsymbol{\Gamma}_{22}, \boldsymbol{\Gamma}_{21}\mathbf{1})$, *where* $\mathbf{y}_2 = \boldsymbol{\eta}_2(\boldsymbol{\Gamma}_{22}, \mathbf{a})$ *is the unique positive solution to*

$$f(\{\mathbf{y}_2\}_i) = \frac{\{\mathbf{y}_2\}_i}{1-\{\mathbf{y}_2\}_i} = \{\boldsymbol{\Gamma}_{22}\mathbf{y}_2 + \mathbf{a}\}_i, \ (i = 1, ..., n-m).$$

PROOF. Clearly $\mathbf{y} = \mathbf{0}$ is a solution to equation (8.11). When $\boldsymbol{\mu} > \mathbf{0}$, equation (8.11) can be rewritten in the form

$$f(y_i) = \frac{y_i}{1-y_i} = \sum_{j=1}^{n} \frac{\sigma_j \lambda_{ij}}{\mu_i} y_j,$$

for $i = 1, ..., n$. If $\rho(\boldsymbol{\Gamma}) \leq 1$, where $\boldsymbol{\Gamma} = (\text{diag}(\boldsymbol{\mu}))^{-1}\boldsymbol{\Lambda}$, it follows from Theorem B.2 that $\mathbf{y} = \mathbf{0}$ is the only non-negative solution.

Suppose either $\mu_i = 0$ for some i or $\boldsymbol{\mu} > \mathbf{0}$ and $\rho(\boldsymbol{\Gamma}) > 1$. If $y_i = 0$ for some i, since $\boldsymbol{\Lambda}$ is non-reducible it follows that $y_i = 0$ for all i. Thus if $\mathbf{y} \neq \mathbf{0}$ is a non-negative solution to equation (8.11), then $\mathbf{y} > \mathbf{0}$.

We now obtain the positive solution, and show uniqueness, for each of the three specified cases.

1. When $\boldsymbol{\mu} > \mathbf{0}$ and $\rho(\boldsymbol{\Gamma}) > 1$ equation (8.11) can be rewritten in the form $(\text{diag}(\mathbf{1}-\mathbf{y}))^{-1}\mathbf{y} = \boldsymbol{\Gamma}\mathbf{y}$. The result then follows from Theorem B.2 part 1 if we take $f(y) = y/(1-y)$, $\mathbf{B} = \boldsymbol{\Gamma}$ and $\mathbf{a} = \mathbf{0}$.

2. Since $\boldsymbol{\mu} = \mathbf{0}$, $\mathbf{y} > \mathbf{0}$ and $\boldsymbol{\Lambda}$ is non-reducible it immediately follows that $\boldsymbol{\Lambda}\mathbf{y} > \mathbf{0}$ so that the only positive solution to equation (8.11) is $\mathbf{y} = \mathbf{1}$.
3. As for case 2, since $\boldsymbol{\Lambda}$ is non-reducible, any solution $\mathbf{y} > \mathbf{0}$ to equation (8.11) has $(\boldsymbol{\Lambda}_{11}\mathbf{y}_1 + \boldsymbol{\Lambda}_{12}\mathbf{y}_2) > \mathbf{0}$. As $\boldsymbol{\mu}_1 = \mathbf{0}$ it then follows from equations (8.11) that $\mathbf{y}_1 = \mathbf{1}$. Then $\mathbf{y}_2$ must satisfy the following equation

$$(\text{diag}(\mathbf{1} - \mathbf{y}_2))^{-1}\mathbf{y}_2 = (\text{diag}(\boldsymbol{\mu}_2))^{-1}\boldsymbol{\Lambda}_{22}\mathbf{y}_2 + (\text{diag}(\boldsymbol{\mu}_2))^{-1}\boldsymbol{\Lambda}_{21}\mathbf{1}, \tag{8.12}$$

where $\boldsymbol{\Lambda}_{21}\mathbf{1} \neq \mathbf{0}$ since $\boldsymbol{\Lambda}$ is non-reducible.

Let $\boldsymbol{\Gamma}_{2j} = (\text{diag}(\boldsymbol{\mu}_2))^{-1}\boldsymbol{\Lambda}_{2j}$ for $j = 1, 2$. In Theorem B.2, take $f(y) = y/(1-y)$, $\mathbf{B} = \boldsymbol{\Gamma}_{22}$ and $\mathbf{a} = \boldsymbol{\Gamma}_{21}\mathbf{1}$. If $\boldsymbol{\Gamma}_{22}$ is non-reducible, since $\boldsymbol{\Gamma}_{21}\mathbf{1} \neq \mathbf{0}$ from the non-reducibility of $\boldsymbol{\Lambda}$, the result follows from Theorem B.2 part 1. When $\mathbf{B} = \boldsymbol{\Gamma}_{22}$ is reducible and is written in normal form, the non-reducibility of $\boldsymbol{\Lambda}$ implies that the first s vector components of $\mathbf{a} = \boldsymbol{\Gamma}_{21}\mathbf{1}$ corresponding to that normal form are all non-zero. Hence from Theorem B.2 part 3, with $\mathbf{B} = \boldsymbol{\Gamma}_{22}$, there is a unique positive solution to equation (8.12) given by $\mathbf{y}_2 = \boldsymbol{\eta}_2(\boldsymbol{\Gamma}_{22}, \boldsymbol{\Gamma}_{21}\mathbf{1})$.

□

8.6 The global asymptotic stability of the equilibria

The local asymptotic stability of the equilibria can be examined in standard fashion by taking a small disturbance from an equilibrium and using a linear approximation to the equations obtained. Since the results also follow immediately from the stronger global stability results derived in this section we merely state them. Attention is restricted to the case when $\boldsymbol{\Lambda}$ is non-reducible. When $\boldsymbol{\mu} > \mathbf{0}$ and $\boldsymbol{\Gamma} \leq 1$ the unique equilibrium $\boldsymbol{\zeta} = \mathbf{0}$ is locally asymptotically stable. In all other cases the zero equilibrium corresponding to no infection is not locally asymptotically stable, but the unique non-zero (endemic) equilibrium $\boldsymbol{\zeta} > \mathbf{0}$ is locally asymptotically stable.

Now consider the global asymptotic stability of the equilibrium $\boldsymbol{\zeta}$. We show that, provided $\mathbf{y}(0) \neq \mathbf{0}$, then $\mathbf{y}(t)$ tends to $\boldsymbol{\zeta}$ as t tends to infinity. Hence the infection is sure to die out when $\boldsymbol{\mu} > \mathbf{0}$ and $\rho\left((\text{diag}(\boldsymbol{\mu}))^{-1}\boldsymbol{\Lambda}\right) \leq 1$, since $\boldsymbol{\zeta} = \mathbf{0}$. In all other cases the infection becomes endemic.

Global asymptotic stability when $\boldsymbol{\mu} > \mathbf{0}$ and $\rho(\boldsymbol{\Gamma}) \leq 1$.

Let $\mathbf{u}$ be the right eigenvector and $\mathbf{v}'$ be the left eigenvector of $\boldsymbol{\Gamma}$ corresponding to $\rho(\boldsymbol{\Gamma})$. When $\rho(\boldsymbol{\Gamma}) \leq 1$ we present a simple proof of the global asymptotic stability based on the function $a(t) = \max_i(y_i(t)/u_i)$ where $u_i = \{\mathbf{u}\}_i$. A proof of this result using a different Liapunov function is given in Lajmanovich and York [L1]. The proof presented is adapted from Rass and Radcliffe [R20].

THEOREM 8.3. *When $\boldsymbol{\mu} > \mathbf{0}$ and $\rho(\boldsymbol{\Gamma}) \leq 1$, $\mathbf{y} = \mathbf{0}$ is globally asymptotically stable on $[0,1]^n$.*

PROOF. Suppose that $y_i(t) > 0$ for at least one i. Define $a(t) = \max_i(y_i(t)/u_i)$. For a given t, choose i such that $y_i(t)/u_i = a(t)$. There may be more than one such i. However, i can be chosen so that, for sufficiently small $\epsilon > 0$, $a(s) = y_i(s)/u_i$ for

$s \in [t, t+\epsilon]$. Then $y_i(t) = u_i a(t)$ and $y_j(t) \le u_j a(t)$ for $j \ne i$. The function $a(t)$ so defined is a continuous function of t with a right hand derivative given by

$$\begin{aligned}
u_i \frac{d_+ a(t)}{dt} &= \frac{dy_i(t)}{dt} \\
&= \mu_i \left[(1 - y_i(t)) \sum_{j=1}^{n} \frac{\sigma_j \lambda_{ij}}{\mu_i} y_j(t) - y_i(t) \right] \\
&= \mu_i \left[(1 - u_i a(t)) \sum_{j=1}^{n} \frac{\sigma_j \lambda_{ij}}{\mu_i} y_j(t) - u_i a(t) \right] \\
&\le \mu_i \left[(1 - u_i a(t)) a(t) \sum_{j=1}^{n} \frac{\sigma_j \lambda_{ij}}{\mu_i} u_j - u_i a(t) \right] \\
&= \mu_i \left[(1 - u_i a(t)) a(t) \rho(\mathbf{\Gamma}) u_i - u_i a(t) \right] \\
&= \mu_i a(t) u_i \left[\rho(\mathbf{\Gamma})(1 - u_i a(t)) - 1 \right] \\
&= -\mu_i a(t) u_i \left[(1 - \rho(\mathbf{\Gamma})) + u_i a(t) \rho(\mathbf{\Gamma}) \right].
\end{aligned}$$

Let $\alpha = (1-\rho(\mathbf{\Gamma}))\min_i(\mu_i)$ and $\theta = \rho(\mathbf{\Gamma})\min_i(u_i\mu_i)$. If $\rho(\mathbf{\Gamma}) < 1$ then $\frac{d_+ a(t)}{dt} \le -\alpha a(t)$, where $\alpha > 0$. If $\rho(\mathbf{\Gamma}) = 1$ then $\frac{d_+ a(t)}{dt} \le -\theta (a(t))^2$, where $\theta > 0$. Therefore $a(t)$ is monotone decreasing and is bounded below by 0, and so must tend to a limit. It is simple to show that this limit is 0.

Consider first the case where $\rho(\mathbf{\Gamma}) < 1$. Let $a(0) = a_0$ and suppose that $a(t) \downarrow a_1 > 0$ as $t \to \infty$. Now $\frac{d_+ a(t)}{dt} \le -\alpha a_1/2$ provided $a(t) > a_1/2$. Thus if $a(t) > a_1/2$ the graph of $y = a(t)$ lies below the line $y = a_0 - \alpha a_1 t/2$. It follows that $a(t)$ will decrease to $a_1/2$ at least by time $\frac{2a_0 - a_1}{\alpha a_1}$. This contradicts the statement that $a(t) \downarrow a_1 > 0$. If $\rho(\mathbf{\Gamma}) = 1$, a similar argument shows that $a(t)$ decreases to $a_1/2$ before time $\frac{4a_0 - 2a_1}{\theta a_1^2}$. Thus in both cases $a(t) \downarrow 0$ as $t \to \infty$ and the theorem is proved.

□

Global asymptotic stability in all other cases.

Now consider the remaining cases when one of the following occurs; either $\mu_i = 0$ for some i or $\boldsymbol{\mu} > \mathbf{0}$ and $\rho(\mathbf{\Gamma}) > 1$. The part of the proof corresponding to the case when $\boldsymbol{\mu} > \mathbf{0}$ is based on proofs given in Lajmanovich and Yorke [L1]; with the entire proof adapted from the more complex proof required for the reducible case given in Rass and Radcliffe [R20].

Suppose that $y_i(0) > 0$ for at least one i. Then it can be shown that $y_i(t) > 0$ for all $i = 1, ..., n$, and all $t > 0$. This is then used to establish global asymptotic stability. The following lemma is a first step in proving this result. Denote $\frac{d^r y(t)}{dt^r}$ by $y^{(r)}(t)$.

LEMMA 8.1. *If* $\mathbf{y}(0) \in [0,1]^n \backslash \{\mathbf{0}\}$, *then* $\exists\, T > 0$ *such that* $y_i(t) > 0$ $i = 1, ..., n$ *for* $t \in (0, T]$.

PROOF. Let $S = \{i | y_i(0) = 0\}$ and $S' = \{i | y_i(0) > 0\}$. If $i \in S'$ it follows by continuity that $\exists\, T_i > 0$ such that $y_i(t) > 0$ for $t \in [0, T_i]$.

Now consider $i \in S$. Then $y_i^{(1)}(0) = \sum_{j=1}^n \{\mathbf{\Lambda}\}_{ij} y_j(0)$. If $\exists\, j \in S'$ such that $\{\mathbf{\Lambda}\}_{ij} > 0$, then $y_i^{(1)}(0) > 0$. If so stop. Otherwise differentiate the ith equation of (8.4). Then

$$y_i^{(2)}(0) = \sum_{j=1}^n \{\mathbf{\Lambda}\}_{ij} y_j^{(1)}(0) = \sum_{j=1}^n \{\mathbf{\Lambda}^2\}_{ij} y_j(0).$$

If $\exists\, j \in S'$ such that $\{\mathbf{\Lambda}^2\}_{ij} > 0$, then $y_i^{(2)}(0) > 0$. If so stop. Otherwise differentiate the ith equation of (8.4) twice and continue. Since $\mathbf{\Lambda}$ is non-reducible, for any $j \in S'$ $\exists\, r \leq n-1$ such that $\{\mathbf{\Lambda}^r\}_{ij} > 0$ and so the process must terminate after at most r steps. Thus for any $i \in S$ there exists an s such that $1 \leq s \leq n-1$ with $y_i^{(m)}(0) = 0$ for $1 \leq m \leq s$ and $y_i^{(s)}(0) > 0$. Here s is the smallest positive integer such that $\{\mathbf{\Lambda}^s \mathbf{y}(0)\}_i > 0$. Hence, for each $i \in S$ $\exists\, T_i > 0$ such that $y_i(t) > 0$ for $t \in (0, T_i]$.

The lemma then follows by taking $T = \min_i T_i$.

□

THEOREM 8.4. *When one of the following occurs, either* $\mu_i = 0$ *for some* i *or* $\boldsymbol{\mu} > \mathbf{0}$ *and* $\rho(\mathbf{\Gamma}) > 1$, *then*

1. $\mathbf{y}(t) > \mathbf{0}$ *for all* $t > 0$ *if* $\mathbf{y}(0) \in [0,1]^n \backslash \{\mathbf{0}\}$.
2. $\mathbf{y} = \boldsymbol{\zeta}$ *is globally asymptotically stable on* $[0,1]^n \backslash \{\mathbf{0}\}$.

PROOF. Observe that here $\boldsymbol{\zeta} > \mathbf{0}$. Let $\zeta_i = \{\boldsymbol{\zeta}\}_i$.

Define $b(t) = \min_i (y_i(t)/\zeta_i)$. For a given t, choose i such that $y_i(t)/\zeta_i = b(t)$. There may be more than one such i. However, i can be chosen so that, for sufficiently small $\epsilon > 0$, $b(s) = (y_i(s)/\zeta_i)$ for $s \in [t, t+\epsilon]$. Then $y_i(t) = \zeta_i b(t)$ and $y_j(t) \geq \zeta_j b(t)$ for $j \neq i$. The function $b(t)$ so defined is a continuous function of t with a right hand derivative given by

$$\begin{aligned}
\zeta_i \frac{d_+ b(t)}{dt} &= \frac{dy_i(t)}{dt} \\
&= (1 - y_i(t)) \sum_{j=1}^n \sigma_j \lambda_{ij} y_j(t) - \mu_i y_i(t) \\
&= (1 - \zeta_i b(t)) \sum_{j=1}^n \sigma_j \lambda_{ij} y_j(t) - \mu_i \zeta_i b(t) \\
&\geq (1 - \zeta_i b(t)) \sum_{j=1}^n \sigma_j \lambda_{ij} \zeta_j b(t) - \mu_i \zeta_i b(t).
\end{aligned}$$

Now consider the cases (i) $\mu_i = 0$ and (ii) $\mu_i > 0$ separately.

(i) From the relation $(1 - \zeta_i) \sum_{j=1}^n \sigma_j \lambda_{ij} \zeta_j = \mu_i \zeta_i$ it follows that $\mu_i = 0$ implies that $\zeta_i = 1$. Thus if $\mu_i = 0$, necessarily $b(t) = y_i(t) \leq 1$ and

$$\frac{d_+b(t)}{dt} \geq (1-b(t))\sum_{j=1}^{n}\sigma_j\lambda_{ij}\zeta_j b(t) = b(t)(1-b(t))\sum_{j=1}^{n}\sigma_j\lambda_{ij}\zeta_j.$$

(ii) If $\mu_i > 0$ and if $b(t) \leq 1$, then

$$\begin{aligned}\zeta_i\frac{d_+b(t)}{dt} &\geq (1-\zeta_i b(t))\sum_{j=1}^{n}\sigma_j\lambda_{ij}\zeta_j b(t) - \mu_i\zeta_i b(t)\\ &= \mu_i\left[(1-\zeta_i b(t))b(t)\frac{\zeta_i}{1-\zeta_i} - \zeta_i b(t)\right],\end{aligned}$$

and so

$$\frac{d_+b(t)}{dt} \geq \frac{\mu_i\zeta_i}{(1-\zeta_i)}b(t)(1-b(t)).$$

Thus in both case (i) and case (ii), if $b(t) \leq 1$,

$$\frac{d_+b(t)}{dt} \geq Ab(t)(1-b(t)),$$

where $A = \min\left(\min_{i:\mu_i=0}\left(\sum_{j=1}^{n}\sigma_j\lambda_{ij}\zeta_j\right), \min_{i:\mu_i>0}\frac{\mu_i\zeta_i}{1-\zeta_i}\right).$

From Lemma 8.1 $\mathbf{y}(t) > \mathbf{0}$ and hence $b(t) > 0$ for $t \in (0,T]$ for some T. Now $\dfrac{d_+b(t)}{dt} > 0$ if $0 < b(t) < 1$ and $\dfrac{d_+b(t)}{dt} \geq 0$ if $b(t) = 1$. It follows that $b(t) \geq \min(1, b(T)) > 0$ for $t > T$ and hence that $\mathbf{y}(t) \geq \min(1, b(T))\boldsymbol{\zeta} > \mathbf{0}$ for $t > T$. Then $\mathbf{y}(t) > \mathbf{0}$ for $t > 0$, which proves part 1 of the theorem.

Let $m(t) = \min(b(t), 1)$. Then $m(t)$ is monotone non-decreasing and bounded above by 1, and so must tend to a limit. We show that this limit is 1. If $b(T) \geq 1$ then $m(t) = 1$ for $t \geq T$ and the limit is 1. Consider the case where $0 < b(T) = b_0 < 1$. Suppose $m(t) \uparrow b_1 < 1$ as $t \to \infty$. Now

$$\frac{d_+m(t)}{dt} = \frac{d_+b(t)}{dt} \geq \theta,$$

where $\theta = A\min[b_0(1-b_0), b_1(1-b_1)]$. So the graph of $y = m(t)$ lies above the line $y = b_0 + \theta(t-T)$. Thus $m(t)$ will increase to $(1+b_1)/2$ at least by time $T + \dfrac{1+b_1-2b_0}{2\theta}$. This contradicts the statement that $m(t) \uparrow b_1 < 1$ as $t \to \infty$. Thus $m(t) \uparrow 1$ as $t \to \infty$.

Define $c(t) = \max_i(y_i(t)/\zeta_i)$. Then $c(t)$ is a continuous function of t. We can choose i so that, for sufficiently small $\epsilon > 0$, $c(s) = (y_i(s)/\zeta_i)$ for $s \in [t, t+\epsilon]$. Then $y_i(t) = \zeta_i c(t)$ and $y_j(t) \leq \zeta_j c(t)$ for $j \neq i$.

We need to consider the cases (i) $\mu_i = 0$ and (ii) $\mu_i > 0$ separately.

(i) If $\mu_i = 0$ then $\zeta_i = 1$ and so $c(t) = y_i(t) \leq 1$ and the right hand derivative of $c(t)$ is given by,

$$\frac{d_+c(t)}{dt} = \frac{dy_i(t)}{dt} = (1-c(t))\sum_{j=1}^{n}\sigma_j\lambda_{ij}y_j(t).$$

In this case if $c(t) \geq 1$, then necessarily $c(t) = 1$ and $\dfrac{d_+c(t)}{dt} = 0$.

(ii) When $\mu_i > 0$ then, if $c(t) \geq 1$,

$$\begin{aligned}
\zeta_i \frac{d_+c(t)}{dt} &= \frac{dy_i(t)}{dt} \\
&= \left[(1 - y_i(t)) \sum_{j=1}^{n} \sigma_j \lambda_{ij} y_j(t) - \mu_i y_i(t) \right] \\
&\leq \left[(1 - \zeta_i c(t)) \sum_{j=1}^{n} \sigma_j \lambda_{ij} \zeta_j c(t) - \mu_i \zeta_i c(t) \right] \\
&= \mu_i \left[(1 - \zeta_i c(t)) c(t) \frac{\zeta_i}{1 - \zeta_i} - \zeta_i c(t) \right] \\
&= -\frac{\mu_i \zeta_i^2}{(1 - \zeta_i)} c(t)(c(t) - 1).
\end{aligned}$$

Hence for $c(t) \geq 1$,

$$\frac{d_+c(t)}{dt} \leq -Bc(t)(c(t) - 1),$$

where $B = \min_i \dfrac{\mu_i \zeta_i}{1 - \zeta_i}$.

From case (ii) it follows that $\dfrac{d_+c(t)}{dt} < 0$ if $c(t) > 1$ and from cases (i) and (ii) it follows that $\dfrac{d_+c(t)}{dt} \leq 0$ if $c(t) = 1$.

Let $M(t) = \max(c(t), 1)$. Then $M(t)$ is monotone non-increasing and bounded below by 1, and so must tend to a limit. We show that this limit is 1. If $c(0) \leq 1$ then $M(0) = 1$ and the limit is 1. Consider the case where $c(0) = c_0 > 1$. Suppose $M(t) \downarrow c_1 > 1$ as $t \to \infty$. Now

$$\frac{d_+M(t)}{dt} = \frac{d_+c(t)}{dt} \leq -\alpha,$$

where $\alpha = Bc_1(c_1 - 1)$. So the the graph of $y = M(t)$ lies below the line $y = c_0 - \alpha t$. Thus $M(t)$ will decrease to $(1 + c_1)/2$ before time $\dfrac{2c_0 - 1 - c_1}{2\alpha}$. This contradicts the statement that $M(t) \downarrow c_1 > 1$. Thus $M(t) \downarrow 1$ as $t \to \infty$.

Now for all i,

$$m(t) \leq b(t) \leq \frac{y_i(t)}{\zeta_i} \leq c(t) \leq M(t).$$

Since $m(t) \to 1$ and $M(t) \to 1$ as $t \to \infty$, it follows that $y_i(t) \to \zeta_i$ as $t \to \infty$ for $i = 1, ..., n$.

□

Theorem 8.3 shows that if $\boldsymbol{\mu} > \mathbf{0}$ and $\rho(\mathbf{\Gamma}) \leq 1$ then, regardless of the amount of initial infection, the epidemic is not sustainable and will eventually die out.

Theorem 8.4 proves that in all other cases, provided there is some initial infection, then an endemic occurs. The proportion of infection at time t, $\mathbf{y}(t)$, tends to the unique positive endemic level as $t \to \infty$.

The non-spatial models considered are for a non-reducible system. Equivalent results have been obtained for a reducible system in Rass and Radcliffe [R20].

8.7 Epidemic models on a finite number of sites

In this section a spatial aspect is introduced, with individuals at a finite number of sites. The approach is essentially the same as that of Chapter 6, Sections 6.1 and 6.2, where the epidemic considered admitted no return to the susceptible state.

The one-type epidemic.

First consider a single type epidemic on N sites. This can be regarded as an N-type non-spatial epidemic by treating the sites as types. In such a model σ_i denotes the size of the population at the ith site and λ_{ij} denotes the rate of infection of a susceptible individual at the ith site by an infectious individual at the jth site. The combined rate at which infected individuals at site i either return to the susceptible state, emigrate or die is μ_i. As in section 8.3 it is assumed that the population size stays constant. Then if $y_i(t)$ represents the proportion of infected individuals at site i, the $y_i(t)$ satisfy the equations.

$$(8.13) \qquad y_i'(t) = (1 - y_i(t)) \sum_{j=1}^{N} \sigma_j \lambda_{ij} y_j(t) - \mu_i y_i(t), \ (i = 1, ..., N).$$

The results for this model follow immediately from those for an N-type model described by equations (8.4), and stated in Theorems 8.3 and 8.4. When $\boldsymbol{\mu} > \mathbf{0}$ and $\rho(\mathbf{\Gamma}) \leq 1$ then the infection will eventually die out at all the sites. In all other cases, provided there is some infection present initially, the infection becomes endemic at all sites. The proportion of infection at site i tends to the unique endemic level $\{\boldsymbol{\zeta}\}_i > 0$ as t tends to infinity.

For most situations the emigration, death and recovery rates would be the same for all sites, so that $\mu_i = \mu^*$ for all $i = 1, ..., N$. Some neat results linking the spatial to non-spatial results are obtained if some additional assumptions are made. The population size and the rates of infection of a susceptible individual at site i are assumed to be site independent. Hence $\sigma_i = \sigma^*$ and $\sum_{j=1}^{N} \lambda_{ij} = \lambda^*$ for all $i = 1, ..., N$. Note that we can think of the infection rate λ_{ij} of a susceptible individual at site i by an infective individual at site j as the product $\lambda_{ij} = \alpha r_{ij}$, where α is the infection rate per contact between a susceptible and infected individual and r_{ij} is the contact rate between an individual at site i and one at site j. The condition on the infection rates essentially stipulates that the total contact rate is the same for all individuals at all sites.

A simple model where these assumptions hold occurs with a circular arrangement of sites with the infection rate dependent on the relative ordered position only. The sites may be considered to be at the integers mod N so that $\lambda_{ij} = \lambda^* p(\min(|i-j|, N - |i-j|)$, where $i, j = 0, 1, ..., N-1$. The assumption is also reasonable with less trivial arrangements of sites, with the contact rates at more extreme sites increased to compensate for less between site contacts.

With these constraints on the parameters, we can prove the result that $y_i(t)$ tends to zero at all sites as t tends to infinity if $\mu^* > 0$ and $\sigma^*\lambda^* \leq \mu^*$. When $\sigma^*\lambda^* > \mu^*$ (which includes the case when $\mu^* = 0$), $y_i(t)$ tends to the unique non-spatial endemic level, $\zeta^* = 1 - \dfrac{\mu^*}{\sigma^*\lambda^*}$ for a non-spatial epidemic with population size σ^*, infection rate λ^* and combined death/emigration/recovery rate μ^*.

In order to prove this result we need only establish two results. First, that when $\mu^* > 0$ the Perron-Frobenius root of $(\text{diag}(\boldsymbol{\mu}))^{-1}\mathbf{\Lambda}$ is $\dfrac{\sigma^*\lambda^*}{\mu^*}$. Second that, when $\sigma^*\lambda^* > \mu^*$, then $\mathbf{y} = \zeta^*\mathbf{1} > \mathbf{0}$ is a solution to

$$\text{diag}(\mathbf{1} - \mathbf{y})\mathbf{\Lambda}\mathbf{y} = \text{diag}(\boldsymbol{\mu})\mathbf{y}, \tag{8.14}$$

(and hence from Theorem 8.2 is the unique such positive solution).

If $\mu^* = 0$ then $\zeta^* = 1$ so that $\mathbf{y} = \zeta^*\mathbf{1} = \mathbf{1}$ satisfies equation (8.14) with $\boldsymbol{\mu} = \mu^*\mathbf{1} = \mathbf{0}$, which completes the proof for this case.

When $\mu^* > 0$, $(\text{diag}(\boldsymbol{\mu}))^{-1}\mathbf{\Lambda}\mathbf{1} = \dfrac{1}{\mu^*}\mathbf{\Lambda}\mathbf{1} = \dfrac{\sigma^*\lambda^*}{\mu^*}\mathbf{1}$. Hence $\mathbf{1}$ is a positive eigenvector of the non-reducible non-negative matrix $(\text{diag}(\boldsymbol{\mu}))^{-1}\mathbf{\Lambda}$ corresponding to the eigenvalue $\dfrac{\sigma^*\lambda^*}{\mu^*}$. From Theorem A.2 part 7, this eigenvalue must therefore be the Perron-Frobenius root of the matrix.

If we take $\mathbf{y} = \zeta^*\mathbf{1}$, where $\zeta^* = 1 - \dfrac{\mu^*}{\sigma^*\lambda^*}$, then

$$\text{diag}(\mathbf{1} - \mathbf{y})\mathbf{\Lambda}\mathbf{y} = (1 - \zeta^*)\zeta^*\mathbf{\Lambda}\mathbf{1} = (1 - \zeta^*)\zeta^*\sigma^*\lambda^*\mathbf{1} = \zeta^*\mu^*\mathbf{1} = \text{diag}(\boldsymbol{\mu})\mathbf{y}.$$

Hence, if $\sigma^*\lambda^* > \mu^* > 0$, $\mathbf{y} = \zeta^*\mathbf{1} > \mathbf{0}$ is a solution to equation (8.14) as required.

The n-type epidemic.

Now consider an n-type epidemic on N sites. Let $\lambda_{ij}(r,s)$ be the rate of infection of a type i susceptible individual at site r by a type j infective individual at site s. Let $\sigma_i(r)$ be the size of the ith population at site r. Let μ_i be the combined death/emigration/recovery rate of an type i infectious individual. Let $y_i(r,t)$ be the proportion of type i infected individuals at site r at time t.

This epidemic can also be treated as a non-spatial $S \to I \to S$ epidemic. In this case there will be nN 'types', which consist of all type and site combinations. Define $y_{(i-1)N+r}(t) = y_i(r,t)$, $\sigma_{(i-1)N+r} = \sigma_i(r)$, $\mu_{(i-1)N+r} = \mu_i$, $i = 1, ...n$, $r = 1, ...N$, and $\lambda_{(i-1)N+r,(j-1)N+s} = \lambda_{ij}(r,s)$, $i = 1, ...n$, $j = 1, ...n$, $r = 1, ...N$ and $s = 1, ..., N$. Here $\sigma_i(r)$ is the population size of type i at site r and $y_i(r,t)$ is the corresponding proportion of infectives at time t. Also $\mu_i(r)$ is the combined death and recovery rate and $\lambda_{ij}(r,s)$ is the infection rate of a type i susceptible at site r by a type j infective at site s. Results for this finite site model are then immediately obtained from Theorems 8.3 and 8.4, as for the one-type model on sites.

We now impose equivalent constraints to those imposed for the one-type model, namely that $\sigma_i(r) = \sigma_i^*$, $\mu_i(r) = \mu_i^*$ and $\sum_{r=1}^{N} \lambda_{ij}(s,r) = \lambda_{ij}^*$. Thus we are assuming that the population sizes and combined recovery, emigration and death rates are site independent. Also that the total of the rates of infection of type i susceptibles at site s by type j infective individuals over all sites is the same for all s.

The equations describing this model are

$$\frac{\partial y_i(s,t)}{\partial t} = \left[(1 - y_i(s,t)) \sum_{j=1}^{n} \sum_{r=1}^{N} \sigma_j^* \lambda_{ij}(s,r) y_j(r,t) - \mu_i^* y_i(s,t)\right].$$

This may be written in matrix form using the Nn combined types so that

$$\frac{d\mathbf{y}(t)}{dt} = \text{diag}(\mathbf{1} - \mathbf{y}(t))\mathbf{\Lambda}\mathbf{y}(t) - \text{diag}(\boldsymbol{\mu})\mathbf{y}(t). \tag{8.15}$$

Here $\mathbf{\Lambda}$ is an $Nn \times Nn$ matrix and $\boldsymbol{\mu}$ is a vector of length Nn given by

$$\mathbf{\Lambda} = \begin{pmatrix} \mathbf{\Lambda}_{11} & \mathbf{\Lambda}_{12} & \dots & \mathbf{\Lambda}_{1n} \\ \mathbf{\Lambda}_{21} & \mathbf{\Lambda}_{22} & \dots & \mathbf{\Lambda}_{2n} \\ \vdots & \vdots & \ddots & \vdots \\ \mathbf{\Lambda}_{n1} & \mathbf{\Lambda}_{n2} & \dots & \mathbf{\Lambda}_{nn} \end{pmatrix} \text{ and } \boldsymbol{\mu} = \begin{pmatrix} \boldsymbol{\mu}_1 \\ \boldsymbol{\mu}_2 \\ \vdots \\ \boldsymbol{\mu}_n \end{pmatrix}$$

where $\{\mathbf{\Lambda}_{ij}\}_{rs} = \sigma_j^* \lambda_{ij}(r,s)$, so that $\mathbf{\Lambda}_{ij}\mathbf{1} = \sigma^* \lambda_{ij}^* \mathbf{1}$, and $\boldsymbol{\mu}_i = \mu_i^* \mathbf{1}$.

When $\mu_i^* > 0$ for all i, then we define $\{\mathbf{\Gamma}^*\}_{ij} = \sigma_j^* \lambda_{ij}^* / \mu_i^*$. The results on global stability are given in Theorem 8.5.

THEOREM 8.5. *If $\mu_i^* > 0$ for all i and $\rho(\mathbf{\Gamma}^*) \leq 1$, then $y_i(r,t) \to 0$ as $t \to \infty$ for all i and r, so that the infection dies out eventually for all types at all sites.*

Consider the case when either $\mu_i^* = 0$ for some i or the following holds; $\mu_i^* > 0$ for all i and $\rho(\mathbf{\Gamma}^*) > 1$. In this case $y_i(r,t)$ tends to the endemic level $\zeta_i^* > 0$ as $t \to \infty$. The endemic level for each type is the same for all sites, and $y_i = \zeta_i^*$ is the unique positive solution to

$$(1 - y_i) \sum_{j=1}^{n} \sigma_j^* \lambda_{ij}^* y_j = \mu_i^* y_i, \ (i = 1, ..., n).$$

PROOF. We first show that when $\mu_i^* > 0$ for all i, so that $\boldsymbol{\mu} > \mathbf{0}$, then $\rho((\text{diag}(\boldsymbol{\mu})^{-1}\mathbf{\Lambda}) = \rho(\mathbf{\Gamma}^*)$.

Let $\mathbf{v}^* > \mathbf{0}$ be the right eigenvector of $\mathbf{\Gamma}^*$ corresponding to $\rho(\mathbf{\Gamma}^*)$ and define $\mathbf{v}' = (v_1^* \mathbf{1}', ..., v_n^* \mathbf{1}')$ where $v_i^* = \{\mathbf{v}^*\}_i$ and the unit vectors are each of length N. Then

$$(\mathrm{diag}(\boldsymbol{\mu}))^{-1}\boldsymbol{\Lambda}\mathbf{v} = \begin{pmatrix} \frac{1}{\mu_1^*}\boldsymbol{\Lambda}_{11} & \frac{1}{\mu_1^*}\boldsymbol{\Lambda}_{12} & \cdots & \frac{1}{\mu_1^*}\boldsymbol{\Lambda}_{1n} \\ \frac{1}{\mu_2^*}\boldsymbol{\Lambda}_{21} & \frac{1}{\mu_2^*}\boldsymbol{\Lambda}_{22} & \cdots & \frac{1}{\mu_2^*}\boldsymbol{\Lambda}_{2n} \\ \vdots & \vdots & \ddots & \vdots \\ \frac{1}{\mu_n^*}\boldsymbol{\Lambda}_{n1} & \frac{1}{\mu_n^*}\boldsymbol{\Lambda}_{n2} & \cdots & \frac{1}{\mu_n^*}\boldsymbol{\Lambda}_{nn} \end{pmatrix} \begin{pmatrix} v_1^*\mathbf{1} \\ v_2^*\mathbf{1} \\ \vdots \\ v_n^*\mathbf{1} \end{pmatrix}$$

$$= \begin{pmatrix} \frac{1}{\mu_1^*}\sum_{j=1}^n v_j^*\boldsymbol{\Lambda}_{1j}\mathbf{1} \\ \frac{1}{\mu_2^*}\sum_{j=1}^n v_j^*\boldsymbol{\Lambda}_{2j}\mathbf{1} \\ \vdots \\ \frac{1}{\mu_n^*}\sum_{j=1}^n v_j^*\boldsymbol{\Lambda}_{nj}\mathbf{1} \end{pmatrix} = \begin{pmatrix} \frac{1}{\mu_1^*}\sum_{j=1}^n \sigma_j^*\lambda_{1j}^* v_j^*\mathbf{1} \\ \frac{1}{\mu_2^*}\sum_{j=1}^n \sigma_j^*\lambda_{2j}^* v_j^*\mathbf{1} \\ \vdots \\ \frac{1}{\mu_n^*}\sum_{j=1}^n \sigma_j^*\lambda_{nj}^* v_j^*\mathbf{1} \end{pmatrix}$$

$$= \begin{pmatrix} \sum_{j=1}^n v_j^*\{\boldsymbol{\Gamma}^*\}_{1j}\mathbf{1} \\ \sum_{j=1}^n v_j^*\{\boldsymbol{\Gamma}^*\}_{2j}\mathbf{1} \\ \vdots \\ \sum_{j=1}^n v_j^*\{\boldsymbol{\Gamma}^*\}_{nj}\mathbf{1} \end{pmatrix} = \begin{pmatrix} \rho(\boldsymbol{\Gamma}^*)v_1^*\mathbf{1} \\ \rho(\boldsymbol{\Gamma}^*)v_2^*\mathbf{1} \\ \vdots \\ \rho(\boldsymbol{\Gamma}^*)v_n^*\mathbf{1} \end{pmatrix} = \rho(\boldsymbol{\Gamma}^*)\mathbf{v}.$$

Hence $\mathbf{v} > \mathbf{0}$ is the eigenvector of $(\mathrm{diag}(\boldsymbol{\mu}))^{-1}\boldsymbol{\Lambda}$ corresponding to the eigenvalue $\rho(\boldsymbol{\Gamma}^*)$ and therefore, from Theorem A.2 part 7, $\rho((\mathrm{diag}(\boldsymbol{\mu}))^{-1}\boldsymbol{\Lambda}) = \rho(\boldsymbol{\Gamma}^*)$.

Hence the conditions that $\mu_i^* > 0$ for all i and $\rho(\boldsymbol{\Gamma}^*) \leq 1$ imply that $\boldsymbol{\mu} > \mathbf{0}$ and $\rho((\mathrm{diag}(\boldsymbol{\mu}))^{-1}\boldsymbol{\Lambda}) \leq 1$. Thus, from Theorem 8.3, $y_i(r,t) \to 0$ as $t \to \infty$ for all i and r and so the infection eventually dies out.

Now consider the remaining cases. When $\mu_i^* = 0$ for some i then $\boldsymbol{\mu}$ has at least one zero entry. When $\mu_i^* > 0$ for all i and $\rho(\boldsymbol{\Gamma}^*) > 1$ this implies that $\boldsymbol{\mu} > \mathbf{0}$ and $\rho((\mathrm{diag}(\boldsymbol{\mu}))^{-1}\boldsymbol{\Lambda}) > 1$. Thus it follows from Theorem 8.4 that $\mathbf{y}(t) \to \boldsymbol{\zeta}$ as $t \to \infty$. It only remains to show that

$$\boldsymbol{\zeta} = \begin{pmatrix} \zeta_1^*\mathbf{1} \\ \vdots \\ \zeta_n^*\mathbf{1} \end{pmatrix},$$

where $\zeta_1^*, ..., \zeta_n^*$ satisfy the equation

$$(8.16) \qquad (1-\zeta_i^*)\sum_{j=1}^n \sigma_j^*\lambda_{ij}^*\zeta_j^* = \mu_i^*\zeta_i^*, \ (i = 1, ..., n).$$

Let $\mathbf{y}' = (\zeta_1^*\mathbf{1}', ..., \zeta_n^*\mathbf{1}')$, with the unit vectors all of length N, where $\zeta_1^*, ..., \zeta_n^*$ are all positive and satisfy equations (8.16). Then

$$\mathrm{diag}(\mathbf{1}-\mathbf{y})\mathbf{\Lambda y} = \begin{pmatrix} (1-\zeta_1^*)\mathbf{\Lambda}_{11} & (1-\zeta_1^*)\mathbf{\Lambda}_{12} & \dots & (1-\zeta_1^*)\mathbf{\Lambda}_{1n} \\ \vdots & \vdots & \ddots & \vdots \\ (1-\zeta_n^*)\mathbf{\Lambda}_{n1} & (1-\zeta_n^*)\mathbf{\Lambda}_{n2} & \dots & (1-\zeta_n^*)\mathbf{\Lambda}_{nn} \end{pmatrix} \begin{pmatrix} \zeta_1^*\mathbf{1} \\ \vdots \\ \zeta_n^*\mathbf{1} \end{pmatrix}$$

$$= \begin{pmatrix} (1-\zeta_1^*)\sum_{j=1}^n \zeta_j^*\mathbf{\Lambda}_{1j}\mathbf{1} \\ \vdots \\ (1-\zeta_n^*)\sum_{j=1}^n \zeta_j^*\mathbf{\Lambda}_{nj}\mathbf{1} \end{pmatrix} = \begin{pmatrix} (1-\zeta_1^*)\sum_{j=1}^n \sigma_j^*\lambda_{ij}^*\zeta_j^*\mathbf{1} \\ \vdots \\ (1-\zeta_n^*)\sum_{j=1}^n \sigma_j^*\lambda_{ij}^*\zeta_j^*\mathbf{1} \end{pmatrix}.$$

Substituting in from equation (8.16), we therefore obtain the result that

$$\mathrm{diag}(\mathbf{1}-\mathbf{y})\mathbf{\Lambda y} = \begin{pmatrix} \mu_1^*\zeta_1^*\mathbf{1} \\ \vdots \\ \mu_n^*\zeta_n^*\mathbf{1} \end{pmatrix} = \mathrm{diag}(\boldsymbol{\mu})\mathbf{y}.$$

Hence $\mathbf{y}' = (\zeta_1^*\mathbf{1}', ..., \zeta_n^*\mathbf{1}')$ satisfies equation (8.14). There is a unique positive solution to this equation and therefore $(\boldsymbol{\zeta})' = (\zeta_1^*\mathbf{1}', ..., \zeta_n^*\mathbf{1}')$. This completes the proof of the theorem.

□

Note that Theorem 8.5 says that the endemic equilibrium, when it exists, is globally asymptotically stable, and that this equilibrium is the same at all sites. The common value at all sites for type i is just ζ_i^*, the endemic equilibrium for a non-spatial epidemic with the following parameters. For type i, the population size is σ_i^* and the combined death/emigration/recovery rate is μ_i^*. The infection rate of a type i susceptible by a type j infective is λ_{ij}^*.

8.8 Saddle point results for spatial models

In this section we consider spatial versions in $\Re^N$ and Z^N of the $S \to I \to S$ model of Section 8.2 in both closed systems and open systems with constant population sizes. The saddle point results of Chapter 7 then enable us to write down the speed of spread of the forward front of this epidemic.

First consider an epidemic in $\Re^N$. Let $y_i(\mathbf{s}, t)$ be the proportion of individuals in the ith population at position $\mathbf{s}$ who were infectious at time t. The equations for an epidemic which starts with a small amount of infection in a compact region amongst the n populations of types at time $t = 0$, are given by

$$\frac{\partial y_i(\mathbf{s},t)}{\partial t} = (1-y_i(\mathbf{s},t))\sum_{j=1}^n \sigma_j\lambda_{ij}\int_{R^N} y_j(\mathbf{s}-r,t)p_{ij}(\mathbf{r})d\mathbf{r} - \mu_i y_i(\mathbf{s},t), \quad (i=1,...,n).$$

Here μ_i is the combined death/emigration/recovery rate for infectious individuals, λ_{ij} is the infection rate of a type i susceptible by a type j infectious individual and $p_{ij}(\mathbf{r})$ is the corresponding contact distribution. Each $y_i(\mathbf{s}, 0)$ has compact support.

We consider the speed of first spread of the epidemic in the forward region. In the forward region we obtain the approximate equations

$$(8.17) \qquad \frac{\partial y_i(\mathbf{s},t)}{\partial t} = \sum_{j=1}^{n} \sigma_j \lambda_{ij} \int_{\Re^N} y_j(\mathbf{s}-r,t) p_{ij}(\mathbf{r}) d\mathbf{r} - \mu_i y_i(\mathbf{s},t), \quad (i=1,...,n).$$

Now consider an epidemic on the integer lattice Z^N. The parameters are the same, but the contact distribution is now discrete. Also we only consider the speed of spread in the direction of one of the N co-ordinate axes. The approximate equations in the forward region are just

$$(8.18) \qquad \frac{\partial y_i(\mathbf{s},t)}{\partial t} = \sum_{j=1}^{n} \sigma_j \lambda_{ij} \sum_{\mathbf{r} \in Z^N} y_j(\mathbf{s}-r,t) p_{ij}(\mathbf{r}) - \mu_i y_i(\mathbf{s},t), \quad (i=1,...,n).$$

The saddle point results of chapter 7 can now be applied to write down the speed of first spread based on either equations (8.17) or (8.18). To use the saddle point method some constraints on the contact distributions and initial infection are needed, which are specified in Chapter 7. In particular it is necessary that the projection of each contact distribution in the specified direction is exponentially dominated in the forward tail. Let Δ be the minimum of the abscissae of convergence of the Laplace transforms of these projected contact distributions. Then the speed of first spread is given by $c = \inf_{\theta \in (0,\Delta)} \frac{\rho(\mathbf{K}(\theta)) - \mu}{\theta}$, where $\mu = \max(\mu_1, ..., \mu_n)$, $\boldsymbol{\mu} = (\mu_i)$, $\mathbf{K}(\theta) = (\sigma_j \lambda_{ij} P_{ij}(\theta)) + \mu \mathbf{I} - \text{diag}(\boldsymbol{\mu})$ and $P_{ij}(\theta)$ is the Laplace transform of the marginal distribution of $p_{ij}(\mathbf{r})$ in the appropriate direction. The same expression for c is obtained for the epidemic in $\Re^N$ or on the integer lattice Z^N.

Note that the speed is only obtained if, in the expression for c, the infimum is achieved for some $\theta \in (0, \Delta)$ and is positive. In particular this can only occur provided $\rho(\mathbf{K}(0)) > \mu$.

When $\boldsymbol{\mu} > \mathbf{0}$ this condition can be shown (as in Chapter 7, Section 7.4) to be equivalent to the condition $\rho(\mathbf{\Gamma}) > 1$, where $\mathbf{\Gamma} = (\text{diag}(\boldsymbol{\mu}))^{-1} \mathbf{\Lambda}$. When $\mu_i = 0$ for some i, then $\{\mathbf{K}(0)\}_{ii} \geq \mu$. If $n > 1$ then, from Theorem A.2 part 5, $\rho(\mathbf{K}(0)) > \{\mathbf{K}(0)\}_{ii} \geq \mu$. If $n = 1$ then $\mu = 0$ and $\rho(\mathbf{K}(0)) = \{\mathbf{K}(0)\}_{11} > 0 = \mu$. Therefore a necessary condition to use the saddle point method to obtain the positive speed of first spread of infection is that either $\mu_i = 0$ for some i, or $\boldsymbol{\mu} > \mathbf{0}$ and $\rho(\mathbf{\Gamma}) > 1$. This is precisely the condition for global asymptotic convergence to an endemic equilibrium for a non-spatial epidemic with populations sizes σ_i, infection rates λ_{ij} and combined emigration/recovery/removal rates μ_i.

The saddle point method can also be used to obtain the speed of first spread of infection in spatial models of open systems when birth and death rates and immigration and emigration balances for susceptible individuals, but infectious individuals have a higher death than birth rate. In the forward front, far from the initial focus of infection, the population size for type i individuals will be approximately constant. Hence the same approximate equations as (8.17) or (8.18) are obtained. If individuals are always born susceptible, then μ_i represents the combined death/emigration/recovery rate for type i infectious individuals. However if the offspring of infectious individuals of type i have probability p_i of being born with the disease, then μ_i is the difference between the combined death/emigration/recovery rate for type i infectious individuals and p_i times the birth rate for these individuals.

The model can be extended to include a latent state and/or a removed state, and leads to the $S \to L \to I \to S$, $S \to I \to R \to S$ and $S \to L \to I \to R \to S$ models. These are the obvious extensions of the $S \to L \to I$, $S \to I \to R$ and $S \to L \to I \to R$ models of Chapter 7. The speed of first spread can be found in an identical manner to the method used in Chapter 7 Section 7.5. For a detailed discussion see Radcliffe and Rass [R11]. That paper also treats spatial epidemic models which allow for sensitive and resistant strains of infection; the equivalent non-spatial model having being analysed in Pinsky and Shonkwiler [P1].

There are at present no exact results for the asymptotic speed of propagation of infection in models which allow a return to the susceptible state.

CHAPTER 9

Contact branching processes

9.1 Introduction

A link exists between stochastic contact birth processes and the spatial form of the deterministic simple $S \to I$ epidemic. Suppose a simple contact birth process, where all individuals are of the same type, starts at time zero with one individual at the origin. After an exponential time it gives rise to a new individual at a random displacement $\mathbf{S}$ relative to the original individual. Individuals act independently. Let $U(t)$ be the random variable giving the position of furthest spread in a specified direction by time t and define $y(s,t) = P(U(t) > s)$. Then it is known that $y(s,t)$ satisfies the equation for the one dimensional spatial $S \to I$ epidemic. However the initial conditions differ, so that the corresponding integral equation for $w(s,t) = -\log(1 - y(s,t))$ is only obtained for $s \geq 0$.

Exact methods developed for epidemic models can be adapted to provide results for the contact birth process model. The asymptotic speed of translation of the distribution function of furthest spread in the contact birth process is obtained, and is shown to be the minimum wave speed c_0 for the corresponding $S \to I$ epidemic. It is also shown that $U(t)/t$ converges in probability to c_0.

A connection also exists between the multi-type contact birth process and the multi-type $S \to I$ epidemic. Exact methods can be used for this process. $U(t)$ can now represent the random variable of furthest spread in a specific direction for a specified type, or for all types. Again results concerning the asymptotic speed of translation and convergence in probability are obtained.

Convergence results had been obtained previously by Mollison [M6] and Biggins [B8] for the one-type contact birth process using probabilistic methods. The almost sure convergence of $U(t)/t$ to c_0 was proved, which implies the weaker convergence in probability result. It may well be possible to use probabilistic methods for multi-type models to obtain similar (or stronger) results to the ones derived in this chapter, although at present such results have not been obtained. Note however that probabilistic methods used for contact birth and branching processes cannot be used to obtain results for deterministic epidemic models without imposing unrealistic initial conditions. For the one-type contact birth process, $y(s,t)$ is a monotone function of s for all t, and $y(s,0) = 1$ for $s < 0$, although the origin of course may be moved. For an epidemic model this would require the epidemic to be on the real line $\Re$, with initially everyone already infected to the left of a specified position.

Section 9.5 considers contact birth-death processes and shows a connection with the spatial $S \to I \to S$ epidemic. In Section 9.6 branching process extensions are discussed. Exact analytic results have not yet been obtained for spatial models with return to the susceptible state. However the saddle point method has been used to obtain the speed of first spread of infection. The saddle point method therefore can be used for these contact birth-death-branching process models to obtain an expression for the speed of translation in a given direction of the extreme of the distribution of furthest spread.

9.2 The simple contact birth process and the McKean connection

In this section the one type contact birth model is described and the McKean connection with the $S \to I$ epidemic is derived. This connection was pointed out by Mollison [M6,M7]. The name comes from a similar connection between the equation for the spread of a gene in a population process and the point of furthest spread in a branching diffusion process, which was first observed by McKean [M1].

Consider a contact birth process in $\Re^N$. A simple contact birth process consists of only one type of individual and individuals act independently. The process commences at time $t = 0$ with a single individual at position $\mathbf{0}$. At a random time T which has a probability density function $\lambda e^{-\lambda t}$, for $0 \leq t < \infty$, each individual gives rise to a further individual at a position $\mathbf{S}$ from the parent, where $\mathbf{S}$ is a vector of random variables. The distribution of $\mathbf{S}$ is referred to as the contact distribution. Let U(t) be the position of the furthest individual from position $\mathbf{0}$ at time t in a given direction. We look at the behaviour of $P(U(t) > s)$.

Define $p(s)$ to be the density function of the projection of the contact distribution in the specified direction. Also let $y(s,t) = P(U(t) > s)$. Consider the time T until the first birth occurs. If $T > t$, then at time t there is still only the original individual at the origin. If the birth occurs at time $\tau < t$, then at time t the furthest spread of individuals is the maximum of the spread from two individuals after a further time $t - \tau$. The first of these individuals is the original individual at the origin, so the probability that the process it generates spreads beyond position s in time $t-\tau$ is $y(s, t-\tau)$. The second individual is the first offspring, which is located at position r in the specified direction (with density $p(r)$). The probability that the process the second individual generates spreads beyond a further $s - r$ in time $t - \tau$ is then $y(s - r, t - \tau)$. So for given τ and r, the probability that the process spreads beyond position s by time t is $1 - (1 - y(s, t - \tau))(1 - y(s - r, t - \tau))$.

By conditioning on the time of first birth, and using the ideas in the previous paragraph, equations can easily be obtained for $y(s,t)$. Therefore

$$y(s,t) = e^{-\lambda t}y(s,0)+\int_0^t \lambda e^{-\lambda\tau}\int_{-\infty}^{\infty} p(r)(1-(1-y(s,t-\tau))(1-y(s-r,t-\tau)))drd\tau.$$

Hence, scaling by $e^{\lambda t}$ and changing the variable in the integral to $u = t - \tau$,

$$y(s,t)e^{\lambda t} = y(s,0) + \int_0^t \lambda e^{\lambda u}\int_{-\infty}^{\infty} p(r)(1 - (1 - y(s,u))(1 - y(s - r,u)))drdu.$$

Differentiating with respect to t we obtain

$$\lambda e^{\lambda t}y(s,t) + e^{\lambda t}\frac{\partial y(s,t)}{\partial t} = \lambda e^{\lambda t}\int_{-\infty}^{\infty} p(r)(1 - (1 - y(s,t))(1 - y(s - r,t)))dr.$$

Since $p(r)$ is a density function, rearrangement of the terms gives

$$\frac{\partial y(s,t)}{\partial t} = \lambda(1-y(s,t))\int_{-\infty}^{\infty} p(r)y(s-r,t)dr. \tag{9.1}$$

Consider equations (3.1) when $n = 1$ and $N = 1$, and drop the subscripts. For an $S \to I$ epidemic $\lambda(\tau) = \lambda$ for all $\tau \geq 0$ and $h(s,t) = \sigma\lambda^*\varepsilon * p^*(s)$. Then the equation for $y(s,t) = \int_0^t I(s,t,\tau)d\tau = 1 - x(s,t)$ is almost the same as equation (9.1), except for the additional term $h(s,t)$ which reflects the effect of the infectives from outside.

An alternative model was given in Chapter 7, Section 7.3, for a one-type $S \to I \to R$ model in $\Re^N$. Here infection is not introduced into an infection-free population at time zero by the introduction of infectives from outside the population. Instead we consider a population which has some infection at time zero. We can consider how infection, which is confined to a bounded region at time zero, spreads over time. Take $N = 1$, $\mu = 0$, $\sigma = 1$. Then the equation for the proportion of individuals at position s who are infected at time t, $y(s,0)$, is identical to equation (9.1). This establishes the McKean connection between the contact birth and the $S \to I$ epidemic processes.

For the contact birth process, $y(s,0) = 0$ for $s \geq 0$ and $y(s,0) = 1$ for all $s < 0$. Since the initial individual does not die it is also true that $y(s,t) = 1$ for all $s < 0$ for every $t \geq 0$. For the model for the $S \to I$ epidemic to give the same initial conditions would therefore require everyone to be infected at time zero to the left of the origin, with no infection elsewhere.

The exact analysis of the epidemic model of Chapters 3 and 5 (which include the $S \to I$ epidemic as a special case) is based on the model when the epidemic is initiated by the introduction of infection from outside at time zero. Specifically the analysis is based on an equation for $w(s,t) = -\log(1-y(s,t))$. For the contact birth process, $w(s,t)$ will be infinite when $s < 0$. Equation (9.1) can only be integrated to give a integro-differential equation for $w(s,t)$ when $s \geq 0$. Then for the contact birth process, for $s \geq 0$ we obtain

$$w(s,t) = \lambda\int_0^t\int_{-\infty}^{s} p(r)\left(1-e^{-w(s-r,\tau)}\right)drd\tau + \lambda t\int_s^{\infty} p(r)dr. \tag{9.2}$$

When $s < 0$, for all $t \geq 0$ $y(s,t) = 1$ and hence $w(s,t)$ is infinite.

The corresponding equation for the one-type $S \to I$ epidemic on the line when $\sigma = 1$ is easily obtained from either of equations (3.1) or (5.1), both of which are for the n-type model with varying infectivity. The equation for $w(s,t)$ for this $S \to I$ model is

$$w(s,t) = \lambda\int_0^t\int_{-\infty}^{\infty} p(r)\left(1-e^{-w(s-r,\tau)}\right)drd\tau + H(s,t),$$

where $H(s,t) = \displaystyle\int_0^t\int_{-\infty}^{\infty}\varepsilon(r)p^*(s-r)\lambda^* e^{-\mu^* u}drdu.$

Since equation (9.2) only holds for $s > 0$, the first integral over r has restricted range and $H(s,t) = \lambda t\int_{-\infty}^{s} p(r)dr$ is not uniformly bounded, the proofs of Chapter

5 do not immediately give the speed of translation for the contact birth process. Some adaptation of the proofs is required.

9.3 The multi-type contact birth process

An n-type version of a contact birth process can also be formulated. The time until a birth event occurs still has an exponential distribution but its parameter will now depend upon the type of individual. When such an event occurs to a type i individual, then there is an associated probability that it will give rise to a type j individual. For this multi-type system $U(t)$ can either be the random variable giving the position of furthest spread of any individual at time t, or giving the position of furthest spread of a specific type of individual at time t. The distribution of $U(t)$ will also depend upon the initial conditions.

Let $y_i(s,t)$ represent the probability that $U(t) > s$ given that the process starts with one type i individual at position $\mathbf{0}$ at time $t = 0$. Then $y_i(s,t)$, $i = 1, ..., n$, can be shown to satisfy the equations obtained for the spatial n-type $S \to I$ deterministic epidemic.

The contact birth process is set up for the case when $U(t)$ gives the position of furthest spread of all types in a specified direction. Let the probability that a type i individual gives birth to an offspring in a time interval $(t, t+\delta t)$ be $\alpha_i \delta t + o(\delta t)$. The probability that this offspring is of type j is q_{ij}, where $\sum_{j=1}^{n} q_{ij} = 1$. The position of such a new individual relative to the parent has contact distribution $p^*_{ij}(\mathbf{s})$ with $p_{ij}(s)$ the marginal contact distribution in the direction of interest.

The time T until the first birth has probability density function $\alpha_i e^{-\alpha_i t}$, $0 \leq t < \infty$. Let $\lambda_{ij} = \alpha_i q_{ij}$. Note that $\alpha_i = \sum_{j=1}^{n} \lambda_{ij}$. Then $y_i(s,t) \equiv 1$ for $s < 0$ and $t \geq 0$. For $s \geq 0$, conditional arguments similar to those for the one-type model of Section 9.2 lead to the following equations for $y_i(s,t)$,

$$(1 - y_i(s,t)) = P(T > t)(1 - y_i(s,0)) + \int_0^t \alpha_i e^{-\alpha_i \tau}(1 - y_i(s, t-\tau)) \sum_{j=1}^{n} q_{ij} \int_{-\infty}^{\infty} (1 - y_j(s-r, t-\tau)) p_{ij}(r) dr d\tau.$$

Thus

$$(1 - y_i(s,t))e^{\alpha_i t} = (1 - y_i(s,0)) + \int_0^t \alpha_i e^{\alpha_i \theta}(1 - y_i(s,\theta)) \sum_{j=1}^{n} \frac{\lambda_{ij}}{\alpha_i} \int_{-\infty}^{\infty} (1 - y_j(s-r,\theta)) p_{ij}(r) dr d\theta. \tag{9.3}$$

Since the $p_{ij}(r)$ are continuous and integrable and $y_i(s,0)$ is continuous for $s \geq 0$, it is easy to show that $y_i(s,t)$ is a continuous function for $s \geq 0$ and $t \geq 0$ and that it is partially differentiable with respect to t. Differentiating equations (9.3) with respect to t, for $s \geq 0$, gives

$$\frac{\partial y_i(s,t)}{\partial t} = (1 - y_i(s,t)) \sum_{j=1}^{n} \lambda_{ij} \int_{-\infty}^{\infty} y_j(s-r, t) p_{ij}(r) dr, \tag{9.4}$$

for $i = 1, ..., n$. This equation also holds for $s < 0$ since then $y_i(s,t) = 0$ so that the right and left hand sides of equations (9.4) are both zero.

Consider the generalisation of the one-type $S \to I \to R$ epidemic of Section 7.3. Let $N = 1$, $\mu_i = 0$ and $x_i(s,t) = 1 - y_i(s,t)$ for all i, so that the model describes the $S \to I$ epidemic on the line $\Re$. If in addition we let $\sigma_i = 1$, then the equation for $y_i(s,t)$ for this epidemic is the same as equations (9.4). This establishes the McKean connection between the n-type contact birth and $S \to I$ spatial epidemic processes.

Equations (9.4) can be rewritten in the form

$$\frac{\partial y_i(s,t)}{\partial t} = (1 - y_i(s,t)) \sum_{j=1}^{n} \lambda_{ij} \left[\int_{-\infty}^{s} y_j(s-r,t) p_{ij}(r) dr + \int_{s}^{\infty} p_{ij}(r) dr \right], \tag{9.5}$$

for $i = 1, ..., n$. Joint continuity of, and the uniform bound for, the partial derivative of $w_i(s,t) = -\log(1 - y_i(s,t))$ with respect to t is immediate. An equivalent equation for $w_i(s,t)$ can then be obtained for $s \geq 0$. When $s < 0$ then $w_i(s,t)$ is infinite. This leads to the equations

$$\begin{aligned} w_i(s,t) &= \sum_{j=1}^{n} \lambda_{ij} \int_0^t \left(\int_{-\infty}^{s} (1 - e^{-w_j(s-r,\tau)}) p_{ij}(r) dr + \int_{s}^{\infty} p_{ij}(r) dr \right) d\tau \\ &= \sum_{j=1}^{n} \lambda_{ij} \int_0^t \int_{-\infty}^{s} (1 - e^{-w_j(s-r,\tau)}) p_{ij}(r) dr d\tau + H_i(s,t), \end{aligned} \tag{9.6}$$

for $s \geq 0$ and $i = 1, ..., n$, where $H_i(s,t) = t \sum_{j=1}^{n} \lambda_{ij} \int_s^\infty p_{ij}(r) dr$.

Equations (9.6) for the contact birth process differ from those for the $S \to I$ epidemic since they only hold for $s \geq 0$. Also conditions imposed on the initial infection for the epidemic model constrained $H_i(s,t)$ to be uniformly bounded over all s and t. For the contact birth process $H_i(s,t)$ is not uniformly bounded. There is also some simplicity for the contact birth process since each $H_i(s,t)$ is monotone in s, as also is each $w_i(s,t)$. This is not true for the $S \to I$ epidemic.

Note that if we consider the furthest spread of a specific type k, rather than any type, then equations may be set up in a similar manner. However, $y_i(s,0) = 0$ for all s if $i \neq k$; and $y_k(s,t) \equiv 1$ for $s < 0$ and $y_k(s,0) = 0$ for $s > 0$. Equations (9.4) then hold for all i and s. Again the McKean connection with the spatial deterministic $S \to I$ epidemic is obtained.

An equation for $w_i(s,t) = -\log(1 - y_i(s,t))$ is now obtained for all s when $i \neq k$. However when $i = k$ the equation is only obtained for $s \geq 0$. When $s < 0$ then $w_k(s,t)$ is infinite. The equivalent equations to (9.6) are then

$$
\begin{aligned}
w_i(s,t) &= \sum_{j\neq k} \lambda_{ij} \int_0^t \int_{-\infty}^{\infty} (1 - e^{-w_j(s-r,\tau)}) p_{ij}(r)\,dr\,d\tau \\
&\quad + \lambda_{ik} \int_0^t \left(\int_{-\infty}^{s} (1 - e^{-w_k(s-r,\tau)}) p_{ik}(r)\,dr + \int_s^{\infty} p_{ik}(r)\,dr \right) d\tau \\
&= \sum_{j\neq k} \lambda_{ij} \int_0^t \int_{-\infty}^{\infty} (1 - e^{-w_j(s-r,\tau)}) p_{ij}(r)\,dr\,d\tau \\
&\quad + \lambda_{ik} \int_0^t \int_{-\infty}^{s} (1 - e^{-w_k(s-r,\tau)}) p_{ik}(r)\,dr\,d\tau + H_i(s,t),
\end{aligned}
\tag{9.7}
$$

for all s if $i \neq k$ but only for $s \geq 0$ when $i = k$. Here $H_i(s,t) = t\lambda_{ik} \int_s^{\infty} p_{ik}(r)dr$.

For both definitions of $U(t)$ the exact methodology developed for the spatial epidemic models of Chapter 3 can be adapted to provide results for the n-type contact birth process. These results are obtained in Section 9.4. We need to use certain properties of the function $g(x) = 1 - e^{-x}$, which are easily established, namely:

1. $0 \leq g(x) \leq x$ for $0 \leq x \leq 1$;
2. $|g(x) - g(y)| \leq |x - y|$ for all $0 \leq x, y \leq 1$.

It is interesting to note that when considering one type spatial models in genetics, equations similar to equations (9.7) arise with $n = 1$ but with $g(x) = 1 - e^{-x}$ replaced by a different function (see Diekmann and Kaper [D9] and Lui [L2,L3]).

9.4 Exact analytic results

A rigorous analytic method was used in Chapter 5 to obtain the speed of propagation of infection for a generalisation of the $S \to I$ epidemic which allowed for the possibility of varying infectivity. The method of proof required the contact distributions to be radially symmetric. This method can be adapted to give a rigorous derivation of the speed of propagation of $y_i(s,t)$ for the contact birth process described in Section 9.3 in the non-reducible case with the projections of the contact distributions symmetric about zero. Note that the contact birth process is non-reducible if $\mathbf{\Lambda}$ is non-reducible, where $\{\mathbf{\Lambda}\}_{ij} = \lambda_{ij} = \alpha_i q_{ij}$. This speed is shown to be c_0, the corresponding speed of propagation of infection and least velocity for which wave solutions exist in the $S \to I$ epidemic if the parameter σ_j is taken to be one for all j and the contact distribution for the epidemic is just the appropriate projection of the contact distribution for the contact-birth process.

The result is not immediate since the equations for $w_i(s,t)$ are not valid for all s, the range of the first spatial integral is restricted and the functions $H_i(s,t)$ are not uniformly bounded. The speed of propagation is also no longer defined in terms of $|s|$. This speed is referred to as the speed of translation of the forward tail of the distribution, rather than the speed of propagation.

Consider first what is meant by the asymptotic speed of translation of the forward tail of the distribution of $U(t)$, where initially there is one type i individual at position zero. For the contact birth process, unlike the epidemic, we need to consider the speed of spread in the positive direction only. Note also that $y_i(s,t) = P(U(t) > s)$ is monotone in s for each t, so that $\sup_{s\geq ct} y_i(s,t)$ and $\inf_{s\leq ct} y_i(s,t)$

are both equal to $y_i(ct,t)$. The conditions for the speed of translation of $y_i(s,t)$ to be c^* are therefore,

1. for any $c > c^*$, $\lim_{t\to\infty} P((U(t)/t) > c) = \lim_{t\to\infty} y_i(ct,t) = 0$;
2. for each c such that $0 < c < c^*$ there exists an $\varepsilon > 0$ and $T > 0$ such that $P((U(t)/t) > c) = y_i(ct,t) \geq \varepsilon$ for all $t \geq T$.

In terms of $w_i(s,t)$, to prove that the speed of translation is c^* we need to show that

1. for any $c > c^*$, $\lim_{t\to\infty} w_i(ct,t) = 0$;
2. for each c such that $0 < c < c^*$ there exists an $\varepsilon > 0$ and $T > 0$ such that $w_i(ct,t) \geq \varepsilon$ for all $t \geq T$.

It is shown that these results hold when the contact distributions are exponentially dominated in the forward tail if we take $c^* = c_0$, the minimum wave speed of the corresponding $S \to I$ epidemic. Therefore the asymptotic speed of translation of the forward tail of the distribution of furthest spread of the contact birth process is c_0.

The result for part 2 is then strengthened. It is shown that for any $c > c_0$ $w_i(ct,t) \to \infty$, and hence $y_i(ct,t) \to 1$, as $t \to \infty$. Together with the result from part 1 this shows that $\lim_{t\to\infty} P((U(t)/t) > c) = 0$ for $c > c_0$ and $\lim_{t\to\infty} P((U(t)/t) > c) = 1$ for $c < c_0$. It is then simple to deduce that $U(t)/t$ converges in probability to c_0.

When at least one contact distribution is not exponentially dominated in the forward tail, the result can be obtained that $\lim_{t\to\infty} P((U(t)/t) > c) = 1$ for any $c > 0$. Then the asymptotic speed of translation may be considered to be infinite.

The proofs are based on equations (9.6) and (9.7) which respectively describe the spread of all types and the spread of a specific type k. The projection of each contact distribution is assumed to be symmetric about zero and, for most of this section, is also assumed to be exponentially dominated in the forward tail. At the end of the section Theorem 9.5 gives the result for the case when at least one contact distribution is not exponentially dominated in the tail.

The same definitions are used for $\mathbf{V}(\lambda)$, Δ_v and $K_c(\lambda)$ as for the epidemic processes, where the value of c has been suppressed in the notation for $\mathbf{V}(\lambda)$. For these branching process models $\{\mathbf{V}(\lambda)\}_{ij} = \lambda_{ij} P_{ij}(\lambda)/(c\lambda)$, Δ_v is the minimum of the abscissae of convergence of the $P_{ij}(\lambda)$ in the right hand half of the complex plane, and $K_c(\lambda) = \rho(\mathbf{V}(\lambda))$. Note that in Chapter 4 it was shown that for each $c > c_0$ there exists a λ such that $K_c(\lambda) \leq 1$. This is required in the proof of Theorem 9.1.

THEOREM 9.1. *When the $p_{ij}(r)$ are all exponentially dominated in the tail, for each of equations (9.6) and (9.7) there exists a monotone (in t) jointly continuous solution $w_i(s,t)$ to the specific system of equations which is unique. This solution is a monotone decreasing function of s and is partially differentiable with respect to t with the partial derivative jointly continuous and uniformly bounded. For any $c > c_0$, $\lim_{t\to\infty} w_i(ct,t) = 0$ for $i = 1, ..., n$.*

PROOF. The proof will be given for the contact birth process described by equations (9.6). For equations (9.7), the proof follows in almost identical fashion.

The existence, uniqueness, continuity and monotonicity can be shown as in Theorem 3.1. Partial differentiability with respect to t and the conditions on the partial derivatives are also easily established. It only remains to show that for any $c > c_0$, $\lim_{t\to\infty} w_i(ct,t) = 0$ for $i = 1, ..., n$. An alternative construction is used.

Consider any $c > c_0$ and choose d such that $c > d > c_0$. Then there exists a $\lambda \in (0, \Delta_v)$ such that $K_d(\lambda) < 1$. Define $x_i(s,t) = w_i(s,t)e^{\lambda(s-dt)}$. Rewriting equations (9.6) in terms of the $x_i(s,t)$ gives, for $s \geq 0$,

(9.8)

$$x_i(s,t) = \sum_{j=1}^{n} \lambda_{ij} \int_0^t \left[e^{\lambda(s-dt)} \int_s^\infty p_{ij}(r)dr \right.$$
$$\left. + \int_0^\infty e^{\lambda(r-d\tau)} \left(1 - e^{-x_j(r,\tau)e^{-\lambda(r-d\tau)}}\right) p_{ij}(s-r)e^{\lambda(s-r)}e^{-\lambda d(t-\tau)}dr \right] d\tau.$$

For $s \geq 0$ define

$$x_i^{(0)}(s,t) = \sum_{j=1}^{n} \lambda_{ij} \int_0^t e^{\lambda(s-dt)} \int_s^\infty p_{ij}(r)drd\tau = \sum_{j=1}^{n} \lambda_{ij} te^{-\lambda dt} e^{\lambda s} \int_s^\infty p_{ij}(r)dr.$$

Note that $x_i^{(0)}(s,t)$ is jointly continuous. Then for $s \geq 0$ and $m = 0, 1, 2, ...$ define

$$x_i^{(m+1)}(s,t) = \sum_{j=1}^{n} \lambda_{ij} \int_0^t \left[e^{\lambda(s-dt)} \int_s^\infty p_{ij}(r)dr \right.$$
$$\left. + \int_0^\infty e^{\lambda(r-d\tau)} \left(1 - e^{-x_j^{(m)}(r,\tau)e^{-\lambda(r-d\tau)}}\right) p_{ij}(s-r)e^{\lambda(s-r)}e^{-\lambda d(t-\tau)}dr \right] d\tau.$$

Now $te^{-\lambda dt}$ is uniformly bounded for all $t \geq 0$ and $e^{\lambda s}\int_s^\infty p_{ij}(r)dr \leq P_{ij}(\lambda)$ for $s \geq 0$ so is also uniformly bounded. Hence we can choose a scalar multiple $\mathbf{D}$ of the right eigenvector of $\mathbf{V}(\lambda)$ corresponding to $K_d(\lambda)$ such that $x_i^{(0)}(s,t) \leq \{\mathbf{D}\}_i$ for all $s \geq 0$ and $t \geq 0$. Define $u_i^{(m)} = \sup_{s\geq 0, t\geq 0} |x_i^{(m)}(s,t) - x_i^{(m-1)}(s,t)|$ and let $\{\mathbf{u}^{(m)}\}_i = u_i^{(m)}$. Using the properties of $g(x) = 1 - e^{-x}$ given at the end of Section 9.3, it is then easily seen that

$$|x_i^{(1)}(s,t) - x_i^{(0)}(s,t)|$$
$$= \sum_{j=1}^{n} \lambda_{ij} \int_0^t \int_0^\infty e^{\lambda(r-dt)} \left(1 - e^{-x_j^{(0)}(r,\tau)e^{-\lambda(r-d\tau)}}\right) p_{ij}(s-r)e^{\lambda(s-r)}e^{-\lambda d(t-\tau)}drd\tau$$
$$\leq \sum_{j=1}^{n} \lambda_{ij} \int_0^t \int_0^\infty e^{\lambda(r-dt)} x_j^{(0)}(r,\tau)e^{-\lambda(r-d\tau)} p_{ij}(s-r)e^{\lambda(s-r)}e^{-\lambda d(t-\tau)}drd\tau$$
$$\leq \sum_{j=1}^{n} \lambda_{ij} D_j \int_0^t \int_0^\infty p_{ij}(s-r)e^{\lambda(s-r)}e^{-\lambda d(t-\tau)}drd\tau$$
$$\leq \sum_{j=1}^{n} \lambda_{ij} D_j \frac{P_{ij}(\lambda)}{d\lambda} = \sum_{j=1}^{n} D_j \{\mathbf{V}(\lambda)\}_{ij},$$

so that $\mathbf{u}^{(1)} \le \mathbf{V}(\lambda)\mathbf{D} = K_d(\lambda)\mathbf{D}$. Note that here $\lambda_{ij} = \alpha_i q_{ij}$.

For $m \ge 1$, since the properties of $g(x) = 1 - e^{-x}$ imply that $|g(x) - g(y)| \le |x - y|$ for all non-negative x, y, we obtain

$$\begin{aligned}|x_i^{(m+1)}(s,t) - x_i^{(m)}(s,t)| &\le \sum_{j=1}^{n} \lambda_{ij} \int_0^t \int_0^\infty e^{\lambda(r-dt)} |(x_j^{(m)}(r,\tau) - x_j^{(m-1)}(r,\tau)| \\ &\quad \times e^{-\lambda(r-d\tau)} p_{ij}(s-r) e^{\lambda(s-r)} e^{-\lambda d(t-\tau)} dr d\tau \\ &\le \sum_{j=1}^{n} u_j^{(m)} \{\mathbf{V}(\lambda)\}_{ij}.\end{aligned}$$

Hence $\mathbf{u}^{(m+1)} \le \mathbf{V}(\lambda)\mathbf{u}^{(m)}$ for $m \ge 1$. Therefore $\mathbf{u}^{(m)} \le (K_d(\lambda))^m \mathbf{D}$ for all $m \ge 1$. Then it immediately follows that, for any integers $m > 0$ and $l \ge 0$ that

$$|x_i^{(l+m)}(s,t) - x_i^{(l)}(s,t)| \le \sum_{r=1}^{m} (K_d(\lambda))^{r+l} \{\mathbf{D}\}_i.$$

Joint continuity of the sequence of functions is easily shown. Since $K_d(\lambda) < 1$, using Cauchy convergence $x_i^{(m)}(s,t)$ converges uniformly for $t \ge 0$ and $s \ge 0$ to a jointly continuous limit $x_i(s,t)$.

It is simple to obtain a uniform upper bound for $x_i^{(m)}(s,t)$. From the result above with $l = 0$,

$$x_i^{(m)}(s,t) \le \sum_{r=0}^{m} (K_d(\lambda))^r \{\mathbf{D}\}_i \le \frac{\{\mathbf{D}\}_i}{1 - K_d(\lambda)}.$$

Since each $x_i^{(m)}(s,t)$ is non-negative and $1 - e^{-x} \le x$ for x non-negative, dominated convergence can be used to show that the constructed solution $x_i(s,t)$, $i = 1, ..., n$, satisfies equations (9.8).

If we now write $w_i(s,t) = x_i(s,t)e^{-\lambda(s-dt)}$, then $w_i(s,t)$ is a continuous solution to equations (9.6). The monotonicity of $w_i(s,t)$ in each of s and t can be established from the construction. It is simple to show iteratively for $m = 0, 1, 2, ...$ that $x_i^{(m)}(s,t)e^{-\lambda(s-dt)}$ is monotone increasing in t and monotone decreasing in s. This result therefore also holds for $w_i(s,t) = x_i(s,t)e^{-\lambda(s-dt)}$. Since there is a unique solution to equations (9.6) our constructed solution is this unique solution.

An upper bound is now obtained for $w_i(s,t)$. From the uniform upper bound on the $x_i^{(m)}(s,t)$ obtained in the construction, it immediately follows that $x_i(s,t) \le \{\mathbf{D}\}_i/(1 - K_d(\lambda))$. Therefore

$$w_i(s,t) \le \frac{D_i}{1 - K_d(\lambda)} e^{\lambda(dt-s)},$$

for all $s \ge 0$ and $t \ge 0$. Now $d < c$, so that

$$w_i(ct,t) \le \frac{D_i}{1 - K_d(\lambda)} e^{\lambda(dt-ct)} = \frac{D_i}{1 - K_d(\lambda)} e^{-\lambda t(c-d)}.$$

The right hand side of this inequality tends to zero as t tends to infinity. Hence for any $c > c_0$, $\lim_{t\to\infty} w_i(ct,t) = 0$ for $i = 1, ..., n$. This completes the proof of the theorem for the speed of spread of all types.

If we consider the speed of spread of a specific type k, the modifications to the proof are straightforward. Equation (9.7) can be written in terms of $x_i(s,t)$. The corresponding sequence $x_i^{(m)}(s,t)$ is now defined, for $s \geq 0$ if $i \neq k$ and for all s otherwise, as follows

$$x_i^{(0)}(s,t) = \lambda_{ik} \int_0^t e^{\lambda(s-dt)} \int_s^\infty p_{ik}(r) dr d\tau = \lambda_{ik} t e^{-\lambda dt} e^{\lambda s} \int_s^\infty p_{ik}(r) dr.$$

Also for $m = 0, 1, 2, ...,$

$$x_i^{(m+1)}(s,t) = \lambda_{ik} \int_0^t e^{\lambda(s-dt)} \int_s^\infty p_{ik}(r) dr d\tau$$
$$+ \lambda_{ik} \int_0^t \int_0^\infty e^{\lambda(r-d\tau)} \left(1 - e^{-x_k^{(m)}(r,\tau)e^{-\lambda(r-d\tau)}}\right) p_{ik}(s-r) e^{\lambda(s-r)} e^{-\lambda d(t-\tau)} dr d\tau$$
$$+ \sum_{j\neq k} \lambda_{ij} \int_0^t \int_{-\infty}^\infty e^{\lambda(r-d\tau)} \left(1 - e^{-x_j^{(m)}(r,\tau)e^{-\lambda(r-d\tau)}}\right) p_{ij}(s-r) e^{\lambda(s-r)} e^{-\lambda d(t-\tau)} dr d\tau.$$

The remainder of the proof follows in identical fashion to the proof for the spread of all types.

□

Theorem 9.1 shows that c_0 provides an upper bound for the speed of translation. In order to prove that c_0 also provides an lower bound, we first prove a stronger result for the contact birth process than was obtained in Lemma 5.6 for the epidemic processes. The proof uses the monotonicity of $w_i(s,t)$ in s for the contact birth process.

LEMMA 9.1. *For any $R > 0$ and $t_0 > 0$, $w_i(R, t_0) > 0$ and hence $w_i(s,t) > 0$ for $(s,t) \in B_R \times [t_0, \infty)$.*

PROOF. For both forms of contact birth process modelled by equations (9.6) and (9.7) there exists a k such that $w_k(s,t)$ is infinite for $s < 0$ and all t. Consider any i. Since the contact birth process is non-reducible there exists a sequence $j_0 = k, j_1, ..., j_r = i$ such that $\lambda_{i_x i_{x+1}} \neq 0$ for all $x = 0, ..., r-1$. A convolution of symmetric densities is a symmetric density. Hence the convolution of $p_{i_x i_{x+1}}(s)$, for $x = 0, ..., r$, is symmetric, so there exists an interval $[A, B] \in \Re_+$ with $0 < A < B$ on which this convolution density is positive. Hence, from equations (9.6) or (9.7), $w_i(B,t) > 0$ for all $t > 0$. Using monotonicity of $w_i(s,t)$ in s this implies that $w_i(s,t) > 0$ for all $s \leq B$ for all $t > 0$.

Again from the non-reducibility of the branching process there exists a sequence $j_0 = i, j_1, ..., j_m = i$ such that $\lambda_{i_x i_{x+1}} \neq 0$ for all $x = 0, ..., m-1$. Hence there exists an interval $[C, D] \in \Re_+$ with $0 < C < D$ for which this convolution density is positive. Hence $w_i(s,t) > 0$ for $s \leq B + D$ and all $t > 0$. This second step may be repeated, so that after r steps we obtain $w_i(s,t) > 0$ for $s \leq B + rD$. The result that $w_i(R, t_0) > 0$ is then obtained by taking $r > (R - B)/D$ and any $t_0 > 0$. The monotonicity of $w_i(s, t)$ in each of s and t then implies that $w_i(s, t) > 0$ for $(s, t) \in B_R \times [t_0, \infty)$.

□

A clever device is now used to enable the results of Chapter 5 concerning the lower bound on the speed of propagation to be used directly to obtain a lower bound on the speed of translation of the contact birth process. New functions $w_i^*(s, t)$ are defined in terms of the $w_i(s, t)$. These new functions satisfy precisely the same system of inequalities obtained in Chapter 5 for the $w_i(\mathbf{s}, t)$, in the special case of an $S \to I$ epidemic on the real line with each $\sigma_i = 1$. The inequalities were required for the proof of Theorem 5.2.

THEOREM 9.2 (A LOWER BOUND FOR THE SPEED OF TRANSLATION). *For each i and each positive $c < c_0$ there exist constants $b_i > 0$ and $T_i > 0$ such that $w_i(ct, t) \geq b_i$ for all $t \geq T_i$.*

PROOF. Consider the solution $w_i(s, t)$ to equations (9.6) or (9.7). Equations (9.6) are obtained when $U(t)$ measures the furthest spread of all types and (9.7) are obtained when $U(t)$ measures the furthest spread of a specific type k. First observe that, from Theorem 9.1, each $w_i(s, t)$ is continuous in s for $s > 0$ and is right continuous for $s = 0$. Similarly from equations (9.7) $w_i(s, t)$ can be shown to be continuous in s, for all s when $i \neq k$ and for $s > 0$ when $i = k$, with $w_k(s, t)$ right continuous at $s = 0$.

When $U(t)$ measures the furthest spread of all types, for each i define $w_i^*(s, t) = w_i(-s, t)$ for $s < 0$ and $w_i^*(s, t) = w_i(s, t)$ otherwise. If $U(t)$ measures the furthest spread of type k, define $w_k^*(s, t) = w_k(-s, t)$ for $s < 0$ and define $w_i^*(s, t) = w_i(s, t)$ for $s \geq 0$ when $i = k$ and for all s when $i \neq k$. Then, for both cases, $w_i^*(s, t)$ is a continuous function of s for all $s \in \Re$ and $i = 1, ..., n$. Also, taking $g(x) = 1 - e^{-x}$,

$$w_i^*(s, t) = \sum_{j=1}^{n} \lambda_{ij} \int_0^t \int_{-\infty}^{\infty} g(w_j^*(s - r, \tau)) p_{ij}(r) dr d\tau + H_i^*(s, t).$$

For the first case, for $s \geq 0$,

$$H_i^*(s, t) = \sum_{j=1}^{n} \lambda_{ij} \int_0^t \int_s^{\infty} e^{-w_j^*(s-r,\tau)} p_{ij}(r) dr d\tau,$$

for all $i = 1, ..., n$. For the second case, for $s \geq 0$,

$$H_i^*(s, t) = \lambda_{ik} \int_0^t \int_s^{\infty} e^{-w_k^*(s-r,\tau)} p_{ik}(r) dr d\tau.$$

In both cases $H_i^*(s, t) = H_i^*(-s, t)$ when $s < 0$ and $H_i^*(s, t) \geq 0$ for all s and $t \geq 0$.

These equations for the $w_i^*(s, t)$ are the same as equations (5.1) for $w_i(s, t)$ for the n-type $S \to I$ epidemic on the line with $\sigma_i = 1$ and a special form for the function $H_i(s, t)$. Since $H_i^*(s, t) \geq 0$ for all s and $t \geq 0$ we obtain the inequalities

$$w_i^*(s, t) \geq \sum_{j=1}^{n} \lambda_{ij} \int_0^t \int_{\Re} \left(1 - e^{-w_j^*(s-r,t-\tau)}\right) p_{ij}(r) dr d\tau, \ (i = 1, ..., n).$$

These are the same inequalities as were obtained in Section 5.4 for the $w_i(\mathbf{s},t)$, in the special case when $N = 1$ and $\sigma_i = 1$ for the $S \to I$ epidemic for which $\gamma_{ij}(\tau) \equiv \sigma_i \lambda_{ij}$.

Theorem 5.2 will then follow immediately for the $w_i^*(s,t)$, provided we can establish the result of Lemma 5.6 for $w_i^*(s,t)$, namely that for any $R > 0$ there exists a $t_0 = t_0(R)$ such that $w_i^*(s,t) > 0$ for $|s| \leq R$ and $t \geq t_0$. Also each $w_i^*(s,t)$ is a jointly continuous function of s and t for all real s and $t \geq 0$.

Now Lemma 9.1 shows that for any $R > 0$ and $t_0 > 0$, $w_i(s,t) \geq w_i(R,t_0) > 0$ for $|s| \leq R$ and $t \geq t_0$. But, for each i and s, either $w_i^*(s,t) = w_i(s,t)$ or $w_i^*(s,t) = w_i(-s,t)$. Therefore $w_i^*(s,t) \geq w_i(R,t_0) > 0$ for $|s| \leq R$ and $t \geq t_0$.

Hence using this result in place of Lemma 5.6, Theorem 5.2 can then be established for $w_i^*(s,t)$ exactly as in Chapter 5, so that for each i and each positive $c < c_0$ there exists a $b_i > 0$ and $T_i > 0$ such that $\inf_{|s| \leq ct} w_i^*(s,t) \geq b_i$ for all $t \geq T_i$.

Having established this result for the $w_i^*(s,t)$, we now consider $w_i(s,t)$ for the contact birth process. Since, for each s and t, either $w_i(s,t) = w_i^*(s,t)$ or $w_i(s,t)$ is infinite, clearly the same values of b_i and T_i can be used for the $w_i(s,t)$ as for the $w_i^*(s,t)$.

Therefore for each i and each positive $c < c_0$ there exists a $b_i > 0$ and $T_i > 0$ such that $\inf_{|s| \leq ct} w_i(s,t) \geq b_i$ for all $t \geq T_i$. From the monotonicity of $w_i(s,t)$ in s the result is obtained that $w_i(ct,t) \geq b_i$ for all $t \geq T_i$.

□

It is now easy to show that the asymptotic speed of translation of the forward tail of the distribution of furthest spread of the process is c_0. An explicit expression for c_0 was obtained for the $S \to I$ epidemic in Section 7.5. By taking $\sigma_i = 1$ for all $i = 1, .., n$ the corresponding expression is obtained for the contact birth process. Define $\mathbf{P}(\theta)$ to be the matrix with ij^{th} entry the Laplace transform of the projection of the contact distribution $p_{ij}(\mathbf{r})$ in direction $\boldsymbol{\xi}$. Then $c_0 = \inf_{\theta > 0} \rho(\mathbf{P}(\theta))/\theta$.

THEOREM 9.3 (THE ASYMPTOTIC SPEED OF TRANSLATION). *If the projections of the contact distributions are symmetric about zero and exponentially dominated in the tail, then the following results hold for each $i = 1, ..., n$.*

1. *For any $c > c_0$, $\lim_{t \to \infty} w_i(ct,t) = 0$.*
2. *For each c such that $0 < c < c_0$ there exists an $\varepsilon > 0$ and $T > 0$ such that $w_i(ct,t) \geq \varepsilon$ for all $t \geq T$.*

PROOF. Parts 1 and 2 follow immediately from Theorem 9.1 and 9.2 respectively.

□

The results of Theorem 9.2 are now strengthened in Lemma 9.2 in order to obtain the result that $U(t)/t$ converges in probability to c_0.

LEMMA 9.2. *For each i and each positive $c < c_0$, $w_i(ct,t) \to \infty$ as $t \to \infty$.*

PROOF. Take any positive $c < c_0$ and choose d such that $c < d < c_0$. From Theorem 9.2 for this d we may choose $b_i > 0$ and $T_i > 0$ such that $w_i(r,\tau) \geq b_i$ for all $r \leq d\tau$ and $\tau \geq T_i$. Let $T = \max_i T_i$. Note that $g(x)$ is a monotone increasing function of x for all $x \geq 0$. Then for any $t > T$,

$$
\begin{aligned}
w_i(ct,t) &\geq \sum_{j=1}^{n} \lambda_{ij} \int_0^t \int_{-\infty}^{\infty} g(w_j(r,\tau))p_{ij}(ct-r)drd\tau \\
&\geq \sum_{j=1}^{n} \lambda_{ij} g(b_j) \int_T^t \int_{-\infty}^{d\tau} p_{ij}(ct-r)drd\tau \\
&= \sum_{j=1}^{n} \lambda_{ij} g(b_j) \int_T^t \int_{ct-d\tau}^{\infty} p_{ij}(u)dud\tau \\
&= \sum_{j=1}^{n} \lambda_{ij} g(b_j) \int_0^{t-T} \int_{ct-d(T+\theta)}^{\infty} p_{ij}(u)dud\theta.
\end{aligned}
$$

Now $ct - d(T+\theta) \leq 0$ if $\theta \geq (ct - dT)/d = (c/d)t - T$ and $p_{ij}(u)$ is symmetric about zero so that $\int_0^\infty p_{ij}(u)du = 1/2$. Therefore for $t > (d/c)T$,

$$
\begin{aligned}
w_i(ct,t) &\geq \frac{1}{2} \sum_{j=1}^{n} \lambda_{ij} g(b_j) \int_{(c/d)t-T}^{t-T} d\theta \\
&= \frac{1}{2}\left(1 - \frac{c}{d}\right) t \sum_{j=1}^{n} \lambda_{ij} g(b_j).
\end{aligned}
$$

The right hand side of this inequality tends to infinity as $t \to \infty$. Hence for any $0 < c < c_0$, $w_i(ct,t) \to \infty$ as $t \to \infty$.

□

Theorem 9.1 and Lemma 9.2 can then be used to show that $U(t)/t$ converges in probability to c_0. Here $U(t)$ is the position of furthest spread, either of all types or of a fixed type k, given there is one type i individual initially. The result holds for all $i = 1, ..., n$.

THEOREM 9.4. *Consider any type i and direction $\boldsymbol{\xi}$. If the projections of the contact distributions are symmetric about zero and exponentially dominated in the tail, then $U(t)/t$ converges in probability to c_0.*

PROOF. Note that $c_0 > 0$. In order to prove that $U(t)/t$ converges in probability to c_0 it is necessary to show that for any $\varepsilon > 0$ and $0 < \delta < 1$ there exists a T such that $P(|(U(t)/t) - c_0| \leq \varepsilon) > 1 - \delta$, for all $t > T$. We need only prove this result for $\varepsilon < c_0$.

Consider any such ε and δ. Now

$$
\begin{aligned}
P\left(\left|\frac{U(t)}{t} - c_0\right| \leq \varepsilon\right) &\geq P\left(c_0 - \varepsilon < \frac{U(t)}{t} \leq c_0 + \varepsilon\right) \\
&= y_i((c_0-\varepsilon)t,t) - y_i((c_0+\varepsilon)t,t).
\end{aligned}
$$

Now $0 < c_0 - \varepsilon < c_0$. From Lemma 9.2, as $t \to \infty$, $w_i((c_0-\varepsilon)t,t)$ tends to infinity and hence $y_i((c_0-\varepsilon)t,t)$ tends to one. Hence there exists a T_1 such that $y_i((c_0-\varepsilon)t,t) > 1 - \delta/2$ for all $t > T_1$. Also $c_0 + \varepsilon > c_0$, so from Theorem 9.1 as t tends to infinity $w_i((c_0+\varepsilon)t,t)$ tends to zero and so $y_i((c_0+\varepsilon)t,t)$ also tends to zero. Hence there exists a T_2 such that $y_i((c_0+\varepsilon)t,t) < \delta/2$ for all $t > T_2$.

Now take $T = \max(T_1, T_2)$. Then $P(|(U(t)/t) - c_0| \leq \varepsilon) > 1 - \delta$ for all $t > T$. Hence $U(t)/t$ converges in probability to c_0.

□

Now consider the case when at least one contact distribution is not exponentially dominated in the forward tail. Theorem 9.5 shows that, for any type i and any $c > 0$, as t tends to infinity $w_i(ct, t)$ tends to infinity and hence $y_i(ct, t)$ tends to one. Hence the asymptotic speed of translation may be regarded as being infinite and for any $c > 0$, $P((U(t)/t) > c)$ tends to one as t tends to infinity.

THEOREM 9.5. *When the $p_{ij}(r)$ are symmetric about zero but not all are exponentially dominated in the tail there exists a monotone (in t) solution $w_i(s,t)$ to equations (9.6) (when considering the furthest spread of all types) or to equations (9.7) (when considering the furthest spread of type k) which is unique. For each i and each $c > 0$, $w_i(ct, t) \to \infty$ as $t \to \infty$.*

PROOF. The existence and uniqueness of the solution $w_i(s,t)$ to either equations (9.6) or (9.7) is proved in an identical fashion to Theorems 2.1 and 3.1 by splitting the ranges of s and t and using contraction arguments.

For the $S \to I$ epidemic, the infection matrix has infinite entries so that $K_c(0)$ is infinite for all $c > 0$. Lemma 5.7 can then be used together with the proof of Theorem 9.2 to show that for each $c > 0$ and each i, there exist a $b_i > 0$ and $T_i > 0$ such that $w_i(ct, t) \geq b_i$. The proof of Lemma 9.2 will then be valid for any $c > 0$ so can be used to strengthen this result and show that $\lim_{t\to\infty} w_i(ct, t) = \infty$.

□

The exact results obtained for the contact birth process in this section require the projection of each contact distribution to be symmetric about zero. The saddle point method of Chapter 7 may be used to obtain results for non-symmetric contact distributions. This method is applied in Sections 9.5 and 9.6 to more complex contact processes which allow for deaths and multiple births. Results are obtained for the speed of spread of $y_i(s,t)$, the forward tail of the distribution of $U(t)$. A natural definition, since $y_i(s,t)$ is monotone decreasing in s for each t, is to consider a small $\eta > 0$ and to find $s(t)$ such that $y_i(s(t), t) = \eta$. This is a minor modification of the saddle point method presented in Chapter 7, and discussed therein. The speed of spread of $y_i(s,t)$ is then $\lim_{t\to\infty} s(t)/t$.

It is easily seen that, once it has been shown by the exact methods that the asymptotic speed of translation is c_0, this immediately implies that $\lim_{t\to\infty} s(t)/t = c_0$. Intuitively one would expect this result to hold.

Consider the results we have obtained when the contact distributions are symmetric for the contact birth process. If $s(t)/t \to c^*$ then for every $\varepsilon > 0$ there exist a T sufficiently large such that

$$y_i((c^* + \varepsilon)t, t) \leq \eta \leq y_i((c^* - \varepsilon)t, t) \tag{9.9}$$

for $t > T$. If $c^* > c_0$ then ε can be chosen so that $c^* - \varepsilon > c_0$ so the right hand side of inequality (9.11) tends to zero as $t \to \infty$ giving a contradiction. Similarly if $c^* < c_0$ then ε can be chosen so that $c^* + \varepsilon < c_0$ and the left hand side of inequality (9.9) tends one as $t \to \infty$ giving a contradiction. Hence $\lim_{t\to\infty} s(t)/t = c_0$. Thus the definition of the speed of propagation given by Aronson and Weinberger [A4,A5], and used for the exact method, gives the same value as the definition used when applying the saddle point method to the contact birth process.

9.5 The multi-type contact birth-death process

A connection is shown to exist between the n-type contact birth-death process and the $S \to I \to S$ epidemic. The saddle point method has been used in Chapter 8 to obtain an expression for the speed of first spread of the $S \to I \to S$ epidemic. It can therefore also be used to obtain the speed of translation of the extreme of the distribution of the contact birth-death process.

Consider the n-type contact birth process of section 9.3, but now individuals can also die. The assumptions are the same as for the contact birth process except that individuals may now die. The probabilities that an individual of type i gives birth or dies in time $(t, t+\delta t)$ are $\lambda_i \delta t + o(\delta t)$ and $\mu_i \delta t + o(\delta t)$ respectively. When a birth event occurs for a type i individual there is a probability q_{ij} that the offspring is of type j. Then the time T until the first birth or death occurs for a type i individual has probability density function $\alpha_i e^{-\alpha_i t}$, for $0 \leq t < \infty$, where $\alpha_i = \sum_{j=1}^n \lambda_i q_{ij} + \mu_i = \lambda_i + \mu_i$.

Consider $U(t)$, the position of furthest spread of all types at time t in a specified direction. By similar conditional arguments to sections 9.2 and 9.3 we can write down equations for $y_i(s,t)$, which give the probability that $U(t) > s$ given that the process starts with one type i individual at $s = 0$ at time $t = 0$.

$$(1 - y_i(s,t)) = P(T > t)(1 - y_i(s,0)) + P(T < t)\frac{\mu_i}{\alpha_i}$$
$$+ \int_0^t \alpha_i e^{-\alpha_i \tau}(1 - y_i(s, t-\tau)) \sum_{j=1}^n \frac{\lambda_i q_{ij}}{\alpha_i} \int_{-\infty}^{\infty} (1 - y_j(s-r, t-\tau)) p_{ij}(r) dr d\tau.$$

Note that $y_i(s,0) = 0$ if $s \geq 0$, and $y_i(s,0) = 1$ if $s < 0$. Then

$$(1 - y_i(s,t))e^{\alpha_i t} = (1 - y_i(s,0)) + (e^{\alpha_i t} - 1)\frac{\mu_i}{\alpha_i}$$
$$+ \int_0^t \alpha_i e^{\alpha_i \theta}(1 - y_i(s,\theta)) \sum_{j=1}^n \frac{\lambda_i q_{ij}}{\alpha_i} \int_{-\infty}^{\infty} (1 - y_j(s-r, \theta)) p_{ij}(r) dr d\theta.$$

Differentiating with respect to t, the following equations are obtained for $i = 1, ..., n$,

$$\frac{\partial y_i(s,t)}{\partial t} = (1 - y_i(s,t)) \sum_{j=1}^n \lambda_i q_{ij} \int_{-\infty}^{\infty} y_j(s-r, t) p_{ij}(r) dr - \mu_i y_i(s,t). \tag{9.10}$$

Since we have considered the position of furthest spread of any type then $y_i(s,0) = 1$ for $s < 0$ and $y_i(s,0) = 0$ for $s \geq 0$. If instead we considered the furthest spread of a specific type k, equations (9.10) are still obtained, but now $y_i(s,0) = 0$ for $i \neq k$ and $y_k(s,0) = 1$ for $s < 0$ and $y_k(s,0) = 0$ for $s \geq 0$.

Equations (9.10) are the same as the equations given in Section 8.8 for $y_i(s,t)$ for the n-type $S \to I \to S$ epidemic on the line (so that $N = 1$) if $\sigma_i \lambda_{ij}$ is replaced by $\lambda_i q_{ij}$ for all $i, j = 1, ..., n$. This establishes the McKean connection between the multi-type contact birth-death and the $S \to I \to S$ processes.

In the forward tail $1 - y_i(s,t) \approx 1$, so the approximate equations obtained from equations (9.10) which are valid in the forward tail are

$$(9.11) \qquad \frac{\partial y_i(s,t)}{\partial t} = \sum_{j=1}^{n} \lambda_i q_{ij} \int_{-\infty}^{\infty} y_j(s-r,t) p_{ij}(r) dr - \mu_i y_i(s,t), \ (i = 1, ..., n).$$

These equations are identical to the equations in the forward front for the corresponding $S \to I \to S$ epidemic, although the initial conditions differ. The conditions are consistent with those assumed in Chapter 7 when obtaining the saddle point result. Hence the asymptotic speed of translation of the forward tail of the distribution of furthest spread (for all types or for a specific type) for the contact birth-death process is

$$c_0 = \max\left(0, \inf_{\theta > 0} \frac{\rho(\mathbf{A}(\theta)) - \mu}{\theta}\right),$$

where $\mu = \max_i(\mu_i)$, $P_{ij}(\theta) = \int_{-\infty}^{\infty} e^{\theta r} p_{ij}(r) dr$ and $\mathbf{A}(\theta)$ is the matrix with ij^{th} entry $\lambda_i q_{ij} P_{ij}(\theta) + \delta_{ij}(\mu_i - \mu)$. Note that δ_{ij} is one if $i = j$ and is zero otherwise.

9.6 Saddle point results for birth-death processes with branching

The generalisation of the contact processes, which have been considered in this chapter, to a contact branching process may be carried out in a number of ways. One-type processes are discussed in Biggins [B8].

The contact birth and contact birth-death processes are special cases of a contact branching process. For a general Markovian contact branching process there is still an exponential time until a birth or death event occurs for each type i individual; however when a birth event occurs the number of offspring produced is a random variable. These offspring may all be of the same type, or the types may differ, so that the number of offspring of different types will have a distribution. The positions of these offspring may all be the same, have independent distributions or have a more general joint distribution. These distributions may depend upon the type of both parent and offspring.

Many variations of such models can be formulated to describe different situations. These models can be specified by equations analogous to equations (9.10). An analysis of these processes leads to essentially the same form of equations in the forward tail as given by equations (9.11). The saddle point method can then be used to obtain the speed of translation of the forward tail for the distribution of furthest spread for general contact branching processes.

Three models are discussed in this section to illustrate the technique. The first and last of these models appear in Radcliffe and Rass [R11]. The second model is a minor modification of the first model. Model 1 is quite simple and has all the offspring of the same type and situated at the same position relative to the parent. Model 2 takes the positions of the offspring relative to the parent to have independent distributions. The final model is more complicated and realistic and allows for the types of offspring to differ.

Note that, although both the contact birth and the contact birth-death processes are special cases of all three models, none of the three branching process models is a special case of either of the other two.

Model 1.

A simple n-type contact branching process with deaths differs from the contact birth-death process by allowing multiple births to occur. For this process, $g_i(M)$ is the probability that a type i individual gives rise to M offspring when a birth event occurs to that individual; q_{ij} is the probability that they are type j (all offspring being restricted to being of the same type); and $p_{ij}(r)$ is the density function of the distance r in a specific direction at which they are all situated relative to the individual giving birth.

Let $U(t)$ represent the furthest spread either of all types or of a specific type k in the specified direction at time t. Define $y_i(s,t)$ to be the probability that $U(t) > s$ when initially there is one type i individual at the origin.

The method used to derive the equations for the contact birth-death process may now be used to obtain the analogous equations for this branching process model. The equations for the $y_i(s,t)$ so obtained are given by

$$\begin{aligned}(1-y_i(s,t)) &= (1-y_i(s,0))e^{-\alpha_i t} + \frac{\mu_i}{\alpha_i}(1-e^{-\alpha_i t}) \\ &\quad + \sum_{j=1}^{n} q_{ij} \int_0^t \lambda_i e^{-\alpha_i \tau}(1-y_i(s,t-\tau)) \\ &\quad \times \sum_{M=1}^{\infty} g_i(M) \int_{-\infty}^{\infty} (1-y_j(s-r,t-\tau))^M p_{ij}(r)\,dr\,d\tau,\end{aligned}$$

where $\alpha_i = (\lambda_i + \mu_i)$.

In a similar manner to Section 9.5 we obtain equations analogous to equations (9.10), namely

$$\frac{\partial y_i(s,t)}{\partial t} = (1-y_i(s,t)) \sum_{j=1}^{n} \lambda_i q_{ij} \int_{-\infty}^{\infty} (1-\pi_i(1-y_j(s-r,t)))p_{ij}(r)\,dr - \mu_i y_i(s,t),$$

for $i = 1,...,n$, where $\pi_i(z) = \sum_{M=1}^{\infty} g_i(M) z^M$.

Consider the speed of spread in the forward tail. In the tail $1 - y_i(s,t) \approx 1$. Also $1-\pi_i(1-z) \approx \pi_i'(1)z$ when z is small. Hence if the contact distributions have small variances, then we obtain the following approximate linear equations in the forward tail:

$$\frac{\partial y_i(s,t)}{\partial t} = \sum_{j=1}^{n} \lambda_i q_{ij} \pi_i'(1) \int_{-\infty}^{\infty} y_j(s-r,t)p_{ij}(r)\,dr - \mu_i y_i(s,t), \quad (i=1,...,n). \tag{9.12}$$

These are the same equations as equations (8.17) for the $S \to I \to S$ epidemic with $\sigma_j\lambda_{ij} = \lambda_i q_{ij}\pi_i'(1)$. Note that $\pi_i'(1)$ is the expected number of offspring of a type i individual when a birth event occurs for that individual.

Equations (9.12) are also the same as equations (9.11) for the contact birth-death process when ${\pi'}_i(1) = 1$ for all i. This condition implies that the number of offspring when a birth event occurs is certain to be one. The contact birth-death process is in fact just a special case of the simple contact branching process described in this section.

Model 2.
As for model 1, at each birth event a type i individual gives rise to a random number of offspring which are all of the same type. The probability that they are of type j is q_{ij}. However they are no longer situated at the same position relative to the parent. Instead the positions are independent random variables. The common density function of the distance r in a specific direction at which an offspring of type j is situated relative to the type i individual giving birth is $p_{ij}(r)$.

Using the same definitions as for model 1, we obtain

$$\begin{aligned}(1-y_i(s,t)) &= (1-y_i(s,0))e^{-\alpha_i t} + \frac{\mu_i}{\alpha_i}(1-e^{-\alpha_i t})\\ &+ \int_0^t \lambda_i e^{-\alpha_i\tau}(1-y_i(s,t-\tau))\\ &\times \sum_{M=1}^{\infty} g_i(M)\sum_{j=1}^{n} q_{ij}\left(\int_{-\infty}^{\infty}(1-y_j(s-r,t-\tau))p_{ij}(r)dr\right)^M d\tau\\ &= (1-y_i(s,0))e^{-\alpha_i t} + \frac{\mu_i}{\alpha_i}(1-e^{-\alpha_i t})\\ &+ \int_0^t \lambda_i e^{-\alpha_i\tau}(1-y_i(s,t-\tau))\\ &\times \sum_{j=1}^{n} q_{ij}\pi_i\left(\int_{-\infty}^{\infty}(1-y_j(s-r,t-\tau))p_{ij}(r)dr\right) d\tau.\end{aligned}$$

Proceeding as in section 9.5 we obtain

$$\begin{aligned}\frac{\partial y_i(s,t)}{\partial t} &= (1-y_i(s,t))\lambda_i\sum_{j=1}^{n} q_{ij}\left(1-\pi_i\left(1-\int_{-\infty}^{\infty} y_j(s-r,t)p_{ij}(r)dr\right)\right)\\ &- \mu_i y_i(s,t),\end{aligned}$$

where $\pi_i(z)$ is the probability generating function of the number of offspring of a type i individual when a birth event occurs.

Consider the speed of spread in the forward tail. In the tail $1-y_i(s,t) \approx 1$. If z is small then $(1-\pi_i(1-z))$ is approximately equal to $\pi_i'(1)z$. Also when the variances of the contact distributions are fairly small, $\int_{-\infty}^{\infty} y_j(s-r,t)p_{ij}(r)dr$ will be small. So we obtain the approximate linear equations

$$(9.13)\quad \frac{y_i(s,t)}{\partial t} = \sum_{j=1}^{n} \lambda_i q_{ij} \pi_i'(1) \int_{-\infty}^{\infty} y_j(s-r,t) p_{ij}(r) dr - \mu_i y_i(s,t), \quad (i = 1, ..., n).$$

These are the same equations as equations (8.17) for the $S \to I \to S$ epidemic with $\sigma_j \lambda_{ij} = \lambda_i q_{ij} \pi_i'(1)$. Again $\pi_i'(1)$ is the expected number of offspring of a type i individual when a birth event occurs for that individual.

The contact birth-death process is a special case of this contact branching process when the number of offspring when a birth event occurs is certain to be one.

Model 3.

Consider a contact branching process where $g_i(M)$ is the number of offspring of a type i individual. The number of offspring of types 1,...,n has a multinomial distribution with parameters M and $q_{i1}, ..., q_{in}$. The joint density function for the contact distribution if the M individuals are of types $j_1, ..., j_M$ is $p_{ij_1,...,j_M}(r_1...., r_M)$ where r_s is the distance of the sth offspring (type j_s) from the type i parent. The marginal distribution of the distance for the sth individual is assumed to depend only on i and j_s, and can be written as $p^*_{ij_s}(r_s)$. With $y_i(s,t)$ and α_i defined as before, we have

$$(1 - y_i(s,t)) = (1 - y_i(s,0))e^{-\alpha_i t} + \frac{\mu_i}{\alpha_i}(1 - e^{-\alpha_i t})$$

$$+ \int_0^t \lambda_i e^{-\alpha_i \tau}(1 - y_i(s, t-\tau)) \sum_{M=1}^{\infty} g_i(M) \sum \frac{M!}{s_1!...s_n!} q_{i1}^{s_1} ... q_{in}^{s_n}$$

$$\times \int_{-\infty}^{\infty} ... \int_{-\infty}^{\infty} p_{ij_1,...,j_M}(r_1...., r_M) \prod_{l=1}^{M} (1 - y_{j_l}(s - r_l, t - \tau)) dr_1...dr_M d\tau,$$

where $j_1, ..., j_M$ consists of a sequence, where 1 occurs s_1 times,...,n occurs s_n times. The summation is over $s_1, ..., s_n$ such that $s_i \geq 0$ and $\sum_{i=1}^{n} s_i = M$.

Again proceeding as in section 9.5 we obtain

$$\frac{\partial y_i(s,t)}{\partial t} = (1 - y_i(s,t))\lambda_i \sum_{M=1}^{\infty} g_i(M) \sum \frac{M!}{s_1!...s_n!} q_{i1}^{s_1} ... q_{in}^{s_n}$$

$$\times \int_{-\infty}^{\infty} ... \int_{-\infty}^{\infty} \left(1 - \prod_{l=1}^{M} (1 - y_{j_l}(s - r_l, t)\right) p_{ij_1,...,j_M}(r_1...., r_M) dr_1...dr_M - \mu_i y_i(s,t).$$

If we consider an approximation in the forward tail where $1 - y_i(s,t) \approx 1$, and assume that the variance-covariance matrix for the joint contact distributions has fairly small entries, we obtain the approximate linear equations

$$\frac{\partial y_i(s,t)}{\partial t} = \lambda_i \sum_{M=1}^{\infty} g_i(M) \sum \frac{M!}{s_1!...s_n!} q_{i1}^{s_1}...q_{in}^{s_n}$$

$$\times \int_{-\infty}^{\infty} ... \int_{-\infty}^{\infty} \sum_{l=1}^{M} y_{j_l}(s-r_l,t) p_{ij_1,...,j_M}(r_1....,r_M) dr_1...dr_M - \mu_i y_i(s,t)$$

$$= \lambda_i \sum_{M=1}^{\infty} g_i(M) \sum \frac{M!}{s_1!...s_n!} q_{i1}^{s_1}...q_{in}^{s_n} \sum_{j=1}^{n} s_j \int_{-\infty}^{\infty} y_j(s-r,t) p_{ij}^*(r) dr - \mu_i y_i(s,t)$$

$$= \lambda_i \sum_{M=1}^{\infty} g_i(M) \sum_{j=1}^{n} M q_{ij} \int_{-\infty}^{\infty} y_j(s-r,t) p_{ij}^*(r) dr - \mu_i y_i(s,t)$$

$$= \sum_{j=1}^{n} \lambda_i q_{ij} \pi_i'(1) \int_{-\infty}^{\infty} y_j(s-r,t) p_{ij}^*(r) dr - \mu_i y_i(s,t).$$

This is the same as equations (8.17) for the $S \to I \to S$ epidemic if we take $\sigma_i \lambda_{ij} = \lambda_i q_{ij} \pi_i'(1)$ and replace $p_{ij}(r)$ by $p_{ij}^*(r)$. Once again the contact birth-death process is a special case of this process when the number of offspring when a birth event occurs is certain to be one.

Since the approximate equations in the forward tail are the same for all three contact branching process models, we can consider all of them simultaneously when obtaining the speed of spread of the forward tail for $y_i(s,t)$. Using the results for the $S \to I \to S$ epidemic on the real line, the speed of translation for all three models is

$$c = \max\left(0, \inf_{\theta>0} \frac{\rho(\mathbf{K}(\theta)) - \mu}{\theta}\right),$$

where $\mathbf{K}(\theta)$ has ij^{th} entry $\lambda_i q_{ij} \pi_i'(1) P_{ij}(\theta) + \delta_{ij}(\mu - \mu_i)$ with $\mu = \max_i \mu_i$ and $P_{ij}(\theta)$ is the Laplace transform of $p_{ij}(r)$ for models 1 and 2 and $p_{ij}^*(r)$ for model 3.

Note that in general for the case $N = 1$, the saddle point method is applied to equations in $y_i(s,t)$ which are not monotone in s. Hence, given a small positive η, it is not possible to uniquely define $s(t)$ such that $y_i(s,t) = \eta$ and hence find the limit of $s(t)/t$ as t tends to infinity. Instead $s(t)$ was defined so that $\int_{s(t)}^{\infty} y_i(s,t) ds = \eta$ and the corresponding limit obtained.

For the contact branching process $\int_s^{\infty} y_i(x,t) dx = \int_s^{\infty} (x-s) f_{U(t)}(x) dx$, where $f_{U(t)}(x)$ is the density function of $U(t)$. This approach would consider the speed at which the weighted tail of the distribution of $U(t)$ first moves out. Because of the monotonicity of $y_i(s,t)$ in s, we can in fact use the simpler approach for the contact branching processes i.e. define $s(t)$ such that $y_i(s,t) = \eta$ and find $\lim_{t\to\infty} s(t)/t$. This was discussed in Chapter 7. The same value is obtained for the speed of translation using either definition of $s(t)$. The forward tail and the weighted forward tail of the distribution of furthest spread move at the same rate in the direction specified, which intuition suggests must happen.

Appendices

Appendix A. Extended Perron-Frobenius theory

This section collects together certain properties of the class $\mathfrak{B}$ of non-negative, non-reducible, finite, square matrices that are used in this book. We first introduce some preliminary notation and definitions.

Let $\mathbf{B} = (b_{ij})$ denote a matrix with ij^{th} element b_{ij} and let $\mathbf{b} = (b_i)$ denote a column vector with i^{th} entry b_i. The ith element of a vector $\mathbf{b}$ is denoted by $\{\mathbf{b}\}_i$ and the ij^{th} element of a matrix $\mathbf{B}$ is denoted by $\{\mathbf{B}\}_{ij}$. We denote a vector or matrix with all elements zero by $\mathbf{0}$ and a vector with all its elements unity by $\mathbf{1}$. Inequalities between matrices or vectors imply the corresponding inequalities between the elements of the matrices or vectors.

A matrix is said to be non-negative if all its elements are non-negative. It is said to be finite if all its elements are finite. A square matrix $\mathbf{B} = (b_{ij})$ is said to be non-reducible if for every $i \neq j$ there exists a distinct sequence $i_i, ..., i_r$ with $i_1 = i$ and $i_r = j$ such that $b_{i_s i_{s+1}} \neq 0$ for $s = 1, ..., (r-1)$. Otherwise $\mathbf{B}$ is called reducible.

When $\mathbf{B}$ is a non-negative, finite, square matrix we denote its Perron-Frobenius root by $\rho(\mathbf{B})$. The Perron-Frobenius root is defined to be the maximum of the moduli of the eigenvalues of $\mathbf{B}$. It is a real eigenvalue of $\mathbf{B}$, and is the eigenvalue with largest real part. This definition is not restricted to the case when $\mathbf{B}$ is non-reducible. When $\mathbf{B}$ is non-reducible the Perron-Frobenius root $\rho(\mathbf{B})$ is simple, but for the reducible case $\rho(\mathbf{B})$ may have multiplicity greater than one.

We need to consider the limit of a non-reducible matrix $\rho(\mathbf{B})$ for situations in which elements of $\mathbf{B}$ may tend to infinity. For simplicity of exposition it is convenient to define $\rho(\mathbf{B}) = \infty$ when $\mathbf{B}$ is a non-negative, non-reducible square matrix with at least one infinite element.

The definition of the Perron-Frobenius root is easily extended to non-reducible, finite, square matrices with non-negative off-diagonal entries. Such a matrix, $\mathbf{A}$ may always be written in the form $\mathbf{A} = \mathbf{B} - c\mathbf{I}$ where $\mathbf{B}$ is a non-negative, square matrix and $\mathbf{I}$ is the identity matrix of the same size. The matrices $\mathbf{A}$ and $\mathbf{B}$ have eigenvalues differing by c and the same eigenvectors. We define $\rho(\mathbf{A}) = \rho(\mathbf{B}) - c$. Hence $\rho(\mathbf{A})$ is the eigenvalue of $\mathbf{A}$ with the largest real part, with corresponding eigenvector the eigenvector of $\mathbf{B}$ corresponding to $\rho(\mathbf{B})$.

Theorem A.1 is the basic Perron-Frobenius theorem for a non-negative, finite, square matrix; some useful results for the non-reducible case being given in Theorem A.2. Corollary A.1 gives certain results for the extended class where only the off-diagonal entries of the matrix are restricted to be non-negative (see the survey on M-matrices of Poole and Boullion [P2]). Since these results may be found in standard texts, or are easily derivable, they are omitted. Useful texts are Berman and Plemmons [B7] and Gantmacher [G1]. For the convenience of the reader, the proofs of Theorems A.1, A.2 and Corollary A.1 are available in pdf format at *http : \\www.maths.qmul.ac.uk\~lr\book.html.*

Non-negative, non-reducible, square matrices whose entries are functions of a single real variable θ are then considered. Let $\mathbf{B}(\theta)$ be such a matrix. In applications, $\{\mathbf{B}(\theta)\}_{ij} = b_{ij}P_{ij}(\theta)$, where (b_{ij}) is a non-negative, non-reducible matrix and $P_{ij}(\theta)$ is the Laplace transform of a contact distribution. Theorem A.3 collects together continuity and convexity results for the Perron-Frobenius root and associated eigenvectors for matrices of this form.

In Theorem A.3, it is necessary to extend the definition of $\rho(\mathbf{B}(\theta))$ to cover certain situations where θ lies in an open ball in the complex plane centred on a real value θ_0. If the radius of the ball is sufficiently small then there is a unique eigenvalue of $\mathbf{B}(\theta)$ with largest real part, which we define to be $\rho(\mathbf{B}(\theta))$. Analyticity results are obtained for $\rho(\mathbf{B}(\theta))$.

THEOREM A.1. *Let* $\mathbf{B}$ *be a non-negative, finite, square matrix and define* $\rho(\mathbf{B})$ *to be the maximum of the moduli of the eigenvalues of* $\mathbf{B}$. *Then* $\rho(\mathbf{B})$ *is a non-negative real eigenvalue of* $\mathbf{B}$.

When $\mathbf{B}$ *is non-reducible, and is not the zero matrix of order* 1, $\rho(\mathbf{B}) > 0$. *There exist a positive right eigenvector and a positive left eigenvector corresponding to* $\rho(\mathbf{B})$.

When $\mathbf{B}$ *is reducible, there exists a right eigenvector* $\mathbf{v} \gneq \mathbf{0}$ *and a left eigenvector* $\mathbf{u} \gneq \mathbf{0}$ *corresponding to* $\rho(\mathbf{B})$. *In addition, when there exists a positive right or left eigenvector corresponding to an eigenvalue* λ *of* $\mathbf{B}$, *then* $\lambda = \rho(\mathbf{B})$.

□

The results for the reducible case may be obtained from those for a non-reducible matrix in a straightforward manner by writing the reducible matrix in normal form. When $\mathbf{B}$ is non-reducible $\rho(\mathbf{B})$ has multiplicity one, however it can have multiplicity greater than one for the reducible case so that the corresponding eigenvectors may not be unique up to a multiple.

Various results for non-negative matrices may be derived from the Perron-Frobenius Theorem. Some results for the non-reducible case which are of particular relevance to this book are collected together in the following theorem.

THEOREM A.2. *The class* $\mathfrak{B}$ *of non-negative, non-reducible, finite, square matrices has the following properties:*

1. *If* $\mathbf{B} \in \mathfrak{B}$ *then* $\rho(\mathbf{B})$ *increases as any element of* $\mathbf{B}$ *increases.*
2. *For any matrix* $\mathbf{B} = (b_{ij})$ *of order* n *in* $\mathfrak{B}$, *and any* $s = 1, ...n$,

$$b_{ss} \leq \rho(\mathbf{B}) \leq \max_i \sum_{j=1}^{n} b_{ij}.$$

3. *Let* $\mathbf{C} = (c_{ij})$ *be a matrix of complex valued elements and* $\mathbf{C}^+ = (|c_{ij}|)$. *If* $\mathbf{C}^+ \leq \mathbf{B}$, *where* $\mathbf{B} \in \mathfrak{B}$, *with strict inequality for at least one element, then for any eigenvalue* μ *of* $\mathbf{C}$, $|\mu| < \rho(\mathbf{B})$.

4. *Define the adjoint of a square matrix of order* 1 *to be the identity matrix of order* 1. *Then for any* $\mathbf{B} \in \mathfrak{B}$

$$\begin{aligned} Adj(\lambda\mathbf{I} - \mathbf{B}) &> \mathbf{0}, \quad for \quad \lambda > \rho(\mathbf{B}), \\ |\lambda\mathbf{I} - \mathbf{B}| &> 0, \quad for \quad \lambda > \rho(\mathbf{B}), \\ |\lambda\mathbf{I} - \mathbf{B}| &= 0, \quad for \quad \lambda = \rho(\mathbf{B}). \end{aligned}$$

5. *When* $\mathbf{B} \in \mathfrak{B}$ *is of order* $n > 1$, *and* $\mathbf{B}^*$ *is any* k*-dimensional principal minor of* $\mathbf{B}$, *where* $k < n$, *then* $\rho(\mathbf{B}^*) < \rho(\mathbf{B})$.
6. *If* $\mathbf{B} \in \mathfrak{B}$ *is of order* $n > 1$, *and* $\mathbf{B}^*$ *is any* $(n-1)$*-dimensional principal minor of* $\mathbf{B}$, *then* $|\lambda\mathbf{I} - \mathbf{B}| < 0$ *for* $\rho(\mathbf{B}^*) < \lambda < \rho(\mathbf{B})$.
7. *If* $\mathbf{B} \in \mathfrak{B}$ *then* $\rho(\mathbf{B})$ *is a simple eigenvalue of* $\mathbf{B}$ *and hence the corresponding right and left eigenvectors are each unique up to a multiple. No other eigenvalue of* $\mathbf{B}$ *has a real, non-negative right or left eigenvector.*
8. *If* $\mathbf{B} \in \mathfrak{B}$ *then* $Adj(\rho(\mathbf{B})\mathbf{I} - \mathbf{B}) > \mathbf{0}$.
9. *If* $\mathbf{B}$ *is a positive square matrix then all eigenvalues of* $\mathbf{B}$ *other than* $\rho(\mathbf{B})$ *have modulus less than* $\rho(\mathbf{B})$.
10. *If* $\mathbf{B} = (b_{ij}) \in \mathfrak{B}$ *and* $b_{ii} > 0$ *for all* i, *then there exists an integer* $s \geq 1$ *such that* $\mathbf{B}^s > \mathbf{0}$. *All eigenvalues of* $\mathbf{B}$ *other than* $\rho(\mathbf{B})$ *have modulus less than* $\rho(\mathbf{B})$.
11. *If* $\mathbf{B} = (b_{ij}) \in \mathfrak{B}$ *is not the* 1×1 *zero matrix, then there exists an integer* $s \geq 1$ *such that*

$$\mathbf{B}^s = \begin{pmatrix} \mathbf{C}_{11} & \mathbf{0} & \dots & \mathbf{0} \\ \mathbf{0} & \mathbf{C}_{22} & \dots & \mathbf{0} \\ \vdots & \vdots & \ddots & \vdots \\ \mathbf{0} & \mathbf{0} & \dots & \mathbf{C}_{kk} \end{pmatrix},$$

where $\mathbf{C}_{ii} > \mathbf{0}$ *with* $\rho(\mathbf{C}_{ii}) = (\rho(\mathbf{B}))^s$ *for all* $i = 1, ..., k$. *Here* $k \geq 1$ *is the number of eigenvalues of* $\mathbf{B}$ *which have modulus equal to* $\rho(\mathbf{B})$.

□

COROLLARY A.1. *Let* $\mathbf{A}$ *be a non-reducible, finite matrix with off diagonal entries non-negative. This matrix may always be written in the form* $\mathbf{A} = \mathbf{B} - c\mathbf{I}$, *where* $\mathbf{B}$ *is non-negative.*

There is a unique eigenvalue, $\rho(\mathbf{A})$, *with largest real part. This eigenvalue is real and simple.*

There exist positive left and right eigenvectors corresponding to $\rho(\mathbf{A})$, *each of which is unique up to a multiple. This is the only eigenvalue of* $\mathbf{A}$ *for which there is a positive eigenvector.*

The eigenvalue $\rho(\mathbf{A})$ *increases as any element of* $\mathbf{A}$ *increases.*

□

Continuity results for the Perron-Frobenius root are also required. A preliminary lemma is first proved concerning the continuity of the eigenvalues of a general matrix. Dieudonné [D11 p.248] uses Rouché's theorem to derive continuity results for the roots of a polynomial, from which equivalent results may be obtained for the eigenvalues of a matrix. This is the approach used in Lemma A.1. An alternative proof of the continuity of the eigenvalues of a matrix as functions of its entries may be found in Ostrowski [O1 p. 282].

Lemma A.1 may then be used to show that if $\mathbf{A} = (a_{ij})$ is a non-negative, non-reducible matrix, then the definition of the Perron-Frobenius root may be extended to matrices whose entries are within an ε distance of those of $\mathbf{A}$, for some $\varepsilon > 0$. Within this region, the Perron-Frobenius root is a continuous function of its entries. The non-reducible matrices of interest have entries which are analytic functions of a single parameter θ.

LEMMA A.1. *Consider an $n \times n$ matrix $\mathbf{A}^* = (a^*_{ij})$ which has (not necessarily distinct) eigenvalues $\mu_1(\mathbf{A}^*), \mu_2(\mathbf{A}^*), ..., \mu_n(\mathbf{A}^*)$ and define*

$$d = \min_{\mu_i(\mathbf{A}^*) \neq \mu_j(\mathbf{A}^*)} |\mu_i(\mathbf{A}^*) - \mu_j(\mathbf{A}^*)|.$$

Let r_i be the multiplicity of $\mu_i(\mathbf{A}^*)$. Then for every $0 < \varepsilon < d/2$ there exists a $\delta > 0$ such that $\mathbf{A} = (a_{ij})$ has exactly r_i eigenvalues within an ε-neighbourhood of $\mu_i(\mathbf{A}^*)$ for each distinct $\mu_i(\mathbf{A}^*)$ for every matrix $\mathbf{A}$ such that $\max_{i,j} |a_{ij} - a^*_{ij}| < \delta$.

When $\mu_i(\mathbf{A}^*)$ has multiplicity 1 then we define $\mu_i(\mathbf{A})$ to be the unique eigenvalue of $\mathbf{A}$ within the ε-neighbourhood of $\mu_i(\mathbf{A}^*)$. Then $\mu_i(\mathbf{A})$ is a jointly continuous function of the entries of $\mathbf{A}$ at $\mathbf{A} = \mathbf{A}^*$, and hence $\mu_i(\mathbf{A}) \to \mu_i(\mathbf{A}^*)$ as $\mathbf{A} \to \mathbf{A}^*$.

PROOF. Rouché's theorem states that if f and g are analytic in an open region $\mathfrak{G}$ of the complex plane and if γ is a Jordan circuit, such that its graph Γ and its interior lie within $\mathfrak{G}$, and if $|g(z)| < |f(z)|$ for all z on Γ, then f and $f + g$ have the same number of zeros within Γ.

Consider the polynomials $f(z) = z^n + \sum_{j=0}^{n-1} b^*_j z^j$ and $h(z) = z^n + \sum_{j=0}^{n-1} b_j z^j$, where $f(z)$ has distinct zeros $\mu_1, ..., \mu_s$ of multiplicities $r_1, ..., r_s$. Define $d = \min_{i \neq j} |\mu_i - \mu_j|$. We use Rouché's theorem to show that for any $0 < \varepsilon < d/2$ there exists a $\delta^* > 0$ such that $h(z)$ has exactly r_i zeros within an ε-neighbourhood of μ_i for $i = 1, ..., s$ provided $\max_{0 \le j \le n-1} |b_j - b^*_j| < \delta^*$.

Consider any ε such that $0 < \varepsilon < d/2$. Define $g(z) = h(z) - f(z)$ and take Γ_i to be the circumference of a ball of radius ε centred on μ_i. Then $f(z)$ is continuous and non-zero on the closed set Γ_i and hence there exists a $k_i > 0$ such that $|f(z)| \ge k_i$ for z on Γ_i. Let $A = \max_{0 \le j \le n-1} |b_j - b^*_j|$. Then $|g(z)| \le A \sum_{j=0}^{n-1} (|\mu_j| + \varepsilon)^j < k_i$ on Γ_i provided $A < k_i / (\sum_{j=0}^{n-1} (|\mu_i| + \varepsilon)^j)$. Let $\delta^* = \min_{1 \le i \le s} k_i / (\sum_{j=0}^{n-1} (|\mu_i| + \varepsilon)^j)$. Then from Rouché's Theorem $h(z) \equiv f(z) + g(z)$ has exactly r_i zeros within an ε-neighbourhood of μ_i for each $i = 1, ..., s$ provided $\max_{0 \le j \le n-1} |b_j - b^*_j| < \delta^*$.

It is simple to use this result to obtain an equivalent result for the distinct eigenvalues $\mu_1, ..., \mu_s$ of a matrix, since the eigenvalues of an $n \times n$ matrix are just the zeros of the characteristic polynomial. Again we let r_i denote the multiplicity of μ_i and let $d = \min_{i \neq j} |\mu_i - \mu_j|$. Let $f(z) = |z\mathbf{I} - \mathbf{A}^*| = z^n + \sum_{j=0}^{n-1} b^*_j z^j$ and $h(z) = |z\mathbf{I} - \mathbf{A}| = z^n + \sum_{j=0}^{n-1} b_j z^j$. Then each b^*_j and b_j are just sums of products of the corresponding entries of the matrices $\mathbf{A}^* = (a^*_{ij})$ and $\mathbf{A} = (a_{ij})$ respectively. Hence each b_j is a jointly continuous function of the entries of $\mathbf{A}$ and $b_j \to b^*_j$ as $\mathbf{A} \to \mathbf{A}^*$. Therefore for any $\delta^* > 0$ there exists a $\delta > 0$ such that $\max_j |b_j - b^*_j| < \delta^*$ provided $\max_{i,j} |a_{ij} - a^*_{ij}| < \delta$. Then it immediately follows from the result for polynomials that for every $0 < \varepsilon < d/2$ there exists a $\delta^* > 0$, and hence a $\delta > 0$, such that exactly r_i eigenvalues of $\mathbf{A}$ lie within an ε-neighbourhood of μ_i for $i = 1, ..., s$, provided $\max_{i,j} |a_{ij} - a^*_{ij}| < \delta$.

Now denote the distinct eigenvalue $\mu_1, ..., \mu_s$ of $\mathbf{A}^*$ by $\mu_1(\mathbf{A}^*), ..., \mu_s(\mathbf{A}^*)$. If $\mu_i(\mathbf{A}^*)$ has multiplicity one then there is exactly one eigenvalue of $\mathbf{A}$ in its ε-neighbourhood, provided $\max_{i,j} |a_{ij} - a^*_{ij}| < \delta$. We define this eigenvalue to be $\mu_i(\mathbf{A})$. Hence for every $0 < \varepsilon < d/2$ there exists a $\delta > 0$ such that $|\mu_i(\mathbf{A}) - \mu_i(\mathbf{A}^*)| < \varepsilon$ provided $\max_{i,j} |a_{ij} - a^*_{ij}| < \delta$, which establishes the continuity and limiting results.

□

We now turn our attention to matrices in the class $\mathfrak{B}$ of non-negative, non-reducible, finite, square matrices of a particular form. The entries of $\mathbf{B}$ are each taken to be analytic functions of a single parameter θ, we therefore write the matrix as $\mathbf{B}(\theta)$. Lemma A.1 implies that $\rho(\mathbf{B}(\theta))$ may be defined for θ in an open neighbourhood of the complex plane centred on θ_0 for each $\theta_1 < \theta < \theta_2$. It is a continuous function of θ in each such neighbourhood and can be shown to be analytic in the (possibly restricted) neighbourhood of each θ_0 for which $trace(Adj(\rho(\mathbf{B}(\theta)\mathbf{I} - \mathbf{B}(\theta)))) \neq 0$. A simple proof of the analyticity is given. It may also be derived from Theorem 9 of Bochner and Martin [B9].

In particular we are interested in results when each $\{\mathbf{B}(\theta)\}_{ij}$ is the Laplace transform of a non-negative function, where these entries all exist in some range $\theta_1 < Re(\theta) < \theta_2$. In this case it is easily seen that the non-zero entries of $\mathbf{B}(\theta)$ are superconvex for θ real in the range (θ_1, θ_2). The convexity of $\rho(\mathbf{B}(\theta))$, for θ real with $\theta_1 < \theta < \theta_2$, may then be established. The method used is a generalisation of that of Kingman [K4] for positive matrices. The proof requires results concerning the class of superconvex functions given in Lemma A.2. A positive function $g(\theta)$, for $\theta_1 < \theta < \theta_2$, is said to be superconvex if $\log(g(\theta))$ is convex in the range, i.e. if $g(x\theta + (1-x)\phi) \leq g(\theta)^x g(\phi)^{1-x}$ for all $\theta_1 < \theta, \phi < \theta_2$ and all $0 \leq x \leq 1$.

LEMMA A.2. *The class $\mathfrak{S}$ of superconvex functions of θ, for $\theta_1 < \theta < \theta_2$, is closed under addition, multiplication and raising to a positive power. In addition if g_n is in the class for $n \geq 1$ and if g_n tends to a limit g, which is a positive function, as n tends to infinity, then g is also in the class.*

PROOF. The closure under multiplication and raising to a positive power and the result on limits follow trivially from the defining inequality for superconvex functions. We only need to consider closure under addition.

We first prove an inequality, which may be derived from Holder's inequality, namely that $1 + u^x v^{1-x} \leq (1+u)^x (1+v)^{1-x}$ for any positive u and v and $0 \leq x \leq 1$. The inequality trivially holds when $u = v$. Now consider the case when $u \neq v$. Define

$$f(x) = 1 + v\left(\frac{u}{v}\right)^x - (1+v)\left(\frac{1+u}{1+v}\right)^x.$$

Then $f(x)$ is continuous with $f(0) = f(1) = 0$, so that a minimum or maximum must occur for some $0 < x < 1$. Now

$$f'(x) = v\left(\frac{u}{v}\right)^x \log\left(\frac{u}{v}\right) - (1+v)\left(\frac{1+u}{1+v}\right)^x \log\left(\frac{(1+u)}{(1+v)}\right).$$

For any x^* such that $f(x^*) = 0$ it is easily seen that $f''(x^*) > 0$. Hence x^* is unique and $f(x)$ is minimised at $x = x^*$. It then follows immediately that $f(x) \leq 0$ for $0 \leq x \leq 1$ and hence $1 + u^x v^{1-x} \leq (1+u)^x(1+v)^{1-x}$ for x in this range.

Let g and h be any two superconvex functions in $\mathfrak{S}$ and consider any $\theta_1 < \theta, \phi < \theta_2$ and $0 \leq x \leq 1$. Using the superconvexity of g and h and the above inequality we obtain

$$\begin{aligned} g(x\theta + (1-x)\phi) + h(x\theta + (1-x)\phi) &\leq g(\theta)^x g(\phi)^{1-x} + h(\theta)^x h(\phi)^{1-x} \\ &= g(\theta)^x g(\phi)^{1-x}\left(1 + \left(\frac{h(\theta)}{g(\theta)}\right)^x \left(\frac{h(\phi)}{g(\phi)}\right)^{1-x}\right) \\ &\leq g(\theta)^x g(\phi)^{1-x}\left(1 + \frac{h(\theta)}{g(\theta)}\right)^x \left(1 + \frac{h(\phi)}{g(\phi)}\right)^{1-x} \\ &= (g(\theta) + h(\theta))^x (g(\phi) + h(\phi))^{1-x}. \end{aligned}$$

Hence the class $\mathfrak{S}$ is closed under addition.

□

THEOREM A.3. *Consider a matrix $\mathbf{B}(\theta) = (b_{ij}(\theta))$, which is not the 1×1 zero matrix, and lies in the class of non-negative, non-reducible, finite, square matrices for θ real with $\theta_1 < \theta < \theta_2$. Let $b_{ij}(\theta)$ be analytic for $\theta_1 < Re(\theta) < \theta_2$ for all i, j. Also define $T(\theta, i; i_1, ..., i_r) = \prod_{s=1}^{r} \{\mathbf{B}(\theta)\}_{i_s i_{s+1}}$, where $i_{r+1} = i_1 = i$ and the sequence $i_1, i_2, ..., i_r$ is distinct. Then the following results hold.*

1. *If θ_0 is real and $\theta_1 < \theta_0 < \theta_2$, then there exists a $\delta_0 > 0$ such that the definition of $\rho(\mathbf{B}(\theta))$ may be extended to complex values of θ in an open ball radius δ_0 centred at θ_0. There exists an open ball centred on θ_0 of radius $0 < \delta^* \leq \delta_0$ in which $\rho(\mathbf{B}(\theta))$ is an analytic function of θ.*
2. *There exists an open ball centred on θ_0 in which $\rho(\mathbf{B}(\theta))$ is defined and is analytic, where the entries of the corresponding right and left eigenvectors, $(\mathbf{u}(\theta))'$ and $\mathbf{v}(\theta)$, with first entries unity and the entries of the corresponding idempotent are also analytic.*
3. *If θ_2 is real and finite and $\lim_{\theta \uparrow \theta_2} \mathbf{B}(\theta)$ exists and is non-reducible, then $\lim_{\theta \uparrow \theta_2} \rho(\mathbf{B}(\theta)) = \rho(\lim_{\theta \uparrow \theta_2} \mathbf{B}(\theta))$. In addition, when $\mathbf{B}(\theta_2)$ has all finite entries, then $\lim_{\theta \uparrow \theta_2} (\mathbf{u}(\theta))' = (\mathbf{u}(\theta_2))'$ and $\lim_{\theta \uparrow \theta_2} \mathbf{v}(\theta) = \mathbf{v}(\theta_2)$. A similar result holds as $\theta \downarrow \theta_1$.*
4. *When $\theta_2 = \infty$, then if there exists at least one integer i and sequence $i_1, i_2, ..., i_r$ for which $\lim_{\theta \to \infty} T(\theta; i; i_1, ..., i_r) = \infty$ then $\lim_{\theta \to \infty} \rho(\mathbf{B}(\theta)) = \infty$. Also when $\theta_2 = \infty$, if $\lim_{\theta \to \infty} T(\theta, i; i_1, ..., i_r) = 0$ for all i and all possible distinct sequences $i_1, ..., i_r$, then $\lim_{\theta \to \infty} \rho(\mathbf{B}(\theta)) = 0$.*
5. *When the eigenvalues of $\mathbf{B}(\theta_0)$ are distinct then there exists an open ball centred at θ_0 in which all the eigenvalues of $\mathbf{B}(\theta)$ are distinct, and the entries of the eigenvalues and the corresponding idempotents (as in part 2) are continuous functions of θ.*

6. *Let each $b_{ij}(\theta)$ be the Laplace transform of a non-negative function, which exists for θ real with $\theta_1 < Re(\theta) < \theta_2$. Then for θ real in this range, the $b_{ij}(\theta)$ which are non-zero are superconvex functions of θ, and hence $\rho(\mathbf{B}(\theta))$ is a superconvex function of θ. It is therefore strictly convex except in the degenerate case when $\rho(\mathbf{B}(\theta))$ is constant, which occurs when each $b_{ij}(\theta)$ is constant.*

PROOF.

1. Let the eigenvalues of $\mathbf{B}(\theta_0)$ be $\mu_1, \mu_2, ..., \mu_n$, where $\mu_1 = \rho(\mathbf{B}(\theta_0))$. Now from Theorem A.1 $\rho(\mathbf{B}(\theta_0))$ is real with $|\mu_j| \le \rho(\mathbf{B}(\theta_0))$ for all $j > 1$. Also from Theorem A.2 part 7 it has multiplicity 1. Hence $Re(\mu_j) < \rho(\mathbf{B}(\theta_0))$ for all $j > 1$. Let $d = \min_{\mu_i \ne \mu_j} |\mu_i - \mu_j|$, $k = \min_{j \ge 2}(\rho(\mathbf{B}(\theta_0)) - Re(\mu_j))$ and $\varepsilon = (1/2)\min(d, k)$. Then for any $i > 1$, if z and w lie respectively in the ε-neighbourhoods of $\rho(\mathbf{B}(\theta_0))$ and μ_i, it immediately follows that $Re(z) > Re(w)$.

   Then from Lemma A.1 and the result above there exists a $\delta > 0$ such that the following result holds. For every $\mathbf{A} = (a_{ij})$ with $\max_{i,j} |a_{ij} - b_{ij}(\theta_0)| < \delta$, $\mathbf{A}$ has a unique eigenvalue in the ε-neighbourhood of $\rho(\mathbf{B}(\theta_0))$, which is the eigenvalue of $\mathbf{A}$ with largest real part which we define to be $\rho(\mathbf{A})$. This eigenvalue has multiplicity 1. Now each $b_{ij}(\theta)$ is a continuous function of θ. Hence there exists a δ_0 with $0 < \delta_0 < \min(\theta_0, \Delta_v - \theta_0)$ such that $\max_{i,j} |b_{ij}(\theta) - b_{ij}(\theta_0)| < \delta$ provided that $|\theta - \theta_0| < \delta_0$. Hence $\rho(\mathbf{B}(\theta))$ has been defined as the unique eigenvalue of largest real part for all complex values of θ such that $|\theta - \theta_0| < \delta_0$.

   Now consider any θ^* such that $|\theta^* - \theta_0| < \delta_0$, so that $\rho(\mathbf{B}(\theta^*))$ has been defined. From Lemma A.1 $\rho(\mathbf{B}(\theta))$ is a continuous function of its entries $b_{ij}(\theta)$, and hence from the continuity of the $b_{ij}(\theta)$ it is a continuous function of θ, at $\theta = \theta^*$. This establishes the existence and continuity of $\rho(\mathbf{B}(\theta))$ in an open ball of radius δ_0 centred on θ_0.

   Finally consider the analyticity of $\rho(\mathbf{B}(\theta))$. Take $\delta_0 > 0$ such that $\rho(\mathbf{B}(\theta))$ is defined for $|\theta - \theta_0| < \delta_0$. Now $Adj(\lambda\mathbf{I} - \mathbf{B}(\theta))$ is a continuous function of λ and θ and, from Theorem A.2 part 8, $Adj(\rho(\mathbf{B}(\theta_0)\mathbf{I} - \mathbf{B}(\theta_0))) > \mathbf{0}$. Also we have shown the continuity of $\rho(\mathbf{B}(\theta))$ at $\theta = \theta_0$. Hence there exists a $0 < \delta_0^* < \delta_0$, such that $Re[Adj(\rho(\mathbf{B}(\theta))\mathbf{I} - \mathbf{B}(\theta))] > \mathbf{0}$ for $|\theta - \theta_0| < \delta_0^*$.

   Consider any $\theta = \theta^*$ in this range. We give a simple proof which shows that $\rho(\mathbf{B}(\theta))$ is differentiable at $\theta = \theta^*$ with continuous derivative, so that $\rho(\mathbf{B}(\theta))$ is an analytic function of θ for $|\theta - \theta_0| < \delta_0^*$.

   Let $\delta\rho = \rho(\mathbf{B}(\theta^* + \delta\theta)) - \rho(\mathbf{B}(\theta^*))$ and $\mathbf{C} = (c_{ij})$, where $c_{ij} = b_{ij}(\theta^* + \delta\theta) - b_{ij}(\theta^*)$. Now

$$
\begin{aligned}
0 &= |\rho(\mathbf{B}(\theta^* + \delta\theta))\mathbf{I} - \mathbf{B}(\theta^* + \delta\theta)| \\
&= |[\rho(\mathbf{B}(\theta^*))\mathbf{I} - \mathbf{B}(\theta^*)] + \delta\rho\mathbf{I} - \mathbf{C}| \\
&= |\rho(\mathbf{B}(\theta^*))\mathbf{I} - \mathbf{B}(\theta^*)| + \delta\rho[trace(Adj(\rho(\mathbf{B}(\theta^*))\mathbf{I} - \mathbf{B}(\theta^*))) + f(\delta\rho)] \\
&- \sum_i \sum_j [c_{ij}\{Adj(\rho(\mathbf{B}(\theta^*))\mathbf{I} - \mathbf{B}(\theta^*))\}_{ij}) + g_{ij}(\delta\rho, \{c_{st}\})] .
\end{aligned}
$$

Here $f(\delta\rho)$ involves first order and higher polynomial terms in $\delta\rho$ and each $g_{ij}(\delta\rho, \{c_{st}\})$ involves first order and higher polynomial terms in $\delta\rho$ and the c_{st}, so that each of these functions tends to zero as $\delta\theta$ tends to zero. Now $|\rho(\mathbf{B}(\theta^*))\mathbf{I} - \mathbf{B}(\theta^*)| = 0$ and $trace(Adj(\rho(\mathbf{B}(\theta^*))\mathbf{I} - \mathbf{B}(\theta^*))) \neq 0$. Hence for $\delta\theta$ sufficiently small,

$$\frac{\delta\rho}{\delta\theta} = \frac{\sum_{i,j}(c_{ij}/\delta\theta)\,[\{Adj(\rho(\mathbf{B}(\theta^*))\mathbf{I} - \mathbf{B}(\theta^*))\}_{ij} + g(\delta\rho, \{c_{st}\})]}{trace(Adj(\rho(\mathbf{B}(\theta^*))\mathbf{I} - \mathbf{B}(\theta^*))) + f(\delta\rho)}.$$

It follows immediately that the derivative $\rho'(\mathbf{B}(\theta^*))$ exists and is given by

$$\rho'(\mathbf{B}(\theta^*)) = \frac{\sum_{i,j} b'_{ij}(\theta^*)\{Adj(\rho(\mathbf{B}(\theta^*))\mathbf{I} - \mathbf{B}(\theta^*))\}_{ij}}{trace(Adj(\rho(\mathbf{B}(\theta^*))\mathbf{I} - \mathbf{B}(\theta^*)))},$$

for all $|\theta^* - \theta_0| < \delta_0^*$. The continuity of the derivative and the analyticity is then immediate.

2. Using part 1 of this theorem, there exists an open ball centred on θ_0 for which $Adj(\rho(\mathbf{B}(\theta))\mathbf{I} - \mathbf{B}(\theta))$ has entries which are analytic functions of θ and which are strictly positive at $\theta = \theta_0$. Hence there exists an open ball centred on θ_0 of radius δ in which $\rho(\mathbf{B}(\theta))$ is defined and is analytic and in which the entries of the adjoint matrix are non-zero. Now the columns of the adjoint matrix are proportional to the right eigenvector of $\mathbf{B}(\theta)$ corresponding to $\rho(\mathbf{B}(\theta))$ and the rows are proportional to the left eigenvector. It therefore follows that, for θ within this open ball centred at θ_0, the right and left eigenvectors, $\mathbf{v}(\theta)$ and $(\mathbf{u}(\theta))'$ can be chosen to have first entry one.

   Partition the matrix $\mathbf{B}(\theta)$ and the right eigenvector $\mathbf{v}(\theta)$ corresponding to $\rho(\mathbf{B}(\theta))$ so that

$$\mathbf{B}(\theta) = \begin{pmatrix} b_{11}(\theta) & (\mathbf{b}_{12}(\theta))' \\ \mathbf{b}_{21}(\theta) & \mathbf{B}_{22}(\theta) \end{pmatrix} \quad \text{and} \quad \mathbf{v}(\theta) = \begin{pmatrix} 1 \\ \mathbf{v}_2(\theta) \end{pmatrix}.$$

   Since $|\rho(\mathbf{B}(\theta))\mathbf{I} - \mathbf{B}_{22}(\theta_0)| = \{Adj(\rho(\mathbf{B}(\theta))\mathbf{I} - \mathbf{B}(\theta))\}_{11} \neq 0$ for all $|\theta - \theta_0| < \delta$, then the entries of $(\rho(\mathbf{B}(\theta))\mathbf{I} - \mathbf{B}_{22}(\theta))^{-1}$ are analytic functions of θ in this open ball. Therefore in this region

$$\mathbf{v}_2(\theta) = (\rho(\mathbf{B}(\theta)\mathbf{I} - \mathbf{B}_{22}(\theta))^{-1}\mathbf{b}_{21}(\theta),$$

   which has entries which are analytic functions of θ. Hence there is a right eigenvector of $\mathbf{B}(\theta)$ corresponding to $\rho(\mathbf{B}(\theta))$ which is analytic, namely the eigenvector with first entry one.

   The proof for the left eigenvector follows in an identical manner. Now the corresponding idempotent $\mathbf{E}(\theta) = \mathbf{v}(\theta)(\mathbf{u}(\theta))'/(\mathbf{u}(\theta))'\mathbf{v}(\theta)$. Also, since $(\mathbf{u}(\theta_0))'\mathbf{v}(\theta_0) > 0$, from the continuity of these eigenvectors there is an open ball centred on θ_0 of radius $\delta^* < \delta$ in which $(\mathbf{u}(\theta))'\mathbf{v}(\theta) \neq 0$. The analyticity of the entries of the idempotent in this open ball is then immediate.

3. Consider a non-reducible, non-negative, finite, square matrix $\mathbf{A}^* = (a_{ij}^*)$ with eigenvalues $\mu_1 = \rho(\mathbf{A}^*), \mu_2, ..., \mu_n$. As in part 1, $\rho(\mathbf{A}^*)$ is a real simple eigenvalue which is the unique eigenvalue of largest real part. Then from

Lemma A.1 there is a $\delta > 0$ so that for any matrix $\mathbf{A} = (a_{ij})$, for which $\max_{i,j} |a_{ij} - a^*_{ij}| < \delta$, there is a unique simple eigenvalue of largest real part which we define to be $\rho(\mathbf{A})$ and $\rho(\mathbf{A}) \to \rho(\mathbf{A}^*)$ as $\mathbf{A} \to \mathbf{A}^*$.

Now, if $\mathbf{B}(\theta_2)$ is finite and non-reducible, as $\theta \uparrow \theta_2$ the entries of $\mathbf{B}(\theta)$ tend to the entries of $\mathbf{B}(\theta_2)$. Hence from the result above $\rho(\mathbf{B}(\theta)) \to \rho(\mathbf{B}(\theta_2))$ as $\theta \uparrow \theta_2$. Therefore $\lim_{\theta \uparrow \theta_2} \rho(\mathbf{B}(\theta)) = \rho(\mathbf{B}(\theta_2))$.

The result that $\lim_{\theta \uparrow \theta_2} \mathbf{v}(\theta) = \mathbf{v}(\theta_2)$ then follows immediately from the definition of $\mathbf{v}(\theta)$ in part 2. The result for the left eigenvector follows in an identical manner.

Now consider the case when $\lim_{\theta \uparrow \theta_2} b_{ij}(\theta) = \infty$ for at least one i, j. Define $\mathbf{B}(\theta_2) = \lim_{\theta \uparrow \theta_2} \mathbf{B}(\theta)$. Suppose that there exists an i such that $\lim_{\theta \uparrow \theta_2} b_{ii}(\theta) = \infty$. By property 2 of Theorem A.2 $\rho(\mathbf{B}(\theta)) \geq b_{ii}(\theta)$ for all $\theta_1 < \theta < \theta_2$. Hence $\lim_{\theta \uparrow \theta_2} \rho(\mathbf{B}(\theta)) = \infty = \rho(\mathbf{B}(\theta_2))$. Now suppose $\lim_{\theta \uparrow \theta_2} b_{ij}(\theta) = \infty$ for some $i \neq j$. By relabelling we may take $i = 1$ and $j = 2$. Since $\lim_{\theta \uparrow \theta_2} \mathbf{B}(\theta)$ is non-reducible, there exists a sequence, which by relabelling may be written as $b_{23}(\theta), ..., b_{k-1,k}(\theta), b_{k1}(\theta)$ with limits all positive as $\theta \uparrow \theta_2$. If $\mathbf{C}(\theta) = (c_{ij}(\theta))$ is the cyclic matrix of order k with $c_{i,i+1}(\theta) = b_{i,i+1}(\theta)$ for $i = 1, ..., (k-1)$, $c_{k1}(\theta) = b_{k1}(\theta)$ and all other entries zero, then $\mathbf{C}(\theta)$ is bounded above by the corresponding k-dimensional principle minor of $\mathbf{B}(\theta)$. By properties 1 and 5 of Theorem A.2 we obtain $\rho(\mathbf{B}(\theta)) \geq \rho(\mathbf{C}(\theta)) = |b_{12}(\theta)b_{23}(\theta)...b_{k1}(\theta)|^{1/k}$ for $\theta_1 < \theta < \theta_2$. As $\theta \uparrow \theta_2$ the right-hand side of this inequality tends to infinity, hence $\lim_{\theta \uparrow \theta_2} \rho(\mathbf{B}(\theta)) = \infty = \rho(\mathbf{B}(\theta_2))$. Thus in all cases when $\lim_{\theta \uparrow \theta_2} \rho(\mathbf{B}(\theta))$ is non-reducible $\lim_{\theta \uparrow \theta_2} \rho(\mathbf{B}(\theta)) = \rho(\mathbf{B}(\theta_2))$.

The proofs when $\theta \downarrow \theta_1$ follow in identical fashion.

4. The first result follows as for the proof of part 3. Since there exists an i and $i_1, ..., i_r$ such that $\lim_{\theta \to \infty} T(\theta, i; i_1, ..., i_r) = \infty$ we take $\mathbf{C}(\theta)$ to be the cyclic matrix of order r with the $(s, s+1)^{th}$ entry $b_{i_s, i_{s+1}}(\theta)$ for $s = 1, ..., r-1$ and $(r, 1)^{th}$ entry $b_{i_r, i_1}(\theta)$. All other entries are zero. Then $\mathbf{C}(\theta)$ is bounded above by a principal minor of $\mathbf{B}(\theta)$ in permuted form. Therefore, as in part 3, $\rho(\mathbf{B}(\theta)) \geq \rho(\mathbf{C}(\theta)) = |b_{i_1,i_2}(\theta)b_{i_2,i_3}(\theta)...b_{i_r,i_1}(\theta)|^{1/r}$. Since the right hand side of the inequalities tends to infinity as $\theta \to \infty$, the first result is immediate, i.e. $\lim_{\theta \to \infty} \rho(\mathbf{B}(\theta)) = \infty$.

Now consider the second part where $\lim_{\theta \to \infty} T(\theta, i; i_1, ..., i_r) = 0$ for all i and all sequences $i_1, ..., i_r$. Take any $\varepsilon > 0$ and consider

$$|z\mathbf{I} - \mathbf{B}(\theta)| = z^n + \alpha_1(\theta)z^{n-1} + ... + \alpha_n(\theta).$$

Here $|\alpha_s(\theta)| \leq \sum |T(\theta, i; i_1, ..., i_s)|$, where the summation is over all i and all distinct sequences $i_1, ..., i_s$ of length s. But $\lim_{\theta \to \infty} T(\theta, i; i_1, ..., i_r) = 0$ for all integers i, all distinct sequences $i_1, ..., i_r$ and all $r = 1, ..., n$. Therefore $\lim_{\theta \to \infty} \alpha_s(\theta) = 0$ for all $s = 1, ..., n$.

Let $g(z, \boldsymbol{\alpha}) = z^n + \{\boldsymbol{\alpha}\}_1 z^{n-1} + ... + \{\boldsymbol{\alpha}\}_n$. If $\boldsymbol{\alpha} = \mathbf{0}$, then the zeros of $g(z, \boldsymbol{\alpha})$ are all zero. Consider $\boldsymbol{\alpha} \in \Re^n$. Using results for the roots of a polynomial contained in the proof of Lemma A.1, there is an open neighbourhood D of $\boldsymbol{\alpha} = \mathbf{0}$ for which all the zeros of $g(z, \boldsymbol{\alpha})$ have modulus less than ε.

Since $\lim_{\theta\to\infty} \alpha_s(\theta) = 0$ for all $s = 1, ..., n$, there exists a θ_0 sufficiently large such that $(\alpha_1(\theta), ..., \alpha_n(\theta)) \in D$ for $\theta > \theta_0$. Hence $\rho(\mathbf{B}(\theta)) < \varepsilon$ for $\theta > \theta_0$. Therefore $\lim_{\theta\to\infty} \rho(\mathbf{B}(\theta)) = 0$ in this case.

5. If the eigenvalues of $\mathbf{B}(\theta_0)$ are distinct then, in a similar manner to the proof of part 1, using Lemma A.1 there exists an open ball centred on θ_0 in which the eigenvalues of $\mathbf{B}(\theta)$ stay distinct and tend to the appropriate eigenvalue of $\mathbf{B}(\theta_0)$ as θ tends to θ_0. The continuity at $\theta = \theta^*$ for any θ^* within this open ball also follows in a similar manner.

   Now consider the idempotent $\mathbf{E}_j(\theta)$ corresponding to $\mu_j(\theta)$. Since the eigenvalues are distinct,

$$\mathbf{E}_j(\theta) = \frac{1}{\prod_{i\neq j}(\mu_j(\theta) - \mu_i(\theta))} \prod_{i\neq j}(\mathbf{B}(\theta) - \mu_i(\theta)\mathbf{I}).$$

   The continuity of the idempotents follows immediately.

6. We first show that the non-zero entries of $\mathbf{B}(\theta) = (b_{ij}(\theta))$ are superconvex by showing that each $\log(b_{ij}(\theta))$ is convex.

   Consider the Laplace transform $A(\theta)$ of a non-negative function $a(x)$ for $x \in \Re$, so that $A(\theta) = \int_{x\in\Re} e^{\theta x} a(x)dx$. Now

$$\frac{d^2 \log(A(\theta))}{d\theta^2} = \frac{A(\theta)A''(\theta) - (A'(\theta))^2}{(A(\theta))^2}.$$

   Here $A''(\theta) = \int_{y\in\Re} y^2 e^{\theta y} a(y)dy$. Then $A(\theta)A''(\theta)$ may be written as

$$\int_{x\in\Re}\int_{y\in\Re} \frac{1}{2}(x^2 + y^2)e^{\theta(x+y)}a(x)a(y)dxdy.$$

   Hence

$$A(\theta)A''(\theta) - (A'(\theta))^2 = \int_{x\in\Re}\int_{y\in\Re} \frac{1}{2}(x - y)^2 e^{\theta(x+y)}a(x)a(y)dxdy > 0.$$

   Hence $\dfrac{d^2 \log(A(\theta))}{d\theta^2} > 0$, so $\log(A(\theta))$ is convex. Hence the non-zero entries of the matrix $\mathbf{B}(\theta)$ are superconvex.

   Now if $b_{ij}(\theta) = 0$ for some θ, it is zero for all θ. Note that the matrix $\mathbf{B}(\theta)$ may have k eigenvalues with modulus equal to $\rho(\mathbf{B}(\theta))$. From Theorem A.2 part 11 there exists an integer $s > 0$ such that

$$(\mathbf{B}(\theta))^s = \begin{pmatrix} \mathbf{A}_{11}(\theta) & \mathbf{0} & \dots & \mathbf{0} \\ \mathbf{0} & \mathbf{A}_{22}(\theta) & \dots & \mathbf{0} \\ \vdots & \vdots & \ddots & \vdots \\ \mathbf{0} & \mathbf{0} & \dots & \mathbf{A}_{kk}(\theta) \end{pmatrix},$$

   where each $\mathbf{A}_{ii}(\theta)$ is a positive matrix with eigenvalue $(\rho(\mathbf{B}(\theta)))^s$ and all other eigenvalues have modulus strictly less than $(\rho(\mathbf{B}(\theta)))^s$. Since the zeros of $\mathbf{B}(\theta)$, and hence of $(\mathbf{B}(\theta))^s$, occur in the same places for all θ, the structure of $(\mathbf{B}(\theta))^s$ will be the same for all θ, so that this result holds for all $\theta_1 < \theta < \theta_2$.

   Now consider

$$\lim_{n\to\infty}\left[\frac{1}{k}trace\left((\mathbf{B}(\theta))^{sn}\right)\right]^{1/sn}.$$

From Lemma A.2 this function is superconvex. It is easily seen that the limit is just $\rho(\mathbf{B}(\theta))$. Hence $\rho(\mathbf{B}(\theta))$ is superconvex.

Now from part 1 and Theorem A.1, $\rho(\mathbf{B}(\theta))$ is analytic and is positive for θ real such that $\theta_1 < \theta < \theta_2$. Since it is superconvex,

$$\rho''(\mathbf{B}(\theta)) \geq (\rho'(\mathbf{B}(\theta)))^2/\rho(\mathbf{B}(\theta)) \geq 0.$$

This inequality is strict if $\rho'(\mathbf{B}(\theta)) \neq 0$. Since the Perron-Frobenius root is analytic, it is therefore strictly convex except in the degenerate case where $\rho(\mathbf{B}(\theta))$ is constant, which occurs when all the entries of $\mathbf{B}(\theta)$ do not depend on θ.

□

Appendix B. Non-negative solutions of a system of equations

In this section we prove theorems concerning the properties of the non-negative solutions of certain systems of equations. These theorems are used throughout the book and are of major importance both for the non-spatial analysis of Chapter 2 and for the spatial analysis in the ensuing chapters. They are also used in Chapter 8 to study the equilibrium solutions of the $S \to I \to S$ epidemic.

Consider the non-spatial analysis of Chapter 2. When $\rho(\mathbf{\Gamma})$ is finite, the final size equations are given by equation (2.19) i.e.

$$-\log(1-y_i) = \sum_{j=1}^{n} \gamma_{ij} y_j + a_i, \quad (i = 1, ..., n). \tag{B.1}$$

The same system of equations arises in the spatial analysis.

Now consider the equilibrium equations for the $S \to I \to S$ epidemic of Chapter 8. These are given by equations (8.11). When the μ_i are non-zero, these may be rewritten in the form

$$\frac{y_i}{1-y_i} = \sum_{j=1}^{n} \frac{\sigma_j \lambda_{ij}}{\mu_i} y_j, \quad (i = 1, ..., n). \tag{B.2}$$

Equations (B.1) and (B.2) are both special cases of the system of equations

$$f(y_i) = \sum_{j=1}^{n} b_{ij} y_j + a_i, \quad (i = 1, ..., n), \tag{B.3}$$

where $f(y)$ is a continuous, strictly monotone increasing function on $[0, 1)$, such that $f(0) = 0$, $f'(0) = 1$ and $f''(y) > 0$, (i.e. $f(y)$ is convex) and $\lim_{y\uparrow 1} f(y) = \infty$.

The intuitive discussion of the final size equations in Chapter 2 using figures 2.1, 2.2 and 2.3 illustrates the possible non-negative solutions for the cases $n = 1$ and 2. The results for general n are similar. These results for equation (B.3) are now proved in two theorems. Let $z = f(y)$ and denote $f^{-1}(z)$ by $g(z)$. Then $g(z)$ is a strictly monotone increasing, concave function on $[0, \infty)$ with $g(0) = 0$, $g'(0) = 1$, $g''(z) < 0$ and $\lim_{z\to\infty} g(z) = 1$. These conditions on $g(z)$ hold for the two cases in which we are interested, i.e. $g(z) = 1 - e^{-z}$ and $g(z) = z/(1+z)$.

If we write $z_i = f(y_i)$, equations (B.3) may be rewritten in terms of g as

$$z_i = \sum_{j=1}^{n} b_{ij} g(z_j) + a_i, \quad (i = 1, ..., n). \tag{B.4}$$

This is the more convenient form for establishing the results. Theorem B.1 proves existence, uniqueness, continuity and differentiability results for solutions to equations (B.4) when the matrix $\mathbf{B} = (b_{ij})$ is a non-negative, non-reducible, finite square matrix. These results are proved in Radcliffe and Rass [R4] Lemma 1 for the special function $g(z) = 1 - e^{-z}$. A similar result appears in Heathcote and Thieme [H2]. The existence part of the proof is related to Theorem 4.11 of Krasnosel'skii [K7]. The notation $\mathbf{x} \not\geq \mathbf{y}$ is used when $\{\mathbf{x}\}_i < \{\mathbf{y}\}_i$ for some i.

Corollary B.1 uses these results to prove equivalent results for the solutions to equations (B.3). Theorem B.2 then extends the results to the case when the matrix $\mathbf{B} = (b_{ij})$ may be reducible. Results for reducible matrices are required since principle minors of non-reducible matrices can be reducible. These arise when partitioning of the infection matrix is necessary. Theorem B.2 gives results for both the non-reducible and reducible case. It is the major theorem in this appendix and is used throughout the monograph.

THEOREM B.1 (NON-NEGATIVE SOLUTIONS OF A SYSTEM OF EQUATIONS INVOLVING A CONCAVE FUNCTION). *Let* $\mathbf{z} = \mathbf{G}(\mathbf{a}, \mathbf{z})$ *be an equation in* $[0, \infty)^n$, *where* $\mathbf{G}(\mathbf{a}, \mathbf{z}) = \mathbf{a} + \mathbf{B}y$ *with* $\{\mathbf{y}\}_i = g(\{\mathbf{z}\}_i)$. *Also let* $g(z)$ *be a strictly increasing, continuous function of* z *for* $z \geq 0$, *with* $g(0) = 0$, $g'(0) = 1$, $\lim_{z\to\infty} g(z) = 1$ *and* $g''(z) < 0$ *for* $z \geq 0$. *Here* $\mathbf{B} = (b_{ij})$ *is a non-negative, non-reducible square matrix and* $\mathbf{a} = (a_i) \geq \mathbf{0}$.

When $\mathbf{a} \neq \mathbf{0}$, *then* $\mathbf{z} = \mathbf{G}(\mathbf{a}, \mathbf{z})$ *has a unique solution* $\mathbf{z} = \mathbf{z}(\mathbf{a})$, *which is positive. If* $\mathbf{a} = \mathbf{0}$, *then* $\mathbf{z} = \mathbf{0}$ *is a solution to* $\mathbf{z} = \mathbf{G}(\mathbf{0}, \mathbf{z})$. *No other solution is possible when* $\rho(\mathbf{B}) \leq 1$, *and we define* $\mathbf{z}(\mathbf{0}) = \mathbf{0}$ *in this case. When* $\mathbf{a} \neq \mathbf{0}$ *and* $\rho(\mathbf{B}) > 1$, *there exists a unique non-zero solution* $\mathbf{z} = \mathbf{z}(\mathbf{0})$ *to* $\mathbf{z} = \mathbf{G}(\mathbf{0}, \mathbf{z})$ *which is positive.*

The solution $\mathbf{z}(\mathbf{a})$ *is a continuous, increasing function of* $\mathbf{a}$ *for* $\mathbf{a}$ *in* $[0, \infty)^n$. *Also* $\mathbf{z}(\mathbf{a})$ *is twice differentiable in* $[0, \infty)^n$ *with*

$$\frac{\partial\{\mathbf{z}(\mathbf{a})\}_i}{\partial\{\mathbf{a}\}_j} > 0 \quad \textit{and} \quad \frac{\partial^2\{\mathbf{z}(\mathbf{a})\}_i}{\partial\{\mathbf{a}\}_j \partial\{\mathbf{a}\}_k} \leq 0 \quad \textit{for all } i, j, k.$$

PROOF. Observe that $\{\mathbf{G}(\mathbf{a}, \mathbf{z})\}_i$ is strictly monotone increasing in a_i and in each $\{\mathbf{z}\}_j$ for which $b_{ij} \neq 0$. It is monotone increasing in $\mathbf{a}$ and $\mathbf{z}$.

First consider the existence of solutions to $\mathbf{z} = \mathbf{G}(\mathbf{a}, \mathbf{z})$. Observe that $\mathbf{z} = \mathbf{0}$ is a solution only if $\mathbf{a} = \mathbf{0}$.

When $\rho(\mathbf{B}) \leq 1$ and $\mathbf{a} = \mathbf{0}$, let $\mathbf{x}' > \mathbf{0}'$ be the left eigenvector of $\mathbf{B}$ corresponding to $\rho(\mathbf{B})$ and let $\mathbf{z}(\mathbf{0})$ be a solution to $\mathbf{z} = \mathbf{G}(\mathbf{0}, \mathbf{z})$. Then $\mathbf{x}'\mathbf{z}(\mathbf{0}) = \rho(\mathbf{B})\mathbf{x}'\mathbf{y}(\mathbf{0})$, where $\{\mathbf{y}(\mathbf{0})\}_i = g(\{\mathbf{z}(\mathbf{0})\}_i)$, so that $\sum_{i=1}^{n} \{\mathbf{x}\}_i(\{\mathbf{z}(\mathbf{0})\}_i - \rho(\mathbf{B})g(\{\mathbf{z}(\mathbf{0})\}_i)) = 0$. Now $g(z) = g'(\zeta)z$ for some ζ such that $0 \le \zeta \le z$ and $g'(\zeta) \le g'(0) = 1$ with equality only if $z = 0$. Thus $g(z) < z$ for $z > 0$, and hence $(z - \rho(\mathbf{B})g(z)) > 0$ for $z > 0$. So the ith term in the summation is positive unless $\{\mathbf{z}(\mathbf{0})\}_i = 0$. Hence $\mathbf{z} = \mathbf{z}(\mathbf{0}) = \mathbf{0}$ is the only solution to $\mathbf{z} = \mathbf{G}(\mathbf{0}, \mathbf{z})$.

Consider next the case $\rho(\mathbf{B}) \le 1$ and $\mathbf{a} \ne \mathbf{0}$. (The construction of the solution in this paragraph also works when $\rho(\mathbf{B}) > 1$ and $\mathbf{a} \ne \mathbf{0}$) Define $\mathbf{u}^{(0)}(\mathbf{a}) = \mathbf{a}$ and $\mathbf{u}^{(N)}(\mathbf{a}) = \mathbf{G}(\mathbf{a}, \mathbf{u}^{(N-1)}(\mathbf{a}))$ for $N = 1, 2, ...$ Now $\mathbf{G}(\mathbf{a}, \mathbf{z})$ is an increasing function of $\mathbf{z}$. Hence $\mathbf{u}^{(1)}(\mathbf{a}) = \mathbf{G}(\mathbf{a}, \mathbf{a}) \ge \mathbf{G}(\mathbf{a}, \mathbf{0}) = \mathbf{a} = \mathbf{u}^{(0)}(\mathbf{a})$ and, using induction, $\mathbf{u}^{(N)}(\mathbf{a})$ is a monotone increasing sequence. The sequence is bounded above since $\mathbf{u}^{(N)}(\mathbf{a}) \le \mathbf{a} + \mathbf{B1}$ for $N \ge 0$. Hence it converges to a limit $\mathbf{z}(\mathbf{a})$ satisfying $\mathbf{z}(\mathbf{a}) = \mathbf{G}(\mathbf{a}, \mathbf{z}(\mathbf{a}))$. Now $\mathbf{a} \ne \mathbf{0}$. Hence for some j, $\{\mathbf{a}\}_j > 0$ and therefore $\{\mathbf{u}^{(0)}(\mathbf{a})\}_j > 0$. This then implies that $\{\mathbf{u}^{(1)}(\mathbf{a})\}_i > 0$ for all i such that $\{\mathbf{u}^{(0)}(\mathbf{a})\}_i > 0$ and/or $\{\mathbf{u}^{(0)}(\mathbf{a})\}_j > 0$ and $b_{ij} \ne 0$. Using the non-reducibility of $\mathbf{B}$ we obtain $\mathbf{u}^{(n-1)}(\mathbf{a}) > \mathbf{0}$ and hence $\mathbf{z}(\mathbf{a}) > \mathbf{0}$. Therefore $\mathbf{z} = \mathbf{z}(\mathbf{a})$ is a positive solution to $\mathbf{z} = \mathbf{G}(\mathbf{a}, \mathbf{z})$.

Suppose $\rho(\mathbf{B}) > 1$ and let $\mathbf{w} > \mathbf{0}$ be the right eigenvector of $\mathbf{B}$ corresponding to $\rho(\mathbf{B})$. Take $0 < \alpha < 1$ such that $\alpha\rho(\mathbf{B}) > 1$. Now $g(z) = g'(\zeta)z$ where $0 < \zeta < z$ and $g'(z) \uparrow 1$ as $z \downarrow 0$. So $g(z) > \alpha z$ for z sufficiently small. Hence $\exists\ \varepsilon$ such that $g(\varepsilon\{\mathbf{w}\}_j) \ge \alpha\varepsilon\{\mathbf{w}\}_j$ for $j = 1, ..., n$ Define $\mathbf{u}^{(0)}(\mathbf{a}) = \varepsilon\mathbf{w}$ and $\mathbf{u}^{(N)}(\mathbf{a}) = \mathbf{G}(\mathbf{a}, \mathbf{u}^{N-1)}(\mathbf{a}))$ for $N = 1, 2, ...$ Then $\mathbf{u}^{(1)}(\mathbf{a}) \ge \mathbf{G}(\mathbf{0}, \varepsilon\mathbf{w}) \ge \alpha\varepsilon\mathbf{B}\mathbf{w} = \alpha\rho(\mathbf{B})\varepsilon\mathbf{w} > \mathbf{u}^{(0)}(\mathbf{a}) > \mathbf{0}$. Using induction, $\mathbf{u}^{(N)}(\mathbf{a})$, for $N \ge 0$, is a monotone increasing sequence of positive vectors. This sequence is bounded above by $\mathbf{a} + \mathbf{B1}$, so that it tends to a limit $\mathbf{z}(\mathbf{a}) > \mathbf{0}$ satisfying $\mathbf{z}(\mathbf{a}) = \mathbf{G}(\mathbf{a}, \mathbf{z}(\mathbf{a}))$. Then $\mathbf{z} = \mathbf{z}(\mathbf{a})$ is a positive solution to $\mathbf{z} = \mathbf{G}(\mathbf{a}, \mathbf{z})$. Note that $\mathbf{z} = \mathbf{0}$ is only also a solution if $\mathbf{a} = \mathbf{0}$.

We now show that if $\mathbf{a} \ne \mathbf{0}$ and/or $\rho(\mathbf{B}) > 1$, then $\mathbf{z}(\mathbf{a})$ is the unique non-zero solution of $\mathbf{z} = \mathbf{G}(\mathbf{a}, \mathbf{z})$. Note that if $\mathbf{a} = \mathbf{0}$ and $\rho(\mathbf{B}) \le 1$ we have already established that no non-zero solution is possible. In addition, using the non-reducibility of $\mathbf{B}$, the only solution of $\mathbf{z} = \mathbf{G}(\mathbf{a}, \mathbf{z})$ with $\{\mathbf{z}\}_i = 0$ for some i is $\mathbf{z} = \mathbf{0}$ (and this is only possible if $\mathbf{a} = \mathbf{0}$). Assume therefore that two distinct positive solutions $\mathbf{z}_1$ and $\mathbf{z}_2$ to $\mathbf{z} = \mathbf{G}(\mathbf{a}, \mathbf{z})$ are possible for some $\mathbf{a} \ne \mathbf{0}$ and/or $\rho(\mathbf{B}) > 1$. Without loss of generality we may assume that $\mathbf{z}_1 \not\ge \mathbf{z}_2$. Define $t_0 = \min_i(\{\mathbf{z}_1\}_i/\{\mathbf{z}_2\}_i)$. Then $0 < t_0 < 1$, $\{\mathbf{z}_1\}_i = t_0\{\mathbf{z}_2\}_i$ for some i and $\mathbf{z}_1 \ge t_0\mathbf{z}_2$. Now $g(z)$ is strictly concave, $g(0) = 0$ and $0 < t_0 < 1$. Hence $g(t_0 z) > t_0 g(z) + (1 - t_0)g(0) = t_0 g(z)$. Using this result together with the monotonicity of $\mathbf{G}(\mathbf{a}, \mathbf{z})$ we obtain $\mathbf{z}_1 = \mathbf{G}(\mathbf{a}, \mathbf{z}_1) \ge \mathbf{G}(\mathbf{a}, t_0\mathbf{z}_2) > t_0\mathbf{G}(\mathbf{a}, \mathbf{z}_2) = t_0\mathbf{z}_2$. Hence $\mathbf{z}_1 > t_0\mathbf{z}_2$, which contradicts the definition of t_0. The uniqueness of the non-zero solution is thus established.

It is simple to establish that $\mathbf{z}(\mathbf{a})$ is monotone increasing in $\mathbf{a}$, using the construction of $\mathbf{z}(\mathbf{a})$ from the existence proofs.

If $\mathbf{a} = \mathbf{0}$, $\mathbf{a}^* \ne \mathbf{0}$ and $\rho(\mathbf{B}) \le 1$, then $\mathbf{z}(\mathbf{a}^*) > \mathbf{0} = \mathbf{z}(\mathbf{0})$. If $\mathbf{a}^* \ge \mathbf{a}$ with $\mathbf{a} \ne \mathbf{0}$ and $\rho(\mathbf{B}) \le 1$ and if $\mathbf{u}^{(0)}(\mathbf{a}) = \mathbf{a}$, $\mathbf{u}^{(0)}(\mathbf{a}^*) = \mathbf{a}^*$, $\mathbf{u}^{(N)}(\mathbf{a}) = \mathbf{G}(\mathbf{a}, \mathbf{u}^{(N-1)}(\mathbf{a}))$ and $\mathbf{u}^{(N)}(\mathbf{a}^*) = \mathbf{G}(\mathbf{a}^*, \mathbf{u}^{(N-1)}(\mathbf{a}^*))$ for $N = 1, 2, ...$, then $\mathbf{u}^{(0)}(\mathbf{a}^*) \ge \mathbf{u}^{(0)}(\mathbf{a})$ and inductively, using the monotonicity of $\mathbf{G}(\mathbf{a}, \mathbf{z})$, $\mathbf{u}^{(N)}(\mathbf{a}^*) \ge \mathbf{u}^{(N)}(\mathbf{a})$ for $N \ge 0$. Hence $\mathbf{z}(\mathbf{a}^*) \ge \mathbf{z}(\mathbf{a})$.

Now consider the case $\mathbf{a}^* \geq \mathbf{a}$ and $\rho(\mathbf{B}) > 1$, and let $\mathbf{u}^{(0)}(\mathbf{a}) = \mathbf{u}^{(0)}(\mathbf{a}^*) = \varepsilon \mathbf{w}$, where ε and $\mathbf{w}$ are defined as in the construction. Also define $\mathbf{u}^{(N)}(\mathbf{a}) = \mathbf{G}(\mathbf{a}, \mathbf{u}^{N-1)}(\mathbf{a}))$ and $\mathbf{u}^{(N)}(\mathbf{a}^*) = \mathbf{G}(\mathbf{a}^*, \mathbf{u}^{N-1)}(\mathbf{a}^*))$ for $N \geq 1$. It immediately follows that $\mathbf{u}^{(N)}(\mathbf{a}^*) \geq \mathbf{u}^{(N)}(\mathbf{a})$ for $N \geq 0$. Hence $\mathbf{z}(\mathbf{a}^*) \geq \mathbf{z}(\mathbf{a})$.

The continuity of $\mathbf{z}(\mathbf{a})$ may be established by contradiction. Now $\mathbf{0} < \mathbf{z}(\mathbf{a}) \leq \mathbf{a} + \mathbf{B1}$, so that $\mathbf{z}(\mathbf{a})$ is bounded for $\mathbf{a}$ in a bounded region of $[0, \infty)^n$. Suppose that $\mathbf{a}^* \neq \mathbf{0}$, then if $\mathbf{z}(\mathbf{a})$ is not continuous at $\mathbf{a} = \mathbf{a}^*$, there exists an $\epsilon > 0$ and a sequence $\{\mathbf{a}^{(N)}\}$ in a bounded region of $[0, \infty)^n$ with $\mathbf{a}^{(N)} \to \mathbf{a}^*$ as $N \to \infty$ and each $\mathbf{z}(\mathbf{a}^{(N)})$ outside a ball of radius ϵ about $\mathbf{z}(\mathbf{a}^*)$. Hence there exists a subsequence of $\{\mathbf{a}^{(N)}\}$, (which we also label $\{\mathbf{a}^{(N)}\}$) which also converges to $\mathbf{a}^*$ and $\mathbf{z}(\mathbf{a}^{(N)}) \to \mathbf{z}$ as $N \to \infty$ with $\mathbf{z} \neq \mathbf{z}(\mathbf{a}^*)$. But $\mathbf{z} = \mathbf{G}(\mathbf{a}^*, \mathbf{z})$ and hence, from uniqueness, $\mathbf{z} = \mathbf{z}(\mathbf{a}^*)$ and a contradiction is obtained.

When $\mathbf{a}^* = \mathbf{0}$ we proceed as above, but observe in addition that $\mathbf{z}(\mathbf{a}^{(N)}) \geq \mathbf{z}(\mathbf{0})$. Hence $\mathbf{z} \geq \mathbf{z}(\mathbf{0})$ and $\mathbf{z} \neq \mathbf{z}(\mathbf{0})$. Since $\mathbf{z} = \mathbf{G}(\mathbf{0}, \mathbf{z})$ a contradiction is again obtained.

Finally we consider the differentiability of $\mathbf{z}(\mathbf{a})$. Suppose that $a_j > 0$. Let $\boldsymbol{\delta} = \delta a_j \mathbf{e}_j$ where $\delta a_j > 0$ and $\mathbf{e}_j$ is a vector with j^{th} element 1 and all other elements 0. Then for $\mathbf{a}$ (and hence $\mathbf{a}+\delta$) in $[0, \infty)^n$, $\mathbf{z}(\mathbf{a}+\boldsymbol{\delta}) - \mathbf{z}(\mathbf{a}) = \boldsymbol{\delta} + \mathbf{BA}^*(\mathbf{z}(\mathbf{a}+\boldsymbol{\delta}) - \mathbf{z}(\mathbf{a}))$, where $\mathbf{A}^*$ is a diagonal matrix with $\{\mathbf{A}^*\}_{mm} = g'(x_m)$ for some x_m in the interval $(\{\mathbf{z}(\mathbf{a})\}_m, \{\mathbf{z}(\mathbf{a}) + \boldsymbol{\delta}\}_m)$. Now

$$\text{(B.5)} \qquad \mathbf{z}(\mathbf{a}) = \mathbf{G}(\mathbf{a}, \mathbf{z}(\mathbf{a})) \geq \mathbf{G}(\mathbf{0}, \mathbf{z}(\mathbf{a})) = \mathbf{BAz}(\mathbf{a}),$$

where $\mathbf{A}$ is a diagonal matrix with $\{\mathbf{A}\}_{mm} = g'(\zeta_m)$ for some $\zeta_m \in (0, \{\mathbf{z}(\mathbf{a})\}_i)$. From Theorem A.2 part 1 and Lemma A.1, if $\mathbf{B}^*$ is a non-negative, non-reducible matrix, then $\rho(\mathbf{B}^*)$ is a continuous strictly increasing function of the elements of $\mathbf{B}^*$. From this result and the fact that $g(z)$ is a strictly decreasing function of z, it follows that $\rho(\mathbf{BA}) > \rho(\mathbf{BA}^*)$. Let $\mathbf{w}'$ be the left eigenvector of $\mathbf{BA}$ corresponding to $\rho(\mathbf{BA})$. Then $\mathbf{w}'\mathbf{z}(\mathbf{a}) \geq \rho(\mathbf{BA})\mathbf{w}'\mathbf{z}(\mathbf{a})$. Therefore $\rho(\mathbf{BA}) \leq 1$, and hence $\rho(\mathbf{BA}^*) < 1$. Therefore, from Theorem A.2 part 4, $(\mathbf{I} - \mathbf{BA}^*)^{-1}$ exists and is positive for $\delta a_j \geq 0$. Therefore $\mathbf{z}(\mathbf{a} + \boldsymbol{\delta}) - \mathbf{z}(\mathbf{a}) = (\mathbf{I} - \mathbf{BA}^*)^{-1}\boldsymbol{\delta}$. Hence

$$\lim_{\delta a_j \downarrow 0} \frac{(\mathbf{z}(\mathbf{a} + \boldsymbol{\delta}) - \mathbf{z}(\mathbf{a}))}{\delta a_j} = \lim_{\boldsymbol{\delta} \downarrow \mathbf{0}} (\mathbf{I} - \mathbf{BA}^*)^{-1} \mathbf{e}_j = (\mathbf{I} - \mathbf{BC})^{-1} \mathbf{e}_j,$$

where $\mathbf{C}$ is a diagonal matrix with $\{\mathbf{C}\}_{mm} = g'(\{\mathbf{z}(\mathbf{a})\}_m)$.

Note that if $\delta a_j < 0$ we can prove the same result by using, in place of equation (B.5), the following:

$$\mathbf{z}(\mathbf{a} + \boldsymbol{\delta}) = \mathbf{G}(\mathbf{a} + \boldsymbol{\delta}, \mathbf{z}(\mathbf{a} + \boldsymbol{\delta})) \geq \mathbf{G}(\mathbf{0}, \mathbf{z}(\mathbf{a} + \boldsymbol{\delta})) = \mathbf{BAz}(\mathbf{a} + \boldsymbol{\delta}),$$

where $\mathbf{A}$ is a diagonal matrix with $\{\mathbf{A}\}_{mm} = g'(\zeta_m)$ for some $\zeta_m \in (0, \{\mathbf{z}(\mathbf{a} + \boldsymbol{\delta})\}_m)$.

Together these prove the existence of $\dfrac{\partial \mathbf{z}(\mathbf{a})}{\partial \{\mathbf{a}\}_j}$ and give the result that

$$\frac{\partial \mathbf{z}(\mathbf{a})}{\partial \{\mathbf{a}\}_j} = (\mathbf{I} - \mathbf{BC})^{-1} \mathbf{e}_j > \mathbf{0} \text{ for } \{\mathbf{a}\}_j > 0$$

Note that for $\{\mathbf{a}\}_j = 0$ we only prove differentiability from the right.

It then follows, by differentiating the identity $(\mathbf{I} - \mathbf{BC})^{-1}(\mathbf{I} - \mathbf{BC}) = \mathbf{I}$, that

$$\frac{\partial^2 \mathbf{z}(\mathbf{a})}{\partial\{\mathbf{a}\}_j \partial\{\mathbf{a}\}_k} = (\mathbf{I}-\mathbf{BC})^{-1}\mathbf{BD}(\mathbf{I}-\mathbf{BC})^{-1}\mathbf{e}_j,$$

where $\mathbf{D}$ is a diagonal matrix with $\{\mathbf{D}\}_{mm} = g''(\{\mathbf{z}(\mathbf{a})\}_m)\frac{\partial\{\mathbf{z}(\mathbf{a})\}_m}{\partial\{\mathbf{a}\}_k}$. Observe that $g''(\{\mathbf{z}(\mathbf{a})\}_m) < 0$, and $\frac{\partial\{\mathbf{z}(\mathbf{a})\}_m}{\partial\{\mathbf{a}\}_k} > 0$ and we have shown that $(\mathbf{I}-\mathbf{BC})^{-1}$ has positive elements. Hence $\frac{\partial^2\{\mathbf{z}(\mathbf{a})\}_i}{\partial\{\mathbf{a}\}_j\partial\{\mathbf{a}\}_k} \le 0$ for all i, j, k.

□

COROLLARY B.1. *Let $\mathbf{B} = (b_{ij})$ be a non-negative $n \times n$ non-reducible matrix and let $a_i \ge 0$ for $i = 1, ..., n$. Take $f(y)$ to be a continuous function of y on $[0, 1)$ such that $f(0) = 0$, $f'(0) = 1$ and $f''(y) > 0$ and $\lim_{y\uparrow 1} f(y) = \infty$.*

Consider the possible solutions of the system of equations

$$f(y_i) = \sum_{j=1}^{n} b_{ij}y_j + a_i, \tag{B.6}$$

for $i = 1, ..., n$. Denote the vector with i^{th} entry y_i by $\mathbf{y}$.

1. *When $\mathbf{a} \neq \mathbf{0}$, then there is a unique solution $\mathbf{y} = \boldsymbol{\eta}(\mathbf{B}, \mathbf{a})$, which is positive.*
2. *If $\mathbf{a} = \mathbf{0}$, then $\mathbf{y} = \mathbf{0}$ is a solution to equations (B.6). No other solution is possible when $\rho(\mathbf{B}) \le 1$, and we define $\boldsymbol{\eta}(\mathbf{B}, \mathbf{0}) = \mathbf{0}$ in this case. When $\mathbf{a} = \mathbf{0}$ and $\rho(\mathbf{B}) > 1$, there exists a unique non-zero solution $\mathbf{y} = \boldsymbol{\eta}(\mathbf{B}, \mathbf{0})$ to equations (B.6) which is positive.*
3. *The solution $\boldsymbol{\eta}(\mathbf{B}, \mathbf{a})$ is a continuous, increasing function of $\mathbf{a}$ for $\mathbf{a}$ in $[0, 1)^n$. Also it is twice differentiable in $[0, 1)^n$ with*

$$\frac{\partial\{\boldsymbol{\eta}(\mathbf{B}, \mathbf{a})\}_i}{\partial\{\mathbf{a}\}_j} > 0 \quad and \quad \frac{\partial^2\{\boldsymbol{\eta}(\mathbf{B}, \mathbf{a})\}_i}{\partial\{\mathbf{a}\}_j\partial\{\mathbf{a}\}_k} \le 0 \quad for\ all\ i, j, k.$$

PROOF. Define $g = f^{-1}$, where the domain of g is taken to be $[0, \infty)$. Then from the conditions on f, it is easily seen that $g(z)$ is a strictly increasing function of z for $z \ge 0$, with $g(0) = 0$, $g'(0) = 1$, $\lim_{z\to\infty} g(z) = 1$ and $g''(z) < 0$ for $z \ge 0$.

Then g satisfies the conditions of Theorem B.1 and a solution $\mathbf{y}$ to equations (B.6) for $\mathbf{y} \in [0, 1)^n$ corresponds to a solution $\mathbf{z} \in [0, \infty)^n$, where $\{\mathbf{z}\}_i = f(\{\mathbf{y}\}_i)$ and $\{\mathbf{y}\}_i = g(\{\mathbf{z}\}_i)$. Results 1 and 2 then follow immediately from Theorem B.1. The result that $\boldsymbol{\eta}(\mathbf{B}, \mathbf{a})$ is a continuous, increasing, twice differentiable function of $\mathbf{a}$ is also immediate. It only remains to show that the first derivative with respect to any entry of $\mathbf{a}$ is positive and all the second derivatives are non-positive.

Now

$$\frac{\partial\{\boldsymbol{\eta}(\mathbf{B}, \mathbf{a})\}_i}{\partial\{\mathbf{a}\}_j} = \frac{\partial g(\{\mathbf{z}(\mathbf{a})\}_i)}{\partial\{\mathbf{a}\}_j} = g'(\{\mathbf{z}(\mathbf{a})\}_i)\frac{\partial\{\mathbf{z}(\mathbf{a})\}_i}{\partial\{\mathbf{a}\}_j}.$$

This is positive since $g(z)$ is a strictly increasing function of z and, from Theorem B.1, the derivative of $\mathbf{z}(\mathbf{a})$ with respect to any entry of $\mathbf{a}$ is positive. Also

$$\frac{\partial^2\{\boldsymbol{\eta}(\mathbf{B},\mathbf{a})\}_i}{\partial\{\mathbf{a}\}_j\partial\{\mathbf{a}\}_k} = g''(\{\mathbf{z}(\mathbf{a})\}_i)\frac{\partial\{\mathbf{z}(\mathbf{a})\}_i}{\partial\{\mathbf{a}\}_j}\frac{\partial\{\mathbf{z}(\mathbf{a})\}_i}{\partial\{\mathbf{a}\}_k} + g'(\{\mathbf{z}(\mathbf{a})\}_i)\frac{\partial^2\{\mathbf{z}(\mathbf{a})\}_i}{\partial\{\mathbf{a}\}_j\partial\{\mathbf{a}\}_k}.$$

Now $g'(z) > 0$ and $g''(z) < 0$. Also, from Theorem B.1, the first derivatives of $\mathbf{z}(\mathbf{a})$ with respect to the entries of $\mathbf{a}$ are positive and the second derivatives are non-positive. Hence the second derivatives of $\boldsymbol{\eta}(\mathbf{B},\mathbf{a})$ with respect to the entries of $\mathbf{a}$ are non-negative. This completes the proof of the corollary.

□

Although we have restricted attention to models where the infection matrix is non-reducible, it is necessary in some places to partition this matrix, which can result in a reducible submatrix. Theorem B.2 therefore not only gives solutions when the matrix $\mathbf{B}$ is non-reducible, but also gives solutions of a particular form in the reducible case. There are other solutions possible when $\mathbf{B}$ is reducible which are irrelevant to the mathematical analysis considered in this monograph. A different solution to the one specified in Theorem B.2 part 3 may be found by taking $\mathbf{y}_i = \mathbf{0}$ for at least one i satisfying $1 \le i \le s$, $\mathbf{a}_i = \mathbf{0}$ and $\rho(\mathbf{B}_{ii}) > 1$. In general there will a multiplicity of such solutions. The solution specified in part 3 of Theorem B.2 is precisely the one required to give the limiting results in part 4 of that theorem.

THEOREM B.2. *Let* $\mathbf{B} = (b_{ij})$ *be a non-negative* $n \times n$ *matrix and let* $a_i \ge 0$ *for all* $i = 1, ..., n$. *Let* $f(y)$ *be a continuous function of* y *on* $[0,1)$ *such that* $f(0) = 0$, $f'(0) = 1$ *and* $f''(y) > 0$ *and* $\lim_{y\uparrow 1} f(y) = \infty$.

Consider the possible solutions of the system of equations

$$f(y_i) = \sum_{j=1}^{n} b_{ij}y_j + a_i, \tag{B.6}$$

for $i = 1, ..., n$, *where* $0 \le y_i < 1$.

The matrix $\mathbf{B}$ *is written in normal form (Gantmacher [G1] p.75) and the* n *dimensional vectors* $\mathbf{a}$ *and* $\mathbf{y}$, *with* $\{\mathbf{a}\}_i = a_i$ *and* $\{\mathbf{y}\}_i = y_i$, *are partitioned so that*

$$\mathbf{B} = \begin{pmatrix} \mathbf{B}_{11} & \mathbf{0} & \cdots & \mathbf{0} & \mathbf{0} & \cdots & \mathbf{0} \\ \mathbf{0} & \mathbf{B}_{22} & \cdots & \mathbf{0} & \mathbf{0} & \cdots & \mathbf{0} \\ \vdots & \vdots & \ddots & \vdots & \vdots & \ddots & \vdots \\ \mathbf{0} & \mathbf{0} & \cdots & \mathbf{B}_{ss} & \mathbf{0} & \cdots & \mathbf{0} \\ \mathbf{B}_{s+1,1} & \mathbf{B}_{s+1,2} & \cdots & \mathbf{B}_{s+1,s} & \mathbf{B}_{s+1,s+1} & \cdots & \mathbf{0} \\ \vdots & \vdots & \ddots & \vdots & \vdots & \ddots & \vdots \\ \mathbf{B}_{g1} & \mathbf{B}_{g2} & \cdots & \mathbf{B}_{gs} & \mathbf{B}_{g,s+1} & \cdots & \mathbf{B}_{gg} \end{pmatrix},$$

$$\mathbf{a} = \begin{pmatrix} \mathbf{a}_1 \\ \mathbf{a}_2 \\ \vdots \\ \mathbf{a}_g \end{pmatrix} \quad \text{and} \quad \mathbf{y} = \begin{pmatrix} \mathbf{y}_1 \\ \mathbf{y}_2 \\ \vdots \\ \mathbf{y}_g \end{pmatrix}.$$

Here $\mathbf{B}_{ii}$ is a non-reducible square matrix of order r_i and $\mathbf{a}_i$ and $\mathbf{y}_i$ are r_i dimensional vectors for $i = 1, ..., g$. In addition, if $s < g$, at least one $\mathbf{B}_{i1}, ..., \mathbf{B}_{i,i-1}$ is non-zero for each i such that $s+1 \leq i \leq g$.

1. *If $\mathbf{a}_i \neq \mathbf{0}$ for all $i = 1, ..., s$, then equations (B.6) have a unique solution $\mathbf{y} = \boldsymbol{\eta}(\mathbf{B}, \mathbf{a}) > \mathbf{0}$.*
2. *When $\mathbf{B}$ is non-reducible (i.e. $s = g = 1$) and $\mathbf{a} = \mathbf{0}$, then equations (B.6) admit the trivial solution $\mathbf{y} = \mathbf{0}$. If $\rho(\mathbf{B}) > 1$ there exists a unique non-trivial solution $\mathbf{y} = \boldsymbol{\eta}(\mathbf{B}, \mathbf{0}) > \mathbf{0}$. When $\rho(\mathbf{B}) \leq 1$ no non-trivial solution exists. In this case we define $\boldsymbol{\eta}(\mathbf{B}, \mathbf{0}) = \mathbf{0}$.*
3. *When $\mathbf{B}$ is reducible with at least one $\mathbf{a}_i = \mathbf{0}$ for $i = 1, ..., s$, there exists a solution $\mathbf{y}$ to equations (B.6) of a particular form. For each $i = 1, ..., s$, this form has $\mathbf{y}_i > \mathbf{0}$ if $\rho(\mathbf{B}_{ii}) > 1$ and/or $\mathbf{a}_i \neq \mathbf{0}$, and $\mathbf{y}_i = \mathbf{0}$ otherwise. Then successively (if $s < g$) for $i = s+1, ..., g$, it has $\mathbf{y}_i > \mathbf{0}$ if $\rho(\mathbf{B}_{ii}) > 1$ and/or $\sum_{j<i} \mathbf{B}_{ij}\mathbf{y}_j + \mathbf{a}_i \neq \mathbf{0}$. Again $\mathbf{y}_i = \mathbf{0}$ otherwise.*

   The solution is the unique solution of this form. We denote it by $\mathbf{y} = \boldsymbol{\eta}(\mathbf{B}, \mathbf{a})$, and partition it so that

$$\boldsymbol{\eta}(\mathbf{B}, \mathbf{a}) = \begin{pmatrix} \boldsymbol{\eta}_1(\mathbf{B}, \mathbf{a}) \\ \boldsymbol{\eta}_2(\mathbf{B}, \mathbf{a}) \\ \vdots \\ \boldsymbol{\eta}_g(\mathbf{B}, \mathbf{a}) \end{pmatrix}.$$

   The components of the solution are specified in terms of solutions based on a non-reducible matrix as follows. For each $i = 1, ..., s$, $\boldsymbol{\eta}_i(\mathbf{B}, \mathbf{a}) = \boldsymbol{\eta}(\mathbf{B}_{ii}, \mathbf{a}_i)$. Then, successively for each $i = s+1, ..., g$, $\boldsymbol{\eta}_i(\mathbf{B}, \mathbf{a}) = \boldsymbol{\eta}(\mathbf{B}_{ii}, \mathbf{b}_i)$, where $\mathbf{b}_i = \sum_{j<i} \mathbf{B}_{ij}\boldsymbol{\eta}_j(\mathbf{B}, \mathbf{a}) + \mathbf{a}_i$.

   Note that if $\mathbf{a} = \mathbf{0}$ and $\rho(\mathbf{B}) \leq 1$ only the trivial solution is possible. In this case $\boldsymbol{\eta}(\mathbf{B}, \mathbf{0}) = \mathbf{0}$.
4. *In all cases, if $\mathbf{b} \geq \mathbf{a} \geq \mathbf{0}$, then $\boldsymbol{\eta}(\mathbf{B}, \mathbf{b}) \geq \boldsymbol{\eta}(\mathbf{B}, \mathbf{a})$. Also, for any $\mathbf{a} \geq \mathbf{0}$,*

$$\lim_{\mathbf{b} \downarrow \mathbf{a}} \boldsymbol{\eta}(\mathbf{B}, \mathbf{b}) = \boldsymbol{\eta}(\mathbf{B}, \mathbf{a}).$$

PROOF. Partition equations (B.6) to correspond to the partitioning of $\mathbf{B}$ and let $\{\mathbf{z}_i\}_j = f(\{\mathbf{y}_i\}_j)$. Then equations (B.6) become

$$\mathbf{z}_i = \begin{cases} \mathbf{B}_{ii}\mathbf{y}_i + \mathbf{a}_i, & \text{for } i = 1, ..., s, \\ \mathbf{B}_{ii}\mathbf{y}_i + \mathbf{a}_i + \sum_{j<i} \mathbf{B}_{ij}\mathbf{y}_j, & \text{for } i = s+1, ..., g. \end{cases} \tag{B.7}$$

1. Let $\mathbf{y} = \boldsymbol{\eta}(\mathbf{B}, \mathbf{a})$ be partitioned as in (B.7) so that $\mathbf{y}_i = \boldsymbol{\eta}_i(\mathbf{B}, \mathbf{a})$. Then we show that the required solution is

$$\boldsymbol{\eta}_i(\mathbf{B}, \mathbf{a}) = \boldsymbol{\eta}(\mathbf{B}_{ii}, \mathbf{c}_i),$$

   where $\mathbf{c}_i = \mathbf{a}_i$ for $i = 1, ..., s$, and then $\mathbf{c}_i$ is defined successively for each $i = s+1, ..., g$ by

$$\mathbf{c}_i = \mathbf{a}_i + \sum_{j<i} \mathbf{B}_{ij}\boldsymbol{\eta}(\mathbf{B}_{jj}, \mathbf{c}_j).$$

In addition the solution is shown to be strictly positive.

It is shown, using induction on k, that $\mathbf{y}_i = \boldsymbol{\eta}(\mathbf{B}_{ii}, \mathbf{c}_i) > \mathbf{0}$ is the unique solution to the subset $i = 1, ..., k$ of equations (B.7). Taking $k = g$ then gives the required result.

Consider any $k \leq s$. For each $i = 1, ..., k$ the corresponding equation from (B.7) involves $\mathbf{y}_i$ only. The result therefore follows immediately from Corollary B.1. Now assume that the result holds for all $1 \leq k \leq K$ where $s \leq K < g$. Consider equation $i = K + 1$ from (B.7). Since the solution to the first K equations of (B.7) is unique and is positive, the solution $\mathbf{y}_{K+1}$ must satisfy the equation

$$\begin{aligned}\mathbf{z}_{K+1} &= \mathbf{B}_{(K+1),(K+1)}\mathbf{y}_{K+1} + \mathbf{a}_{K+1} + \sum_{j<(K+1)} \mathbf{B}_{(K+1),j}\boldsymbol{\eta}_j(\mathbf{B}_{jj}, \mathbf{c}_j) \\ &= \mathbf{B}_{(K+1),(K+1)}\mathbf{y}_{K+1} + \mathbf{c}_{K+1}.\end{aligned} \tag{B.8}$$

Now the normal form requires the existence of a $j < K + 1$ for which $\mathbf{B}_{(K+1),j} \gneq \mathbf{0}$. Also $\boldsymbol{\eta}(\mathbf{B}_{jj}, \mathbf{c}_j) > \mathbf{0}$. Hence $\mathbf{c}_{K+1} \gneq \mathbf{0}$. Then from Corollary B.1 there is a unique solution $\mathbf{y}_{K+1} = \boldsymbol{\eta}(\mathbf{B}_{(K+1),(K+1)}, \mathbf{c}_{K+1})$ to (B.8) which is positive. Therefore $\mathbf{y}_i = \boldsymbol{\eta}(\mathbf{B}_{ii}, \mathbf{c}_i) > \mathbf{0}$, for $i = 1, ..., K + 1$, is the unique solution to the first $K + 1$ of equations (B.7). By induction the result then holds for all $i = 1, ..., g$.

2. Since $\mathbf{B}$ is non-reducible, the result follows immediately from Corollary B.1.
3. As in part 1 of this proof, we show that the required solution is

$$\mathbf{y}_i = \boldsymbol{\eta}_i(\mathbf{B}, \mathbf{a}) = \boldsymbol{\eta}(\mathbf{B}_{ii}, \mathbf{c}_i),$$

where $\mathbf{c}_i$ is defined as in that proof. The proof follows in almost identical fashion. However the solution $\mathbf{y}_i = \boldsymbol{\eta}(\mathbf{B}_{ii}, \mathbf{c}_i)$ is no longer always positive.

Consider a solution to the first k of equations (B.7). These equations only involve $\mathbf{y}_i$ for $i = 1, ..., k$. When $1 \leq k \leq s$, each equation involves a single $\mathbf{y}_i$ only. From Corollary B.1 if $\mathbf{c}_i = \mathbf{a}_i$ is non-zero then the solution to equation i of (B.7), $\mathbf{y}_i = \boldsymbol{\eta}(\mathbf{B}_{ii}, \mathbf{c}_i)$, is unique and is strictly positive. When $\mathbf{a}_i = \mathbf{0}$, then $\mathbf{y}_i = \mathbf{0}$ is a solution. Another solution is only possible when $\rho(\mathbf{B}_{ii}) > 1$. In this case there is a unique non-zero solution $\mathbf{y}_i = \boldsymbol{\eta}(\mathbf{B}_{ii}, \mathbf{c}_i) > \mathbf{0}$. When $\rho(\mathbf{B}_{ii}) \leq 1$ then $\boldsymbol{\eta}(\mathbf{B}_{ii}, \mathbf{c}_i) = \mathbf{0}$. Here $\mathbf{c}_i = \mathbf{a}_i$. Hence for each $i = 1, ..., k$ there is a unique solution to equation i of (B.7) which has $\mathbf{y}_i > \mathbf{0}$ if $\rho(\mathbf{B}_{ii}) > 1$ and $\mathbf{y}_i = \mathbf{0}$ otherwise, namely $\mathbf{y}_i = \boldsymbol{\eta}(\mathbf{B}_{ii}, \mathbf{c}_i)$.

Now assume that, for all $1 \leq k \leq K$ with $s \leq K < g$ that the following result holds. For such a k, $\mathbf{y}_i = \boldsymbol{\eta}(\mathbf{B}_{ii}, \mathbf{c}_i)$ for $i = 1, ..., k$ is the unique solution of the subset of the first k of equations (B.7) which has $\mathbf{y}_i > \mathbf{0}$ if $\rho(\mathbf{B}_{ii}) > 1$ and/or $\sum_{j<i} \mathbf{B}_{ij}\mathbf{y}_j + \mathbf{a}_i \neq \mathbf{0}$ and $\mathbf{y}_i = \mathbf{0}$ otherwise.

Now consider equations $i = 1, ..., K+1$ of (B.7). To be a solution of the required form, from the inductive hypothesis with $k = K$, necessarily $\mathbf{y}_i = \boldsymbol{\eta}(\mathbf{B}_{ii}, \mathbf{c}_i)$ for $i = 1, ..., K$, and $\mathbf{y}_{K+1}$ satisfies equation (B.8). This equation has a unique positive solution $\mathbf{y}_{K+1} = \boldsymbol{\eta}(\mathbf{B}_{(K+1),(K+1)}, \mathbf{c}_{(K+1),(K+1)}) > \mathbf{0}$ if $\rho(\mathbf{B}_{(K+1),(K+1)}) > 1$ and/or $\mathbf{c}_{K+1} = \sum_{j<K+1} \mathbf{B}_{(K+1),j}\mathbf{y}_j + \mathbf{a}_{(K+1)} \neq \mathbf{0}$.

Otherwise $\mathbf{y}_{K+1} = \boldsymbol{\eta}(\mathbf{B}_{(K+1),(K+1)}, \mathbf{c}_{(K+1),(K+1)}) = \mathbf{0}$. The result then holds for $k = K+1$ and hence for all $1 \leq k \leq g$.

The result is then immediate.

4. The solutions may be written in the form

$$\boldsymbol{\eta}_i(\mathbf{B}, \mathbf{a}) = \boldsymbol{\eta}(\mathbf{B}_{ii}, \mathbf{c}_i) \text{ and } \boldsymbol{\eta}_i(\mathbf{B}, \mathbf{b}) = \boldsymbol{\eta}(\mathbf{B}_{ii}, \mathbf{d}_i),$$

where $\mathbf{c}_i = \mathbf{a}_i$ and $\mathbf{d}_i = \mathbf{b}_i$ for $i \leq s$. When $s < i \leq g$, then

$$\mathbf{c}_i = \mathbf{a}_i + \sum_{j<i} \mathbf{B}_{ij}\boldsymbol{\eta}(\mathbf{B}_{jj}, \mathbf{c}_j) \text{ and } \mathbf{d}_i = \mathbf{b}_i + \sum_{j<i} \mathbf{B}_{ij}\boldsymbol{\eta}(\mathbf{B}_{jj}, \mathbf{d}_j). \tag{B.9}$$

The result may easily be shown successively for each $i = 1, ..., g$. Consider any $i \leq s$. Then $\mathbf{d}_i = \mathbf{b}_i \geq \mathbf{a}_i = \mathbf{c}_i$. Hence $\mathbf{d}_i \downarrow \mathbf{c}_i$ as $\mathbf{b} \downarrow \mathbf{a}$. It follows immediately from Corollary B.1 that $\boldsymbol{\eta}(\mathbf{B}_{ii}, \mathbf{d}_i) \geq \boldsymbol{\eta}(\mathbf{B}_{ii}, \mathbf{c}_i)$ and that

$$\lim_{\mathbf{d}_i \downarrow \mathbf{c}_i} \boldsymbol{\eta}(\mathbf{B}_{ii}, \mathbf{d}_i) = \boldsymbol{\eta}(\mathbf{B}_{ii}, \mathbf{c}_i).$$

Therefore for each $i = 1, ..., s$, $\boldsymbol{\eta}_i(\mathbf{B}, \mathbf{b}) \geq \boldsymbol{\eta}_i(\mathbf{B}, \mathbf{a})$ and

$$\lim_{\mathbf{b} \downarrow \mathbf{a}} \boldsymbol{\eta}_i(\mathbf{B}, \mathbf{b}) = \boldsymbol{\eta}_i(\mathbf{B}, \mathbf{a}).$$

Now assume that these results hold for all i such that $1 \leq i \leq K < g$, where $K \geq s$. Consider $i = K+1$ where $\mathbf{c}_{K+1}$ and $\mathbf{d}_{K+1}$ are specified by (B.9). From our assumption it is easily seen that $\mathbf{d}_{K+1} \geq \mathbf{c}_{K+1}$ and $\mathbf{d}_{K+1} \downarrow \mathbf{c}_{K+1}$ as $\mathbf{b} \downarrow \mathbf{a}$. The result for $i = K+1$ then follows immediately from Corollary B.1. Hence the result holds for $i = K+1$ and therefore by induction holds for all $1 \leq i \leq g$. This completes the proof.

□

A theorem is now proved showing that $\boldsymbol{\eta}(\mathbf{B}, \mathbf{a}) \uparrow \boldsymbol{\eta}(\mathbf{A}, \mathbf{a})$ as $\mathbf{B} \uparrow \mathbf{A}$ where $\mathbf{a} \geq \mathbf{0}$ and both $\mathbf{A}$ and $\mathbf{B}$ are non-negative non-reducible finite matrices. An equivalent result is obtained for the case when $\mathbf{A}$ may have some infinite entries. These results are required in Chapter 5 when deriving results concerning the asymptotic shape of infection.

THEOREM B.3. *Let* $\mathbf{B} = (b_{ij})$ *be a finite, non-negative, non-reducible* $n \times n$ *matrix and let* $\mathbf{a} = (a_i) \geq \mathbf{0}$. *Also take* $f(y)$ *to be a continuous function of* y *on* $[0, 1)$ *such that* $f(0) = 0$, $f'(0) = 1$ *and* $f''(y) > 0$ *and* $\lim_{y \to 1} f(y) = \infty$. *When* $\rho(\mathbf{B}) > 1$ *and/or* $\mathbf{a} \neq \mathbf{0}$, *define* $y_i = \eta_i(\mathbf{B}, \mathbf{a})$, *for* $i = 1, ..., n$, *to be the unique positive solution, as obtained in Theorem B.2, to equations (B.6), so that*

$$f(\eta_i(\mathbf{B}, \mathbf{a})) = \sum_{j=1}^{n} b_{ij}\eta_j(\mathbf{B}, \mathbf{a}) + a_i, \tag{B.10}$$

for $i = 1, ..., n$, *where* $0 < \eta_i(\mathbf{B}, \mathbf{a}) < 1$. *When* $\rho(\mathbf{B}) \leq 1$ *and* $\mathbf{a} = \mathbf{0}$, *then, from Theorem B.2, equation (B.6) only admits the zero solution so that* $\eta_i(\mathbf{B}, \mathbf{a}) = 0$ *for all* i. *Define* $\boldsymbol{\eta}(\mathbf{B}, \mathbf{a}) = (\eta_i(\mathbf{B}, \mathbf{a}))$.

Then $\rho(\mathbf{B})$ is a continuous, strictly increasing function of the entries of $\mathbf{B}$. Also $\boldsymbol{\eta}(\mathbf{B},\mathbf{a})$ has entries which are increasing functions of the elements of $\mathbf{B}$.

Take $\mathbf{A}$ to be a non-negative, non-reducible matrix. Then, for $\mathbf{B}$ non-negative with $\{\mathbf{B}\}_{ij} = 0$ if $\{\mathbf{A}\}_{ij} = 0$, the following limiting results hold as $\mathbf{B} \uparrow \mathbf{A}$; there being four cases corresponding to different structures for $\mathbf{A}$,

1. *When $\mathbf{A}$ is a finite matrix with $\rho(\mathbf{A}) \le 1$ and $\mathbf{a} = \mathbf{0}$, then $\rho(\mathbf{B}) \uparrow \rho(\mathbf{A})$ as $\mathbf{B} \uparrow \mathbf{A}$ and $\boldsymbol{\eta}(\mathbf{B},\mathbf{0}) = \mathbf{0}$ for all $\mathbf{B} \le \mathbf{A}$.*
2. *When $\mathbf{A}$ is a finite matrix and $\rho(\mathbf{A}) > 1$ and/or $\mathbf{a} \ne \mathbf{0}$, then $\rho(\mathbf{B}) \uparrow \rho(\mathbf{A})$ and $\boldsymbol{\eta}(\mathbf{B},\mathbf{a}) \uparrow \boldsymbol{\eta}(\mathbf{A},\mathbf{a}) = (\eta_i(\mathbf{A},\mathbf{a}))$ as $\mathbf{B} \uparrow \mathbf{A}$. Here $y_i = \eta_i(\mathbf{A},\mathbf{a})$ is the unique positive solution to*

$$f(y_i) = \sum_{j=1}^{n} \{\mathbf{A}\}_{ij} y_j + a_i, \quad (i = 1, ..., n). \tag{B.11}$$

3. *When $\mathbf{A}$ has at least one infinite entry in each row, then $\rho(\mathbf{B}) \to \infty$ and $\boldsymbol{\eta}(\mathbf{B},\mathbf{a}) \uparrow \mathbf{1}$ as $\mathbf{B} \uparrow \mathbf{A}$.*
4. *When $\mathbf{A}$ has infinite entries in some but not all rows, by re-ordering if necessary $\mathbf{A}$ may be written in the form*

$$\mathbf{A} = \begin{pmatrix} \mathbf{A}_{11} & \mathbf{A}_{12} \\ \mathbf{A}_{21} & \mathbf{A}_{22} \end{pmatrix},$$

where $(\mathbf{A}_{21}\ \mathbf{A}_{22})$ has at least one infinite entry in each row, and $\mathbf{A}_{11}$ is $m \times m$ finite matrix which is written in normal form and $\mathbf{A}_{12}$ is finite. Then $\rho(\mathbf{B}) \to \infty$ as $\mathbf{B} \uparrow \mathbf{A}$. Also $\eta_i(\mathbf{B},\mathbf{a}) \uparrow 1$ for $i = m+1, ..., n$ and $\eta_i(\mathbf{B},\mathbf{a}) \uparrow \eta_i^*(\mathbf{A}_{11},\mathbf{b})$ for $i = 1, ..., m$ as $\mathbf{B} \uparrow \mathbf{A}$, where $\{\mathbf{b}\}_i = b_i = a_i + \{\mathbf{A}_{12}\mathbf{1}\}_i$ and $y_i = \eta_i^*(\mathbf{A}_{11},\mathbf{b})$ is the unique positive solution to

$$f(y_i) = \sum_{j=1}^{m} \{\mathbf{A}_{11}\}_{ij} y_j + b_i, \quad (i = 1, ..., m). \tag{B.12}$$

PROOF. First observe that, from Lemma A.1, $\rho(\mathbf{B})$ is a continuous function of the entries of $\mathbf{B}$ and from Theorem A.2 part 1 it is a strictly increasing function of the entries of $\mathbf{B}$. This establishes the first result.

Next consider any finite matrix $\mathbf{B}^* \ge \mathbf{B}$. Then $\mathbf{B}^*$ is non-reducible and

$$\begin{aligned} f(\eta_i(\mathbf{B}^*,\mathbf{a})) &= \sum_{j=1}^{n} \{\mathbf{B}^*\}_{ij} \eta_j(\mathbf{B}^*,\mathbf{a}) + a_i \\ &= \sum_{j=1}^{n} \{\mathbf{B}\}_{ij} \eta_j(\mathbf{B}^*,\mathbf{a}) + c_i, \end{aligned}$$

where $c_i = a_i + \sum_{j=1}^{n} \{\mathbf{B}^* - \mathbf{B}\}_{ij} \eta_j(\mathbf{B}^*,\mathbf{a}) \ge a_i$. Hence from Corollary B.1, if $\mathbf{c} = (c_i)$, then

$$\eta_i(\mathbf{B}^*,\mathbf{a}) = \eta_i(\mathbf{B},\mathbf{c}) \ge \eta_i(\mathbf{B},\mathbf{a}).$$

Therefore $\boldsymbol{\eta}(\mathbf{B},\mathbf{a})$ has entries which are increasing functions of the elements of $\mathbf{B}$.

The four cases are now considered.

1. The result that $\rho(\mathbf{B}) \uparrow \rho(\mathbf{A})$ as $\mathbf{B} \uparrow \mathbf{A}$ is immediate since $\rho(\mathbf{B})$ has been shown to be an increasing, continuous function of the entries of $\mathbf{B}$ and $\rho(\mathbf{A})$ is finite. Therefore $\rho(\mathbf{B}) \leq \rho(\mathbf{A}) \leq 1$. Hence, from Theorem B.2, $\boldsymbol{\eta}(\mathbf{B}, \mathbf{0}) = \mathbf{0}$ for all $\mathbf{B} \leq \mathbf{A}$.
2. Since $\mathbf{A}$ is finite, as for case 1 $\rho(\mathbf{B}) \uparrow \rho(\mathbf{A})$ as $\mathbf{B} \uparrow \mathbf{A}$.

   If $\rho(\mathbf{A}) > 1$ then the continuity of the Perron-Frobenius root implies that we need only consider $\lim_{\mathbf{B}\uparrow\mathbf{A}} \boldsymbol{\eta}(\mathbf{B}, \mathbf{a})$ for $\mathbf{B}$ with $\rho(\mathbf{B}) > 1$.

   Now, taking $\mathbf{B}$ so that $\rho(\mathbf{B}) > 1$ if $\mathbf{a} = \mathbf{0}$, $\boldsymbol{\eta}(\mathbf{B}, \mathbf{a}) > \mathbf{0}$. It is also bounded above by the unit vector, and is an increasing function of the entries of $\mathbf{B}$, so that $\boldsymbol{\eta}(\mathbf{B}, \mathbf{a})$ must tend to a limit $\boldsymbol{\phi} = (\phi_i) > \mathbf{0}$ as $\mathbf{B} \uparrow \mathbf{A}$. From equations (B.6), $f(\eta_i(\mathbf{B}, \mathbf{a})) \leq \{\mathbf{B1}\}_i + a_i \leq \{\mathbf{A1}\}_i + a_i$. So in fact $\phi_i \leq K_i < 1$, where $f(K_i) = \{\mathbf{A1}\}_i + a_i$.

   Let $\mathbf{B} \uparrow \mathbf{A}$ in equations (B.10). Then

$$f(\phi_i) = \sum_{j=1}^{n} \{\mathbf{A}\}_{ij} \phi_j + a_i,$$

   for $i = 1, ..., n$, where $0 < \phi_i < 1$ for all i. But, from Theorem B.2, there is a unique positive solution to equation (B.11). Hence $\boldsymbol{\phi} = \boldsymbol{\eta}(\mathbf{A}, \mathbf{a})$, which completes the proof of case 2.
3. Since $\mathbf{B} \uparrow \mathbf{A}$, we may consider $\mathbf{B} \geq \mathbf{B}_0$, where $\mathbf{B}_0$ is non-negative and non-reducible with $\{\mathbf{B}_0\}_{ij} \neq 0$ if $\{\mathbf{A}\}_{ij} \neq 0$. We can choose a pair i, j such that $\{\mathbf{A}\}_{ij} = \infty$. Since $\mathbf{A}$ is non-reducible there exists a sequence $j = i_1, i_2, ..., i_s = i$ with $\{\mathbf{B}_0\}_{i_k i_{k+1}} > 0$ for $k = 1, ..., s-1$. Hence from Theorem A.2 parts 5 and 1,

$$\begin{aligned}\rho(\mathbf{B}) &\geq \left(\{\mathbf{B}\}_{ij} \prod_{k=1}^{s-1} \{\mathbf{B}\}_{i_k i_{k+1}} \right)^{1/s} \\ &\geq \left(\{\mathbf{B}\}_{ij} \prod_{k=1}^{s-1} \{\mathbf{B}_0\}_{i_k i_{k+1}} \right)^{1/s}.\end{aligned}$$

   The right hand side of this inequality tends to infinity as $\mathbf{B} \uparrow \mathbf{A}$. Hence $\rho(\mathbf{B}) \to \infty$ as $\mathbf{B} \uparrow \mathbf{A}$.

   We may therefore choose $\mathbf{B}_0$, with the entry corresponding to the chosen infinite entry of $\mathbf{A}$ sufficiently large such that $\rho(\mathbf{B}_0) > 1$. Then from the second part of this theorem and Theorem B.2, $\boldsymbol{\eta}(\mathbf{B}, \mathbf{a}) \geq \boldsymbol{\eta}(\mathbf{B}_0, \mathbf{a}) > \mathbf{0}$. Hence from equations (B.6),

$$f(\eta_i(\mathbf{B}, \mathbf{a})) > b_{ij} \eta_j(\mathbf{B}_0, \mathbf{a}) \tag{B.13}$$

   for all i, j.

   Now for each i we can choose a j such that $\{\mathbf{A}\}_{ij}$ is infinite. Hence $f(\eta_i(\mathbf{B}, \mathbf{a})) \to \infty$, and therefore $\eta_i(\mathbf{B}, \mathbf{a}) \to 1$, for all $i = 1, ..., n$ as $\mathbf{B} \uparrow \mathbf{A}$.
4. The proof that $\rho(\mathbf{B}) \to \infty$ as $\mathbf{B} \uparrow \mathbf{A}$ follows as in part 3.

   As in part 3, we may consider the limit as $\mathbf{B} \uparrow \mathbf{A}$ for $\mathbf{B} \geq \mathbf{B}_0$, where $\rho(\mathbf{B}_0)) > 1$. Then, from the second part of this theorem and from Theorem

B.2, $\boldsymbol{\eta}(\mathbf{B}, \mathbf{a}) \geq \boldsymbol{\eta}(\mathbf{B}_0, \mathbf{a}) > \mathbf{0}$. From inequality (B.13), as in part 3, $\eta_i(\mathbf{B}, \mathbf{a}) \uparrow 1$, for all $i = m+1, ..., n$ as $\mathbf{B} \uparrow \mathbf{A}$.

Now consider the first m equations of equations (B.10).

$$f(\eta_i(\mathbf{B}, \mathbf{a})) = \sum_{j=1}^{n} b_{ij}\eta_j(\mathbf{B}, \mathbf{a}) + a_i, \ \ (i = 1, ..., m).$$

Now it has already been shown that $\eta_i(\mathbf{B}, \mathbf{a})$ is an increasing function of the entries of $\mathbf{B}$. Since $\eta_i(\mathbf{B}, \mathbf{a})$ is bounded above by 1, it must therefore tend to some limit ϕ_i, with $0 < \phi_i \leq 1$, as $\mathbf{B} \uparrow \mathbf{A}$. As in part 3, it is easily seen that in fact $\phi_i \leq K_i$ for $i = 1, ..., m$, where $f(K_i) = \{\mathbf{A}_{11}\mathbf{1} + \mathbf{A}_{12}\mathbf{1}\}_i + a_i$.

Taking the limit as $\mathbf{B} \uparrow \mathbf{A}$ in the first m equations of (B.10) gives

$$f(\phi_i) = \sum_{j=1}^{m} \{\mathbf{A}_{11}\}_{ij}\phi_j + b_i, \ \ (i = 1, ..., m).$$

Now $\mathbf{A}_{11}$ is written in normal form as in Theorem B.2 part 3. Partition $\mathbf{b}$ in the corresponding manner. Then the non-reducibility of $\mathbf{A}$ implies that $\mathbf{b}_j \neq \mathbf{0}$ for $j = 1, ..., s$. Hence from Theorem B.2 part 3, equations (B.12) have a unique positive solution. Therefore $\phi_i = \eta_i^*(\mathbf{A}_{11}, \mathbf{b})$, which completes the proof of case 4 and hence the proof of the theorem.

$\square$

Bibliography

[A1] R. M. Anderson and R. M. May, *Infectious Diseases of Humans: Dynamics and Control*, Oxford University Press, Oxford, 1991.

[A2] T. M. Apostol, *Mathematical Analysis*, Addison-Wesley, Reading Massachusetts, 1974.

[A3] D. G. Aronson, *The asymptotic speed of propagation of a simple epidemic*, Non-linear Diffusion, Research Notes in Mathematics, Vol. 14, Pitman, London, 1977, pp. 1–23.

[A4] D. G. Aronson & H. F. Weinberger, *Nonlinear diffusions in population genetics, combustion, and nerve pulse propagation*, Partial Differential Equations and Related Topics, Lect. Notes Math., Vol. 446, Springer-Verlag, Berlin, 1975, pp. 5–49.

[A5] D. G. Aronson & H. F. Weinberger, *Multidimensional nonlinear diffusion arising in population genetics*, Adv. Math. **30** (1978), 33–76.

[A6] G. Aronsson & I. Melander, *A deterministic model in biomathematics. Asymptotic behavior and threshold conditions*, Math. Biosci. **49** (1980), 207–222.

[A7] C. Atkinson & G. E. H. Reuter, *Deterministic epidemic waves*, Math. Proc. Cambridge Philos. Soc **80** (1976), 315–330.

[B1] N. T. J. Bailey, *The Mathematical Theory of Infectious Diseases and its Applications*, Griffin, London, 1975.

[B2] N. T. J. Bailey, *The Biomathematics of Malaria*, Griffin, London, 1982.

[B3] A. D. Barbour, *The uniqueness of Atkinson and Reuter's epidemic waves*, Math. Proc. Cambridge Philos. Soc **82** (1977), 127–130.

[B4] R. G. Bartle, *The Elements of Real Analysis*, Wiley, New York, 1964.

[B5] M. S. Bartlett, *Deterministic and stochastic models for recurrent epidemics*, Proc. Third Berkeley Symp. on Math. Stat. and Prob. **4** (1956), 81–109.

[B6] M. S. Bartlett, *Stochastic Population Models in Ecology and Epidemiology*, Wiley & Methuen, New York & London, 1960.

[B7] A. Berman & R. J. Plemmons, *Non-negative Matrices in the Mathematical Sciences*, Academic Press, New York & London, 1979.

[B8] J. D. Biggins, *The asymptotic shape of the branching random walk*, Adv. Appl. Prob. **10** (1978), 62–84.

[B9] S. Bochner & W. T. Martin, *Several Complex Variables*, Princeton University Press, Princeton, 1948.

[B10] M. Bramson, *Convergence of solutions of the Kolmogorov equation to travelling waves*, Mem. Amer. Math. Soc., Vol. 44, No. 285, American Mathematical Society, Providence, Rhode Island, 1983.

[B11] K. J. Brown & J. Carr, *Deterministic epidemic waves of critical velocity*, Math. Proc. Cambridge Philos. Soc **81** (1977), 431–433.

[B12] S. N. Busenberg, M. Iannelli & H. R. Thieme, *Global behavior of an age-structured epidemic model*, SIAM J. Math Anal. **22** (1991), 1065–1080.

[C1] V. Capasso, *Mathematical Structures of Epidemic Systems, Lecture Notes in Biomathematics 97*, Springer-Verlag, Berlin, Heidelberg, 1993.

[C2] J. S. Clark, *Why do trees migrate so fast: Confronting theory with dispersal biology and the paleorecord*, American Naturalist. **152** (1998), 204–224.

[C3] K. L. Cooke & J. A. Yorke, *Some equations modelling growth processes and gonorrhea epidemics*, Math. Biosci. **16** (1973), 75–101.

[C4] J. T. Cox & R. Durrett, *Limit theorems for the spread of epidemics and forest fires*, Stochastic Processes Appl. **30** (1988), 171–191.

[C5] P. Creegan & R. Lui, *Some remarks about the wave speed and travelling wave solutions of a nonlinear integral operator*, J. Math. Biol. **20** (1984), 59–68.

[D1] H. E. Daniels, *The deterministic spread of a simple epidemic*, Perspectives in Probability and Statistics: Papers in Honour of M. S. Bartlett, Distributed for Applied Probability Trust by Academic Press, London, 1975, pp. 373–386.

[D2] H. E. Daniels, *The advancing wave in a spatial birth process*, J. Appl. Prob. **14** (1977), 689–701.

[D3] M. H. Delange, *Généralisation du théorème de Ikehara*, Ann. Sci. Ecole Norm. Sup. (3) **71** (1954), 213–242.

[D4] O. Diekmann, *Limiting behaviour in an epidemic model*, Nonlinear Analysis, Theory and Applications **1** (1977), 459–470.

[D5] O. Diekmann, *Thresholds and travelling waves for the geographical spread of infection*, J. Math Biol. **6** (1978), 109–130.

[D6] O. Diekmann, *On a nonlinear integral equation arising in mathematical epidemiology*, Differential Equations and Applications, North-Holland, Amsterdam, 1978, pp. 133–140.

[D7] O. Diekmann, *Run for your life. A note on the asymptotic speed of propagation of an epidemic*, J. of Differential Equations **33** (1979), 58–73.

[D8] O. Diekmann & J. A. P. Heesterbeek, *Mathematical Epidemiology of Infectious Diseases*, Wiley, Chichester, England, 2000.

[D9] O. Diekmann & H. G. Kaper, *On the bounded solutions of a nonlinear convolution equation*, Nonlinear Analysis,Theory and Applications **2** (1978), 721–737.

[D10] O. Diekmann & R. Montijn, *Prelude to Hopf bifurcation in an epidemic model: Analysis of a characteristic equation associated with a nonlinear Volterra integral equation*, J. Math Biol. **14** (1982), 117–127.

[D11] J. Dieudonné, *Foundations of Modern Analysis*, Academic Press, New York, 1969.

[D12] R. Durrett, *Lecture Notes on Particle Systems and Percolation*, Wadsworth, Belmont, Calif, 1988.

[D13] R. Durrett, *Crabgrass, measles, and gypsy moths: an introduction to modern probability*, Bull. Amer. Math. Soc. **18** (1988), 117–143.

[E1] M. Essén, *Studies on a convolution inequality*, Ark. Math. **5** (1963), 113–152.

[F1] W. Feller, *An Introduction to Probability Theory and its Applications, Vol. 2*, Wiley, New York, 1966.

[F2] P. C. Fife, *Asymptotic states for equations of reaction and diffusion*, Bull. Amer. Math. Soc. **84** (1978), 693–726.

[F3] P. C. Fife, *Mathematical aspects of reacting and diffusing systems*, Lect. Notes in Biomath. No. 28, Springer-Verlag, Berlin, 1979.

[F4] R. A. Fisher, *The wave of advance of advantageous genes*, Ann. Eugenics **7** (1937), 355–369.

[F5] W. E. Fitzgibbon, M. E. Parrott & W. F. Webb, *Diffusive epidemic models with spatial and age dependent heterogeneity*, Discrete and Continuous Dynamical Systems **1** (1995), 35–57.

[G1] F. R. Gantmacher, *The Theory of Matrices, Vol. 2*, Chelsea, New York, 1959.

[G2] D. A. Griffiths, *Multivariate birth-and-death processes as approximations to epidemic processes*, J. Appl. Prob. **10** (1973), 15–26.

[G3] G. Gripenberg, *On some epidemic models*, Quart. Appl. Math. **39** (1981), 317–327.

[H1] K. P. Hadeler, *Diffusion in Fisher's population model*, Rocky Mountain J. Math. **11** (1985), 39–45.

[H2] H. W. Hethcote & H. R. Thieme, *Stability of the endemic equilibrium in epidemic models with subpopulations*, Math. Biosci. **75** (1985), 205–227.

[K1] D. G. Kendall, *Deterministic and stochastic epidemics in closed populations*, Proc. Third Berkeley Symp. Math. Statist. & Prob., Vol. 4, Berkeley and Los Angeles: Univ. California Press, California, 1956, pp. 149–165.

[K2] D. G. Kendall, *Contribution to the discussion in Bartlett, measles periodicity and community size*, J. R. Statist. Soc. A **120** (1957), 48–70.

[K3] D. G. Kendall, *Mathematical models of the spread of infection.*, Mathematics and Computer Science in Biology and Medicine, H.M.S.O., London, 1965, pp. 213-225.

[K4] J. F. C. Kingman, *A convexity property of positive matrices*, Quart. J. Math Oxford (2) **12** (1961), 283-284.

[K5] A. N. Kolmogoroff, I. G. Petrovsky & N. S. Piscounoff, *Étude de l'équation de la diffusion avec croissance de la quantité de matière et son application a un problème biologique*, Bull. Univ. Moscow (ser. intern) A **1(6)** (1937), 1-25.

[K6] M. Kot, M. A. Lewis & P. van den Driessche, *Dispersal data and the spread of invading organisms*, Ecology **77** (1996), 2027–2042.

[K7] M. A. Krasnosel'skii, *Positive Solutions of Operator Equations*, Groningen, Noordhoff, 1964.

[L1] A. Lajmanovich & J. A. Yorke, *A deterministic model for gonorrhea in a nonhomogeneous population*, Math. Biosci. **28** (1976), 221–236.

[L2] R. Lui, *A nonlinear integral operator arising from a model in population genetics, I. Monotone initial data*, SIAM J. Math. Anal. **13** (1982), 913–937.

[L3] R. Lui, *A nonlinear integral operator arising from a model in population genetics, II. Initial data with compact support*, SIAM J. Math. Anal. **13** (1982), 938–953.

[L4] R. Lui, *Existence and stability of travelling wave solutions of a nonlinear integral operator*, J. Math. Biol. **16** (1983), 199–220.

[L5] R. Lui, *A nonlinear integral operator arising from a model in population genetics, III. Heterozygote inferior case*, SIAM J. Math. Anal. **16** (1985), 1180–1206.

[L6] R. Lui, *A nonlinear integral operator arising from a model in population genetics, IV. Clines*, SIAM J. Math. Anal. **17** (1986), 152–168.

[L7] R. Lui, *Biological growth and spread modeled by systems of recursions, I. Mathematical theory*, Math. Biosci. **93** (1989), 269–295.

[L8] R. Lui, *Biological growth and spread modeled by systems of recursions, II. Biological theory*, Math. Biosci. **93** (1989), 297–312.

[M1] H. P. McKean, *Application of Brownian motion to the equation of Kolmogorov-Petrovskii-Piscunov*, Commun. Pure Appl. Maths. **28** (1975), 323–331.

[M2] J. A. J. Metz, D. Mollison & F. van den Bosch, *The dynamics of invasion waves.*, The Geometry of Ecological Interactions: Simplifying Spatial Complexity, Cambridge University Press, Cambridge, 2000, pp. 482–512.

[M3] H. D. Miller, *A convexity property in the theory of random variables defined on a finite Markov chain*, Ann. Math. Stat. **32** (1961), 1260–1270.

[M4] D. Mollison, *Possible velocities for a simple epidemic*, Adv. in Appl. Prob. **4** (1972), 233–257.

[M5] D. Mollison, *The rate of spatial propagation of simple epidemics*, Proc. Sixth Berkeley Symp. Math. Statist. & Prob., Vol. 3, Berkeley and Los Angeles: Univ. California Press, California, 1972, pp. 579–614.

[M6] D. Mollison, *Spatial contact models for ecological and epidemic spread*, J. R. Statist. Soc. B. **39** (1977), 283–326.

[M7] D. Mollison, *Markovian contact processes*, Adv. Appl. Prob. **10** (1978), 85–108.

[M8] J. D. Murray, *Mathematical Biology*, Springer-Verlag, Berlin, 1989.

[O1] A. M. Ostrowski, *Solutions of Equations and Systems of Equations*, Academic Press, New York, 1960.

[P1] P. Pinsky & R. Shonkwiler, *A gonorrhea model treating sensitive and resistant strains in a multigroup population*, Math. Biosci. **98** (1990), 103–126.

[P2] G. Poole & T. Boullion, *A survey on M-matrices*, SIAM Review **16** (1974), 419–427.

[R1] J. Radcliffe & L. Rass, *Wave solutions for the deterministic non-reducible n-type epidemic*, J. Math. Biol. **17** (1983), 45–66.

[R2] J. Radcliffe & L. Rass, *The spatial spread and final size of models for the deterministic host-vector epidemic*, Math. Biosci. **70** (1984), 123–146.

[R3] J. Radcliffe & L. Rass, *The uniqueness of wave solutions for the deterministic non-reducible n-type epidemic*, J. Math. Biol. **19** (1984), 303–308.

[R4] J. Radcliffe & L. Rass, *The spatial spread and final size of the deterministic non-reducible n-type epidemic*, J. Math. Biol. **19** (1984), 309–327.

[R5] J. Radcliffe & L. Rass, *Saddle-point approximations in n-type epidemics and contact birth processes*, Rocky Mountain J. Math. **14** (1984), 599–617.

[R6] J. Radcliffe & L. Rass, *The rate of spread of infection in models for the deterministic host-vector epidemic*, Math. Biosci. **74** (1985), 257–273.

[R7] J. Radcliffe & L. Rass, *The asymptotic speed of propagation of the deterministic non-reducible n-type epidemic*, J. Math. Biol. **23** (1986), 341–359.

[R8] J. Radcliffe & L. Rass, *The effect of reducibility on the deterministic spread of infection in a heterogeneous population*, Lecture Notes in Pure and Applied Mathematics, Vol 131, Mathematical Population Dynamics, Marcel Dekker, New York, 1991, pp. 93–114.

[R9] J. Radcliffe & L. Rass, *Reducible epidemics: choosing your saddle*, Rocky Mountain J. Math. **23** (1993), 725–752.

[R10] J. Radcliffe & L. Rass, *Multitype contact branching processes*, Lecture Notes in Statistics, Vol 99, Branching Processes: Proceedings of the First World Congress, Springer-Verlag, New York, 1995, pp. 169–179.

[R11] J. Radcliffe & L. Rass, *Spatial branching and epidemic processes*, Mathematical Population Dynamics: Analysis of Heterogeneity, Vol 1, Theory of Epidemics, Wuertz, Winnipeg, Canada, 1995, pp. 147–170.

[R12] J. Radcliffe & L. Rass, *The asymptotic behavior of a reducible system of nonlinear integral equations*, Rocky Mountain J. Math. **26** (1996), 731–752.

[R13] J. Radcliffe & L. Rass, *Discrete time spatial models arising in genetics, evolutionary game theory and branching processes*, Math. Biosci. **140** (1997), 101–129.

[R14] J. Radcliffe & L. Rass, *Spatial Mendelian games*, Math. Biosci. **151** (1998), 199–218.

[R15] J. Radcliffe & L. Rass, *Convergence results for contact models in genetics and evolutionary game theory*, J. Biol. Systems **6** (1998), 411–426.

[R16] J. Radcliffe & L. Rass, *Strategic and genetic models of evolution*, Math. Biosci. **156** (1999), 291–307.

[R17] J. Radcliffe, L. Rass & W. D. Stirling, *Wave solutions for the deterministic host-vector epidemic*, Math. Proc. Cambridge Philos. Soc. **91** (1982), 131–152.

[R18] L. Rass & J. Radcliffe, *The asymptotic spatial behaviour of a class of epidemic models*, Proceedings of the seventh international colloquium on differential equations, VSP, Utrecht, The Netherlands, 1997, pp. 355–362.

[R19] L. Rass & J. Radcliffe, *The derivation of certain pandemic bounds*, Math. Biosc. **156** (1999), 147–165.

[R20] L. Rass & J. Radcliffe, *Global asymptotic convergence results for multitype models*, Int. J. Appl. Math. and Comp. Sci. **10** (2000), 63–79.

[R21] M. G. Roberts, *Stability in a two host epidemic model*, J. Math. Biol. **14** (1982), 71–75.

[S1] E. Seneta, *Non-negative Matrices*, Allen and Unwin, London, 1973.

[S2] N. Shigesada & K. Kawasaki, *Biological Invasions: Theory and Practice*, Oxford University Press, Oxford, New York, Tokyo, 1997.

[S3] C. P. Simon & J. A. Jacquez, *Reproduction numbers and the stability of equilibria of SI models for heterogeneous populations*, SIAM J. Appl. Math. **52** (1992), 541–576.

[T1] H. R. Thieme, *A model for the spread of an epidemic*, J. Math. Biol. **4** (1977), 377–351.

[T2] H. R. Thieme, *The asymptotic behavior of solutions of non-linear integral equations*, Math. Zeitschrift **157** (1977), 141–154.

[T3] H. R. Thieme, *Asymptotic estimates of the solutions of nonlinear integral equations and asymptotic speeds for the spread of populations*, J. Reine. Angew. Math. **306** (1979), 94–121.

[T4] H. R. Thieme, *Density-dependent regulation of spatially distributed populations and their asymptotic speed of spread*, J. Math. Biol. **8** (1979), 173–187.

[T5] H. R. Thieme, *Renewal theorems for some mathematical models in epidemiology*, J. Integral Eqn. **8** (1985), 185–216.

[T6] H. R. Thieme, *Asymptotic proportionallity (weak ergodicity) and conditional asymptotic equality of solutions to time-heterogeneous sublinear difference and differential equations*, J. Diff. Eqn. **73** (1988), 237–268.

[U1] K. Uchiyama, *Spatial growth of a branching process of particles living in* $\Re^d$, Ann. Prob. **10** (1982), 896–918.

[V1] F. van den Bosch, J. A. J. Metz & O. Diekmann, *The velocity of spatial population expansion*, J. Math. Biol. **28** (1990), 529–565.

[V2] A. I. Volpert, V. A. Volpert & V. A. Volpert, *Traveling wave solutions of parabolic systems, Translations of Mathematical Monographs Vol. 140*, American Mathematical Society, Providence, Rhode Island, 1994.

[W1] H. F. Weinberger, *Asymptotic behavior of a model in population genetics*, Lecture Notes in Math., Vol 648, Nonlinear Partial Differential Equations and Applications, Springer, Berlin, 1978, pp. 47–96.

[W2] H. F. Weinberger, *Long-time behavior of a class of biological models*, SIAM J. Math. Anal. **13** (1982), 353–396.

Index